T0201389

Analysis of Poverty Data by Small Area Estimation

WILEY SERIES IN SURVEY METHODOLOGY
Established in part by Walter A. Shewhart and Samuel S. Wilks

A complete list of the titles in this series appears at the end of this volume.

Analysis of Poverty Data by Small Area Estimation

Edited by

Monica Pratesi
University of Pisa, Italy

This edition first published 2016.

Registered office
John Wiley & Sons Ltd, The Atrium, Southern Gate, Chichester, West Sussex, PO19 8SQ, United Kingdom

For details of our global editorial offices, for customer services and for information about how to apply for permission to reuse the copyright material in this book please see our website at www.wiley.com.

Library of Congress Cataloging-in-Publication Data applied for

A catalogue record for this book is available from the British Library.

ISBN: **9781118815014**

Set in 10/12pt, TimesLTStd by SPi Global, Chennai, India.
Printed and bound in Singapore by Markono Print Media Pte Ltd

1 2016

Contents

Foreword

Poverty and living conditions are always at the forefront of analyses and discussions carried out by international and national organizations, governments and researchers from all over the world. All of them agree that the intervention policies to fight against poverty and to improve the quality of life should be specifically designed and implemented at a local level, because the phenomena are heterogeneous and have multiple and different characteristics in the different territorial areas. Obviously, local governments play a fundamental role in implementing actions, but, to do that, they need statistical information (data) to understand the situation and to be able to evaluate the impact of their actions. On the other hand, the stakeholders and citizens are interested in and able to judge the economic situation and the quality of life at a local level and are interested in better understanding the effect of policies on their own territory.

However, usually, the data on income, poverty and quality of life are not available at a local level. In fact, the main sources of statistical data in these fields are from sample surveys that cannot support reliable estimation at a local level because their sample sizes are too small. The problem could be overcome by increasing the sample sizes, but in many practical situations cost–benefit analysis excludes it as a time-consuming and unaffordable solution.

The key solution in order to be able to comply with the information need for measuring poverty at a local level is the use of Small Area Estimation (SAE) methods that researchers and National Statistical Offices of various countries are developing and implementing. This is confirmed by the large amount of literature on these local estimates resulting from many projects, conferences and books in the last decade.

This book provides a very comprehensive and detailed source of information to construct such a key solution; it explains clearly the use of SAE methods efficiently adapted to the distinctive features (identification of relative poverty indicators, classification of statistical units, specific sample design of the surveys, characteristics of panel surveys, etc.) of poverty data coming from surveys and administrative archives. All of these complications add up to make the use of SAE methods a difficult and challenging problem that this book ably and comprehensively tackles.

The book, after having discussed the definition(s) of the poverty indicators and data collection and data integration methods to obtain reliable estimations of them, describes and reviews the advanced methods and techniques recently developed and applied to SAE of poverty, addressing the distinctive features mentioned before (impact of sampling designs, etc.). Then, the book presents the SAE models as applied to poverty. In the extensive literature, there are many methods developed and they are often specified to solve the particular estimation problems for the case under study. However, their presentation in the book has been able to single out and address the main general issues in the estimation of poverty at a local level, such as the erroneous specification of the models and the robustness of the estimations, the use of spatio-temporal models, the estimation of distribution function of income and inequalities, and so on. Each chapter of the book describes insights, introduces methodology, and outlines the cutting-edge necessary for effective estimation and analysis of poverty indicators at a local level. Very interesting advanced new methodologies and new challenges to be faced are presented. All of this makes this book very timely.

One of the particular attractive features of this book is that it is about both theoretical and practical methods and analysis. It does not simply discuss the methodological tools that can be applied in an idealized setting, but also discusses the issues which all applied statisticians and the National Statistical Offices have to face to produce an estimation of poverty indicators at a local level. The practical aspects of the estimation methods are discussed in many of the chapters and, in a specific way, the last three chapters are devoted to the presentation of the procedures used in the EU, USA and Chile, discussing also the quality of the obtained results. Moreover, most of the chapter authors have supported the methods concerning data analysis and models by presenting specific scripts that are also described and written in SAS or R software in an Appendix available on the book's website.

Put together, the attractive features of this book make it a genuinely valuable and very useful book for all the researchers from academia and statistical offices, concerned with the measuring of poverty indicators at a local level and with the survey methodology. Surely this book will stimulate further important research in the field.

Luigi Biggeri
Emeritus Professor of Economic Statistics, University of Florence, Italy
Past President, Italian National Statistical Institute (Istat)

Preface

All over the world, fighting against poverty is assuming a more and more central role and recent radical economic and social transformations have caused a renewed interest in this sector. Such interest is due not only to economic factors but also to issues related to the quality of life and to the protection of social cohesion. This growing attention has strongly reinforced the need to look at poverty as the result of a chain of processes linked together. In this approach, poverty represents not only a problem but also the symptom of the ineffectiveness of the policies to reinforce resilience and to protect against vulnerabilities. Because of this role, it deserves special attention.

These aspects have led to deep modifications in the data provided in this field and in the definition of a set of comparable and readable poverty indicators. Particularly, the demand for poverty and living conditions data, referring to local areas and/or subpopulations, has become urgent. Policy makers and stakeholders need to know the indicators and their spatial distribution at regional and subregional levels. This is important for formulating and implementing policies, distributing resources and measuring the effect of local policy actions.

Income and living conditions surveys are thus conducted all over the world in order to gather a large amount of information on the classic income and consumption, but also on other related monetary and non monetary aspects of living conditions. But those surveys may not support a reliable estimation at the level of a local area because area-specific sample sizes are often too small to provide direct estimates with acceptable variability. In addition, data based statistics on poverty and living conditions are becoming more and more common, and integration of survey and administrative data can raise many distinct issues.

As a result, the statistics produced are so strongly conditioned by this largely diversified demand and supply of data that researchers and National Statistical Offices of many countries, in order to be able to comply with the information need, began to set up a complex system of Small Area Estimation (SAE) methods based on an integrated set of information whose design, implementation and maintenance require a strong methodological effort.

Apart from the difficulties typical of social economic data, such as the qualitative nature of many variables and the high concentration of quantitative variables, small area methods for poverty indicators are indeed characterized by some additional peculiarities that often make it impossible or inefficient to make use of classical small area models proposed in the literature.

In particular we refer to the following:

a) The definition of poverty is neither obvious nor unique, because the list of possible options is quite large (monetary poverty, non monetary poverty, multidimensional poverty) and its choice depends on the phenomenon for which we are interested in collecting the data. Absolute poverty and relative poverty are both valid concepts.[1] Here we refer to relative poverty.

b) The identification of relative poverty indicators and of significant auxiliary data to proxy them is a topic for research itself. Among these, the geography of the country of interest and its subdivision in areas and regions appear to be crucial in poverty studies. In the choice of the proxies also the availability of a source of data of sufficient quality and the possibility of integrating existing data is important. This is especially true at a local level.
c) Typological classifications of the statistical units (households, individuals, social services users) are very important tools to define the estimation domains and to design an efficient integration of survey and administrative data sources. However, harmonized hierarchical nomenclatures are usually not available for a certain definition of statistical unit, or they do exist but are so subjective that they cannot be considered as standard. The dialogue between survey data archives and administrative data archives is not easy and requires statistical matching and data integration.
d) The effect of poverty on a person or a household is directly related to the duration of their poverty and to its persistency. Often the surveys on income and living conditions are panel surveys composed by several waves and this allows for the exploration of the duration of poverty. In this context the issue of estimating sampling error of cumulative and longitudinal poverty indicators from panel data is crucial, especially at subnational level where the sample size can be small.
e) The impact of survey sampling design in SAE of poverty indicators has not yet been completely explored. There are issues to be addressed on the effect of the different sampling designs on the model-based estimates, also in comparison with classical design-based methods. This opens the discussion on which estimation method is preferable in what context.
f) In many circumstances the use of the so-called model assisted and 'model based' methods is considered a standard procedure in SAE. Sometimes there is the obvious consequence that the peculiarities of the methods in benchmarking to estimates for larger areas, their resistance to outliers, their behavior when the auxiliary data are temporal and/or spatial data are not discussed. Special issues arise when the data are skewed, the interest is on complex poverty indicators derived from the income distribution, and the covariates are measured with error. This has evident implications in terms of the quality of the obtained estimates especially from the point of view of Official Statistical Agencies.
g) At least when using geographically referred units, there often exist particular auxiliary variables requiring *ad hoc* procedures to be used in the fitting of a SAE model. Spatial data sets can be fruitfully used in poverty mapping. Nevertheless, extracting the interesting and useful patterns from spatial data sets is more difficult than extracting the corresponding patterns from traditional numeric and categorical data. This is due to the complexity of spatial data types, spatial relationships, and spatial autocorrelation.

As far as we know, in the current literature there exists no comprehensive source of information regarding the use of SAE methods adapted to these distinctive features of poverty data coming from surveys and administrative archives. This book may serve to fill this gap.

[1] The concept of absolute poverty is that there are minimum standards (monetary and non monetary) below which no one anywhere in the world should ever fall. Relative poverty refers to a standard of living which is defined in terms of the society in which an individual lives and which therefore differs between areas in countries and over time.

It contains 20 chapters, the first one of which can be considered as an introductory chapter reviewing the problem and perspective of SAE applied to poverty (Chapter 1. Introduction on measuring poverty at local level using small area estimation methods), and the remaining 19 are divided into six parts:

I. *Definition of indicators and data collection and integration methods* (Chapter 2. Regional and local poverty measures; Chapter 3. Administrative and survey data collection and integration; Chapter 4. Small area methods and administrative data integration).
These chapters provide an overview of the basic tools used in the definitions of poverty and of local poverty indicators, including some practical and theoretical considerations regarding the usage of income and consumption surveys and their integration with administrative data files to produce local poverty measures, in the attempt to address issues (a)–(c) previously described. Attention is then focused on the use of administrative data that in the last few years have evolved from a simple backup source to a very relevant element in ensuring the coverage of a list of units.

II. *Impact of sampling design, weighting and variance estimation* (Chapter 5. Impact of sampling designs in small area estimation with applications to poverty measurement; Chapter 6. Model-assisted methods for small area estimation of poverty indicators; Chapter 7. Variance estimation for cumulative and longitudinal poverty indicators from panel data at regional level).
These chapters review advanced methods and techniques recently developed in the survey data analysis literature as applied to SAE of poverty, in an attempt to address the distinctive features (d)–(e) described above. Some interesting proposals arise from the studies aiming at evaluating the impact of sampling design and model assisted estimation.These studies, together with design-based cumulation techniques for variance estimation, have received a lot of attention in recent years due to the growing demand for reliable small-area statistics needed for formulating policies and programs.

Chapters 8–20 are devoted to SAE methods. SAE models as applied to poverty are indeed many and often specified to solve the particular estimation problems for the case under study. However, there are some general themes that can be singled out in addressing issues (f) and (g) previously described. Each chapter is classified under only one theme, but even then some of them cross-cut more than one theme: to facilitate the reader they are assigned to the theme that can be considered as prevalent. The resulting classification is:

III. *Small area estimation modeling and robustness* (Chapter 8. Models in small area estimation when covariates are measured with error; Chapter 9. Robust domain estimation of income-based inequality indicators; Chapter 10. Nonparametric regression methods for small area estimation).
In some situations the erroneous specification of a model and/or errors in the covariates can result in biased estimators. These chapters describe the use of traditional and more recent SAE methods able to recover these problems and provide good robustification tools as applied to poverty data.

IV. *Spatio-temporal modeling of poverty* (Chapter 11. Area level spatio-temporal small area estimation models; Chapter 12. Unit level spatio-temporal models; Chapter 13. Spatial information and geoadditive small area models).

The temporal and spatial dimensions of poverty are often included in modeling the indicators. There are specific models for statistical units equal to areas (area level models) and models for statistical units equal to households or individuals (unit level models). Additionally, the usefulness of spatial data as the main auxiliary variables for geographically coded units is assessed through empirical evidence.

V. *Small area estimation of the distribution function of income and inequalities* (Chapter 14. Model-based direct estimation of a small area distribution function; Chapter 15. Small area estimation for lognormal data; Chapter 16. Bayesian Beta regression models for the estimation of poverty and inequality parameters in small areas; Chapter 17. Empirical Bayes and hierarchical Bayes estimation of poverty measures for small areas).
The models presented above are applied to carry out a wide range of operations on survey data to estimate many poverty indicators. Auxiliary variables are retrieved from many kinds of mixed sources. However, the particular nature of the target parameters and the availability of *a priori* information allow for different formalization of the problem. These chapters address the estimation of the distribution function of income and inequalities under the frequentist and the Bayesian approach.

VI. *Data analysis and applications* (Chapter 18. Small area estimation using both survey and census unit record data: links, alternatives, and the central roles of regression and contextual variables; Chapter 19. An overview of the U.S. Census Bureau's Small Area Income and Poverty Estimates Program; Chapter 20. Poverty mapping for the Chilean comunas).
The chapters of the last part of the book provide examples of the procedures used in the European Union and United States by the Official Statistical Agencies and traditionally by the World Bank, discussing also the quality of the obtained results. An appraisal is provided of indirect estimates used in the Small Area Income and Poverty Estimates (SAIPE) program, both traditional and model-based, that are used because direct area-specific estimates may not be reliable due to small area-specific sample sizes. A wide application of SAE methods in a developing country, Chile, conclude the book.

The book is completed by an Appendix (Chapter 21. Appendix on Software and Codes Used in the Book) describing scripts written in SAS or R software, that are available on the book's website. Most of the methods concerning data analysis and models are supported by scripts written by the chapter authors. The Appendix is intended to provide guidance on how to use these scripts for actually implementing the advanced methods covered in the book.

The volume originates from a selection of the methodological results obtained during the development of several research projects,[2] which intended to bring together the expertise of academics and of specialists from National Statistical Offices to increase the dissemination of

[2] We refer mainly to SAMPLE (Small Area Methods for Poverty and Living Condition Estimates) and to AMELI (Advanced Methodology for European Laeken Indicators) projects which were financially supported by the European Commission within the 7th Framework Programme. The complete set of project results are available via the homepages (http://www.sample-project.eu and https://www.uni-trier.de/index.php?id=40263&L=2). Another fundamental program which motivated some of the results collected here is the U.S. Census Bureau SAIPE program. It provides annual estimates of income and poverty statistics for all school districts, counties, and states of the U.S. (www.census.gov/did/www/saipe).

the most recent survey data analysis methods in the poverty sector. It also collects the content of many presentations on this topic from international conferences on SAE.[3]

Although the present book can serve as a supplementary text in graduate seminars in survey methodology, the primary audience is researchers having at least some prior training in sampling methods and survey data analysis. Since it contains a number of review chapters on several specific themes in survey research, it will be useful to researchers actively engaged in organizing, managing and conducting poverty mapping who are looking for an introduction to advanced techniques from both a practical and a methodological perspective.

Finally, this book aims at stimulating research in this field and, for this reason, we are aware that it cannot be considered as a comprehensive and definitive reference on the methods that can be used in poverty mapping, since many topics were intentionally omitted. However, it reflects, to the best of my judgement, the state of the art on several crucial issues.

Monica Pratesi
Pisa, Italy

[3] The reference is mainly to the set of conferences held in Jyväskylä, Finland (2005), Pisa, Italy (2007), Alicante, Spain (2009), Trier, Germany (2011) and Bangkok, Thailand (2013). Their declared aim was to develop an information network of individuals and institutions involved in the use and production of small area estimates and also poverty mapping. These conferences were organized with the support of the National Statistical Offices of the hosting country and were often supported by the IASS (International Association of Survey Statisticians) as satellite conferences of the ISI (International Statistical Institute) World Congresses.

Acknowledgements

The editing of the book was conducted within the research infrastructure InGRID (Inclusive Growth Research Infrastructure Diffusion; https://inclusivegrowth.be/), which is financially supported by the European Commission within the 7th Framework Programme under Grant Agreement no. 312691. Thanks are due to Liz Wingett, Prachi Sinha Sahay, Lincy Priya, Richard Davies and Jo Taylor of John Wiley & Sons, Ltd for editorial assistance, and to Alistair Smith of Sunrise Setting Ltd for assistance with LaTeX. Finally, I am grateful to the chapter authors for their diligence and support for the goal of providing an overview of such an active research field, and I would like to thank Luigi Biggeri, Emeritus Professor of Economic Statistics at the University of Florence, for his advice and suggestions during the implementation phase of the project.

About the Editor

Monica Pratesi is Professor of Statistics at the University of Pisa. She has taught several statistics-related courses at the Universities of Florence, Bergamo and at the University of Pisa, where now she is holder of the Jean Monnet Chair "Small Area Methods for Monitoring of Poverty and Living Conditions in EU" (sampleu.ec.unipi.it). Her main research fields include small area estimation, inference in elusive populations, nonresponse in telephone and Internet surveys, and design effect in fitting statistical models. She has been involved in the management of several research projects related to these fields, as the Eframe project (www.eframeproject.eu) and the InGRID project (https://inclusivegrowth.be), and she coordinated a collaborative project on Small Area Methodologies for Poverty and Living Conditions Estimates (S.A.M.P.L.E. project) funded by the European Commission in the 7th Framework Programme.

List of Contributors

Serena Arima, Department of Methods and Models for Economics Territory and Finance, University of Rome La Sapienza, Rome, Italy

Wesley W. Basel, Social, Economic, and Housing Statistics Division, U.S. Census Bureau, Washington, USA

William R. Bell, Research and Methodology Directorate, U.S. Census Bureau, Washington, USA

Emily Berg, Department of Statistics, Iowa State University, Ames, USA

Gianni Betti, Department of Economics and Statistics, University of Siena, Siena, Italy

Chiara Bocci, IRPET-Regional Institute for Economic Planning of Tuscany, Florence, Italy

Jay F. Breidt, Department of Statistics, Colorado State University, Fort Collins, USA

Jan Pablo Burgard, Department of Economics and Social Statistics, University of Trier, Trier, Germany

Carolina Casas-Cordero Valencia, Instituto de Sociología y Centro de Encuestas y Estudios Longitudinales, Universidad Católica de Chile, Santiago, Chile

Ray Chambers, Centre for Statistical and Survey Methodology, University of Wollongong, Wollongong, Australia

Hukum Chandra, Indian Agricultural Statistics Research Institute, New Delhi, India

Alessandra Coli, Department of Economics and Management, University of Pisa, Pisa, Italy

Paolo Consolini, ISTAT, Italian National Staistical Institute, Rome, Italy

Antonella D'Agostino, Department of Business and Quantitative Studies, University of Naples "Parthenope", Naples, Italy

Gauri S. Datta, Department of Statistics, University of Georgia, Athens, USA

Marcello D'Orazio, ISTAT, Italian National Statistical Institute, Rome, Italy

Jenny Encina, Inter-American Development Bank, Washington, DC, USA

Marià Dolores Esteban, Centro de Investigación Operativa, Universidad Miguel Hernández de Elche, Elche, Spain

Enrico Fabrizi, DISES, Università Cattolica del S. Cuore, Piacenza, Italy

Maria Rosaria Ferrante, Dipartimento di Scienze Statistiche "Paolo Fortunati", Università di Bologna, Bologna, Italy

Francesca Gagliardi, Department of Economics and Statistics, University of Siena, Siena, Italy

Caterina Giusti, Department of Economics and Management, University of Pisa, Pisa, Italy

Stephen J. Haslett, Institute of Fundamental Sciences, Massey University, Palmerston North, New Zealand and Statistical Consulting Unit, The Australian National University, Canberra, Australia

Partha Lahiri, Joint Program in Survey Methodology and Department of Mathematics, University of Maryland, College Park, USA

Risto Lehtonen, Department of Social Research, University of Helsinki, Helsinki, Finland

Achille Lemmi, Department of Economics and Statistics and Honorary Fellow ASESD Tuscan Universities Research Centre "Camilo Dagum", University of Siena, Siena, Italy

Brunero Liseo, Department of Methods and Models for Economics Territory and Finance, University of Rome La Sapienza, Rome, Italy

Jerry J. Maples, Center for Statistical Research and Methods, U.S. Census Bureau, Washington, USA

Stefano Marchetti, Department of Economics and Management, University of Pisa, Pisa, Italy

Isabel Molina, Department of Statistics, Universidad Carlos III de Madrid, Madrid, Spain

Domingo Morales, Centro de Investigación Operativa, Universidad Miguel Hernández de Elche, Elche, Spain

Ralf Münnich, Department of Economics and Social Statistics, University of Trier, Trier, Germany

Laura Neri, Department of Economics and Statistics, University of Siena, Siena, Italy

Jean D. Opsomer, Department of Statistics, Colorado State University, Fort Collins, USA

Maria Chiara Pagliarella, Department of Economics and Statistics, University of Siena, Siena, Italy

Tomasz Panek, Warsaw School of Economics, Warsaw, Poland

Agustín Pérez, Centro de Investigación Operativa, Universidad Miguel Hernández de Elche, Elche, Spain

Alessandra Petrucci, Department of Statistics, Informatics, Applications, University of Florence, Florence, Italy

Monica Pratesi, Department of Economics and Management, University of Pisa, Pisa, Italy

M. Giovanna Ranalli, Dipartimento di Scienze Politiche, Università degli Studi di Perugia, Perugia, Italy

Jon. N. K. Rao, School of Mathematics and Statistics, Carleton University, Ottawa, Canada

Nicola Salvati, Department of Economics and Management, University of Pisa, Pisa, Italy

Renato Salvatore, Department of Economics and Jurisprudence, University of Cassino and Southern Lazio, Cassino (FR), Italy

Carlo Trivisano, Dipartimento di Scienze Statistiche "Paolo Fortunati", Università di Bologna, Bologna, Italy

Nikos Tzavidis, Department of Social Statistics and Demography, University of Southampton, Southampton, UK

Ari Veijanen, Statistics Finland, Finland

Vijay Verma, Department of Economics and Statistics, University of Siena, Siena, Italy

Li-Chun Zhang, S3RI/University of Southampton, Southampton, UK and Statistics Norway, Oslo, Norway

Thomas Zimmerman, Department of Economics and Social Statistics, University of Trier, Trier, Germany

1

Introduction on Measuring Poverty at Local Level Using Small Area Estimation Methods

Monica Pratesi and Nicola Salvati
Department of Economics and Management, University of Pisa, Pisa, Italy

1.1 Introduction

All over the world, fighting against poverty is assuming a more and more central role and recent radical economic and social transformations have caused a renewed interest in this field. Poverty is a complex concept. As a consequence, the focus should not be only on monetary poverty, but also on the larger concept of well-being, which preliminarily includes the definition and measure of the following aspects: capability of income production, being involved in a satisfying job, being in good health, living in an adequate house, achieving a proper level of education, having good social relations, and so on. These characteristics require poverty to be defined in a multidimensional setting.

Given that, the reduction of the risk of becoming poor can be achieved only through a very wide range of policy actions and tools: from the mere monetary transfer to a varied supply of social services.

Local governments play a fundamental role in implementing actions to provide help to vulnerable people. By means of providing social services and transfers in kind, Local Governmental Agencies (LGAs) are able to adapt their service supply to multiple and different needs. The governance of local areas must be concerted and shared creating a virtuous pool of governmental and not governmental actors and agencies.

So the policy makers need to know the situation as it is and the impact of their actions at this local level and also stakeholders and citizens are interested in better understanding the effect of policies on their own territory.

Analysis of Poverty Data by Small Area Estimation, First Edition. Edited by Monica Pratesi.

Companion Website: www.wiley.com/go/pratesi/poverty

However the main sources of statistical data on monetary and non-monetary poverty are from sample surveys on income and living conditions. These rarely give credible estimates at sub-regional and local level. From this comes the importance of the Small Area Estimation (SAE) methods for measuring poverty at local level. This is confirmed also by the large amount of literature on these local estimates resulting from many projects, conferences and books in the last decade.

This chapter has a twofold scope. It serves as necessary background to introduce the book as it constitutes also a useful preparation to the specific methodologies described in each chapter, and a common reference for the notation to use. We start from the definition of poverty indicators and the problem of their estimation (Section 1.2), to present then the main issues related to the data as data integration and data quality that are cross-cutting the methodologies presented in the book (Section 1.3). Section 1.4 reviews the model-assisted and model-based methods used in the book and also gives advice and recommendations on the previous issues.

1.2 Target Parameters

1.2.1 Definition of the Main Poverty Indicators

In order to monitor the process of social inclusion, a list of 18 indicators monitoring poverty and social exclusion was proposed in 2001 (Atkinson *et al.* 2002). The list is constantly modified and complemented. It contains both indicators based on household incomes (monetary indicators) and indicators based on non-monetary symptoms of poverty (non-monetary indicators). Among poverty indicators, the so-called Laeken indicators are very often used to target poverty and inequalities. They are a core set of statistical indicators on poverty and social exclusion agreed by the European Council in December 2001, in the Brussels suburb of Laeken, Belgium.

Referring to the monetary poverty and starting from the Income distribution the most frequently used indicators are the average mean of the equalized income, the Head Count Ratio (HCR) and the Poverty Gap (PG). The HCR measures the incidence of poverty and it is the percentage of individuals of households under a poverty line, that can be defined at national or regional level. For example, the European Commission fix it as 60% of the median value of the equivalized income distribution. The PG index measures the intensity of poverty, that are the depth of poverty by considering how far, on average, the poor are from that poverty line.

Formally, the incidence of poverty or HCR and the PG can be obtained by the generalized measures of poverty introduced by 1984. Denoting the poverty line by t, the Foster-Greer-Thorbecke (FGT) poverty measures are defined as:

$$F(\alpha, t) = \frac{1}{N} \sum_{j=1}^{N} \left(\frac{t - y_j}{t} \right)^{\alpha} \mathrm{I}(y_j \leq t). \tag{1.1}$$

Here y is a measure of income for individual/household j, N is the number of individuals/households and α is a "sensitivity" parameter. Setting $\alpha = 0$ defines the HCR, $F(0, t)$, whereas setting $\alpha = 1$ defines the PG, $F(1, t)$.

The HCR indicator is a widely used measure of poverty. The popularity of this indicator is due to its ease of construction and interpretation, even if it has some limitations. As it assumes that all poor individuals/households are in the same situation, the easiest way of reducing its value is by implementing actions to target benefits to people who are just below the poverty line. In fact, they are the ones who are the cheapest to move across the line. Hence, policies based on the headcount index might be not completely effective, as they are not based on the exam of the whole income distribution. For this reason, estimates of the PG indicator are important. The PG can be interpreted as the average shortfall of poor people. It shows how much would have to be transferred to all the poor to bring their expenditure up to the poverty line.

Together with the above indicators, the average value of the distribution of the household income is also important. This is especially true when the level of income is modest and the distribution of income has a long tail. In this case the median value on which the poverty line is computed is expected to be low and the HCR tends to be low as well. Also the PG can lose its relevance, giving a misleading indication of the deprivation of the population under study.

In many cases these measures are considered as a starting point for more in depth studies of poverty and living conditions. In fact, analyses are done using also non-monetary indicators in order to give a more complete picture of poverty and deprivation (Cheli and Lemmi, 1995). In addition, as poverty is a question of graduation, the set of indicators is generally enlarged with other indicators belonging to vulnerable groups, from which it can be likely to move towards the status of poverty (see Chapter 2 of this book). The spatial distribution of these poverty indicators is a feature of high interest. It can be illustrated and represented by building poverty maps. Poverty maps can be constructed using censuses, surveys, administrative data and other data. Here we refer to poverty mapping to visualize the spatial distribution of poverty indicators. This is particularly useful, as it is shown in Chapter 2, to monitor the localization of poverty and the individuation of the most vulnerable areas.

1.2.2 Direct and Indirect Estimate of Poverty Indicators at Small Area Level

The estimates of the different poverty indicators at area level can be done under the design-based (Hansen *et al.* 1953; Kish 1965; Cochran 1977), model-assisted (Särndal *et al.* 1992) and model based approach (Gosh and Meeden, 1997, Valliant *et al.* 2000; Rao 2003), as direct or indirect small area estimates. The direct estimates are produced under the design-based approach using only data coming from one survey, the indirect estimates use auxiliary information (variables) to improve the quality and accuracy of survey estimates or to break down the known values referred to larger areas by using regression-type models. All these estimates belong to the broad class of Small Area Estimation (SAE) methods.

Let us start introducing the notation we use in this chapter and in particular in the review of the small areas model-assisted and model-based methods. Consider that a population U of size N is divided into D non-overlapping subsets U_d (domains of study or areas) of size N_d, $d = 1, ..., D$. We index the population units by j and the small areas by d, the variable of interest is y_{jd}, $\mathbf{x}_{jd}$ is a vector of p auxiliary variables. We assume that $\mathbf{x}_{ij}$ contains 1 as its first component. Suppose that a sample s is drawn according to some, possibly complex, sampling design such that the inclusion probability of unit j within area d is given by π_{jd}, and that area-specific samples $s_d \subset U_d$ of size $n_d \geq 0$ are available for each area. Note that non-sample areas have $n_d = 0$, in which case s_d is the empty set. The set $r_d \subseteq U_d$ contains the $N_d - n_d$ indices of the non-sampled units in small area d.

Values of y_{jd} are known only for sampled values while for the p-vector of auxiliary variables it is assumed that area level totals $\mathbf{X}_d$ or means $\bar{\mathbf{X}}_d$ or individual values $\mathbf{x}_{jd}$ are accurately known from external sources.

The straightforward approach to calculate FGT poverty indicators referring to the areas of interest is to compute direct estimates. For each area, direct estimators use only the data referring to the sampled households, since for these households the information on the household income is available.

The direct estimators of the FGT poverty indicators are of the form:

$$F_d^{dir}(\alpha, t) = \frac{1}{\sum_{i \in s_d} w_{jd}} \sum_{j \in s_d} w_{jd} \left(\frac{t - y_{jd}}{t} \right)^{\alpha} I(y_{jd} < t), \qquad d = 1, \ldots, D, \tag{1.2}$$

where w_{jd} is the sampling weight (inverse of the probability of inclusion) of household j belonging to area d and $\sum_{i \in s_d} w_{jd} = N_d$. In the same way, the mean of the household equivalized income in each small area can be computed as:

$$m_d^{dir} = \frac{1}{\sum_{i \in sd} w_{jd}} \sum_{j \in s_d} w_{jd} y_{jd}, \qquad d = 1, \ldots, D. \tag{1.3}$$

When the sample size in the areas of interest is limited, estimators such as (1.2) and (1.3) cannot be used. In fact the size is too small to obtain acceptable statistical significance of the direct estimates obtained under the sample design. Then the purely design-based solution and the usage of direct estimates often implies the increase of the sample size, oversampling of the studied domains. If oversampling is done, credible estimates can be obtained with appropriate direct estimators and the SAE problem is solved. Nevertheless, in many practical situations oversampling is far from being an option as cost–benefit analysis excludes it as a time-consuming and unaffordable solution.

In these cases, model-assisted and model-based SAE techniques need to be employed. Therefore, the estimation of poverty indicators (target parameters) at local level is computed with indirect methods by using auxiliary variables, usually coming from administrative data available also at local area level. The relationship between the target parameters and the auxiliary variables is described by a suitable model. Considering Särndal *et al.* (1992) we clarify that in this context a model consists of "some assumptions of relationship, unverifiable but not entirely out of place, to save survey resources or to bypass other practical difficulties".

Under these approaches it is useful to express the mean and the FGT indicators for the small area d as shown in the following.

The population small area mean can be written as:

$$m_d = N_d^{-1} \left(\sum_{j \in s_d} y_{jd} + \sum_{j \in r_d} y_{jd} \right). \tag{1.4}$$

Since the y values for the r_d non-sampled units are unknown, they need to be predicted.

The FGT poverty indicators in small area d can be written as:

$$F_d(\alpha, t) = N_d^{-1} \left(\sum_{j \in s_d} z_{jd}(\alpha, t) + \sum_{j \in r_d} z_{jd}(\alpha, t) \right), \tag{1.5}$$

where

$$z_{jd}(\alpha, t) = \left(\frac{t - y_{jd}}{t} \right)^{\alpha} I(y_{jd} < t). \tag{1.6}$$

Also the z values for the r_d non-sampled units are unknown, and they need to be predicted on the basis of the predicted y values.

The prediction of the y is generally based on a set of auxiliary variables following a regression model. In this perspective, the model-based methodologies allow for the construction of efficient estimators and their confidence intervals by borrowing the strength through use of a suitable model.

The prediction process can encounter inadequacies, difficulties, and problems due both to the characteristics of the available data and the specification and fitting of the SAE model. These issues depend on the amount and the extent of the information on the study variable and on the auxiliary information, and on the typology of the study variable we are interested in. Other problems are linked to the specification of the model as the under/over shrinkage effect of the variability of the estimates between the areas, the modeling of the spatial relationships among the areas and/or the units and the treatment of out-of-sample areas (see Section 1.3).

1.3 Data-related and Estimation-related Problems for the Estimation of Poverty Indicators

The data-related problems are faced when preparing the data information available to set up the estimation phase.

There are various sample surveys, both at EU and country level, on household income, consumption, labor force and living conditions that can be used to compute direct estimates of poverty and related indicators. However, these surveys have at least two limitations: (i) problems of incoherent definitions may rise, because no single data source is able to cover all the aspects; and (ii) the estimates are accurate only at the level of large areas, because the sample is sized at regional level (e.g., in Italy not at province and municipality level).

To overcome the first limitation, it is necessary to check the coherence among the different definitions of the target variables and to improve their comparability, as well as to integrate the micro data coming from different surveys and other data sources to increase the accuracy of the direct estimations.

The second limitation means that the survey data do not support reliable estimation at the level of a local area because sample sizes are often too small to provide direct estimates with acceptable variability (as measured by the coefficient of variation). Sometimes, these estimates could be obtained with larger samples, oversampling the areas of interest, but increasing also the survey costs, and this is not a generally feasible solution to the problem.

When the administrative register data are used as covariate in the SAE model, it is frequently necessary to integrate data coming from different administrative sources in order to derive more adequate auxiliary variables and more accurate and complete final statistics. This is not a straightforward procedure, as it is shown in Chapters 3 and 4 of this book. The keyword is the harmonization of the registers in such a way that information from different sources and observed data should be consistent and coherent.

Other data-related problems arise when indirect methods based on sample surveys are used:

(i) *The out-of-sample areas*. The estimation of target parameters at local area use both the data collected by the related survey and the auxiliary variables data available at that area level. Frequently, for some or many areas the values of the study variable are not available, and obviously the SAE have to face with this situation, that is known as the problem of out-of-sample areas or domains.

(ii) *The benchmarking*. Often the target parameters to be estimated at area level are to be related with known values referred to larger areas we want to break down with the estimation models. Once obtained, the small area estimates should be consistent with already known values for larger areas. Benchmarking is the consistency of a collection of small area estimates with a reliable estimate obtained according to ordinary design-based methods for the union of the areas. The population counts or the values of the target parameters in larger areas serve as a benchmark accounting for under coverage or over coverage and underreporting of the small area target values. Realignment of the small area estimates with the known values is an automatic result of the application of some small area methods. This is also particularly important for National Statistical Institutes to ensure coherence between small area estimates and direct estimates produced at higher level planned domains. In Section 1.4 we examine the methods from this perspective giving advice and warnings about their features and impact on the estimates, guiding the reader to other chapters of the book.

(iii) *The excess of zero values*. The excess of significant zero values in the data requires a preliminary investigation to formulate a model of behavior for the study variable in the population. There are many practical situations where the study variable can be conceptualized as skewed and strictly positive: in a population of individuals income and consumption follow those models. The problem of the zero excess emerges in situations where the target variable is not only skewed and strictly positive, but defined over the whole positive axis, zero included. Also, when analyzing significant variables to build up poverty indicators it is likely to be in the presence of survey data where there are many zero values of that variable for many sampled households. We refer here to the case of negative income values that are substituted by zero values. A high frequency of zeros can occur also when the study variable is a characteristic of the households, such as presence of households not able to keep their home adequately warm or with arrears on utility bills in a local area where living conditions are acceptable. In this case the problem is different and should be treated under the umbrella of SAE for a rare population.

(iv) *The outlier*. Outlier detection in the study variable have always been an interesting challenge when examining data to prepare the estimation of small area target parameters. If they are significant and not to be eliminated cleaning up the data set, they require methods that are robust against their effect on the validity of the small area model.

There are solutions described in recent literature to deal with the problem of excess of zeros and with the estimation in the presence of outliers which we will mention in Section 1.4 and they also are presented in the following chapters.

Part III of this book contains chapters devoted to the design-based estimation of poverty indicators and on related themes. Particularly Chapter 5 provides evidence on the effect of the sample design on SAE methods. Chapter 6 shows applications of the design-based framework to SAE and Chapter 7 illustrates the cumulation of panel data to estimate the sampling variance.

The estimation-related problems are inherent to the selected SAE model and its specification and fitting procedure. They produce an effect on the set of small area estimates affecting their heterogeneity and the meaning of their relation with other variables:

(v) *The shrinkage effect*. The SAE estimates can often be motivated from both a Bayesian and a frequentist point of view, can be obtained using the theory of best linear unbiased prediction (BLUP) or empirical best linear unbiased prediction (EBLUP) or under non-parametric and semi-parametric approaches using also M-quantile models. The chapters of Part III and Part V of this book show many of these models and present simulation studies and application to real poverty data. Nevertheless, there are situations where the models have the tendency for under/over-shrinkage of small area estimators. In fact, it is often the case that, if we consider a collection of small area estimates, they misrepresent the variability of the underlying "ensemble" of population parameters. In other words, the expected sampling variance of the set of predictions is less than the expected sampling variance of the ensemble of the true Small Area parameters (see Rao, 2003, section 9.6 for a discussion of this problem and also of adjusted predictors).

(vi) *The spatial modeling*. In recent years there have been significant developments in model-based small area methods that incorporate spatial information in an attempt to improve the efficiency of small area estimates by borrowing strength over space. The possible gains from modeling the correlations among small area random effects used to represent the unexplained variation of the small area target quantities are examined and compared with other parametric and non parametric approaches. The reader can find a review of spatio-temporal models in the chapters of Part IV. In Chapters 11, 12 and 13 there are examples of how these spatial models perform when estimation is for out-of-sample areas that is areas with zero sample, and issues related to estimation of mean squared error (MSE) of the resulting small area estimators are discussed. The emphasis is on point prediction of the target area quantities, and mean square error assessments. However, these alternative small area models using data with geographical information have to be studied also with reference to their performance whenever the Modifiable Area Unit Problem (MAUP) occurs.

(vii) *The Modifiable Area Unit Problem*. The MAUP appears when analyzing the relation (spatial or not) between variables. It is a potential source of error that can affect spatial studies, which utilize aggregate data sources and also the SAE results. The result can be diverse when the same relation is measured on different areal units. This can give misleading results in the specification of SAE models and affect the quality of the small area estimates. A simple strategy to deal with the problem of MAUP in SAE is to undertake analysis at multiple scales or zones. In Section 1.4 we will indicate some preliminary results on the scale effect of MAUP when obtaining small area estimates.

1.4 Model-assisted and Model-based Methods Used for the Estimation of Poverty Indicators: a Short Review

1.4.1 Model-assisted Methods

In the last 30 years mixture modes of making inference have become common in survey sampling: in many cases design-based inference is model assisted. Also in the SAE context the model-assisted approach has become popular and in this section we briefly review the most common estimators under this approach.

Among design-based methods assisted by the specification of a model for the study variable there are three families of methods that have been recently applied in poverty mapping: Generalized Regression (GREG) estimators; pseudo-EBLUP estimators; and M-quantile weighted estimators.

The GREG approach can be used to estimate several poverty indicators. With reference to the estimation of the small area mean, the estimators under this approach share the following structure:

$$\hat{m}_d^{GREG} = \sum_{j\in U_d} \hat{y}_{jd} + \sum_{j\in s_d} w_{jd}(y_{jd} - \hat{y}_{jd}), \tag{1.7}$$

where w_{jd} is the sampling weight of unit j within area d that is the reciprocal of the respective inclusion probability π_{jd}. Different GREG estimators are obtained in association with different models specified for assisting estimation, that is for calculating predicted values $\hat{y}_{jd}, j \in U_d$. In the simplest case a fixed effects regression model is assumed: $E(y_{jd}) = \mathbf{x}_{jd}^T\boldsymbol{\beta}, \forall j \in U_d, \forall d$ where the expectation is taken with respect to the assisting model. Lehtonen and Veijanen (1999) introduce an assisting two-level model where $E(y_{jd}) = \mathbf{x}_{jd}^T(\boldsymbol{\beta} + \mathbf{u}_d)$, which is a model with area-specific regression coefficients. In practice, not all coefficients need to be random and models with area-specific intercepts mimicking linear mixed models may be used (Lehtonen *et al.* 2003). In this case the GREG estimator takes the form of (1.7) with $\hat{y}_{jd} = \mathbf{x}_{jd}^T(\hat{\boldsymbol{\beta}} + \hat{\mathbf{u}}_d)$. Estimators $\hat{\boldsymbol{\beta}}$ and $\hat{\mathbf{u}}$ are obtained using generalized least squares and restricted maximum likelihood methods (Lehtonen and Pahkinen, 2004). See Chapter 6 of this book.

Under the pseudo-EBLUP approach the estimators are derived taking into account the sampling design both via the sampling weights and the auxiliary variables in the models. The estimators of the area mean proposed by Prasad and Rao (1999) and You and Rao (2002) are based on the assumption of a population nested error regression model and it is also assumed that the sampling design is ignorable given the auxiliary variables included in the model. As for the error terms it is assumed that $u_d \overset{i.i.d.}{\sim} N(0, \sigma_u^2)$ and $e_{ij} \overset{i.i.d.}{\sim} N(0, \sigma_e^2)$.

By combining a Hájek type direct estimator of $\bar{m}_d$ defined as $\bar{y}_{dw} = \sum_{j\in s_d} \breve{w}_{jd} y_{jd}$ where $\breve{w}_{jd} = w_{jd}\left(\sum_{j\in s_d} w_{jd}\right)^{-1}$, and the nested error regression model, Prasad and Rao (1999) obtain the following aggregated area level model:

$$\bar{y}_{dw} = \bar{\mathbf{x}}_{dw}^T\boldsymbol{\beta} + v_d + \bar{e}_{dw}, \tag{1.8}$$

with $\bar{e}_{dw} = \sum_{j\in s_d} \breve{w}_{jd} e_{jd}$ and $\bar{\mathrm{X}}_{dw} = \sum_{j\in s_d} \breve{w}_{jd} \mathrm{x}_{jd}$.

The design consistent pseudo-EBLUP estimator $\hat{\eta}_{dw}$ of the d th area mean is then given by:

$$\hat{\eta}_{dw} = \hat{\gamma}_{dw}\bar{y}_{dw} + (\bar{\mathbf{X}}_d - \hat{\gamma}_{dw}\bar{\mathbf{x}}_{dw})^T\hat{\boldsymbol{\beta}}_w, \tag{1.9}$$

where $\hat{\gamma}_{dw} = \hat{\sigma}_u^2(\hat{\sigma}_u^2 + \hat{\sigma}_e^2\delta_d)^{-1}$, $\delta_d = \sum_{j\in s_d} \breve{w}_{jd}^2$ and

$$\hat{\boldsymbol{\beta}}_w(\hat{\sigma}_u^2, \hat{\sigma}_e^2) = \left(\sum_{d=1}^{D}\sum_{j\in s_d} \breve{w}_{jd}\mathbf{x}_{jd}(\mathbf{x}_{jd} - \hat{\gamma}_{dw}\bar{\mathbf{x}}_{dw}^T)\right)^{-1}\left(\sum_{d=1}^{D}\sum_{j\in s_d} \breve{w}_{jd}(\mathbf{x}_{jd} - \hat{\gamma}_{dw}\bar{\mathbf{x}}_{dw}^T y_{jd})\right). \tag{1.10}$$

The variance components (σ_u^2, σ_e^2) can be estimated using for example, Restricted Maximum Likelihood (REML) or the fitting-of-constants method. Both Prasad and Rao (1999) and You and Rao (2002) provided formulae for the model-based MSE associated with the pseudo-EBLUP estimators of the area mean. Jiang and Lahiri (2006) noted that these estimators are not second-order correct. Torabi and Rao (2010) derived a second order unbiased predictor for the pseudo-EBLUP estimator (1.9).

An alternative family of model-assisted small area estimators is based on the M-quantile methodology (Chambers and Tzavidis, 2006), see Chapter 9 of this book. Recently, under this model, Fabrizi *et al.* (2014a) proposed a design consistent estimator of area-specific poverty indicators using the Rao–Kovar–Mantel estimator of the distribution function of income F_i (Rao *et al.* 1990) defined as:

$$\hat{F}_d^{WMQ/RKM} = N_d^{-1}\left[\sum_{j\in s_d} w_{jd} I(y_{jd}\leq t) + \sum_{j\in U_d} I(\mathbf{x}_{jd}^T\hat{\boldsymbol{\beta}}_{w\bar{\theta}_d}\leq t) - \sum_{j\in s_d} w_{jd} I(\mathbf{x}_{jd}^T\hat{\boldsymbol{\beta}}_{w\bar{\theta}_d}\leq t)\right], \tag{1.11}$$

where $\hat{\boldsymbol{\beta}}_{wq}$ is a design consistent estimator of $\boldsymbol{\beta}_q$. In the application of M-quantile regression to SAE, Chambers and Tzavidis (2006) characterize the variability across the population, beyond what is accounted for by the model covariates, by using the so-called M-quantile coefficients of the population units. For unit j in area d, this coefficient is the value θ_{jd} such that $Q_{\theta_{jd}}(y_{jd}|\mathbf{x}_{jd}) = y_{jd}$, where $Q_q(y_{jd}|\mathbf{x}_{jd})$ is the conditional M-quantile that is assumed to be a linear function of the auxiliary information. The authors observe that if a hierarchical structure does explain part of the variability in the population data, units within areas defined by this hierarchy are expected to have similar M-quantile coefficients. Average area coefficients $\bar{\theta}_d$ may be calculated and this represents an alternative approach to estimating area random effects without the need for using parametric assumptions.

More specifically, the weighted M-quantile-based small area estimator of the mean from (1.11) is:

$$\hat{\bar{m}}_d^{WMQ} = \int t d\hat{F}_d^{WMQ/RKM}(t) = \frac{1}{N_d}\sum_{j\in s_d} w_{jd} y_{jd} + \left(\frac{1}{N_d}\sum_{j\in U_d}\mathbf{x}_{jd}^T - \frac{1}{N_d}\sum_{j\in s_d} w_{jd}\mathbf{x}_{jd}^T\right)\hat{\boldsymbol{\beta}}_{w\bar{\theta}_d}. \tag{1.12}$$

The M-quantile method can be also used for estimating the HCR and the PG. Using t to denote the poverty line, different poverty indicators are defined by the area-specific mean of the variable derived:

$$f_{jd}(\alpha, t) = \left(\frac{t - y_{jd}}{t}\right)^{\alpha} I(y_{jd}\leq t), d = 1, \ldots, D; \quad j = 1, \ldots, N_d. \tag{1.13}$$

The population-level small area-specific poverty indicator can be decomposed as:

$$F_d(\alpha, t) = N_d^{-1}\left[\sum_{j\in s_d} f_{jd}(\alpha, t) + \sum_{j\in r_d} f_{jd}(\alpha, t)\right]. \tag{1.14}$$

The first component in (1.14) is observed in the sample, whereas the second component has to be predicted by using the M-quantile model. Tzavidis *et al.* (2014) propose a non-parametric approach by using a smearing-type estimator. More specifically:

$$F_d(\alpha, t) = N_d^{-1}\left[\sum_{j\in s_d} f_{jd}(\alpha, t) + \sum_{j\in r_d} E(f_{jd}(\alpha, t))\right]. \tag{1.15}$$

For simplicity let us focus on the simplest case when $\alpha = 0$. An estimator of $F_d(0,t)$ is obtained by substituting an estimator of $E(f_{jd}(\alpha,t))$ in (1.15) leading to

$$\hat{F}_d(0,t) = N_d^{-1}\left[\sum_{j\in s_d} w_{jd}f_{jd}(0,t) + \frac{1}{\sum_{j\in s_d} w_{jd}}\sum_{k\in r_d}\sum_{j\in s_d} w_{jd}I(\mathbf{x}_{kd}^T\hat{\boldsymbol{\beta}}_{w\bar{\theta}_d} + \hat{e}_{jd} \le t)\right], \quad (1.16)$$

where $\hat{e}_{jd}$s are the estimated residuals from the M-quantile fit. The same approach can be followed to estimate $\hat{F}_d(1,t)$ or any other of the FGT poverty measures.

For the estimation of the variance of the M-quantile (MQ) predictors see Fabrizi *et al.* (2014a) where two alternative estimators of the variance of the MQ predictors are proposed.

Even if the use of design consistent estimators in SAE is somewhat questionable because of the small sample sizes in some or all of the areas, as Pfeffermann noted (Pfeffermann, 2013), the families of methods we have described above offer generally design consistent estimators.

The three approaches previously described give partial solutions to the problems listed in Section 1.3: they give practical solutions to benchmarking, they deal with the presence of outliers, the estimates that they provide are differently affected by the shrinkage effect, and they all offer out-of-sample predictions.

Also to protect against possible model failures, *benchmarking* procedures make the total of small area estimates match a design consistent estimate for a larger area. With respect to benchmarking, all the families of methods offer a solution.

There are two kinds of benchmarked estimators: estimators that are internally benchmarked (or self-benchmarked) and those that are externally benchmarked. Self-benchmarked predictors are the GREG estimator and the pseudo-EBLUP introduced by You and Rao (2002). The externally benchmarked ones are more common under the model-based approach. For a recent review see Wang *et al.* (2008).

The GREG procedure uses the higher level totals as auxiliary data in calculating survey weights, thereby adjusting the lower level weights so that the total and subtotal estimates are consistent (see also Smith and Hidiroglou, 2005). In addition, the weights that are used for direct estimation using survey data in GREG expression are often constructed using calibration methods, Often benchmarking to auxiliary totals is used together with weight equalization. Benchmarking (forcing certain estimates to match known totals) has been shown to reduce variances for statistics correlated with the auxiliary characteristics, and weight equalization (forcing the weights within higher-level units to be equal) has been shown to further reduce variances for statistics measured on the higher-level units (Lehtonen and Veijanen, 1999). The pseudo-EBLUP estimators satisfy the benchmarking property without any adjustment in the sense that they add up to the direct survey regression estimator when aggregated over the areas. A drawback of this type of self-benchmarked estimators is that they force the use of the same auxiliary information used for the direct usually GREG-type estimator also for the model-based small area predictors, whereas it could be very profitable to allow for different auxiliary variables at the small area level. Coming to the M-quantile approach note that expression (1.12) has a GREG-type form. This is the basis to see that the MQ predictors do not satisfy the benchmarking property as it is shown in Fabrizi *et al.* (2014b). Here the authors propose a method of constraining M-quantile regression. It can be applied to obtain benchmarking MQ small area estimates.

The treatment of the *outliers* is not the focus of the estimators of GREG type nor of those under the pseudo-EBLUP approach, while the weighted M-quantile approach this issue.

There are studies under the AMELI (2008) project that illustrate the behavior of the GREG-like estimators in the presence of different models of outlier-contamination of the observed data. The results show that even if a robust method of fitting the logistic mixed model was not available, the poverty rate estimators are fairly robust: this happens both under a simple random sampling design and under a complex sampling design (AMELI, 2008, Deliverable 2.2). To deal with outliers, Beaumont and Alavi (2004) use the weighted generalized M-estimation technique to reduce the influence of units with large weighted population residuals. With respect to the empirical pseudo best approach recalled before there is no contribution addressing the robustification of the estimates against the presence of outliers. Jiang *et al.* (2011) relaxed some of the classical EBLUP model to obtain robust-model based predictors. These relaxations may work also under the pseudo-EBLUP approach but until now no evidence of it has been produced. The AMELI project provides evidence also on the behavior of the Empirical Best Predictor type estimator based on a logistic mixed model. This estimator is least affected by contaminations when the data come from a simple random sample but it is not based on the pseudo-EBLUP approach. As it concerns the M-quantile estimator with respect to GREG-S popular in small area literature (see Rao, 2003, section 2.5), note that: (*i*) the use of an area-specific coefficient ($\bar{\theta}_d$) in M-quantile regression accounts for area characteristics not explained by the auxiliary variables; and (*ii*) the use of M-estimation offers outlier robust estimation. Specifically, the recourse to M-quantile regression reduces the impact that outlier observations have on the estimated regression coefficients and thereby on the small area means.

The models which are assisting the estimation under the design-based approach can have have the tendency for *under/over-shrinkage* of small area estimators.

The desirable property of neutral shrinkage is not achieved under the pseudo-EBLUP approach. In this case it is reasonable that the over-shrinking behavior of the Empirical Best predictors is confirmed. The understatement of extreme values, referred to as over-shrinkage in this context, is problematic when the goal is the description of the overall distribution among areas. However this tendency can be adjusted (see EURAREA, 2001, section B.3) and it is likely that the adjustment can work even under the pseudo-EBLUP approach, but up to now no evidence of it has been produced.

The tendency of GREG estimators is similar to that of direct estimators and in contrast to that of the over-shrinking empirical Bayes (EB) predictors, as the results of the EURAREA project have shown. The behavior of M-quantile-based predictors is then more similar to that of direct estimators and GREG. Fabrizi *et al.* (2014b) propose an adjustment of the benchmarked MQ predictors in order to obtain estimators with approximately neutral shrinkage. This adjustment parallels the one used to adjust EB predictors (Rao, 2003, see Section 9.6). They extend the methodology of Fabrizi *et al.* (2014b) to obtain estimates that enjoy "ensemble" properties, that is properties related to the estimation of a functional of an ensemble of parameters (Frey and Cressie, 2003). An ensemble of estimators is said to be neutral with respect to shrinkage if the variance of the ensemble of the parameters can be unbiasedly estimated by the variance of the ensemble of the estimators. This guarantees a correct representation of the geographical variation of the variable in question. Otherwise, this geographical variation may be over- or underestimated. Neutral shrinkage is important when small area estimators are used to create "maps".

For the set $E = \{d|n_d = 0\}$ of *the out-of-sample areas*, that is areas where $n_d = 0$, the GREG-like estimators cannot be computed. The pseudo-EBLUP approach provides predictors under the specified models which are likely to underestimate the variability of th estimates among areas. Consistently with Chambers and Tzavidis (2006), the small area estimator $\hat{\hat{m}}_d^{WMQ}$ can be defined as $N_d^{-1} \sum_{j \in U_d} \mathbf{x}_{jd}^T \hat{\boldsymbol{\beta}}_{w0.5}$, that is a synthetic estimator based on the weighted M-regression.

1.4.2 Model-based Methods

The most popular method used for model-based SAE employs linear mixed models. In the general case such a model has the form:

$$y_{jd} = \mathbf{x}_{jd}^T \boldsymbol{\beta} + u_d + e_{jd}, \tag{1.17}$$

where u_d is the area-specific random effect and e_{jd} is an individual random effect. The empirical best linear unbiased predictor (EBLUP) of m_d (Henderson, 1975, Rao, 2003, chapter 7) is then

$$\hat{\hat{m}}_d^{LM} = N_d^{-1} \left[\sum_{j \in s_d} y_{jd} + \sum_{j \in r_d} \{ \mathbf{x}_{jd}^T \hat{\boldsymbol{\beta}} + \hat{u}_d \} \right], \tag{1.18}$$

where $\hat{\beta}$, $\hat{u}_d$ are defined by substituting an optimal estimator for the covariance matrix of the random effects in (1.17) in the best linear unbiased estimator of $\boldsymbol{\beta}$ and the BLUP of u_d, respectively. A widely used estimator of the MSE of the EBLUP is based on the approach of Prasad and Rao (1990). This estimator accounts for the variability due to the estimation of the random effects, regression parameters, and variance components.

Models presented in Parts IV and V of this book rely on and often enlarge the assumptions of this popular approach: Chapter 8 introduces the issue of measurement error in the covariates; Chapter 10 extends it to a non-parametric regression environment; and Chapters 11, 12 and 13 extend it to take into account spatial and temporal correlations and the characteristics of geographical patterns.

Assuming model (1.17) on the logarithmically transformed values of income y_{jd}, the most widely used method for small area poverty mapping is the so-called World Bank (WB) or Elbers, Lanjouw and Lanjouw (ELL) method (Elbers *et al.* 2003). Chapter 18 describes links, alternatives and models used under this approach. The model is fitted to clustered survey data from the population of interest, with the random effects in the model corresponding to the cluster used in the survey design. Once the model has been estimated using the survey data, the ELL method uses the following bootstrap population model to generate L synthetic censuses:

$$y_{jd}^* = \mathbf{x}_{jd}^T \hat{\boldsymbol{\beta}} + u_d^* + e_{jd}^*, u_d^* \sim N(0, \hat{\sigma}_u^2), e_{jd}^* \sim N(0, \hat{\sigma}_e^2) \tag{1.19}$$

For each draw, using the synthetic values of the welfare variable y_{jd}^*, values of the poverty indicators of interest for the different small areas are calculated. These are averaged over the L Monte Carlo simulations to produce the final estimates of the poverty quantities, with the simulation variability of these estimates used as an estimate of their uncertainty.

Molina and Rao (2010) point out that when small areas and clusters coincide, in the simplest case of estimating a small area mean, the ELL method leads to a synthetic regression estimator that, in many cases, could be less efficient than the alternative model-based estimators. Molina and Rao (2010) propose a modification of the ELL method (the empirical best predictor (EBP) method) introducing random area effects (rather than random cluster effects) into the linear regression model for the welfare variable, and also simulated out-of-sample data by independent drawings the conditional distribution of the out-of-sample data, given the sample data. Deeper insights on this and recent enhancements of the method are described in Chapter 17.

An alternative approach to EBLUP has been discussed in Chandra and Chambers (2005) and it is based on the use of model-based direct estimation (MBDE) within the small areas. In this case an estimate for a small area of interest corresponds to a weighted linear combination of the sample data for that area, with weights based on a population level version of the linear mixed model. These weights "borrow strength" via this model, which includes random area effects. Provided the assumed small area model is true, the EBLUP is asymptotically the most efficient estimator for a particular small area. In practice however the "true" model for the data is unknown and the EBLUP can be inefficient under misspecification. In such circumstances, Chandra and Chambers (2005) note that MBDE offers an alternative to potentially unstable EBLUP. In particular, MBDE is easy to implement, produces sensible estimates when the sample data exhibit patterns of variability that are inconsistent with the assumed model (e.g., contain too many zeros) and generates robust MSE estimates. The MBDE is presented in Chapter 14 of this book for the estimation of the Cumulative Distribution Function.

A different approach has been proposed in the literature for further robustification of the inference by relaxing some of the model assumptions. This approach is based on M-quantile regression (Breckling and Chambers, 1988). It provides a "quantile-like" generalization of regression based on influence functions (Breckling and Chambers, 1988). A linear M-quantile regression model is one where the qth M-quantile $Q_q(y_{jd}|\mathbf{x}_{jd})$ of the conditional distribution of y given x satisfies:

$$Q_q(y_{jd}|\mathbf{x}_{jd}) = \mathbf{x}_{jd}^T\boldsymbol{\beta}_q. \tag{1.20}$$

That is, it allows a different set of regression parameters for each value of q. For specified q and continuous influence function ψ, an estimate $\hat{\boldsymbol{\beta}}_q$ of $\boldsymbol{\beta}_q$ can be obtained via an iterative weighted least squares algorithm.

As stated in the previous section, extending this line of thinking to SAE, Chambers and Tzavidis (2006) observed that if variability between the small areas is a significant part of the overall variability of the population data, then units from the same small area are expected to have similar M-quantile coefficients. In particular, when (1.20) holds, and $\boldsymbol{\beta}_q$ is a sufficiently smooth function of q, these authors suggest a predictor of m_j of the form:

$$\hat{m}_d^{MQ} = N_d^{-1}\left[\sum_{j\in s_d} y_{jd} + \sum_{j\in r_d} \hat{Q}_{\bar{\theta}_d}(y_{jd}|\mathbf{x}_{jd})\right], \tag{1.21}$$

where $\hat{Q}_{\bar{\theta}_d}(y_{jd}|\mathbf{x}_{jd}) = \mathbf{x}_{jd}^T\hat{\boldsymbol{\beta}}_{\bar{\theta}_d}$ and $\bar{\theta}_d$ is an estimate of the average value of the M-quantile coefficients of the units in area d. Typically this is the average of estimates of these coefficients for sample units in the area. When there is no sample in the area, we can form a "synthetic" M-quantile predictor by setting $\bar{\theta}_d = 0.5$. Tzavidis *et al.* (2010) refer to (1.21) as the "naïve" M-quantile predictor and note that this can be biased and they propose a bias adjusted M-quantile predictor of m_d.

The M-quantile small models are used also for estimating the poverty indicators such as HCR and PG (Tzavidis *et al.* 2014) by using a smearing-type estimator (Duan, 1983). A small area estimator of the HCR is obtained as:

$$\hat{F}_d(0,t) = N_d^{-1}\left[\sum_{j\in s_d} f_{jd}(0,t) + \hat{E}[f_{jd}(0,t)]\right] \tag{1.22}$$

where

$$\hat{E}[f_{jd}(0,t)] = \int I(\mathbf{x}_{jd}^T\hat{\boldsymbol{\beta}}_{\bar{\theta}_d} + \hat{e}_{jd} \leq t)d\hat{F}(\hat{e}) = n^{-1}\sum_{k\in r_d}\sum_{j\in s_d} I(\mathbf{x}_{kd}^T\hat{\boldsymbol{\beta}}_{\bar{\theta}_d} + \hat{e}_{jd} \leq t)$$

with the distribution function estimated as $\hat{F}(\hat{e}) = n^{-1}\sum_{j=1}^{n} I(\hat{e}_j \leq e)$. The same approach can be used to estimate the PG indicator or any other of the FGT poverty measures.

Under the model-based approach many of the problems listed in Section 1.3 have a solution, for example all of them offer out-of-sample predictors. Among the other issues we focus here on the excess of zero values in the data and in the treatment of geographic information and spatial data.

Model-based estimators usually do not have the *benchmarking* property under a complex sampling design. Given a small area estimator, that does not show the benchmarking property, a first simple way of achieving benchmarking is by a ratio type adjustment. Externally benchmarked predictors are obtained through an a-posteriori adjustment of model-based predictors. Among the others, Pfeffermann and Barnard (1991) propose an externally restricted benchmarked estimator of small area means. This is constructed under an area linear mixed model for a continuous response variable.

Many variables of interest in economics surveys on poverty and living conditions are semicontinuous in nature, that is they either take a single fixed value (typically 0, zero) or they have a continuous, often skewed, distribution on the positive real line. They present an *excess of zero values*. A semicontinuous variable is quite different from one that has been left censored or truncated, because the zeros are valid self-representing data values, not proxies for negative or missing responses. A two-part random effects model (Olsen and Schafer, 2001) is widely used for SAE with zero-inflated variables, see for example, Pfeffermann *et al.* (2008) and Chandra and Sud (2012). Chandra and Chambers (2014) propose a SAE method for semicontinuous variables under a two part random effects model. The issues which arise when the data are lognormal are discussed in Chapter 15.

In poverty studies observations that are spatially close may be more alike than observations that are further apart. One approach for incorporating spatial information in *spatial modeling* and in a small area regression model is to assume that the model coefficients themselves vary spatially across the geography of interest and/or the random effects of the model be correlated. Both EBLUP predictors and MQ predictors can be extended to include the effect of the spatial characteristics of the data. These extensions can be applied to poverty studies (see SAMPLE, 2008, deliverables), but are not reviewed in this book.

When geography is included as auxiliary information in modeling, the spatial correlation and the consequent correlation between the random effect in the EBLUP model require the extension of the EBLUP estimator to the Spatial Empirical Best Linear Unbiased Predictor (SEBLUP) estimator (Petrucci and Salvati, 2006, Pratesi and Salvati, 2009).

Under the MQ approach the reference to the Geographically Weighted Regression (GWR) (Brundson *et al.* 1996) helps in modeling spatial variation. This uses local rather than global parameters in the regression model. That is, a GWR model assumes spatial non-stationarity of the conditional mean of the variable of interest. Salvati *et al.* (2012) propose an M-quantile GWR model, that is a local model for the M-quantiles of the conditional distribution of the outcome variable given the covariates. This approach is semi-parametric in that it attempts to capture spatial variability by allowing model parameters to change with the location of the units, in effect by using a distance metric to introduce spatial non-stationarity into the mean structure of the model. The model is then used to define a predictor of the small area characteristic of interest. As a consequence, it integrates the concepts of bias-robust SAE and borrowing strength over space within a unified modeling framework. By construction, the model is a local model and so can provide more flexibility in SAE, particularly for out-of-sample small area estimation, that is areas where there are no sampled units. For the estimation of the variance of the predictors see Chambers *et al.* (2011, 2014).

When studying the spatial distribution of local poverty indicators obtained by SAE methods, it can be relevant to consider the possible effect of the MAUP. This is a source of statistical bias that can radically affect the results of statistical analysis. It affects results when point-based measures of spatial phenomena (e.g., population density) are aggregated into larger areas. The resulting summary values (e.g., totals, rates, proportions) are influenced by the choice of the boundaries of the areas. For example, point-based census or survey data may be aggregated into census enumeration districts, or post-code areas, or any other spatial partition (thus, the "areal units" are "modifiable").

The topic has not yet been treated explicitly in the current literature on SAE. The only empirical study is due to Pratesi and Petrucci (2014) who studied the scale effect on SAE predictors by a simulation experiment. They provide evidence to assess the robustness of SAE methods to different scale of aggregation of the point-based measures inside the pre-defined small areas (domains) of interest. The rationale of this simulation study is to verify to what extent we can aggregate the individual values inside the small areas and still have an acceptable accuracy of the estimate of the small area parameter. Under this simulation experiment, methods that are naturally robust to outliers and not linked to distributional assumption on the study variable as M-quantile methods perform better than the alternative methods for SAE and are found to be resilient to changing scale of analysis. This is likely due to the fact that the changes in geography do not affect the M-quantile coefficients at area level.

References

AMELI: Advanced Methodology for European Laeken Indicators 2008 Project no. SSH-CT-2008-217322. FP7-SSH-2007-1.

Atkinson AB, Cantillon B, Marlier E, and Nolan B 2002 *Social Indicators: The EU and Social Inclusion*. Oxford: Oxford University Press.

Beaumont JF and Alavi A 2004 Robust Generalized Regression Estimation. *Survey Methodology* **30**, 195–208.

Breckling J and Chambers R 1988 M-quantiles. *Biometrika* **795**(4), 761–771.

Brundson, C, Fotheringham AS, and Charlton M 1996 Geographically weighted regression: a method for exploring spatial nonstationarity. *Geographical Analysis* **28**, 281–298.

Chambers R, Chandra H, and Tzavidis N 2011 On bias-robust mean squared error estimation for pseudo-linear small area estimators. *Survey. Methodology*, **37**, 153–170.

Chambers R and Tzavidis N 2006 M-quantile models for Small Area Estimation. *Biometrika*, **93**, 255–268.

Chambers R, Chandra H, Salvati N, and Tzavidis N 2014 Outlier robust small area estimation. *Hournal of the Royal Statistical Society: Series B* **76**(3), 47–69.

Chandra H and Chambers R 2014 Comparing EBLUP and C- EBLUP for Small Area Estimation. *Biometrical Journal*, doi: 10.1002/bimj.201300233.

Chandra H and Chambers R 2005 Small area estimation for semicontinuous data. *Statistics in Transition*, **7**, 637–648.

Chandra H and Sud UC 2012 Small area estimation for zero-inflated data. *Communications in Statistics — Simulation and Computation* **41**(5), 632–643.

Cheli B and Lemmi A 1995 A totally fuzzy and relative approach to the multidimensional analysis of poverty. *Economic Notes* **24**, 115–134.

Cochran WG 1977 *Sampling Techniques*, 3rd ed. New York: John Wiley & Sons, Inc.

Duan N 1983 Smearing estimate: a non parametric retransformation method. *Journal of the American Statistical Association* **78**, 605–610.

Elbers C, Lanjouw JO, and Lanjouw P 2003 Micro-level estimation of poverty and inequality. *Econometrica* **71**, 355–364.

EURAREA 2001 Programme funded by Eurostat under the Fifth Framework (FP5) Programme of the European Union.

Fabrizi E, Giusti C, Salvati N, and Tzavidis N 2014a Mapping average equivalized income using robust small area methods. *Papers in Regional Science* **93**, 685–701.

Fabrizi E, Salvati N, Pratesi M, and Tzavidis N 2014b Outlier robust model-assisted small area estimation. *Biometrical Journal* **56**, 157–175.

Foster J, Greer J, and Thorbecke E 1984 A class of decomposable poverty measures. *Econometrica* **52**, 761–766.

Frey J and Cressie N 2003 Some results on constrained Bayes estimators. *Statistics and Probability Letters* **65**, 389–399.

Gosh M and Meeden G 1997 *Bayesian Methods for Finite Population Sampling*. London: Chapman & Hall.

Hansen MH, Hurwitz WN and Madow WG 1953 *Sample Survey Methods and Theory*. New York: John Wiley & Sons, Inc.

Henderson C 1975 Best linear unbiased estimation and prediction under a selection model. *Biometrics* **31**, 423–447.

Jiang J and Lahiri P 2006 Estimation of finite population domain means: a model-assisted empirical best prediction approach. *Journal of the American Statistical Association* **101**, 301–311.

Jiang J, Nguyen T, and Rao JS 2011 Best predictive small area estimation. *Journal of the American Statistical Association* **106**, 732–745.

Kish L 1965 *Survey Sampling*, New York: John Wiley & Sons, Inc.

Lehtonen R and Pahkinen E 2004 *Practical Methods for Design and Analysis of Complex Surveys* Chichester, England: John Wiley & Sons, Ltd.

Lehtonen R, Särndal CE, and Veijanen A 2003 The effect of model choice in estimation for domains, including small domains. *Survey Methodology* **29**, 33–44.

Lehtonen R and Veijanen A 1999 Domain estimation with logistic generalized regression and related estimators. IASS Satellite Conference on Small Area Estimation. Riga: Latvian Council of Science, 121–128.

Molina I and Rao JNK 2010 Small area estimation of poverty indicators. *The Canadian Journal of Statistics* **38**, 369–385.

Olsen MK and Schafer JL 2001 A two-part random-effects model for semicontinuous longitudinal data. *Journal of the American Statistical Association* **96**, 730–745.

Petrucci A and Salvati N 2006 Small area estimation for spatial correlation in watershed erosion assessment. *Journal of Agricultural, Biological, and Environmental Statistics* **11**, 169–182.

Pratesi M and Salvati N 2009 Small area estimation in the presence of correlated random area effects. *Journal of Official Statistics* **25**(1), 37–53.

Pfeffermann D and Barnard CH 1991 Some new estimators for small-area means with application to the assessment of farmland values. *Journal of Business & Economics Statistics* **9**, 73–84.

Pfeffermann D, Terryn B, and Moura FAS 2008 Small area estimation under a two-part random effects model with application to estimation of literacy in developing countries. *Survey Methodology* **34**, 235–249.

Pfeffermann D 2013 New important developments in small area estimation. *Statistical Science* **28**(1), 40–68.

Prasad NGN and Rao JNK 1990 The estimation of the mean squared error of small-area estimators. *Journal of the American Statistical Association* **85**, 163–171.

Prasad NGN and Rao JNK 1999 On robust small area estimation using a simple random effects model. *Survey Methodology* **25**, 67–72.

Pratesi M and Petrucci A 2014 Methodological and operational solutions to gaps and issues on methods for producing agricultural and rural statistics at small domains level and on methods for aggregation, disaggregation and integration of different kinds of geo-referenced data for increasing the efficiency of agricultural and rural statistics. In *FAO* report.

Rao JNK 2003 *Small Area Estimation*. Hoboken, New York: John Wiley & Sons, Inc.

Rao JNK, Kovar JG, and Mantel HJ 1990 On estimating distribution functions and quantiles from survey data using auxiliary information. *Biometrika* **77**, 365–375.

Salvati N, Tzavidis N, Pratesi M, and Chambers R 2012 Small area estimation via M-quantile geographically weighted regression. *TEST* **21**, 1–28.

SAMPLE: Small Area Methods for Poverty and Living Condition Estimates 2008 Project no. SSH-CT-2008-217565. FP7-SSH-2007-1.

Särndal CE, Swensson B, and Wretman J 1992 *Model Assisted Survey Sampling*. New York: Springer-Verlag.

Smith P and Hidiroglou M 2005. Benchmarking through calibration of weights for microdata. *Working Papers and Studies*. Luxembourg: Office for Official Publications of the European Communities.

Torabi M and Rao JNK 2010 Mean squared error estimators of small area means using survey weights. *The Canadian Journal of Statistics* **38**, 598–608.

Tzavidis N, Marchetti S, and Chambers R 2010 Robust prediction of small area means and distributions. *Australian and New Zealand Journal of Statistics* **52**, 167–186.

Tzavidis N, Marchetti S, and Donbavand S 2014 Outlier robust semi-parametric small area methods for poverty estimation. In *Poverty and Social Exclusion, New Methods and Analysis* edited by Gianni Betti and Achille Lemmi, Chapter 15, 283–300. Routledge, Abingdon, Oxon and simultaneously New York.

Valliant R, Dorfman AH, and Royall RM 2000 *Finite Population Sampling and Inference: A Prediction Approach*. New York: John Wiley & Sons, Inc.

Wang J, Fuller WA, and Qu Y 2008 Small area estimation under a restriction. *Survey Methodology* **34**, 29–36.

You Y and Rao JNK 2002. A pseudo-empirical best linear unbiased prediction approach to small area estimation using survey weights. *The Canadian Journal of Statistics* **30**, 431–439.

Part I

Definition of Indicators and Data Collection and Integration Methods

2

Regional and Local Poverty Measures

Achille Lemmi[1] and Tomasz Panek[2]
[1]*Department of Economics and Statistics and Honorary Fellow ASESD Tuscan Universities Research Centre "Camilo Dagum", University of Siena, Siena, Italy*
[2]*Warsaw School of Economics, Warsaw, Poland*

2.1 Introduction

Combating poverty and social exclusion is one of the main targets of social policy conducted by the EU and its Member States (Maastricht Treaty). Reduction of poverty and social exclusion along with sustainable economic growth and increasing employment are considered as main areas of interest of the European Commission and are fundamental parts of the Lisbon Strategy. Likewise, in a revised version of the Lisbon Strategy social inclusion is still considered as a strategic area for the EU. In 2010 the Council of Europe enacted five major goals of the Europe 2020 Strategy. One of the five goals is to promote social inclusion, in particular, by reducing poverty by lifting at least 20 million individuals out of the poverty by 2020 (Copeland and Daly, 2012).

In spite of leaving large autonomy in the ways of combating poverty and social exclusion to EU Member States, the European Commission stresses the necessity of obtaining internationally comparable results of the undertaken social policies in this area in each country. In order to monitor the process of social inclusion, a list of 18 indicators monitoring poverty and social exclusion was proposed in 2001 (Atkinson *et al.*, 2002). These indicators are known as "Laeken" indicators.[1] The list is constantly modified and complemented.[2] It contains both

[1] A set of indicators was established at the European Council in the Brussels suburb of Laeken in Belgium.
[2] This list is developed by the Indicators Sub-Group of the Social Protection Committee (SPC). An updated list of indicators adopted in September 2009 by the SPC is on the Commission's website: http://ec.europa.eu/eurostat/ramon/nomenclatures/index.cfm?TargetUrl=LST_CLS_DLD&StrNom=NUTS_33&StrLanguageCode=EN

Analysis of Poverty Data by Small Area Estimation, First Edition. Edited by Monica Pratesi.

Companion Website: www.wiley.com/go/pratesi/poverty

indicators based on household incomes (monetary poverty indicators) and indicators based on non-monetary symptoms of social exclusion (indicators of non-income social exclusion dimensions). At the same time, the European Commission decided to launch a new survey aimed at measuring incomes and living conditions in the EU Member States (EU Statistics on Income and Living Conditions – EU-SILC). The EU-SILC was meant to be coordinated by the Eurostat and provide internationally comparable results (Wolf *et al.*, 2010). The EU-SILC is used to calculate basic indicators of poverty and social exclusion.

Goals formulated in the EU and national social policies distinctly indicate the need for analyzing poverty at regional and local levels. Regional differences and marginalization of certain EU regions have recently become one of the main areas of interest of the EU integration policies. Constant monitoring of poverty at a regional level is needed in order to adequately allocate EU funds aimed at combating poverty and social exclusion and assess the effectiveness of their spending. This chapter presents the different definitions of poverty, social exclusion and appropriate indicators in Sections 2.2, 2.3 and 2.4. The multidimensional and fuzzy nature of the poverty and of poverty indicators is described in Sections 2.5 and 2.6. The chapter finally shows and comments on an example of mapping of the regional distribution of many indicators in Europe.

2.2 Poverty – Dilemmas of Definition

The very first step to measure poverty should be providing a definition of the phenomenon in question. The choice of the specific definition of poverty directly influences outcomes of the measurement (Hagenaars, 1986). Depending on the chosen definition of poverty different social groups or various regions in regional analysis may be seen as poverty-stricken. At the same time, the way of defining poverty affects the allocation of EU regional policy funds as well as the way of creating social policy programs aimed at curbing poverty.

All definitions of poverty in the literature are focused on the inability to meet basic needs at a satisfactory level (Drewnowski, 1997).

Until the end of 1960s the basic needs approach was a leading approach used in poverty analysis. Poverty was seen as a situation in which incomes are lower than the ones required to meet the basic needs. This approach to measuring poverty based on monetary indicators, whose foundation was set forth by the School of Welfare Economics (Marshall, 1920), dominated in nearly all research into this phenomenon until the 1970s. Therefore the concept of poverty based on the level of income required for the meeting of basic needs is referred to as a monetary poverty or income poverty.

Gradually, the range of basic needs covered by the poverty category broadened. Along with the broadening of the basic needs scope, the viewpoint that the identification of impoverished persons exclusively on the basis of monetary categories is sufficient, began to meet with considerable criticism (Abel-Smith and Townsend, 1965). It was accompanied at the same time by moving from the concept to understand poverty as a lack of financial resources to satisfy basic needs (the basic needs approach) toward the inability to perform the functions of life, resulting not only from the lack of financial resources but also social and personal determinants that influence the conduction of valuable life (the capabilities approach; Sen, 1985). As a result, poverty is often confused with social exclusion.

Social exclusion is generally defined as a process in which individuals or social groups are restrained from full participation in substantial areas of social, cultural, economic and political life of the society in which they live (Silver, 1994). The dimensions of social exclusion often reinforce one another, and consequently, lead to even deeper marginalization of individuals. The notion of social exclusion is not restricted only to the lack of material resources. It also refers to other constrains that block individuals (families, households, social groups) from living in the way which is accepted in the country in which they live. Identifying poverty with social exclusion results in examining this phenomenon in terms of the inability to access something not only for financial reasons, not limited exclusively to the availability of goods and services meeting basic needs.

Social exclusion should not be considered as a synonym to poverty. The inability to meet basic needs may be identified as poverty only if it is caused by the lack of adequate material resources. Moreover, social exclusion is not always caused by poverty. Thus, poverty may be regarded as a financial dimension of social exclusion.

In this chapter an economic definition of poverty is used. Poverty would imply a situation where an individual (a person, a family, a household) does not have sufficient financial resources (both cash in the form of current income, income from previous periods and accumulated non-cash assets) to satisfy basic needs on an acceptable level.

2.3 Appropriate Indicators of Poverty and Social Exclusion at Regional and Local Levels

We begin this chapter by identifying special features and requirements of the system of indicators of poverty and social exclusion appropriate for use at the regional level. Specifically, the requirement is to identify whether, and if so in what manner, indicators appropriate for the regional level may differ from the indicators designed primarily for the national level.

Indicators of poverty and social exclusion of course have an important territorial dimension, pointing to the need to take account of regional and local differences. In an ideal context, one may seek to give regional breakdown on all indicators. That is, one may introduce regional analysis within each of the indicator fields, for instance producing poverty rates by NUTS classification,[3] urban–rural classification, and so on. However, simply the introduction of more extensive breakdown is neither possible because of data limitations, nor sufficient in itself.

Some of the Laeken indicators may be suitable for regional application; others may be suitable after modification; while some may not be appropriate for the purpose.

For the monitoring of poverty at the regional and local levels, the starting point of course is to specify a set of indicators in this area. It is also absolutely necessary to extend the list of Laeken indicators to non-income poverty indicators (i.e., material deprivation indicators).

2.3.1 *Adaptation to the Regional Level*

Henceforth these indicators have been applied at the national level. It is necessary to adapt them for regional application, taking into account any differences in the requirements, but equally important, differences in the practical situation. As in the case of regional adaptation of all other indicators, it is necessary to focus on the more basic among this set of indicators.

[3] NUTS (Nomenclature of Statistical Territorial Units) classification is a single, coherent system for dividing up the EU territory for statistical purposes. The NUTS classification is available at: http://www.europa.euint/corom/eurostat/ramon/nuts/splash_ regions.html.

This is because of the substantially increased data requirements when the results have to be geographically disaggregated.

Detailed disaggregation of the indicators by age, gender and other characteristics – simultaneously with disaggregation by geographical region – has to be severely restricted where the information comes from sample surveys of limited size, as is the case in most Member States lacking income registers. Broad classification, such as distinguishing children and old persons, may be possible, but even that has to be subsidiary to the need for adequate regional breakdown.

Certain more complex monetary poverty and monetary poverty inequality measures – measures which are more sensitive to details and irregularities of the empirical income distribution – are less suited for disaggregation to small populations and small samples. Indicators such as the Gini coefficient and even decile ratios (S80/S20)[4] may be too demanding at say NUTS 2[5] level.

The above considerations apply, though to a lesser extent, to Laeken indicators such as "Indicator 4: Relative median at-risk-of-poverty gap, by age and gender" and "Indicator 13: At-risk-of-poverty rate before social transfers, by age and gender".

The level of income poverty is determined by the chosen poverty line. By choosing different poverty lines, different numerical values are obtained, and to some extent each such figure provides additional information. It is for this reason that the Laeken list includes Indicator 11 "Dispersion around the at-risk-of-poverty threshold (illustrative values)", meaning poverty rates defined using 40, 50, 60 and 70% of the national mean as the poverty line. Incorporation of the effect of choosing different poverty lines is also important when we move down to the regional level. However, again in view of small sample sizes, it is desirable to avoid producing separate figures, any real differences between which may be overwhelmed by sampling variability and other errors in the data. Rather, it is more useful to consolidate such separate figures into a single (or at most a very small set of) more robust measure(s) if possible. The ideas explained later (see Chapter 19 in particular) of "consolidating" the measures – such as in the form of suitably weighted averaging over different numerical measures computed with different thresholds and levels of poverty lines – could be applied. In specific terms, a single measure based on suitable consolidation over, say, 50, 60 and 70% of median poverty lines, would be preferable to separate indicators such as Laeken Indicators 1 and 11. As a consequence, for the purpose of regional indicators, the focus has to be primarily on ordinary poverty rates for the total population, and possibly some special groups such as children and the elderly.

In addition, it is also necessary to consider whether there is need for addition to the existing indicators developed primarily for application at national level – region-specific indicators able to capture aspects which are essentially regional. It is possible that a more diverse "portfolio of indicators" is required for the purpose of addressing concerns of regional policy and research.

Perhaps the most important of these is simply the mean income levels of the regions, the dispersion among which provides a measure of regional disparities. General entropy measures such as GE(0) and GE(1) may also be useful because they can be decomposed into within and between region components.

It is clear from all the considerations described above, that for indicators at the regional level as well as at local level, in comparison with those at country level, there has to be less stress on

[4] Ratio of total income received by the 20% of the country's population with the highest income (top quintile) to that received by the 20% of the country's population with the lowest income (lowest quintile).

[5] NUTS 2 are basic regions for the purposes of the regional policy and in general contain between 800 000 and 3 million people.

monetary indicators and consequently increased stress on non-monetary indicators (material deprivation indicators). In 2010 the Indicators sub-group (ISG) of the EU Social Protection Committee (SPC) proposed measures aimed at monitoring progress of social integration within the EU. Finally, in June 2010 the Employment, Social Policy, Health and Consumer Affairs Council (EPSCO) accepted the proposal of the SPC to adopt two measures as benchmarks for the assessment of the process of realization of the "Europe 2020" strategy in the fields of poverty reduction (Bradshaw and Mayhew, 2011). These measures are:

- incidence of monetary poverty;
- incidence of material deprivation.

The proposed system of indicators marks a significant step toward a comprehensive assessment of poverty as it incorporates both monetary and non-monetary (material deprivation) indicators of poverty. Our proposal just indicates the necessity of taking into account both current monetary incomes and past incomes (in the form of accumulated assets) when analyzing the ability to meet one's needs. Moreover, the proposed system corresponds to the economic definition of poverty proposed in this chapter, according to which any individual should be considered poverty-stricken if the individual is both monetary impoverished and materially deprived.

2.4 Multidimensional Measures of Poverty

Many researchers have postulated the necessity of treating poverty as a multidimensional phenomenon. Townsend was one of the first researchers to single out the imperfection inherent in identifying poverty exclusively on the basis of the current income criterion. He proposed for poverty analyses to incorporate dwelling conditions, affluence, education, as well as professional and financial resources (Abel-Smith and Townsend, 1965; Townsend, 1979). A broader look at the problem of poverty than just through the prism of income (expenditures) was also presented, among others, by Atkinson and Bourguignon (1982), Hagenaars (1986), Sen (1985), Whelan *et al.* (2001), Bourguignon and Chakravarty (2003), Tsui (2002), Betti *et al.* (2006), Deutsch and Silber (2005), and Alkire and Foster (2008). The authors of a report containing recommendations for the European Union on indicators of poverty and social exclusion also point to the multidimensional nature of the concept of poverty (Atkinson *et al.*, 2002).

Nowadays the multidimensional nature of poverty is a widely recognized fact, not only by the international scientific community, but also by many official statistical agencies (e.g., Eurostat, Istat) and by international institutions (United Nations, World Bank). This fact implies a more complete and realistic vision of this phenomenon and also an increased complexity at both the conceptual and analytical levels. Such a complexity determines the need for adequate tools of analysis and the availability of statistical data that have to be adequate too, complete and reliable.

2.4.1 *Multidimensional Fuzzy Approach to Poverty Measurement*

The multidimensional approach to poverty is focused not only on the current households income but also on the inability to fulfill certain needs, which is caused by inadequacy in the current income as well as the past incomes and accumulated assets measured in non-monetary terms (such as durable goods, apartment, etc.).

One of the multidimensional approaches to poverty measurement is based on the theory of fuzzy sets. This method was utilized in the empirical part of this chapter. Thanks to the fuzzy sets theory the dichotomous distinction between poverty-stricken and non-poor individuals can be avoided. Poverty is not defined in terms of presence or absence in the subset of poor individuals but as a matter of degree of belonging to this subset.

The fuzzy approach considers poverty as a matter of degree rather than an attribute that is simply present or absent for individuals in the population. In this case, two additional aspects have to be introduced:

1. The choice of membership functions, that is quantitative specification of individuals' or households' degrees of poverty and material deprivation;
2. The choice of rules for the manipulation of the resulting fuzzy sets, as complements, intersections, union and aggregation.

An early attempt to incorporate the concept of poverty as a matter of degree at methodological level was made by Cerioli and Zani (1989) who drew inspiration from the theory of Fuzzy Sets initiated by Zadeh (1965). They proposed the introduction of a transition zone $(z_1 - z_2)$ between the two states, a zone over which the membership function declines from 1 to 0 linearly.

Cheli and Lemmi (1995) in their *Totally Fuzzy and Relative* (TFR) approach attempted to overcome the limits of Cerioli and Zani membership function, that is, the arbitrary choice of the two threshold values and the linear form of the function within such values. They defined the membership function as the distribution function $F(y_i)$ of income, normalized (linearly transformed) so as to equal 1 for the poorest and 0 for the richest person in the population. Betti and Verma (1999) modified the membership function version proposed by Chelli and Lemmi taking the membership function as the normalized (linearly transform) Lorenz curve of income $L(F(y_i))$. Finally, Betti *et al.* (2006) combined their previous proposals into the *Integral Fuzzy and Relative* (IFR) approach.

Fuzzy incidence indicators defined under the TFR approach overlook the second basic aspect of poverty analysis, namely poverty depth. The necessity of also taking poverty depth into consideration in multidimensional analyses of poverty has been postulated by many researchers (see, e.g., Shorrocks and Subramanian, 2004). Panek (2010) proposed to extend the IFR approach by incorporating two additional indicators, namely the *Fuzzy Monetary Depth* (FMD) and the *Fuzzy Supplementary Depth* (FSD) indicators.

Further Panek and Zwierzchowski (2014) expanded the IFR approach by introducing two other fuzzy indicators, aimed at measuring poverty intensity and poverty severity. In this chapter we focus on the poverty incidence and depth.

2.4.2 *Fuzzy Monetary Depth Indicators*

2.4.2.1 Fuzzy Monetary Depth Indicator

The starting point to define a FMD indicator, corresponding to the monetary poverty gap index, is the calculation of an individual monetary poverty gap ratio for each individual:

$$v_i = \frac{y^* - y_i^e}{y^*}, \qquad i = 1, 2, \ldots, n, \tag{2.1}$$

with the monetary non-poor individuals (for which $y_i^e \geq y^*$, where y_i^e is the equivalent disposable income of the ith individual and y^* is the monetary poverty line) v_i being assigned the value of zero.

In the next step, the degree of the lack of monetary poverty gap (monetary non-poverty gap score) is defined for each individual:

$$d_i = 1 - v_i, \quad i = 1, 2, \dots, n_{mp}. \tag{2.2}$$

The increase of d_i shows the decrease of monetary poverty gap, that is the increase of income of the poor individual.

The FMD indicator is defined, similarly to the fuzzy monetary incidence (FMI) indicator, as the linear combination of the (1-F^{MD}) function and the (1-L^{MD}) function. The (1-F_i^{MD}) for the ith individual is the proportion of individuals whose monetary non-poverty gap score is higher (who are not as poor or better off) than the individual concerned within the population of impoverished:

$$\lambda_i(v) = FMD_i = (1 - F_i^{MD})^\beta = \left(\frac{\sum_{\gamma=i+1}^{n_{mp}} w_\gamma}{\sum_{\gamma=1}^{n_{mp}} w_\gamma} \right)^\beta, \quad i = 1, 2, \dots, n_{mp}, \tag{2.3}$$

where F_i^{MD} is the value of the distribution function $F(d_i)$ of the monetary non-poverty gap score for the ith individual, w_γ is the weight of the ith individual of rank γ in ascending monetary non-poverty gap score distribution, and β is a parameter.

The (1-L_i^{MD}) is the share of the total monetary non-poverty gap score assigned to all individuals whose monetary non-poverty gap score is higher (who are not as poor or are better off) than the individual concerned within the population of impoverished:

$$\lambda_i(v) = FMD_i = (1 - L_i^{MD})^\beta = \left(\frac{\sum_{\gamma=i+1}^{n_{mp}} w_\gamma d_\gamma}{\sum_{\gamma=1}^{n_{mp}} w_\gamma d_\gamma} \right)^\beta, \quad i = 1, 2, \dots, n_{mp}, \tag{2.4}$$

where L_i^{MD} is the value of the Lorenz curve of the monetary non-poverty gap score $L(F(d_i))$ for the ith individual.

Finally, the membership function to the subset of monetary impoverished with regard to the monetary poverty gap, for the ith individual, is defined as a combination of (2.3) and (2.4):

$$\lambda_i(v) = FMD_i = (1 - F_i^{MD})^{\beta-1}(1 - L_i^{MD}), \quad i = 1, 2, \dots, n_{mp}. \tag{2.5}$$

The overall (for the population in question) FMD indicator, which corresponds to the monetary poverty gap index, is calculated as follows:

$$FMD = \frac{\sum_{i=1}^{n_{um}} \lambda_i(v) \cdot w_i}{\sum_{i=1}^{n} w_i}. \tag{2.6}$$

The parameter β in (2.5) is estimated so that the value of the FMD indicator (for the entire population) is equal to the monetary poverty gap index.

2.4.2.2 Fuzzy Supplementary Depth Indicator

In addition to the monetary (current income) variable, poverty in the multidimensional approach is also explained by non-monetary variables which represent accumulated assets (income from previous periods and non-cash assets). The starting point for including non-monetary variables in poverty analysis is the selection of variables that may be treated as material deprivation symptoms and grouping them into deprivation dimensions (Whelan *et al.*, 2001). An alternative approach may be defining the dimensions of material deprivation in the first step and then choosing the appropriate material deprivation symptoms for each dimension. The next step is to assign numerical values to each deprivation symptom's ordered categories. Then it is necessary to weight the deprivation symptoms scores in order to construct composite indicators and to scale the measures. Since in the EU-SILC survey there has been data on material deprivation symptoms measured on a dichotomous scale, the modified method of calculation of material deprivation indices proposed by Panek (2010) was employed.

A fuzzy measure of the material deprivation depth – *Fuzzy Supplementary Depth* (FSD) – will be defined in a similar step method as a fuzzy measure of the monetary depth. For each dimension of material deprivation we define a variable which assumes values equal to the number of material deprivation symptoms within that dimension ($z_h = 0, 1, \ldots, k_h$). Then numerical values (ranks) are assigned to this variable ($c_h = 1, 2, \ldots, (k+1)_h$) after arranging the values of this variable from the most materially deprived ($c_h = 1$) to the least materially deprived ($c_h = k+1$) situation. The indicator of material deprivation gap for every materially deprived individual and dimension is defined as:

$$x_{h,i} = \frac{(c_h = (k+1)_h - 1) - (c_{h,i} - 1)}{c_h = (k+1)_h - 1}, \quad h = 1, 2, \ldots, m; \quad i = 1, 2, \ldots, n_{md}, \tag{2.7}$$

where $c_h = (k+1)_h$ is the minimal rank assigned to the value of the hth variable, for which material deprivation in the hth dimension is not found.

Next, we define for every materially deprived individual a variable measuring the lack of material deprivation gap for each of the defined dimensions of material deprivation, using the following formula:

$$s_{h,i} = 1 - x_{h,i}, \quad h = 1, 2, \ldots, m; \quad i = 1, 2, \ldots, n_{md}. \tag{2.8}$$

The increase in value of a measure given by (2.8) indicates an improvement of material situation of a given individual. Next, we determine the non-material deprivation gap score (lack of material deprivation gap score) for materially deprived individuals (assessment of the degree of material deprivation gap for materially deprived) for each material deprivation dimension:

$$g_{h,i} = 1 - \frac{1 - F(s_{h,i})}{1 - F(1)}, \quad h = 1, 2, \ldots, m; \quad i = 1, 2, \ldots, n_{md}, \tag{2.9}$$

where $s_{h,i}$ is the value of the lack of material deprivation gap score for the hth dimension and the ith materially deprived individual, $F(s_{h,i})$ is the value of the cumulative distribution function of the lack of material deprivation gap sore, regarding the *h*th deprivation dimension, for the *i*th materially deprived individual, and $F(1)$ is the value of the cumulative distribution function of the lack of material deprivation score that equals 1 for the *h*th dimension and the *i*th materially deprived individual (value of the function that indicates the highest material deprivation gap for materially deprived in the *h*th dimension).

The non-material deprivation gap scores of the deprived individuals (2.9) will be aggregated over the defined dimensions in order to obtain the overall individual lack of material deprivation gap score for every materially deprived:

$$g_i = \frac{\sum_{h=1}^{m} g_{h,i}}{m}, \quad i = 1, 2, \ldots, n_{md}. \tag{2.10}$$

Next, we can define a membership function to the set of materially deprived with respect to the material deprivation gap for every materially deprived individual:

$$\lambda_i(x) = (1 - F_i^{SD})^{\beta' - 1}(1 - L_i^{SD}), \quad i = 1, 2, \ldots, n_{md}, \tag{2.11}$$

where

$$(1 - F_i^{SD})^{\beta'} = \left(\frac{\sum_{\gamma=i+1}^{n_{md}} w_\gamma}{\sum_{\gamma=1}^{n_{md}} w_\gamma} \right)^{\beta'}, \quad i = 1, 2, \ldots, n_{md}, \tag{2.12}$$

and

$$(1 - L_i^{SD})^{\beta'} = \left(\frac{\sum_{\gamma=i+1}^{n_{md}} w_\gamma g_\gamma}{\sum_{\gamma=1}^{n_{md}} w_\gamma g_\gamma} \right)^{\beta'}, \quad i = 1, 2, \ldots, n_{md}, \tag{2.13}$$

where F_i^{SD} is the value of the distribution function of the lack of material deprivation gap score ($F(g_i)$) given in (2.10) for the *i*th materially deprived individual, and L_i^{SD} is the value of the Lorenz curve of the lack of material deprivation gap score ($L(F(g_i))$) for the *i*th materially deprived individual.

The value of F_i^{SD} for the *i*th materially deprived individual is the proportion of materially deprived individuals who have a higher lack of material deprivation gap score (who are less materially deprived) than the individual concerned. The value of L_i^{SD}, for the *i*th materially deprived individual, is the share of the total lack of material deprivation gap score assigned to all materially deprived individuals with higher lack of material deprivation gap score than the materially deprived individuals concerned.

By aggregation of values of the membership function given in (2.11) we define a FSD index, which is a measure of the risk of material deprivation gap for materially deprived:

$$FSD = \frac{\sum_{i=1}^{n_{md}} \lambda_i(x) \cdot w_i}{\sum_{i=1}^{n_{md}} w_i}. \tag{2.14}$$

The value of the parameter β' in (2.11) is estimated so that the value of the FSD index is equal to the value of the material deprivation depth index (Panek, 2014). The estimated value of β' may then be used to calculate values of individual membership functions of all materially deprived individuals to the set of materially deprived with regard to the material deprivation gap in all defined dimensions:

$$\lambda_i(x_h) = (1 - F_{h,i}^{SD})^{\beta'-1}(1 - L_{h,i}^{SD}), \quad h = 1, 2, \ldots, m; \; i = 1, 2, \ldots, n. \tag{2.15}$$

The formula given in (2.15) is aggregated over the entire analyzed population resulting in FSD indices for each of the defined dimensions of material deprivation:

$$FSD_h = \frac{\sum_{i=1}^{n_{md}} \lambda_i(x) \cdot w_i}{\sum_{i=1}^{n_{md}} w_i}, \quad h = 1, 2, \ldots, m. \tag{2.16}$$

2.5 Co-incidence of Risks of Monetary Poverty and Material Deprivation

The risk of poverty is more intense when it jointly applies to monetary poverty and material deprivation (Betti and Verma, 2008). Such a risk of poverty is defined as *manifest poverty risk*.

The degree of manifest poverty risk incidence for the *i*th individual is defined as the minimal value of two membership functions – a membership function to the set of monetary impoverished and the membership function to the set of materially deprived (Betti *et al*., 2006):

$$m_i^I = \min(\lambda_i(y_i^e), \lambda_i(x)), \quad i = 1, 2, \ldots, n. \tag{2.17}$$

The degree of manifest poverty risk depth is defined similarly, as the minimal value of functions given in (2.5) and (2.11):

$$m_i^D = \min(\lambda_i(v), \lambda_i(x)), \quad i = 1, 2, \ldots, n. \tag{2.18}$$

By aggregating the formulas given in (2.17) and (2.18) we obtain the fuzzy manifest poverty incidence and fuzzy manifest poverty depth indicators for the entire analyzed population:

$$M^I = \frac{\sum_{i=1}^{n} m_i^I \cdot w_i}{\sum_{i=1}^{n} w_i}, \tag{2.19}$$

and

$$M^D = \frac{\sum_{i=1}^{n} m_i^D \cdot w_i}{\sum_{i=1}^{n} w_i} \tag{2.20}$$

2.6 Comparative Analysis of Poverty in EU Regions in 2010

2.6.1 Data Source

The empirical analyses conducted in this chapter are based on the data from an EU survey (EU-SILC) carried out in 2010. The main objective of EU-SILC is to supply EU comparable data on the income, poverty, social exclusion and living conditions of the population of the EU Members States.

The survey is based on representative random samples of households and individuals aged 16 and above, who are members of drawn households, for each EU Member State. The survey results are weighted in order to represent the size and structure of the entire population of households and citizens for each EU Member State. The total sum of weights corresponds to the total number of households and individuals for each country.

2.6.2 Object of Interest

In this chapter an object of interest from the point of view of poverty analysis is defined as a person (not as a household). As a consequence, all measures and indicators are calculated for the population of persons. However, the identification of impoverished persons is conducted on the basis of identification of impoverished households, as all members of impoverished households are considered to be impoverished. This approach is adopted to analyze both the monetary poverty and non-monetary poverty (material deprivation). In the case of monetary poverty analysis, every person is assigned an equivalent disposable income of the household to which that person belongs. It is also assumed that every member of a household is characterized by the same material deprivation symptoms as its household.

Household income is defined as yearly household equivalent disposable income in the last calendar year preceding the survey. The equivalent disposable incomes were calculated by dividing disposable household income by the Organisation for Economic Co-operation and Development (OECD) modified equivalence scales. The disposable income is defined as a sum of net monetary income gained by all households' members.

In order to guarantee a comparability of incomes for various EU countries and eliminate differences of price levels between countries, all monetary incomes expressed in national currencies were divided by Purchasing Power Parities (PPP) indicators. Thus, all monetary incomes are quoted in the Purchasing Power Standard (PPS) which is an agreed, artificial common reference currency used in the EU for international comparisons.

2.6.3 Scope and Assumptions of the Empirical Analysis

The empirical comparative analysis was conducted for the EU Member States and EU regions on NUTS 2 level. The interregional comparisons within the EU based on the results of the EU-SILC study come across many practical obstacles. The EU-SILC data concerning some Member States available for scientific research do not allow the region in which the studied households reside[6] to be identified. Furthermore, the countries that have available data enabling the households to be identified by region, often provide region codes only on NUTS 1 level. As a result, due to the inaccessibility of data, the interregional comparisons carried out in the study do not include all Member States.

Within the fuzzy sets approach a calibration of poverty indicators is required (see Section 2.4). In our analysis the parameters of the indicators were calculated using classical measures. The classical monetary poverty measures (monetary headcount ratio and monetary poverty gap) were calculated using a common EU monetary poverty line instead of the national poverty lines[7] to all countries. The adopted monetary poverty line was determined at such a level, so that for the adopted material deprivation threshold 20 million people in the EU would be in poverty, that is, they would be both monetary poor and materially deprived (Panek and Zwierzchowski, 2014). This threshold for 2010 was equal to 6354 euro per year, which was 54% of the median equivalent income quoted in PPS.

The use of a common monetary poverty line for all EU countries provides, first of all, comparable results of analyzes of monetary poverty between EU Member States and their regions. EU Member States are treated as components of a larger structure like the EU.

In the calculation of classical non-monetary poverty (material deprivation) the indicators are based on those recommended by the EU material deprivation symptoms[8] (material deprivation headcount ratio and material deprivation gap index; see Panek, 2014).

2.6.4 Risk of Monetary Poverty

The fuzzy monetary poverty incidence indicator (FMI) assumed the value of 14.2% in the EU in 2010 (Figure 2.1).[9]

[6] This applies to the Netherlands, Germany, and Great Britain.

[7] The EU common monetary poverty line is calculated as 60% of the median of joint household equivalent income distribution in all EU Member States. National household equivalent incomes are expressed in PPS.

[8] A person, whose household has at least 4 out of 12 symptoms, is considered to be materially deprived (Panek and Zwierzchowski, 2014).

[9] The values of fuzzy monetary poverty incidence indicators are presented in Panek and Zwierzchowski (2014), Table A10. A list of acronyms of countries and regions is given on the website: http://epp.eurostat.ec.europa.eu/portal/page/portal/nuts_nomenclature/introduction.

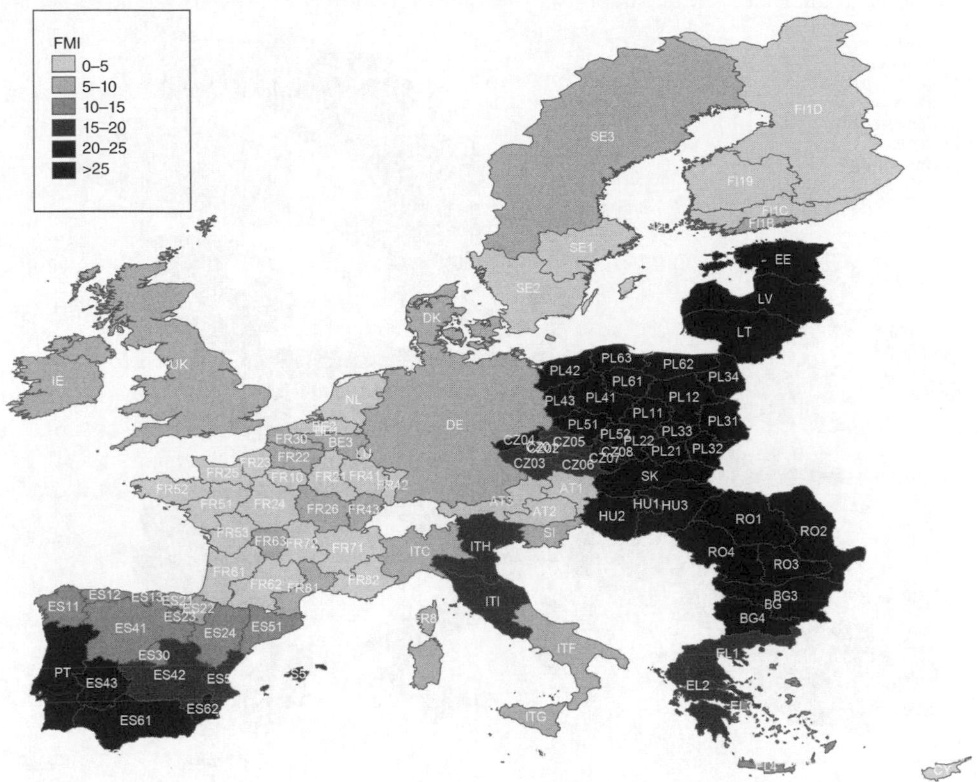

Figure 2.1 Fuzzy monetary poverty incidence indicators in the EU countries and regions in 2010

For the EU Member States the highest values of the FMI indicator were observed in 2010 in: Romania (66.7%), Bulgaria (45.3%), Latvia (43.6%), Lithuania (43.1%), Hungary (37.4%), and Poland (32.2%)[10].

Among regions at the NUTS 2 level for which the required data were available, the following were marked with the highest values of the FMI indicator in 2010 (Figure 2.1): Latvia (43.6%), Lithuania (43.1%), and Lubelskie and Swietokrzyskie voivodships in Poland (36.0 and 35.6%, respectively).

The countries with the highest values of the FMD indicator in 2010 (Figure 2.2) were Romania (88.8%), Bulgaria (78.1%), Hungary (75.7%), Latvia (75.5%), and Lithuania (75.0%). At NUTS 2 level the highest fuzzy monetary poverty depth is observed in Latvia (75.5%), the Lubelskie and Swietokrzyskie voivodships in Poland (76.0% and 75.3%), and Lithuania (75.0%).

[10] The values of fuzzy monetary poverty indicators are presented in Panek and Zwierzchowski (2014),Table A10 and Table A11.

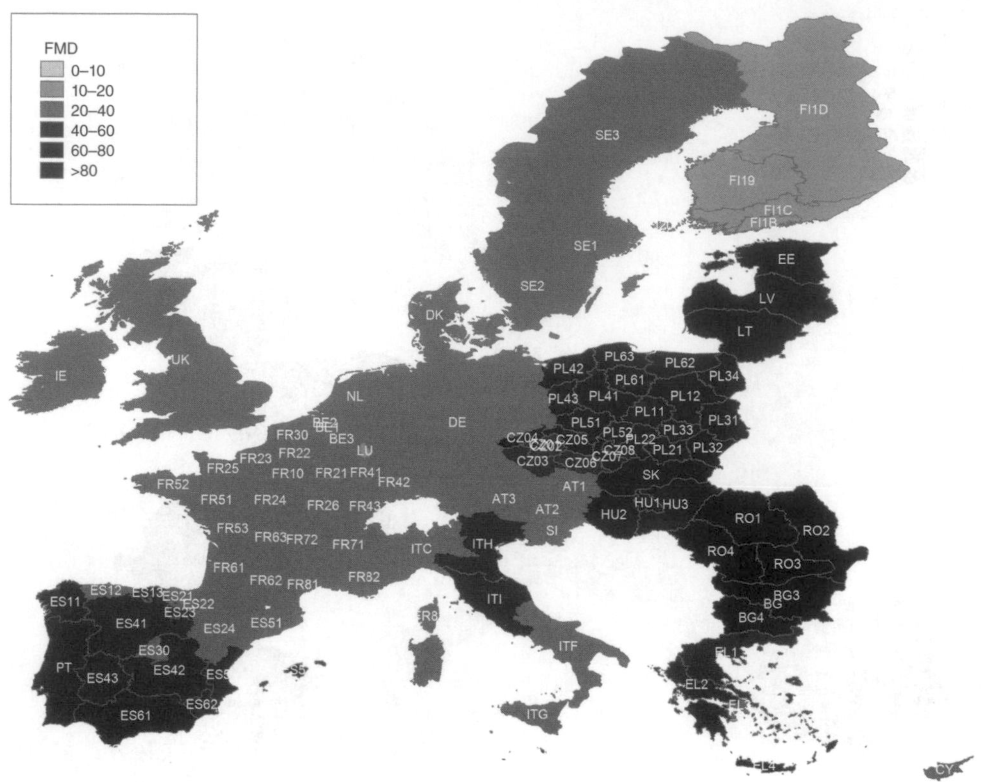

Figure 2.2 Fuzzy monetary poverty depth indicators in the EU countries and regions in 2010

2.6.5 *Risk of Material Deprivation*

For the purpose of the empirical analysis in this chapter, the following dimensions and symptoms of poverty were defined:[11]

1. Equipment of households in durables – symptoms relate to the lack of possession of widely desired durables because of lack of resources:
   (a) Lack of a telephone.
   (b) Lack of a color TV.
   (c) Lack of a computer.
   (d) Lack of a washing machine.
   (e) Lack of a car.
2. Housing facilities and deterioration – symptoms relate to the absence of basic housing facilities and to serious problems with the dwelling:
   (a) Leaky roof, damp walls/floors/foundation, or rot in window frames or floor.

[11] The scope of data in EU-SILC does not allow to distinguish the dimensions of material deprivation closely linked to the need groups of households.

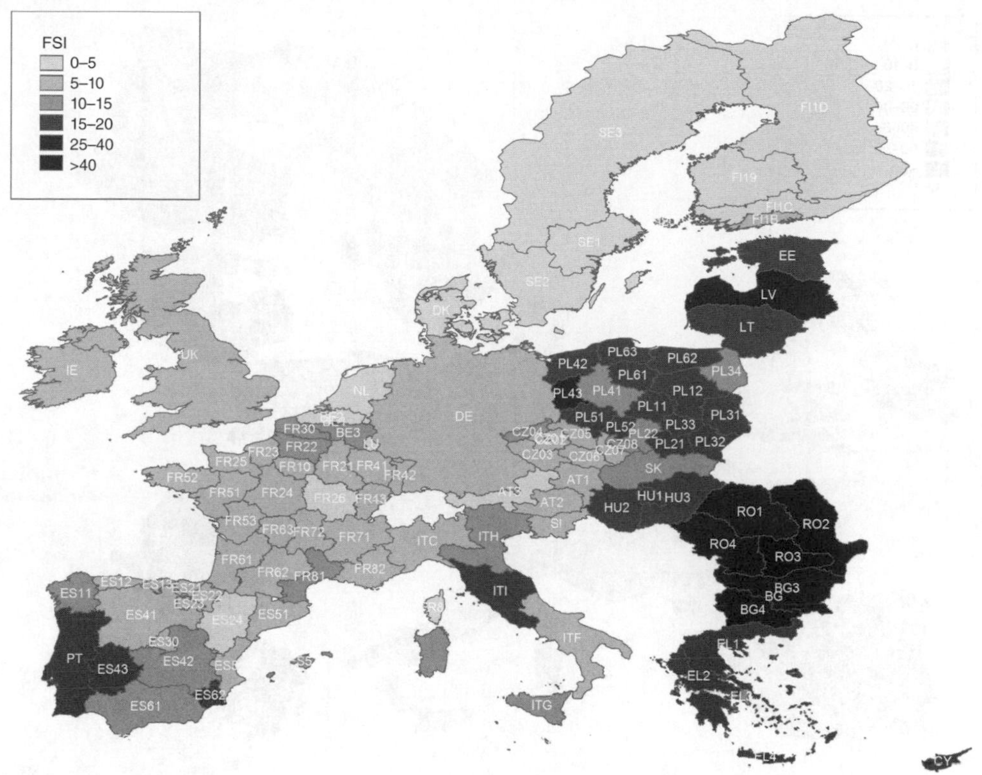

Figure 2.3 Fuzzy supplementary incidence indicators in the EU countries and regions in 2010

(b) A bath or shower in dwelling.
(c) An indoor flushing toilet for sole use of a household.

3. Basic life style – symptoms relate to the lack of ability to afford most basic requirements:
   (a) Paying for one week annual holiday away home.
   (b) Eating a meal with meat, chicken, fish (or vegetarian equivalent) every second day.
   (c) Keeping home adequately warm.
   (d) Ability to pay for unexpected expenses.
   (e) The household has been in arrears during the last 12 months due to rent for accommodation, mortgage repayments, utility bills, or other loan payments.
4. Health care – symptoms relate to the necessity of refraining from basic health care due to financial reasons:
   (a) During the last 12 months a member of the household refrained from visiting a physician due to financial reasons.
   (b) During the last 12 months a member of the household refrained from visiting a dentist due to financial reasons.

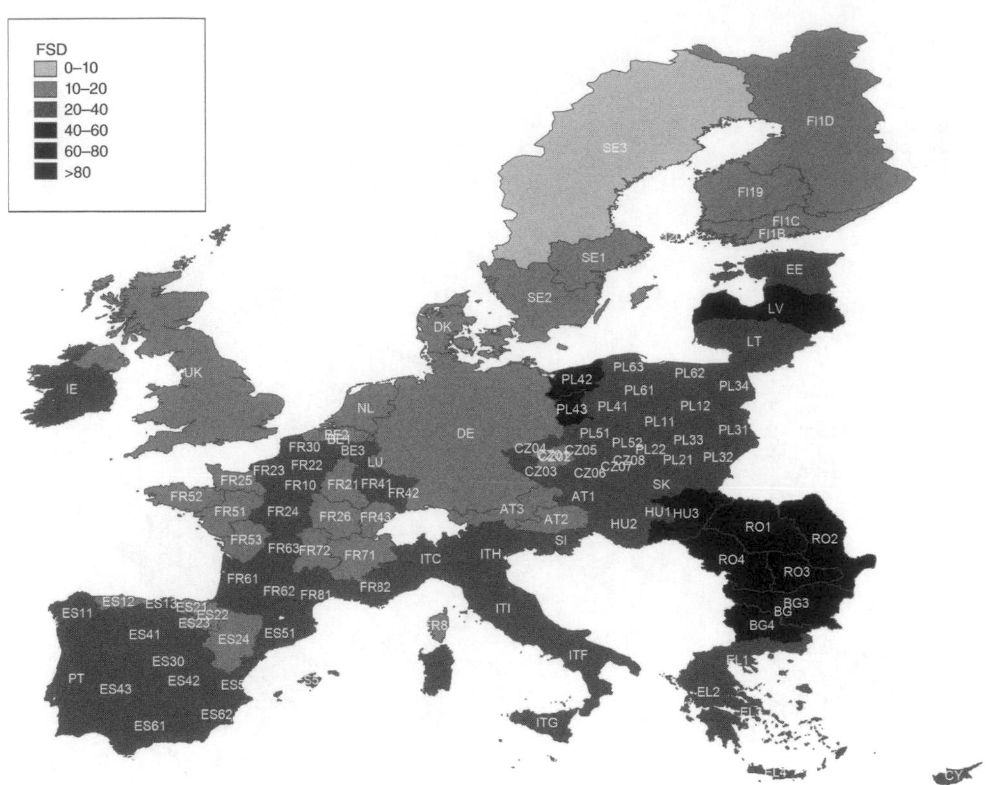

Figure 2.4 Fuzzy supplementary depth indicators in the EU countries and regions in 2010

The Fuzzy Supplementary Incidence Indicators (FSI) index was calibrated so that it was equal to the headcount material deprivation ratio, for the whole EU (11.4%).[12] Therefore, the FSI value is slightly lower than the FMI value for the whole EU. The highest values of the FSI were observed in 2010 in the following countries (Figure 2.3): Romania (43.2), Bulgaria (37.1%), Latvia (36.6%), and Hungary (21.3%). The FSI in Poland was equal to 16.6% and was one of the highest among EU Member States. The lowest values of the fuzzy material deprivation incidence measure were observed in Sweden (3.1%), Luxembourg (3.4%), Finland (4.4%), and Denmark (4.4%).

At the NUTS 2 level, among regions for which the required information was available, the following were observed to have the highest values of the FSI in 2010 (Figure 2.3): Latvia (36.6%), and the Lubuskie (26.6%), Zachodniopomorskie (22.8%) and Lodzkie (21.2%) voivodships in Poland. However, we have good reason to believe, that the majority of regions at the NUTS 2 level in Romania, Bulgaria and some regions in Hungary would have been listed among those with the highest values of the FSI, had the required data been available.

[12] The values of fuzzy supplementary poverty incidence indicators are presented in Panek and Zwierzchowski (2014), Table A10.

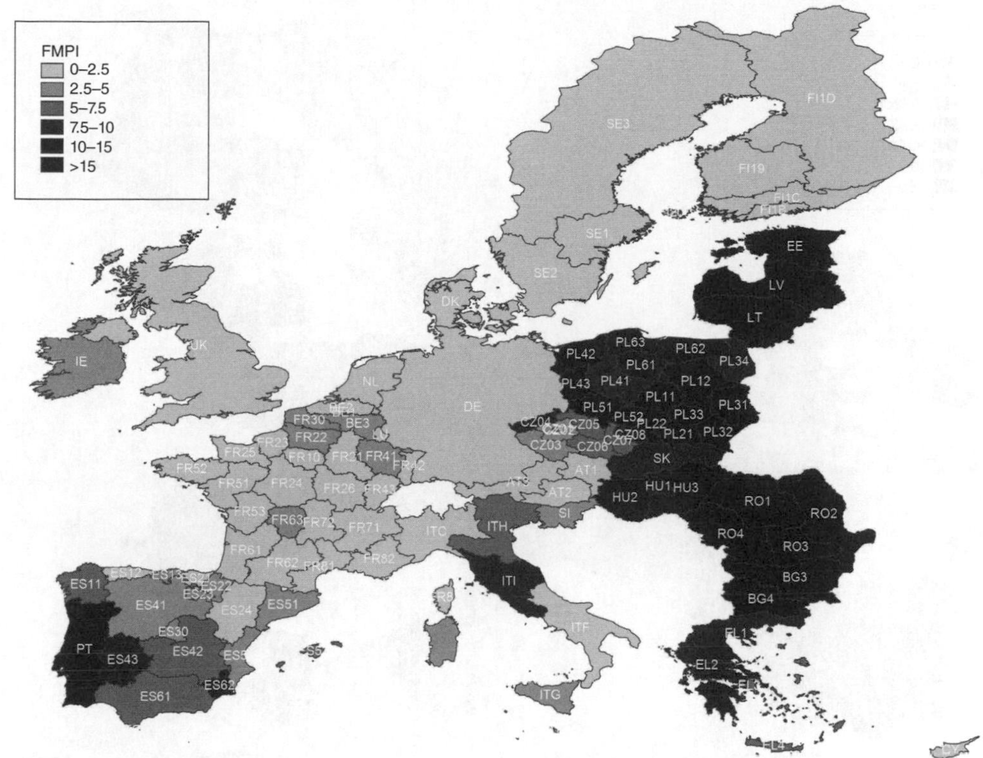

Figure 2.5 Fuzzy manifest poverty incidence indicators in the EU countries and regions in 2010

Countries marked with the highest material deprivation depth in 2010 (Figure 2.4)[13] were Romania (61.2%), Bulgaria (57.7%), and Latvia (55.9%). At the NUTS 2 level the following regions were marked with the highest values of the FSD: Latvia (55.9%), and the Lubuskie (45.9%) and Zachodniopomorskie (42.2%) voivodships in Poland.

2.6.6 *Risk of Manifest Poverty*

The EU countries with the highest risk of manifest poverty incidence in 2010 (Figure 2.5)[14] were Romania (39.5%), Bulgaria (29.6%), Latvia (27.5%), Lithuania (16.9%), and Hungary (16.8%).

A comprehensive international comparison at the NUTS 2 level is not possible due to lack of necessary data for the majority of countries. At the NUTS 2 level, among regions for which the required data were available, the following regions had the highest values of the manifest poverty incidence measure: Latvia (27.5%), the Lubuskie (20.7%) and Zachodniopomorskie (17.5%) voivodships in Poland, and Lithuania (16.9%).

[13] The values of fuzzy supplementary poverty depth indicators are presented in Panek and Zwierzchowski (2014), Table A11.

[14] The values of fuzzy manifest poverty incidence indicators are presented in Panek and Zwierzchowski (2014), Table A10.

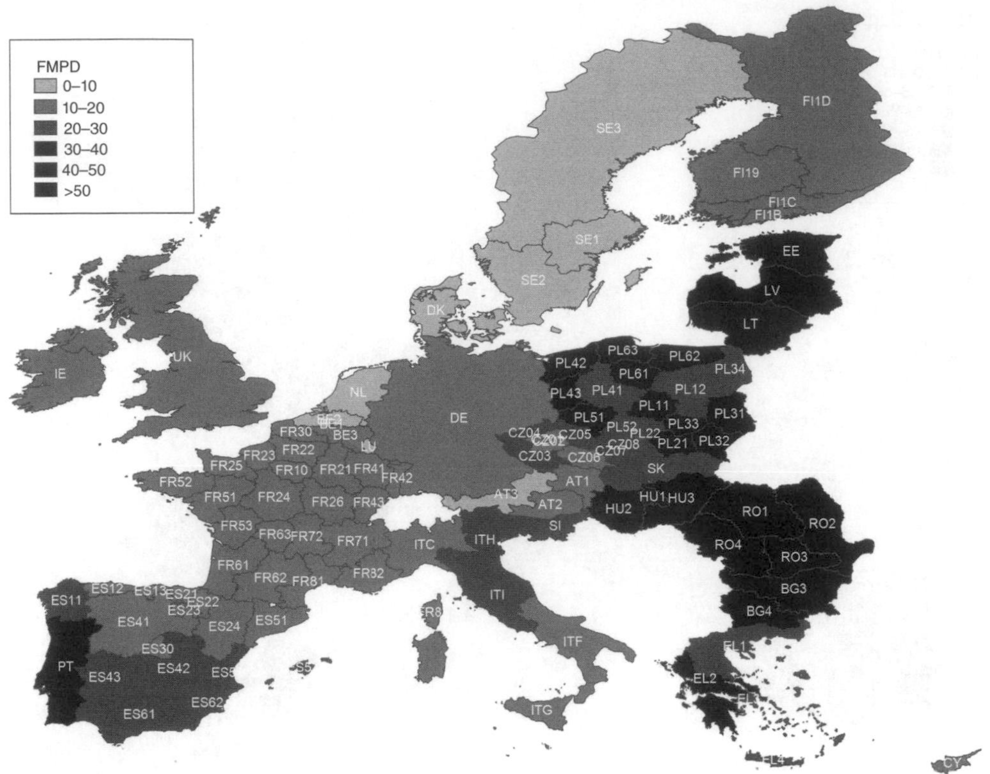

Figure 2.6 Fuzzy manifest poverty depth indicators in the EU countries and regions in 2010

The highest rates of manifest poverty depth were observed in the same three countries with the highest rates of manifest poverty incidence (Figure 2.6),[15] in Romania (60.5%), Bulgaria (55.2%), and Latvia (52.2%). The regions with the highest manifest poverty depth at the NUTS 2 level were Latvia (52.2%), and the Lubuskie (43.7%) and Zachodniopomorskie (39.9%) voivodships in Poland.

2.7 Conclusions

Our major goal was to propose a methodology of poverty measurement for use at the regional and, applying small area estimation, at local level. We proposed the comprehensive multidimensional approach to poverty measurement, which is focused not only on the current households income (monetary poverty) but also on the past incomes and accumulated assets measured in non-monetary terms (material deprivation).

Our methodology was applied for identification of the poorest and most vulnerable countries and regions within the EU in 2010. Furthermore, the results obtained within this approach are fully comparable between EU Member States and regions.

[15] The values of fuzzy manifest poverty depth indicators are presented in Panek and Zwierzchowski (2014), Table A11.

References

Abel-Smith B. and Townsend P. (1965) *The Poor and the Poorest,* Occasional Papers and Social Administration, 17, Bell & Sons, London.

Alkire S. and Foster J. (2008) *Counting and Multidimensional Poverty Measurement*, OPWI Working Paper Series, University of Oxford, Oxford.

Atkinson A.B. and Bourguignon F. (1982) The comparison of multidimensional distribution of economic status, *Review of Economic Studies*, **49**, pp. 183–201.

Atkinson A.B., Cantillon B., Marlier E., and Nolan B. (2002*) Social Indicators: The EU and Social Inclusion*, Oxford University Press, Oxford.

Betti G., Cheli B., Lemmi A., and Verma V. (2006) Multidimensional and longitudinal poverty: an integrated fuzzy approach, in A. Lemmi and G. Betti (eds) *Fuzzy Set Approach to Multidimensional Poverty Measurement*, Springer, New York, pp. 111–137.

Betti G. and Verma V. (1999) *Measuring the Degree of Poverty in a Dynamic and Comparative Context: A Multi-dimensional Approach Using Fuzzy Set Theory*, Proceedings, ICCS-VI, 11, pp. 289–301, 29–31 August 1999, Lahore, Pakistan.

Betti G. and Verma V. (2008) Fuzzy measures of the incidence of relative poverty and deprivation: a multi-dimensional perspective, *Statistical Methods and Applications*, **17**(2), pp. 225–250.

Bourguignon F. and Chakravarty S.R. (2003) The measurement of multidimensional poverty, *Journal of Economic Inequality*, **1**(1), pp. 25–49.

Bradshaw J. and Mayhew E. (2011) *The Measurement of Extreme Poverty in the European Union*, European Commission, Directorate-General for Employment, Social Affairs and Inclusion.

Cerioli A. and Zani S. (1989) A fuzzy approach to the measurement of poverty, in C. Dagum and M. Zenga (eds) *Income and Wealth Distribution, Inequality and Poverty. Studies in Contemporary Economics*, Springer Verlag, Berlin, pp. 272–284.

Cheli B. and Lemmi A. (1995) A totally fuzzy and relative approach to the multidimensional analysis of poverty, *Economic Notes*, **24**(1), pp. 115–134.

Copeland P. and Daly M. (2012) Varieties of poverty reduction: inserting the poverty and social exclusion target into Europe 2020, *Journal of European Social Policy*, **22**(3), pp. 273–287.

Deutsch J. and Silber J. (2005) Measuring multidimensional poverty: an empirical comparison of various approaches, *Review of Income and Wealth*, **51**(1), pp. 145–174.

Drewnowski J. (1997) Poverty: its meaning and measurement, *Development and Change*, **8**, pp. 183–208.

Hagenaars A.J.M. (1986) *The Perception of Poverty*, North-Holland, Amsterdam.

Marshall A. (1920) *Principles of Economics*, 8th ed., McMillan, London.

Panek T. (2010) Multidimensional approach to poverty measurement: fuzzy measures of the incidence and the depth of poverty, *Statistics in Transition*, **11**(2), pp. 361–379.

Panek T. (2014) Methodology of analyzing poverty, in J. Czapiński and T. Panek (eds) *Social Diagnosis 2013. The Objective and Subjective Quality of Life in Poland*, The Council for Social Monitoring, Warsaw, pp. 479–489.

Panek T. and Zwierzchowski J. (2014) *Comparative Analysis of Poverty in the EU State Members and Regions*, Warsaw School of Economics Press, Warsaw.

Sen A. (1985) *Commodities and Capabilities*, North-Holland, Amsterdam.

Shorrocks A. F. and Subramanian S. (2004) *Fuzzy Poverty Indices*, University of Essex, Colchester.

Silver H. (1994) Social exclusion and social solidarity: three paradigms, *International Labour Review*, **133**(5–6), pp. 531–578.

Townsend P. (1979) *Poverty in the United Kingdom*, Penguin Books, London.

Tsui K.Y. (2002) Multidimensional poverty indices, *Social Choice and Welfare*, **19**(1), pp. 69–93.

Whelan C.T., Layte R., Maitre B., and Nolan B. (2001) Income, deprivation and economic strain: an analysis of the European Community Household Panel, *European Sociological Review*, **17**(4), pp. 357–372.

Wolf P., Montaigne F., and Gonzales G.R. (2010) Investing in statistics, in A. B. Atkinson and E. Marlier (eds) *Income Living Conditions in Europe*, Office for the Official Publications of the European Communities, Luxembourg, pp. 37–56.

Zadeh L.A. (1965) Fuzzy sets, *Information and Control*, **8**, pp. 338–353.

3

Administrative and Survey Data Collection and Integration

Alessandra Coli[1], Paolo Consolini[2] and Marcello D'Orazio[2]
[1]*Department of Economics and Management, University of Pisa, Pisa, Italy*
[2]*ISTAT, Italian National Statistical Institute, Rome, Italy*

3.1 Introduction

Poverty studies rely on the analysis of individual data collected by sample surveys, census and registers. In Europe, the EU Statistics on Income and Living Conditions (EU-SILC) instrument represents one of the most relevant sources of data for calculating poverty and social exclusion indicators (e.g., Laeken indicators). EU-SILC covers a large spectrum of aspects such as housing, health, education, labor, and income. Nevertheless, these statistics provide high quality information only at the country or regional level. In Italy, for example, the sample size (about 26,000 households) aims at providing accurate estimates at regional level (NUTS 2 level).[1] However, in Italy as in most EU countries, local governments beneath NUTS 2 level are also in charge of significant social policies concerning health, education, disability, or housing. Such governments cannot rely on official statistics for the planning and evaluation of their policies. In fact, most current Eurostat and national surveys are planned as to assure accuracy at NUTS 2 or 3 levels (e.g., Labor Force Survey, Household Budget Survey, Multipurpose Survey). Increasing accuracy would imply larger samples, that is, more financial resources for data collection, which is not a viable solution in times of budget constraints. The exploitation of all available information at the local level may represent a valid alternative in order to produce direct estimates or model-based estimates using appropriate small area estimation techniques (see Chapter 4). In both cases the integration of micro data from different data sources represents a crucial step.

[1] European official nomenclature of territorial units for statistics. In Italy, the NUTS 2 level corresponds to the 20 administrative regions (http://epp.eurostat.ec.europa.eu).

Analysis of Poverty Data by Small Area Estimation, First Edition. Edited by Monica Pratesi.

Companion Website: www.wiley.com/go/pratesi/poverty

One of the main limitations of official statistics on poverty is represented by inaccuracy of estimates related to non-planned domains. A system of poverty indicators should cover all crucial factors of poverty and social exclusion. Since no single data source is able to cover all such aspects, incoherence problems may emerge, especially among indicators on interrelated phenomena. For example, when evaluating economic well-being, it is necessary to look at households' income, consumption and wealth simultaneously (Stiglitz Commission, 2009) but unfortunately micro data on these phenomena do not always convey a coherent view. This limit is particularly relevant in Italy, where micro data on income, consumption and wealth originate from different sample surveys. In this case, the integration of data sources could help to improve the quality of the final statistical outputs.

Integration of data can represent a viable solution to overcome specific relevance, inaccuracy and incoherence problems like the ones discussed above. In a larger perspective, data integration may represent a valid solution for increasing the quality (especially relevance and completeness) of official statistics as a whole.

In general, data can originate from different sources, in this chapter the focus is mainly on two: data originating from surveys (on sample basis or not) and databases set up and maintained by government agencies or private companies, denoted as Administrative Registers. National Statistical Institutes (NSIs) or private companies are in charge of collecting survey data, usually on a sample basis, for statistical purposes, that is, for exploring a particular phenomenon (e.g., Labor Force Survey, Household Budget Survey, etc.). Government agencies or other institutions collect data for administrative purposes, most of the times to respond to requirements for registering particular events (e.g., births and deaths), administering benefits (e.g., pensions) or duties (e.g., taxation).

Until recently, increasing the relevance and the completeness of the final statistical outputs (e.g., estimates of parameters for a higher number of domains, etc.) meant changing the planning of surveys (e.g., increasing the sample size or the number of variables). In this setting, the possibility of reusing survey data in order to respond to new users' needs was quite limited. On the other hand, the production of statistical outputs entirely based on administrative registers was limited because of the typical defects of administrative data: (i) differences in the definition of the variables with respect to the ones used for statistical purposes; (ii) differences in the definition of the unit of analysis; and (iii) differences populations (population covered by the register does not coincide with the target population of the statistical analysis).

In the last few decades the idea of producing new statistical outputs by integrating survey and administrative data emerged. Initially, integration was mainly conceived as a tool for improving survey-based statistics: administrative data mainly supported the data editing and processing steps (imputation of missing and invalid values, auxiliary variables for estimation purposes, fitting of small area models, etc.). Recently, NSIs have been experiencing a more general approach to integration processes: integration is a tool for exploiting all the available data sources in order to derive more accurate and complete final statistics. The basic idea is that of building a unique data source, which integrates data from all the available data sources (administrative registers and surveys), thus permitting to overcome the drawbacks related to the use of a single source (coverage, definition of variables, etc.).

The aim of this chapter is to present the main integration methods and to show how these can be fruitfully used to improve the quality of statistics for poverty and well-being studies. The chapter is organized as follows. Section 3.2 describes the theoretical aspects of the main integration methods, focusing on each method's objectives and main issues. Section 3.3 argues the advantages of micro-data integration to produce poverty and well-being indicators both at country level and for small geographical areas. Finally, Section 3.4 contains some concluding remarks on the issues discussed in the chapter.

3.2 Methods to Integrate Data from Different Data Sources: Objectives and Main Issues

The integration of different data sources can have different goals depending on the type of data sources at hand. In a statistical framework the integration of two administrative registers is performed to derive a wider register, in terms of variables and coverage of the target population. The resulting register can be used for the direct production of statistics or as a "support" for one or more sample surveys, at different phases: as a sampling frame; as a source of additional auxiliary information to be used in the survey data treatment phase (e.g., editing and imputation) or in the data processing (auxiliary variables are the basis of post-stratification of survey data or for the calibration of survey weights) or to perform ad hoc analyses (e.g., small area estimation, see Chapter 4). On the contrary, the integration of two data sources originating from two independent sample surveys is usually performed to study the relationship of variables that were not jointly observed in a single survey.

In statistics, the integration techniques can be divided in two broad classes of methods: (i) record linkage (RL); and (ii) statistical matching (SM) (or data fusion).

3.2.1 Record Linkage

RL techniques aim at identifying pairs of records, in different data sources, which refer to the same entity (person, household, enterprise, etc.). The term "record linkage" in this context is used in its wider meaning including *exact matching* or *merging*, that is, integration based on unit identifiers supposed to be free of errors (e.g., Personal Identification Number, VAT code, etc.). The exact matching nowadays can be performed in an automated manner on very large data sources, given the computing power and storage made available by modern computers. When units' identifiers are affected by errors or when they are not available, then the linkage is based on the unit indicators – typically text strings such as name, surname, address, and so on – denoted as *key variables*. A perfect agreement of key variables should correspond to a match while partial agreements need to be further processed. RL can be *deterministic* or *probabilistic*. In the first case the *ad hoc* rules are set to manage the results of comparison and provide matched and unmatched pairs. In probabilistic RL the decision whether a pair of records refer to the same entity or not is based on statistical models.

In the deterministic RL all the possible pairs of records obtained by comparing records from one data source (A) with those from the other one (B), are divided in *matches* (units referring to the same entity, usually denoted as M) and *not matches* (U) according to rules specified in advance concerning the similarity of the values presented by key variables being compared. Typically, in deterministic RL applications a pair of records is considered a match if perfect agreement is observed for each key variable (or just on a subset of the most important key variables). In some applications rules account also for non-perfect agreement, providing the threshold for similarity measures adopted to determine how close two records are according to the values presented by the key variables (usually similarity measures are scaled to range from 0 to 1, which denotes perfect agreement).

A formal definition of the problem of modern probabilistic RL is introduced by Fellegi and Sunter (1969). They assume that all the possible pairs of records, obtained by comparing records from A with those from B, can be divided in matches and not matches but the group membership is a latent (unobserved) binary variable that represents the objective of the inference. In this setting, the vector $\gamma^{(a,b)}$ resulting from the comparison of the key variables for records a and b ($a \in A$, $b \in B$), will follow a different distribution depending on the fact the pair is a match (belongs to M) or not (is in U). In particular by comparing the probabilities by means of the log-likelihood it is possible to derive a weight for each possible pair of units ratio:

$$W_{ab} = \log R(\gamma^{(a,b)}) = \frac{m(\gamma^{(a,b)})}{u(\gamma^{(a,b)})} = \frac{Pr(\gamma^{(a,b)}|M)}{Pr(\gamma^{(a,b)}|U)} \tag{3.1}$$

The Fellegy–Sunter (FS) procedure provides a rule to determine the matching status of a pair according to the estimated weight $\widehat{W}_{ab}$; three different subsets can be identified: *links*, *non-links*, and *possible links*. The presence of the last subset is due to the fact that the probability distributions of matches and non-matches show a partial overlapping. The set of the possible links should be reviewed by experienced staff to ascertain which pairs are a link and which are not.

The original FS procedure is based on some critical assumptions:

(i) in order to estimate the parameters of the two distributions M and U, independence is assumed between the variables $\gamma^{(a,b)}$ given the latent matching status;
(ii) the variables $\gamma^{(a,b)}$ are dichotomous (1 if the key variables are equal and 0 otherwise);
(iii) the variables $\gamma^{(a,b)}$ do not account for eventual rare categories of the key variables; agreement on rare categories is more important than agreement on frequent categories in deciding whether a pair refers to the same unit;
(iv) the procedure does not end with a set of unique links, in practice, the same unit can be linked with two distinct units of the other data source. To derive a set of unique links it is necessary to process further the set of linked units by means of an optimization procedure.

As far as point (i) is concerned, the FS procedure provides closed form expressions for estimating parameters when there are just three key variables. Jaro (1989) suggested a procedure based on the EM algorithm to estimate the parameter for an unlimited number of key variables. This method permits to improve the procedure for defining the threshold aimed at identifying links, not-links, and possible links. Some authors suggest to bypass the conditional independence assumption by using opportune models (Thibaudeau, 1989, 1993; Winkler, 1993).

To overcome the problem of dichotomous comparison variables, different string comparators can be employed to derive measures of partial agreement, for example the Jaro–Winkler method based on the number of common characters and number of transpositions or Levenstein distance (Jaro, 1972; Winkler 1990). It is worth remembering that before comparing indicators it would be preferable to pre-process them in order to harmonize classifications, categories, and so on.Usually the text strings should be checked to remove special characters, to convert case; in some circumstances, it may be necessary to split a variable value in two or more different variables or, on the contrary, to join different values. Some further preprocessing may include the correction of known errors. In general, the preprocessing steps strongly depend on the country specific language.

The FS approach, based on a likelihood ratio, does not account for whether the agreement is on rare or frequent categories of a variable; for this reason some enhancements have been proposed by Yancey (2000) and Zhu *et al.* (2009).

A general solution to some drawbacks posed by the FS approach is provided by applying a Bayesian approach as suggested by Tancredi and Liseo (2011).

In practice, the application of RL procedures may become highly computationally demanding in the presence of very large data sources: all the records in a data source should be compared with the ones in the other one. For this reason it would be preferable to reduce the number of comparisons by imposing some constraints (*blocking*) which permit just comparisons of records belonging to the same block. Blocks can be created by considering key indicators assumed to be free of errors in both of the data sources, for example gender, place of birth, first character of the name or surname, and so on. Fernandez (2009, 2011) provides a comprehensive review of the blocking methods.

Alternative approaches to RL are based on the application of methods developed for classification; a vast range of *machine learning* methods can be applied (Support Vector Machine, classification trees, clustering methods, etc.). For the main details see Winkler (2000), Elfeky *et al.* (2003), Sariyar and Borg (2010), and Christen (2012).

The analysis of a data set obtained by integrating two or more data sources requires a certain care. In particular the reliability of the final results will depend on the extent of possible linkage errors. There can be two errors: *false match*, that is link of a pair of units that refer to two different entities; and *false non-match*, that is records referring to the same entity that were not linked. In the first case the analysis of variables observed on false matches is likely to introduce bias into the final statistics. In particular, the risk is that of underestimating the strength of the relationship between variables not observed in the same data source. For instance, Neter *et al.* (1965) showed that false matches introduce bias in the results of record check studies (i.e. comparisons of the value of the same variable observed in different data sources). Various papers (Scheuren and Winkler, 1993, 1997; Lahiri and Larsen, 2000, 2005; Chambers, 2009) deal with fitting regression models involving variables coming from two data sources that were linked; they propose different methods to improve the estimation of the parameters of the model in order to compensate the potential bias due to the presence of linkage errors. Generally speaking, when analysing integrated data sources provided by RL procedures, it may be useful to include into the analysis the estimated probabilities of being a match or, when non-probabilistic linkage procedures are considered, the similarity between the key variables.

The false non-match errors introduce missing values (the new variables added after integration are missing because the unit has not been linked) that need additional treatment (imputation or data estimation in the presence of missing values) in order to derive the required statistics.

For these reasons before using the integrated data set for producing results it would be preferable to assess the incidence of linkage errors (similar measures are adopted in other fields, e.g. *positive predicted values* and *sensitivity*, generally used in the epidemiological field, or *recall* and *precision*, normally used in the information retrieval framework). Unfortunately such evaluation is not simple and requires an additional effort. A possible strategy consists in a repetition of the RL application starting from a sample of pairs of units already processed by staff with specific expertise, which therefore is supposed to provide error free matches; unfortunately this clerical review is costly and burdensome. Alternative approaches consist in estimating the error rates by using opportune statistical models as suggested, for instance, in Belin and Rubin (1995), whose procedure requires the availability of a sample of pairs whose status (link, not link) is known without errors. Torelli and Paggiaro (1999) suggest an EM estimation approach that does not require the availability of a training sample.

A comprehensive overview on the theory and of the practical aspects of an RL procedure are provided by the final reports of EU projects "CENEX on Integration of Survey and Administrative Data" (Eurostat, 2009) and "ESSnet on Data Integration" (Eurostat, 2011).

It is worth noting that practical applications of RL are based on the combination of different steps of deterministic and probabilistic RL. Usually a first step of exact matching or deterministic RL is performed (i.e., exact matching is performed when just a subset of units presents the unique unit identifier) then the probabilistic RL is performed on the remaining records not matched in the first steps. Finally, the possible links are submitted to clerical revision.

3.2.2 *Statistical Matching*

SM (or data fusion) refers to a series of methods that integrate the available data sources with the aim of studying the relationship between variables not jointly observed in a single source. In the typical framework, the available data sources, A and B, share a set of variables X, while the variables Y are observed just in A and the variables Z are observed just in B. The objective of SM consists in investigating the relationship between Y and Z. Usually A and B are relatively small independent samples selected from the same target population, therefore the chance of observing the same unit in both of them is close to zero, unless an unequal probability sampling design has been used such that the most important units are included in the sample with probability close or equal to one (e.g., surveys with a take-all stratum).

In SM the integration does not necessarily involve the creation of a "fused" data source; in fact, in some cases, the objective of SM may consist in estimating one or more parameters; this goal can be achieved by applying appropriate statistical methods on the available data sources without integrating them at micro level (for details see D'Orazio *et al.*, 2006).

SM at micro level provides a synthetic or fused data set in which all the needed variables are available; the term "synthetic" is used to stress that its records are artificial: the values of some variables are not really observed because they are derived (imputed) according to the chosen SM methods. The synthetic data set can be created by (i) concatenating A and B and then filling in the missing values, or (ii) by choosing just one of the available data sources, say A, and then filling it in with the values of the missing variable(s), that is Z, by using the information available in the other one (B in this case). In this latter case, A plays the role of recipient, while B is the donor. The choice of the recipient is crucial; it, once filled in with the missing variables, will be the basis of further statistical analyses; an obvious decision seems that of considering the larger one because it would provide results that are more accurate. Unfortunately, in SM such a way of working may provide inaccurate results, especially when the sizes of the two data sources are very different ($n_{ric} \gg n_{don}$); the risk is that of obtaining imputed valued whose distribution does not reflect the original one (estimated from the donor data set).

Methods commonly used to perform SM at micro level are strictly related to those used for imputing missing values in a survey. Donor-based methods are frequently applied. A donor for each unit in the recipient data set A can be picked up at random (random hot deck); usually, before the donor selection, units in both A and B are grouped in classes according to one or more X variables: donors are randomly selected in the same class to which the recipients belong. Alternatively, it can be considered the closest donor in B according to a distance computed on a suitable subset of the common variables (X) (nearest neighbor hot deck). In the latter approach a crucial decision concerns the selection of the subset of the X variables involved in the computation of distance (matching variables). Different studies show that choosing too many matching variables may affect negatively the matching results; in particular, the marginal distribution of the variable imputed in A may not reflect the one observed in B. When dealing with large data sources it may be convenient and more efficient to divide units in A and B into donation classes according to the values of some categorical common variables (e.g., for a male in A the closest donor male in B will be searched).

The nearest neighbor hot deck can be performed under some particular constraints: for instance, it may be decided that a donor should be used just once, subject to the additional constraint of minimizing the overall matching distance. In this case, the selection of the donors requires the solution of an optimization problem (transportation problem). Such a constrained matching returns an overall matching distance greater than the one in the unconstrained case, but it tends to better preserve the marginal distribution of the variable imputed in the synthetic data set. A different set of constraints can be considered when survey weights are introduced (Barr and Turner, 1981).

A parametric approach to SM requires the specification of a model, then its parameters are estimated and, when the synthetic data set is required, a predictive approach is used to impute the missing values (in the concatenated file or just in the recipient, depending on the choice made). Commonly used parametric micro methods are conditional mean matching and draw based on the predictive distribution. In the first case each missing value is substituted by the corresponding expected value given the observed X variables; regression models are frequently used. Unfortunately such a method provides imputed values without variation with respect to the expected ones (Little and Rubin, 2002), for this reason it is preferable to impute values obtained as random draws from the predictive distribution; in the case of regression models, this approach results in imputing values obtained as the expected one plus a random residual (stochastic regression imputation). The main issues when dealing with a parametric approach to SM concern the specification of the model and the final imputed values: a wrong specification of a model will provide unreliable results; the imputed values are artificial values and are not really observed. For these reasons sometimes, a nonparametric imputation step follows the parametric approach; the resulting SM procedure is said to be mixed. The proposed mixed approaches for SM are based essentially on predictive mean matching imputation methods (see D'Orazio *et al.*, 2006, sections 2.5 and 3.6). In the case of continuous X, Y and Z, a general procedure consists of the following two steps:

- Step 1. Choice of a model $Z = g(X; \theta)$ and then estimation of its parameters (θ). Then, the estimated model is used to derive "artificial" values, $\tilde{z}_a$ of Z in A (predicted values, or predicted plus a random error term).
- Step 2. For each record in A the closest record in B is selected according to a distance $d_{ab}(\tilde{z}_a, z_b)$ computed considering artificial and truly observed values of Z.

  The closest record in B donates to A the value of Z observed on it.

Similar procedures are proposed by Rubin (1986), Singh *et al.* (1993), Moriarity and Scheuren (2001, 2003). The mixed approach avoids the computation of distances on several common variables, whereas variables with low predictive power on the target variable may influence negatively the distances. Moreover, the imputed values are really observed values and not artificial ones. More in general, the mixed approach joins the advantages of both the parametric and nonparametric approaches: the model is more parsimonious while the nonparametric step offers "protection" against model misspecification.

It is important to stress that all the matching applications that use just the set of variables common to both the data sources, assume implicitly that the relationship between the variables not jointly observed is completely explained by these common variables. In other words, in the basic SM framework (just A and B are available, A contains X and Y while B observes X and Z) all the SM methods (parametric, nonparametric, and mixed) that use X to match A and B, implicitly assume the conditional independence of Y and Z given X. The conditional independence assumption means that the joint probability density function for X, Y and Z can be factorized in the following manner:

$$f(x, y, z) = f(y, z|x)f(x) = f(y|x)f(z|x)f(x) \tag{3.2}$$

This assumption is particularly strong and seldom holds in practice. If the conditional independence assumption does not hold then results of SM derived under it will not be valid.

It is possible to skip the conditional independence assumption if some additional auxiliary information can be exploited in the matching application. The term auxiliary information here is used in a wider meaning referring to one or more of the following elements.

a) A third data source C where (X, Y, Z) or just (Y, Z) are jointly observed (e.g., small survey, past survey or census data, administrative register, etc.).
b) Estimates related to the parameters of (Y, Z) or $(Y, Z)|X$ (e.g., estimate of the covariance σ_{YZ}, of the correlation coefficient ρ_{YZ}, or of the partial correlation coefficient $\rho_{YZ|X}$, a contingency table of $Y \times Z$, etc.).
c) A priori knowledge of the investigated phenomenon, which allows to identify some logical constraints on the parameters' values. For instance, when dealing with categorical variables there may be some *structural zeros*, that is events that cannot happen because they are impossible, that is $Pr(Y = y, Z = z) = 0$.

The exploitation of auxiliary information in the SM application depends on the type of information and on the desired outputs. The case (a) can be easily exploited in most of the SM. For the main details on the usage of auxiliary information in the SM see D'Orazio *et al.* (2006) and Singh *et al.* (1993).

An alternative approach to SM consists of exploring the uncertainty in estimating the parameters of interest that is determined by the matching framework. This approach can be applied when the objective is macro. In practice, uncertainty analysis does not end with a unique estimate of the unknown parameter characterizing the joint probability density function for (X, Y, Z); on the contrary, it identifies an interval of plausible values for it. For instance, when dealing with categorical variables, the intervals of plausible values for the probabilities in the table $Y \times Z$ are derived by means of the Fréchet bounds once conditioning on X (D'Orazio *et al.*, 2006):

$$P_{j,k}^{(low)} \leq P_{Y=j,Z=k} \leq P_{j,k}^{(up)} \tag{3.3}$$

with

$$P_{j,k}^{(low)} = \sum_i P_{X=i}\, max\{0; P_{Y=j|X=i} + P_{Z=k|X=i} - 1\}, \qquad j = 1, \cdots, J; k = 1, \cdots, K \quad (3.4)$$

$$P_{j,k}^{(up)} = \sum_i P_{X=i}\, min\{P_{Y=j|X=i}; P_{Z=k|X=i}\}, \qquad j = 1, \cdots, J; k = 1, \cdots, K \quad (3.5)$$

where I, J, and K are the categories of X, Y, and Z, respectively.

Most of the SM techniques assume that A and B are random samples of independent and identically distributed (i.i.d.) observations selected from the same infinite population. Unfortunately, in most of the cases the available data come from complex sample surveys carried out on the same finite population. In this framework, the set of the values assumed by X, Y, and Z for each unit in the finite population are usually viewed as fixed values and not outcomes of random variables. The randomness is introduced by the probability criterion used to select the sample from the population. In practice, in this case the inference is based on the *sampling design* (involving stratification and/or clustering) used to draw the sample. However, when dealing with data from complex sample surveys, apart from the sampling design, the analyst should consider that: (i) the data sources are usually affected by *non-observation errors* (*unit and item nonresponse*; *out-of-scope* units) which determine a reduction of the sample size with respect to the planned size; (ii) some observed values may be affected by measurement errors (not detected by checks); (iii) the final weights w_α associated with the available units are the direct weights corrected to compensate for unit nonresponse, frame undercoverage and to reproduce known population totals (benchmarks) concerning some important auxiliary variables; and (iv) the design variables may be partially available due to the risk of disclosure.

All these issues pose several problems when applying SM. There are a few SM methods that explicitly take into account the sampling design and the design weights: Renssen's approach based on *weights' calibrations* (Renssen, 1998) and Rubin's *file concatenation* (Rubin, 1986). A comparison between these approaches can be found in D'Orazio *et al.* (2010; see also D'Orazio, 2011 and D'Orazio *et al.*, 2012).

The Rubin's approach consists in concatenating the surveys $A \cup B$ and then computing a new inclusion probability for each unit in it:

$$\pi_{A \cup B, c} = \pi_{A,c} + \pi_{B,c} - \pi_{A \cap B, c} \cong \pi_{A,c} + \pi_{B,c} \qquad (3.6)$$

(with $c \in A \cup B$). The approximation holds when the chance of a unit of being included in two independent surveys related to the same finite population is close to zero (usually the samples are relatively small if compared with the population size). Unfortunately, such an approach handles theoretical samples and does not account for reduction of sample size because of unit nonresponse. Moreover, it may be quite difficult to derive the inclusion probabilities $\pi_{A \cup B, c}$ (for the main details on these problems see Ballin *et al.*, 2008). Once the probabilities are estimated, the problem remains of filling in the missing values in the concatenated file (Z is missing in data originating from A and Y is missing in data from B).

Renssen's approach is based on a series of *calibration* steps of the survey weights in A and in B. Calibration is a technique very common in sample surveys for deriving new weights, as close as possible to the starting ones, which fulfill a series of constraints concerning totals (benchmarks) for a set of auxiliary variables (for further details see Särndal and Lundström, 2005). Renssen's procedure is designed to estimate the two-way contingency table $Y \times Z$

starting from the data of two independent complex sample surveys carried out on the same finite population (macro approach). It permits to exploit eventual auxiliary information represented by a third data source C in which (X, Y, Z) or simply (Y, Z) are available. In practice the procedure consists of two steps: (1) harmonization of the marginal/joint distribution of the matching variables in A and in B; and (2) estimation of the two-way contingency table $Y \times Z$ using auxiliary information provided by a third data source C by means of (a) *incomplete two-way stratification*, or (b) *synthetic two-way stratification*.

In order to estimate the table $Y \times Z$, Renssen's approach considers linear probability models; the fitted models permit a regression imputation of the missing variables at micro level. When all the variables (X, Y, Z) are categorical the predicted values of Y in the synthetic data preserve the marginal distribution of the imputed Z variable and the joint distribution $X \times Z$. The main disadvantage is that the imputed value of Z for each unit in A corresponds to a vector of estimated probabilities. Unfortunately, linear probability models present several drawbacks (negative or greater than one estimated probabilities; heteroskedasticity, and residuals not normally distributed). For these reasons, such predictions should be used carefully. When Y and Z are continuous variables, Renssen (1998) suggests to create the synthetic file by concatenating A and B and then filling in the missing values with the predictions provided by the models or by using a mixed procedure. The resulting concatenated data set should preserve the marginal distributions of all the variables.

It is worth noting that the analyses that can be performed on a synthetic data set are strictly related to the objectives of SM and to the applied SM methods. In particular, given that the matching objective is that of exploring the relationship between Y, X and Z, analyses involving different sets of variables in most of the cases will still be based on the Conditional Independence (CI) assumption (independence given the variables being used in the integration), and therefore they will be unreliable if the CI does not hold.

A problem, similar to SM, deals with the integration of two independent surveys: the survey A consisting in a large sample providing data just on the X variables, and B which contains data on Y and X but on a smaller sample ($n_A \gg n_B$). This scenario, according to Kim and Rao (2012), is that of *nonnested two-phase sampling*. The target of the inference is that of imputing Y in A by exploiting the relationship between Y and X, in order to derive more accurate estimates concerning Y. The techniques used to solve the problem sometimes are different. In some cases, imputation methods are applied (*mass imputation*: the file to fill in is larger than the donor). This situation is however strictly related to other situations frequently encountered in small area estimation problems, microsimulation scenarios, and so on (Haslett *et al.*, 2009). The Kim and Rao (2012) approach is interesting; they suggest a method to create a synthetic data set by filling in A with imputed values of Y, in order to estimate the total amount of Y at population level or domain level. The imputed values are derived by using appropriate models and the resulting estimator is a particular *projection estimator*. Unfortunately, the proposed approach cannot be used to estimate quantiles of a continuous variable.

3.3 Administrative and Survey Data Integration: Some Examples of Application in Well-being and Poverty Studies

The traditional approach of economists to the measurement of poverty is to use measures of income or consumption. In particular, the poverty approach aims at subdividing the population into poor and non-poor defined in relation to some chosen poverty lines that represent a percentage (generally 50, 60, or 70%) of the media or the median of the equivalent income distribution. Behind such measures lies the concept of utility, or welfare, which people are assumed to derive from income and consumption. Over the years, scholars have suggested broader sets of measures for assessing poverty than just income and consumption. This multidimensional approach (see SAMPLE,[2] deliverable 4 for a review of the main approaches) reflects the opinion that well-being depends not only on material living standards, but also on other aspects of life. Enjoying good health, gaining access to an appropriate education and training system, having a good job, living in a context of social relations based on trust and environmental care are relevant dimensions of well-being. Today, the multidimensional nature of well-being and poverty is a widely recognized fact, not only by the scientific community, but also by many official statistical agencies and international institutions. In December 2001, at the Laeken European Council, EU Heads of State and Government endorsed a first set of 18 common statistical indicators of social exclusion and poverty (Laeken indicators), indicators that were refined later by the European Commission Social Protection Committee (2001). The Laeken indicators cover four important dimensions of social inclusion (financial poverty, employment, health, and education), which highlight the multidimensionality of the phenomenon. Such indicators are regularly produced for every EU country on a comparable basis. EU-SILC and Labor force surveys are the main data sources used to calculate them.

Official statistics also disseminate macro indicators on people's economic well-being within the National Accounts (NAs) framework. These statistics are particularly relevant when the purpose is to compare material living conditions in time and space also taking into account economic resources not captured by surveys such as social transfers in kind or income from hidden economy. Indeed, the OECD better life index makes use of Houschold Net Adjusted Disposable Income and Household Financial Wealth (both calculations are based on the NAs database) for measuring the economic well-being domain of the index (OECD, 2014).

Although the quality of official statistics on the discussed topics is continuously improving, some information is still lacking. We focus here on two major problems: (i) the incoherence of micro data statistics on income, consumption and wealth, which prevents national accountants from properly measuring disparities in economic well-being; and (ii) the inaccuracy of available statistics on poverty and well-being at the local level.

3.3.1 Data Integration for Measuring Disparities in Economic Well-being at the Macro Level

NAs describe the economy of a country (a region or groups of countries) in a given period recording how production is generated (Gross Domestic Product, GDP), how income originating in production flows to the economic agents, and how the economic agents allocate income to consumption, saving and investment. Finally, NAs include balance sheet accounts thus providing a measure of the country's wealth at a given date.

According to their economic behavior, elementary economic agents are grouped together into homogeneous sectors, called institutional sectors. The Households sector's accounts

[2] Small Area Methods for Poverty and Living condition Estimates (SAMPLE), EU 7FP project (www.sample-project.eu).

provide useful information to measure people's economic well-being. In fact, they provide fully consistent and coherent data on households' income, consumption, and wealth. This is a very desirable property when the purpose is to consider the three aggregates jointly (Stiglitz Commission, 2009, recommendation 3). Unfortunately, NAs allow calculating average values only, without providing any detail on the distribution of income, consumption, and wealth among different households groups. The Eurostat statistics database publishes NAs statistics on an annual and quarterly basis, at the national and local levels (regions NUTS 2) for all Member States.

Differently from NAs, micro sources allow users to analyze the distributions of income, consumption, and wealth across the population, for various subgroups, and over time. Nevertheless, micro sources fail in detecting all people's economic resources such as social transfers in kind or income from hidden economy. Furthermore, micro data on income consumption and wealth are seldom coherent, unless they all come from one single source. In short, macro data assure comprehensiveness, inner coherence and international comparability but cannot provide information on the distribution of economic resources among people; micro data allow users to calculate distribution indicators but do not cover all people's economic resources nor do they always provide coherent information on households' income consumption and wealth.

In order to overcome the NAs' limits, as early as the 1990s, the European System of Accounts suggested partitioning the Household sector into sub-sectors according to the family's largest source of income (Eurostat, 1995). Recently, the OECD-Eurostat Expert Group on Disparities in a NAs framework proposed to group households according to the households' income quintile or to socio-demographic characteristics of the household (Fesseau and Mattonetti, 2013).

Whatever the criterion, sub-sectoring Households' NAs means breaking down NAs items according to distribution indicators from micro-data sources. To this end, it is necessary to use several micro-data sources (at least more than one), given the high detail of information required to break down all the different NAs' items. A major problem is that micro data sources often suffer a certain degree of incoherence, which could bring, at a later stage, to an inconsistent allocation of NAs items among the Household sub-sectors.

Suppose for instance that data source A collects micro data on households' disposable income and its components (earned income, pensions and other transfers net of taxes and social contributions) whereas source B collects data on households' consumption expenditure by kind of products. Using source A it could be possible to calculate distribution indicators to allocate NAs income items among the households' sub-sectors. Conversely, source B should provide information for allocating consumption expenditure by Household sub-sectors and consumption categories. However, what if A and B do not provide consistent information? It could happen that Household sub-sectors record a value of consumption expenditure not compatible with their disposable income. In order to avoid the problem, we should match individual micro data *before* calculating breaking-down indicators for the NAs items.

One possible strategy consists in recovering income data from administrative tax registers rather than from income surveys. RL could be used to merge income and consumption records, taking advantage of the partial overlap between the two data sets in terms of observed units. Indeed, record linkage precisely aims at linking a single individual's records between two or more files making use of a proper identifier. This strategy has the indisputable strength of linking information actually observed on the same unit (except in the case of errors in the identifier variables, see Section 3.2.1). However, the use of administrative data also has disadvantages. One problem is that tax data contain incomplete information on the bottom tail of the income distribution since low incomes are typically tax-exempt. Furthermore, units differ in the two data sources: the household for the consumption survey, and the tax unit for the administrative data source.

One viable alternative consists of merging income and consumption micro data coming from two different sample surveys. In this context, SM seems appropriate. This procedure finds one donor unit in the consumption data set for each unit of the income data set. The final purpose is the construction of a complete file containing all the variables collected by the two data sources. This approach is particularly useful when no alternative data source providing detailed data on both household income and consumption is available (Coli and Tartamella, 2010). However, it is worth noting, that SM should make use of auxiliary information in order to skip the conditional independence assumption (see Section 3.2.2).

3.3.2 Collection and Integration of Data at the Local Level

People living in close geographic areas can experience different quality of life standards. In fact, whilst some well-being dimensions concern the citizens' individual characteristics, others have much more to do with the city they live in. Differences among the regions of the same country can be just as important as differences between countries (OECD, 2014). This depends on many causes, one of which is the existence of different local welfare systems within the same territory. In fact, some well-being domains correspond to typical areas of intervention of the local welfare policies. In most countries, local governments are in charge of significant social policies on health, education, disability, or housing. Each local government establishes the level and composition of its social protection expenditure. For example, in Italy, in 2009, funding of social services ranged from 25 euros per inhabitant in Calabria to 304 euros per inhabitant in the Province of Trento (Istat, 2013); some local governments supported polices in favor of elderly people, others decided to support younger generations, or concentrated funds on housing policies. Local policy makers need accurate and up-to-date information on the needs expressed by people living in their territory. To this end, it is essential to measure well-being where it is being considered. Furthermore, local governments need up-to-date data on the results of their policies in order to evaluate the effectiveness of policies. As already stressed, the lack of accurate and timely indicators at the local level is one of the major limits of current official statistics. Many regions and cities in Europe and beyond have launched their own measurement initiatives to bring data and policy together for improving the quality of life of their communities (OECD, 2014).

The SAMPLE project contributed to research of this topic, pursuing two main objectives: (i) develop new indicators to better investigate inequality and poverty with special attention to social exclusion and deprivation; and (ii) develop models and implement procedures for estimating these indicators and their corresponding accuracy measures at the small area level. SAMPLE produced prototype indicators both on monetary and non-monetary aspects of poverty, at NUTS 3 and LAU 1[3] geographical domains within the Tuscany region. In

[3] To meet the demand for statistics at local level, Eurostat has set up a system of Local Administrative Units (LAUs) compatible with NUTS. The upper LAU level (LAU level 1, formerly NUTS level 4) is defined for most, but not all of the countries. The lower LAU level (LAU level 2, formerly NUTS level 5) consists of municipalities or equivalent units in the 27 EU Member States (http://epp.eurostat.ec.europa.eu).

order to estimate indicators for such unplanned domains, it was necessary to enlarge the EU-SILC sample (EU-SILC oversampling, see Box 3.1) and further integrate such statistics with administrative data from local governance institutions and stakeholders.

Box 3.1 shows the results of the record linkage procedure set up in order to match survey and administrative data. Both data sources refer to individuals resident in the Province of Pisa. Among the main advantages, there is the possibility of correcting administrative indicators for the self-selection bias. In fact, most of the administrative data produce biased estimates because they only refer to people eligible for obtaining a service (i.e., Medicare, pensions, enrolment in social programs). Additionally, matched data allow administrative indicators to be read taking into account the characteristics of the family to which the individuals belong. More often, administrative archives do not collect information on the individual's family. This is a big limitation when the purpose is to measure individual material well-being, which greatly depends on the economic well-being of the individual's family.

Box 3.1: Merging Sample and Administrative Data

The SAMPLE project funded an oversampling of the SILC survey for the Province of Pisa (PI-SILC) in 2008. Sampled households were augmented from 162 to 818. The oversampling made it possible to attempt a RL with individual data from the local Revenue Agency (RA) registers.[4] The RA registers covered all individuals resident in the Province of Pisa who submitted a taxpayer declaration in 2008. In particular, we gained access to tax declarations submitted by employees, pensioners, temporary workers, cooperative workers (the "730" tax returns register), and tax declarations submitted by self-employed ("Unico p.f." tax returns register). Conversely, we were not able to gain access to the tax declarations submitted by employers on behalf of their employees. Therefore, we had to restrict our analysis on a subset of taxpayers.

The merging is aimed at extending the PI-SILC survey records with variables taken from the RA database (for details on the application, see Coli, 2012). Among the main advantages, there is the possibility of grossing up RA-based indicators using PI-SILC sample weights. Additionally, matched data allow RA indicators to be read taking into account the characteristics of the households to which the individuals belong. Relying on RA indicators only, individuals could be classified as poor or marginalized, in spite of actually living in a wealthy family.

The PI-SILC target population corresponds to all residents in the Province of Pisa at the date of the interview (about 406 000 inhabitants). The RA population is only part of the PI-SILC target population. However, the RA database includes many more records (about 187 000) than the PI-SILC database (about 1800), the latter being a sample of the population.

We pre-processed data in order to detect and correct errors, and in order to harmonize the content and the categories of common variables. Furthermore, we selected the PI-SILC individuals who declared to have earned a taxable amount of income in 2007 (1210). In fact, only these individuals are expected to find a match in the RA registers.

The personal items used as match key-variables were: birth month, birth year, gender, birth place (municipality), place of residence (municipality), and nationality.

[4] Chapter 4 focuses on the link between the PI-SILC records and the Provincial Jobcenter database records.

As a first stage, the key-variables were used to classify the PI-SILC and RA units into homogeneous strata. Within each couple of homogeneous strata, we ran a further deterministic linkage procedure taking into account income-related variables. In particular, for each PI-SILC unit we picked out the RA unit sharing approximately the same income level (a difference of no more than 100 euros was allowed) and the same kind of earned income (from employed income, self-employed income, or pensions). When we found more than one match, we chose the one with the lowest difference in terms of gross income.

Using the record linking procedure, we found a link for 411 PI-SILC units out of 1246 (about 33%). This not entirely satisfactory result is due to various reasons: (i) the PI-SILC oversampling involved only part of the municipalities of the Province of Pisa; (ii) the RA database includes only a subset of the tax payers; and (iii) for confidentiality reasons, we could use only "weak" identifiers.

In order to check the quality of the matching, we compared the distributions of some RA variables in the donor (RA database) and in the matched file (Rässler, 2002). RA marginal distributions seem well preserved in the matched sample when considering demographic variables. On the contrary, the distribution of taxable income changes significantly. Comparing the distribution of taxable income by category (Table 3.1) it appears that the linking procedure failed in detecting self-employed and property income earners or in detecting the richest ones among them. In fact, the proportion of self-employed income and property income is considerably lower in the matched sample.

Table 3.1 Distribution of taxable income by categories in the RA register and in the matched sample, year 2007

Data source	Compensation of employees (%)	Self-employed income (%)	Property income (%)	Taxable income (%)
RA register	77.22	11.40	8.63	100.00
Matched sample	93.09	3.36	2.60	100.00

Source: Coli (2012).

Successively, using matched data we calculated mean and quintiles for per capita taxable income from the RA data source. First, we used un-weighted values, then we applied the indirect sampling methodology, that is we used the PI-SILC sampling weights and the link matrix generated by the matching procedure (for details see SAMPLE, deliverable 11). Results (Table 3.2) show that correcting for self-selection bias leads to a general increase in taxable income level.

Table 3.2 RA per capita taxable income (in euros), Province of Pisa, year 2007

	Un-weighted values	Weighted values
Mean	19 878	21 561
1st quintile	12 138	13 369
Median	17 826	19 774
3rd quintile	25 233	27 295

Source: SAMPLE, deliverable 11 (www.sample-project.eu).

3.4 Concluding Remarks

Official statistics currently provide many indicators on poverty and well-being, however these indicators cannot always answer the users' needs. In this chapter, we focused in particular on two limits. On the micro side (Laeken indicators), there is a problem of inaccuracy for the estimate of poverty indicators on unplanned domains (e.g., small geographical areas). On the macro side (NAs indicators on economic well-being), distributional measures of income, consumption, and wealth would be extremely useful to assess disparities among different groups of households but this preliminary requires micro statistics on income, consumption, and wealth to be coherent. Integration of micro data can help overcome both these limitations as discussed in the chapter.

The outcome of the integration procedures depends to a great extent on the quality of the data sources to be integrated and, in particular, on the accuracy of the variables used in the integration process. In RL applications, the integration is simple and straightforward if the two data sources have a units' identifier recorded without errors. If the identifier is not available, the success of the RL procedure (deterministic, probabilistic, or a combination of them) is heavily based on the availability of key variables being "powerful" (e.g., name, surname, etc.) in identifying units. It will be difficult for complex RL procedures to be successful when one or both of the data sources lack powerful key variables. Similarly, it is difficult to obtain good results when the RL is performed on key variables, which are highly affected by errors; the pre-processing steps and the string comparators can compensate for some typographical errors but cannot remediate for poorly recorded values.

A similar reasoning applies to SM applications where, however, the key role is played by the assumption of conditional independence of Y and Z given the common X variables. This assumption is seldom valid in real situations; therefore, SM results based on it will be unreliable when it does not hold. In this context, a SM application will be successful if it can exploit some powerful auxiliary information concerning the relationship between Y and Z.

In this way, the integration step should be considered as an integral part of a wider statistical process. This means that data collection (survey or Administrative register) for a specific purpose should be planned in order to provide the information also needed for a further integration process. If RL has to be applied, it is important to design the collecting process in order to provide an error free identification code for all the units. If this is not possible, for example for confidentiality reasons or other issues, then it is important to collect powerful key variables observed without errors and with the characteristics necessary to use them in the RL, so avoiding a long pre-processing step. In the case of SM, it is important to design the surveys involved in the matching, in order to collect the auxiliary information needed to bypass the CI assumption. For example, to analyze the relationship between income and consumption using matched data, it is advisable to plan the income survey with questions on expenditure or the expenditure survey with questions on income.

Obviously, designing the statistical process having integration in mind is feasible when the integration is performed by the agency which produces the data sources involved in integration; on the contrary, it may be difficult when integration is done with data obtained from other agencies. In such cases, the possibility of improving the characteristics of the data sources are conditioned by the possibility to intervene to modify the process that produces them. In some cases, some formal or informal agreements between the involved agencies can help in solving the problem.

Official statisticians are taking important steps toward the adoption of data integration as a step in the statistical production process. Currently, European statistical production is moving from a system anchored to a one-to-one link of separate surveys with their own specific outputs to an integrated system, including different data sources (surveys but also administrative) that are combined for producing various statistical outputs. For example, Eurostat is re-shaping its system of social statistics to a modular architecture with modules: (i) focusing on specific domains; (ii) autonomous to a high extent, in that there may be some interaction between modules, but the greater interaction and integration occurs within each module; and (iii) hierarchically nested, so that each module can be further decomposed into finer modules (Van der Valk, 2012 & Eurostat, 2002). This type of architecture is based on the re-usability of the modules in the development of different surveys (Reis, 2013). To this end, surveys will be designed in order to maximize the value added of a further integration of data. For example, they will share, to the maximum extent possible, statistical frames, definitions, survey methods, classifications and output categories. Each module will be assigned the mode of data collection which best fits the kind of data to be collected. For example, it is advisable to collect data on income via tax registers rather than interviewing people. Once the best mode of collection is selected for each module, that mode will be used for whatever statistical product (instrument) which uses the module (Van der Valk, 2013).

In this scenario, the policy of the European NSIs is to devote resources to linking, matching and reconciliation procedures rather than to direct data collection. In the last decade, more and more NSIs improved the use of administrative data as a primary source of information, restricting the use of sample surveys for in-depth analysis on specific topics of interest. However, administrative data will not be able to fully replace surveys, although there will be a significant reduction in costs and statistical burdens. In the future, the statistical production will have to rely on a combination of survey and administrative data. The integrated system has the potential of building on the strength of each type of source in order to mitigate the weakness of the other sources, thus improving the robustness of official statistics (Kloek and Vâju, 2012).

References

Ballin, M., Di Zio, M., D'Orazio, M., Scanu, M., and Torelli, N. (2008) File concatenation of survey data: a computer intensive approach to sampling weights estimation, *Rivista di Statistica Ufficiale*, **2-3**, pp. 5–12. November 23, 2014. http://www.istat.it/it/files/2011/05/2_3_20081.pdf.

Barr, R.S. and Turner, J.T. (1981) Microdata file merging through large-scale network technology, *Mathematical Programming Studies*, **15**, pp 1–22.

Belin, T.R. and Rubin, D.B. (1995) A method for calibrating false-match rates in record linkage, *Journal of the American Statistical Association*, **90**, pp. 694–707.

Chambers, R. (2009) Regression analysis of probability-linked data, *Official Statistics Research Series*, **4**.

Christen, P. (2012) *Data Matching, Concepts and Techniques for Record Linkage, Entity Resolution, and Duplicate Detection*. Springer, New York.

Coli, A. (2012) Matching survey and administrative data for studying poverty at the local level, *Rivista Internazionale di Scienze Sociali*, 2012 vol 1- pp. 24–32.

Coli, A. and Tartamella, F. (2010) Income and consumption expenditure by households groups in National accounts, Report 334, Department of Statistics and Mathematics Applied to Economics, University of Pisa.

D'Orazio, M. (2011) Statistical matching when dealing with data from complex survey sampling, Eurostat Report of WP1 of ESSnet project on Data Integration, pp. 33–37.

D'Orazio, M., Di Zio, M., and Scanu, M. (2006) *Statistical Matching, Theory and Practice*. John Wiley & Sons, Ltd, Chichester.

D'Orazio, M., Di Zio, M., and Scanu M. (2010) Old and new approaches in statistical matching when samples are drawn with complex survey designs. Proceedings of the 45th Riunione Scientifica della Società Italiana di Statistica, Padua, Italy, June 16–18, 2010.

D'Orazio, M., Di Zio, M., and Scanu, M. (2012) Statistical matching of data from complex sample surveys, European Conference on Quality in Official Statistics (Q2012), Athens, Greece, May 29–June 1, 2012.

Elfeky, M., Verykios, V., Elmagarmid, A.K., and Ghanem, T.M. (2003) Record linkage: a machine learning approach, a toolbox, and a digital government Web service, Computer Science Technical Reports, Department of Computer Science, Purdue University.

European Commission Social Protection Committee (2001) Report on indicators in the field of poverty and social exclusion. Technical report.

Eurostat (1995) System of National and Regional Accounts, Luxembourg.

Eurostat (2002) Handbook on Social Accounting Matrices and Labour Accounts, Luxembourg.

Eurostat (2009) State of the art on statistical methodologies for integration of surveys and administrative data, Report of WP1, CENEX on Integration of Survey and Administrative Data. November 23, 2014. http://cenex-isad.istat.it.

Eurostat (2011) State of the art on statistical methodologies for data integration, Report of WP1, ESSnet on Data Integration. November 23, 2014. http://www.cros-portal.eu/content/data-integration-1.

Fellegi, I.P. and Sunter, A.B. (1969) A theory for record linkage, *Journal of the American Statistical Association*, **64**, pp. 1183–1210.

Fernandez, G. (2009) Blocking procedures, in Eurostat, Report of WP1, CENEX on Integration of Survey and Administrative Data, pp. 12–16.

Fernandez, G. (2011) Efficient blocking, in Eurostat, Report of WP1, ESSnet on Data Integration, pp. 32–35.

Fesseau, M. and Mattonetti, M.L. (2013) Distributional Measures Across Household Groups in a National Accounts Framework: Results from an Experimental Cross-country Exercise on Household Income, Consumption and Saving, OECD Statistics Working Papers, No. 2013/04, OECD Publishing.

Haslett, S., Jones, G., Noble, A., and Ballas, D. (2009) More for Less? Comparing small area estimation, spatial microsimulation, and mass imputation. Proceedings of the Section on Survey Research Methods – JSM 2010. Vancouver, British Columbia, July 31-August 5, 2010.

Istat (2013) Indagine sugli interventi e i servizi sociali dei comuni singoli e associati, datawarehouse I.Stat.

Jaro, M.A. (1972) UNIMATCH: a computer system for generalized record linkage under conditions of uncertainty, Proceedings of the 1972 Spring Joint Computer Conference, American Federation of Information Processing Societies, New York, USA, May 16–18, 1972, pp. 523–530.

Jaro, M.A. (1989) Advances in record-linkage methodology as applied to matching the 1985 Census of Tampa, Florida, *Journal of the American Statistical Association*, **84**, pp. 414–420.

Kim, J.K., and Rao, J.N.K. (2012) Combining data from two independent surveys: a model-assisted approach, *Biometrika*, **99**, 85–100.

Kloek, W. and Vâju, S. (2012) The use of administrative data in integrated statistics. Proceedings of the NTTS 2013 Conference, Brussels, Belgium, March 5–7, 2013.

Lahiri, P. and Larsen, M.D. (2000) Model based analysis of records linked using mixture models. Proceedings of the Section on Survey Research Methods, American Statistical Association, Indianapolis, Indiana, August 13-17, 2000. pp. 11–19.

Lahiri, P. and Larsen, M.D. (2005) Regression analysis with linked data, *Journal of the American Statistical Association*, **100**, pp. 222–230.

Little, R.J.A. and Rubin, D.B. (2002) *Statistical Analysis with Missing Data*, 2nd Edition. John Wiley & Sons, Ltd, Chichester.

Moriarity, C. and Scheuren, F. (2001) Statistical matching: a paradigm for assessing the uncertainty in the procedure, *Journal of Official Statistics*, **17**, pp. 407–422.

Moriarity, C. and Scheuren, F. (2003) A note on Rubin's statistical matching using file concatenation with adjusted weights and multiple imputation, *Journal of Business and Economic Statistics*, **21**, pp. 65–73.

Neter, J., Maynes, S. and Ramathan, R. (1965) The effect of mismatching on the measurement of response error, *Journal of the American Statistical Association*, **60**, pp. 1005–1027.

OECD (2014) How's Life in Your Region? Measuring Regional and Local Well-being for Policy Making. OECD Publishing, Paris.

Rässler, S. (2002) *Statistical Matching: a Frequentist Theory, Practical Applications and Alternative Bayesian Approaches*, Lecture notes in statistics, 168. Springer, New York.

Reis, F. (2013) Links between centralisation of data collection and survey integration in the context of the industrialisation of statistical production. Proceedings of UNECE Seminar on Statistical Data Collection, Geneva, Switzerland, September 25–27, 2013.

Renssen, R.H. (1998) Use of statistical matching techniques in calibration estimation, *Survey Methodology*, **24**, pp. 171–183.

Rubin, D.B. (1986) Statistical matching using file concatenation with adjusted weights and multiple imputations, *Journal of Business and Economic Statistics*, **4**, pp. 87–94.

Sariyar, M. and Borg, A. (2010) The RecordLinkage package: detecting errors in data, *The R Journal*, **2**, pp. 61–67.

Särndal, C. E. and Lundström, S. (2005) *Estimation in Surveys with Nonresponse*. John Wiley & Sons, Ltd, Chichester.

Scheuren, F. and Winkler, W.E. (1993) Regression analysis of data files that are computer matched, *Survey Methodology*, **19**, pp. 39–58.

Scheuren, F. and Winkler, W.E. (1997) Regression analysis of data files that are computer matched - part II, *Survey Methodology*, **23**, pp. 157–165.

Singh, A.C., Mantel, H., Kinack, M., and Rowe, G. (1993) Statistical matching: use of auxiliary information as an alternative to the conditional independence assumption, *Survey Methodology*, **19**, pp. 59–79.

Stiglitz Commission (2009) Report on the measurement of economic performance and social progress. November, 23 2014. http://www.ofce.sciences-po.fr/pdf/dtravail/WP2009-33.pdf

Tancredi, A. and Liseo, B. (2011) A hierarchical Bayesian approach to matching and size population problems, *The Annals of Applied Statistics*, **5**, pp. 1553–1585.

Thibaudeau, Y. (1989) Fitting log-linear models when some dichotomous variables are unobservable. Proceedings of the Section on Statistical Computing, American Statistical Association, August 1-6, Washington, pp. 283–288.

Thibaudeau, Y. (1993) The discrimination power of dependency structures in record linkage, *Survey Methodology*, **19**, pp. 31–38.

Torelli, N. and Paggiaro, A. (1999) La Stima della Quota di Errori in Procedure di Abbinamento Esatto. Atti del Convegno SIS 99: Verso i censimenti del 2000. *Proceedings of the Conference of the Italian Statistical Society* - SIS "Verso i censimenti del 2000", Udine, Italia, 7-9 giugno 1999.

Van der Valk, J. (2012) Introducing modularity in order to improve quality and efficiency. Proceedings of the 7th Workshop on LFS Methodology, Madrid, Spain, May 10–11, 2012.

Van der Valk, J. (2013) Modernisation of Social Statistics – A Possible Future of the European Social Surveys, Mimco.

Winkler, W.E. (1990) String comparator metrics and enhanced decision rules in the Fellegi–Sunter model of record linkage. Proceedings of the Section on Survey Research Methods, American Statistical Association, Anaheim, CA, 1990-8-6, pp. 354–359.

Winkler, W.E. (1993) Improved decision rules in the Fellegi–Sunter model of record linkage. Proceedings of the Section on Survey Research Methods, American Statistical Association, pp. 274–279.

Winkler, W.E. (2000) Machine learning, information retrieval, and record linkage. Proceedings of the Section on Survey Research Methods, American Statistical Association, Indianapolis, Indiana, August 13-17, 2000. pp. 20–29.

Yancey, W.E. (2000) Frequency-dependent probability measures for record linkage, Statistical Research Report Series, no. RR2000/07, U.S. Bureau of the Census, Washington, DC.

Zhu, V.J., Overhage, M.J., Egg, J., Downs, S.M., and Grannis, S.J. (2009) An empiric modification to the probabilistic record linkage algorithm using frequency-based weight scaling, *Journal of the American Medical Informatics Association*, **16**, pp. 738–745.

4

Small Area Methods and Administrative Data Integration

Li-Chun Zhang[1] and Caterina Giusti[2]
[1]*S3RI/University of Southampton, Southampton, UK and Statistics Norway, Oslo, Norway*
[2]*Department of Economics and Management, University of Pisa, Pisa, Italy*

4.1 Introduction

The literature of Small Area Estimation (SAE) is dominated by sample-survey-based applications, where administrative register data are used as covariates in the models. Register-based statistics, however, are becoming more and more common, and integration of survey and administrative data can raise many distinct issues. Compared with sample survey data, an important advantage of the register data is that statistics can be produced at much more detailed aggregation levels. For instance, register-based census have been carried out in a number of European countries. Statistical measures of uncertainty, however, are rarely produced for register-based statistics, partly due to a lack of theoretical developments, partly due to the complexity of the errors involved (Zhang 2012). These issues are particularly relevant also for the production of poverty and well-being indicators. In many countries data coming from large sample surveys, such as the Labour Force Survey (LFS) or the EU-SILC (EU Statistics on Income and Living Conditions), can be complemented or integrated with data coming from several population registers to obtain either more accurate estimates at the local level, or multidimensional indicators that could not have been produced using each source on its own.

We shall characterize the settings of SAE which involve administrative data according to (a) whether there are relevant additional sample survey data present and, if so, (b) how the target measure is related to the available data in the two sources.

The term register-based is often understood to refer to statistics that are produced by tabulation of statistical registers processed purely from administrative data. It may be that no

Analysis of Poverty Data by Small Area Estimation, First Edition. Edited by Monica Pratesi.

Companion Website: www.wiley.com/go/pratesi/poverty

additional survey data are available at all, such as when health statistics are exclusively compiled based on clinical or pharmaceutic records. Sometimes, relevant survey data are available, but are only used 'indirectly' to define the processing rules and/or to assess the accuracy of the register data but not to adjust them. For example, analysis of past census and sample survey household data may help to define the 'rules' by which administrative data are combined to construct the statistical register households in a register-based census (Zhang 2011). As another example, to size the inter-regional over-/under-coverage of the register-based population enumeration, additional questions were administered in the Norwegian LFS in the 4th quarter of 2011. Table 4.1 cross-tabulates the 1029 persons (out of over 20 000 LFS respondents) who have reported a different address in the LFS to the Population Register. Further analysis reveals two causes that dominate the discrepancy. For instance, out the 150 persons in column I, 100 have permanently moved to another address and 32 of the rest are students. The former demonstrates the time lag that exists in many registers, which will be discussed in Section 4.2.2. The latter reflects potential relevance error, as the population registration abides by different regulations compared with the traditional census residence definition. We discuss the relevance error in Section 4.3.2.

It is important to emphasize that combining data from multiple sources is generally necessary for the register-based statistics. In particular, integration with one or several base registers (Wallgren and Wallgren 2006), including Population Register, Business Register and Immobility Register of building, property and land, is almost always required to obtain the target population frame and to improve the data quality. For instance, part of the input to register-based education statistics are university exam results, which may be utilized for deriving, say, the variable highest level of education. Matching these data to the Population Register is necessary in order to verify the in-scope target population, and it helps to combine multiple exam results of the same person and check their plausibility, and so on, and it enables these data to be 'linked' with other relevant sources of educational data for the stated purpose. Chapter 3 of this book focuses on methods to integrate survey and register data with the aim of producing measures of well-being at the macro or micro level.

A different setting is when sample survey data are available and considered to provide the target measure. This is by and large the commonly treated scenario in the SAE literature, where

Table 4.1 Region of residence by register enumeration and LFS

Region	Region in register enumeration								
in LFS	Missing	I	II	III	IV	V	VI	VII	Total
Missing	1	31	10	29	22	40	20	19	172
I	0	116	15	24	14	8	5	8	190
II	0	5	53	1	1	3	1	1	65
III	3	11	1	90	4	2	2	4	117
IV	1	4	0	2	94	5	1	5	112
V	0	6	2	7	14	118	6	3	156
VI	0	5	2	7	2	11	58	10	95
VII	1	3	0	4	2	4	5	103	122
Total	5	150	73	135	131	151	78	134	1029

Source: Internal quality report, Norwegian register-based Census (2011).

the administrative data supply the covariates of the model. A special situation arises when the administrative source contains a proxy to the target survey measure. To distinguish it from the other auxiliary variables, we loosely define a proxy measure to be a variable that is similar in definition and has the same support compared with the target variable. For instance, while variables such as age, sex, education, income, and so on are auxiliary variables to the binary unemployment status, the binary register-based job-seeker status is a proxy measure. But the job-seeker status is not a proxy variable to the activity status defined (employed, unemployed, inactive) because the two have different support.

Statistical registers are rich sources of concurrent proxy measures that have complete (or virtually complete) coverage. A proxy variable is typically the most powerful among all the auxiliary variables for regression modelling of sample survey data. But it also enables another perspective, namely to adjust the relevance error inherent in the proxy variable that has complete coverage by the target survey variable that is only available for a sample of the population, where the sample size is typically small in the SAE context.

In short, the two settings to be considered in more detail are: Section 4.2, register-based SAE without any survey data; and Section 4.3, SAE based on integration of sample survey and register data. The topics will be organized by the nature of the predominant error of concern—see Zhang (2012) for a total-error framework for data integration. These include sampling error, measurement error arising from progressive administrative data, coverage error, relevance error and probability linkage error. The discussions will be focused on their relevance to the different settings of SAE, in such a way that the general concepts and issues involved are applicable to the production of poverty and well-being indicators.

The key message we wish to convey is that there are considerable challenges for SAE based on administrative data integration beyond conventional regression modelling of the sample survey data. With or without sample surveys, there are many important areas of application for administrative data, including the production of poverty and well-being indicators, but also plenty of problems that need to be solved.

4.2 Register-based Small Area Estimation

4.2.1 *Sampling Error: A Study of Local Area Life Expectancy*

When the target parameter is of a theoretical nature, such as life expectancy, disease prevalence or well-being, the actual finite-population can be conceived as a sample from an appropriately defined infinite theoretical or super-population, and the register-based direct domain (or sub-population) counts the direct sample estimates. Despite the total sample size being very large, the effects of sampling errors will be noticeable at very detailed levels.

For example, domain mortality rates are needed in order to calculate life expectancy (Chiang 1984) at a disaggregated level. Let the domains be classified according to: (1) area, denoted by $i = 1, \ldots, m$ for fixed m; (2) sex, denoted by $j = 1$ for male and $j = 2$ for female; and (3) age, denoted by $a = 0, \ldots, 99$. (The few people over 99 are grouped with those of the age 99.) Suppose that the domain-specific number of deaths over a given period is available in the relevant administrative register, denoted by y_{ija}. Since the domain population size is not constant over the period during which the death records are accumulated, some kind of a hypothetical *equivalent* population size is needed, denoted by n_{ija}.

Poisson distribution of y_{ija} seems natural, with parameter $\lambda_{ija} = n_{ija}\tau_{ija}$, where τ_{ija} is the theoretical domain mortality rate. A *direct* estimator of τ_{ija} is then given by

$$\hat{\tau}_{ija} = y_{ija}/n_{ija}$$

But $\hat{\tau}_{ija}$ can be highly unstable, yielding many extreme mortality rates in the smallest population domains. A simple alternative is the *synthetic* estimator given by

$$\hat{\tau}^S_{ija} = \left(\sum_{i=1}^{m} y_{ija}\right) / \left(\sum_{i=1}^{m} n_{ija}\right) = \xi_{ja}$$

yielding a single mortality rate for each sex-age group without any between-area variation.

SAE of the domain mortality rates is thus needed. On the one hand, this should reduce the large variance of the direct estimators in order to bring stability over time and to avoid an implausibly huge range of the local area life expectancy. On the other hand, this should avoid the evident over-smoothing of the synthetic estimator.

Let each sex-age group form a cohort. Initially, within each cohort $h = (j, a)$, the observed area-specific mortality rates can be smoothed to yield estimates of $\{\tau_{hi}; i = 1, \ldots, m\}$. Repeating the same procedure separately in each cohort yields then the estimates of all $\{\tau_{hi}; i = 1, \ldots, m \text{ and } h = 1, \ldots, H\}$. This *basic* smoothing approach has a theoretical drawback because the estimates of $V(\tau_{hi})$ do not necessarily vary smoothly over the 'neighbouring' cohorts for fixed i. The neighbouring cohorts may be the neighbouring age groups, either for a given sex or when both sexes are considered pairwise. Yet there appears to be no reason *a priori* why one should expect such 'jerky' dispersions.

A variance-component model for domain relative risk was developed in ESSnet SAE (2011, pp. 86–109) to address the problem. As commonly found in the literature of disease mapping (e.g. Rao 2003, section 9.5), let $\theta_{hi} = \tau_{hi}/\xi_h$ be the *relative risk* (RR), or the standardized mortality rate (SMR), where ξ_h denotes the cohort mortality rate that is calculated using the data from all the areas and treated as fixed. The variance-component model is given as

$$\begin{aligned} &y_{hi}\,|\theta_{hi} \sim \text{Poisson}\,(\lambda_{hi}) \quad \text{where} \quad \lambda_{hi} = \mu_{hi}\theta_{hi} \quad \text{and} \quad \mu_{hi} = n_{hi}\xi_h \\ &\theta_{hi} = \psi_h\psi_{hi} \quad \text{where} \quad E(\psi_h) = E(\psi_{hi}) = 1 \quad \text{and} \quad V(\psi_h) = \sigma^2_\psi \quad \text{and} \quad V(\psi_{hi}) = \sigma^2_h \end{aligned} \tag{4.1}$$

and ψ_h and ψ_{hi} are independent of each other, and ψ_{hi} and ψ_{hj} are independent of each other for $i \neq j$. The domain RR θ_{hi} is then the product of a cohort-level random effect ψ_h and a cohort-domain-level random effect ψ_{hi}. The basic separate-cohort smoothing approach corresponds to the special case of $\sigma^2_\psi = 0$. In general $V(\psi_{hi}) = \sigma^2_h$ is allowed to vary across the cohorts, possibly with a functional expression. A special case is $\sigma^2_h = \sigma^2$, which is referred to as the *variance homogeneity* assumption. Notice that the data across all the domains will be used to estimate the model (4.1).

The variance-component model (4.1) was applied in a case study of the life expectancy across the Norwegian municipalities $i = 1, \ldots, m$. An approach based on moving-average with different choices of the window width was explored for the estimation of σ^2_h, in addition to the variance homogeneity assumption. Some results for the quantiles of the resulting estimated life expectancy across the over 400 municipalities are given in Table 4.2. For the present context, we notice only that the direct register-based estimates have an implausibly huge range across

Table 4.2 Quantiles of estimated life expectancies across the municipalities. Basic: Separate basic smoothing in each cohort. Neighbour: Neighbouring variance homogeneity assumption and moving average variance estimator of bandwidth 50. Global: Variance homogeneity assumption

Sex	Method	Minimum	25% Quantile	Median	75% Quantile	Maximum
Male	Direct Estimator	62.2	76.7	78.0	79.1	83.6
	Basic	77.6	77.9	78.0	78.0	78.8
	Neighbour	76.9	77.6	78.0	78.4	79.8
	Global	75.1	77.3	78.0	78.6	80.7
Female	Direct Estimator	75.2	81.8	82.8	83.9	88.3
	Basic	82.1	82.5	82.6	82.6	82.9
	Neighbour	81.1	82.4	82.6	82.9	84.1
	Global	79.9	82.1	82.6	83.3	85.3

Source: ESSnet SAE (2011, p. 101).

the municipalities. The basic cohort-specific smoothing more or less wipes out all the potential between-area variation. The estimated cohort variance components $\hat{\sigma}_h^2$ are especially jerky for the lower age groups, whilst being zero in many cohorts which causes over-shrinkage in those cohorts. The variance-component model provides an attractive alternative for the smoothing of the direct register-based estimates. In practice, the appropriate degree of smoothing may be decided together with the demographers.

4.2.2 Measurement Error due to Progressive Administrative Data

It is customary that sample survey or census data are collected over a specified period of time for the field operation, after which the observations become static and no longer change. This is not the case with administrative data. Most administrative data are event-triggered/-based, such as birth or death, taking an exam or graduation, paying tax, hiring an employee, and so on. The mandatory registration of the event is often self-administered. Delays and mistakes are not avoidable entirely, whether by allowance or negligence. Measures from the administrative sources are thus progressive in the sense that the 'observed' value for a given statistical reference time point t may differ depending on the measurement time point $t + s$, for $s \geq 0$, where s may be referred to as the measurement delay.

As explained in Zhang (2012), the progressiveness of the input data source can affect either the measurement (i.e. variables) of the secondary integrated data or the representation (i.e. units and population frame), depending on the process of data integration. For example, a delay in the reporting of a new employee may cause a measurement error of the register-based activity status, as along as the target population is defined by the Population Register which nevertheless contains the person. Whereas a delay in the registration of a newly finished building may cause an under-coverage error in the register-based Building Statistics.

Under-coverage caused by reporting delay has been studied for example for epidemiological, insurance, and product warranty applications. Hedlin *et al.* (2006) applied a log-linear type of model to estimate the reporting delays for the introduction of birth units to the Business Register. Linkletter and Sitter (2007) used a non-parametric method to estimate and adjust for delays in Natural Gas Production reports in Texas. Similar issues can arise in SAE applications, such as utilizing the VAT register for short-term business statistics, where the numerous industrial groups (e.g. at the 5-digit NACE level) form the domains of interest.

We are currently unaware of any established register-based SAE application that deals with the coverage error arising from progressive administrative data. Coverage errors in the context of population size estimation will be discussed in Section 4.3.1, where the setting involves coverage or post-enumeration surveys. Here we focus on the measurement error.

Zhang and Fosen (2012) study the progressive measurement errors in the register-based small-area employment rate. Let $y_k(t; t+s)$ be the binary employment status of person k at statistical time point t, which is available at the measurement time point $t+s$, for $s \geq 0$. Take any two measurement time points $(t+r, t+s)$ where $0 \leq r < s$. Person k is said to have a *delayed entry* between $t+r$ and $t+s$ if $y_k(t; t+s) \neq y_k(t; t+r)$ due to updates between r and s. Person k is said to have a *recurred entry* between $t+r$ and $t+s$ if $y_k(t; t+r) = y_k(t; t+s)$ despite there being updates between $t+r$ and $t+s$ concerning person k. For the register-based production at time $t+s_0$, the recurred entries between t and $t+s_0$ are *ignorable* progressive data, but the delayed entries are *non-ignorable*.

Let $y_k(t)$ be the true register-status of person k for time t, based on ideal error-free input administrative data. To simplify the matter we shall assume that

$$y_k(t) = \lim_{s \to \infty} y_k(t; t+s)$$

Let $N_{ab}(t+r, t+s)$ be the number of persons with $y_k(t; t+r) = a$ and $y_k(t; t+s) = b$ for $a, b = 0, 1$. Let $t+s_0$ be the production time point. A simple selection model of the measurement mechanism (i.e. binary classification) at s_0 is given by

$$p_1 = P[y_k(t; t+s_0) = 1 | y_k(t) = 1] = \lim_{s \to \infty} \frac{N_{11}(t+s_0, t+s)}{N_{11}(t+s_0, t+s) + N_{01}(t+s_0, t+s)}$$

$$p_0 = P[y_k(t; t+s_0) = 1 | y_k(t) = 0] = \lim_{s \to \infty} \frac{N_{10}(t+s_0, t+s)}{N_{00}(t+s_0, t+s) + N_{10}(t+s_0, t+s)}$$

The classification is assumed to be independent and identically distributed across the units.

Zhang and Fosen (2012) explore the measurement errors using historic data from the Norwegian Employer Employee Register (NEER). Put

$$a_s = N_{01}(t, t+s)/N_{11}(t, t) \qquad \text{and} \qquad b_s = N_{10}(t, t+s)/N_{11}(t, t)$$

which gives, respectively, the relative increase and decrease of the register-based employment rate due to the delayed entries between measurement time points t and $t+s$.

Table 4.3 shows the historic values of a_s and b_s in the NEER for reference time points in years 2002, 2004 and 2006, respectively. The first measurement delay $s = 140$ corresponds roughly to the actual production time point $t+s_0$ of the register-based employment rate. It can be seen that delayed entries may keep arriving a long time after that. Only b_s seems to have converged after about 6 years (say, for $t > 2190$) for the reference year 2002. Convergence does not seem to be the case for the other series. Nevertheless, one can get a feeling of the level of the classification probabilities (p_1, p_0) on substituting $s = 2555$ for $s = \infty$ for reference year 2002, which gives $\hat{p}_1 = 1 - 0.053/1.052 = 0.950$ and $\hat{p}_0 = 0.576 \cdot 0.030/$

Table 4.3 Historic data in the NEER. Statistical reference time point (t) in week 45 of 2002, 2004 and 2006. Measurement delay (s) in days after the reference time point (t). Increase (a_s) and decrease (b_s) due to delayed entries

Measurement	Reference time point (t)					
	Year 2002		Year 2004		Year 2006	
delay (s)	a_s	b_s	a_s	b_s	a_s	b_s
140	.043	.014	.031	.025	.041	.027
365	.070	.036	.044	.036	.056	.037
548	.080	.040	.051	.041	.064	.041
730	.084	.041	.055	.043	.068	.042
1095	.089	.042	.060	.045	.070	.044
1460	.091	.043	.062	.046		
1825	.094	.043	.063	.047		
2190	.095	.044				
2555	.096	.044				

Source: Zhang and Fosen (2012, Table 2, p. 99). Reproduced with permission of Indian Agricultural Statistics Research Institute.

$(1 - 0.576 \cdot 1.052) = 0.044$, that is 5% misclassification for those with $y_k(t) = 1$ and 4.4% for those with $y_k(t) = 0$.

To see the relevance for SAE, consider the following simple model. Let θ_i be the theoretical area mean, and let y_{ij} be the error-free binary register variable for unit j in area i, where $i = 1, \ldots, m$ and $j = 1, \ldots, N_i$. Let x_{ij} be the observed binary register variable. The fixed reference and production time points $(t, t + s_0)$ are dropped to simplify the notation. Put

$$P(x_{ij} = 1|y_{ij}) = \begin{cases} p_{i1} & \text{if } y_{ij} = 1 \\ p_{i0} & \text{if } y_{ij} = 0 \end{cases}$$

$$y_{ij}|\theta_i \sim \text{ Bernoulli } (\theta_i)$$

$$\theta_i = \theta + u_i$$

where $E(u_i) = 0$ and $V(u_i) = \sigma_u^2$. Let $\bar{y}_i = \sum_{j=1}^{N_i} y_{ij}/N_i$. We have

$$\bar{y}_i = \theta_i + e_i = \theta + u_i + e_i$$

where $E(e_i|u_i) = 0$, and $V(e_i|u_i) = \theta_i(1 - \theta_i)/N_i$, and $Cov(u_i, e_i) = 0$. Let $\bar{x}_i = \sum_{j=1}^{N_i} x_{ij}/N_i$. Let $\lambda_i = p_{i1} - p_{i0}$. We have

$$\bar{x}_i = p_{i0} + \lambda_i \theta + \lambda_i u_i + b_i \tag{4.2}$$

where $E(b_i|u_i) = 0$ and $V(b_i|u_i) = V(x_{ij}|\theta_i)/N_i = \theta_i(1 - \theta_i)\lambda_i^2 + \theta_i p_{i1}(1 - p_{i1}) + (1 - \theta_i) p_{i0}(1 - p_{i0})$. The expected true area mean $\bar{y}_i$ conditional on the observed $\bar{x}_i$ is

$$E(\bar{y}_i|\bar{x}_i) = \frac{\bar{x}_i \theta_i p_{i1}}{\theta_i p_{i1} + (1 - \theta_i)p_{i0}} + \frac{(1 - \bar{x}_i)\theta_i(1 - p_{i1})}{\theta_i(1 - p_{i1}) + (1 - \theta_i)(1 - p_{i0})}$$

which is not equal to $\bar{x}_i$ unless $(p_{i1}, p_{i0}) = (1, 0)$, that is no measurement error.

The random effect u_i represents the heterogeneity across the areas. It is of the order $O_p(1)$. The random error b_i arises from the within-area individual variations and the measurement errors. It is of the order $O_p(1/\sqrt{N_i})$. The model (4.2) differs from the model of Fay and Herriot (1979) in that the sampling variance $V(b_i|\theta_i)$ depends on the mean parameter θ_i. It is nevertheless easier to handle compared with an alternative generalized linear mixed model. Because we are dealing with population registers, N_i is usually large enough to warrant a normal approximation to the distribution of b_i, as long as θ_i is not very close to either 0 or 1.

Zhang and Fosen (2012) explore in addition a model for the register-based change. They carry out sensitivity analysis of the level and change estimates of small-area employment rates, using alternative values of (p_{i1}, p_{i0}) under the simplification that $(p_{i1}, p_{i0}) = (p_1, p_0)$ does not vary over time. However, as the historic data in Table 4.3 suggest, the measurement error mechanism does vary over time, and may well be expected to vary across the areas. For future research it will be interesting to develop more elaborate models, which allow for differential error mechanisms both over time and across the population domains.

Finally, we notice that provided relevant sample survey data are available, such as the LFS in the context of Employment Statistics, alternative approaches are possible based on administrative and survey data integration, as we shall discuss later in Section 4.3.2.

4.3 Administrative and Survey Data Integration

4.3.1 Coverage Error and Finite-population Bias

A register has under-coverage (or missing enumeration) of the target population if there exist population units that are not enlisted in the register; it has over-coverage (or erroneous enumeration) if there are units in the register that do not belong to the target population.

An important SAE application that involves the coverage error is list enumeration adjustment using separate coverage surveys. The list may be the census or Population Register enumeration. See Hogan (1993) for an early account of the methodology in the US. See Nirel and Clickman (2009) for a recent and more comprehensive review of the uses of sample surveys in censuses.

A potential adaption to the production of poverty indicators can be as follows. On the one hand, one has from the tax authority some past, say previous-year, income, based on which it is possible to classify the 'poor' persons. Such a 'poor' person, however, may be erroneous in the current situation, if his/her income has changed to above the poverty threshold in the meantime; whereas, a 'not-poor' person may be a missing enumeration in the reverse case. On the other hand, one has a sample of the current population, where the poverty status can be correctly classified, but for the nonrespondents. In other words, one recognizes that the survey enumeration has under-coverage.

Disregarding several additional issues including sampling design, weighting and imputation for missing data, matching and processing of datasets, a stylized population size estimator can be given as

$$\hat{N} = n(1 - \theta)x/m$$

where x is the list population count, and n an independent under-coverage survey (U-sample) count, and m the number of enumerations in both the list and the U-sample. The parameter

θ is the proportion (or probability) of erroneous list enumeration, and is traditionally estimated using a separate O-sample drawn from the list enumeration. Wolter (1986) details the underlying assumptions without explicit reference to $1 - \theta$, but the account can be rephrased to acknowledge the additional O-sample. See Zhang (2015) for modelling approaches to the over- and under-coverage errors, where the adjustment requires only the U-sample.

A key difficulty from the SAE perspective arises when the coverage samples are not large enough to support the direct estimator $\hat{N}_i$ in all the local areas $i = 1, \ldots, m$. In particular, when a multi-stage sampling design is used, there may be many areas that are not represented in the sample. To focus on the key issue at hand, put, to start with,

$$\tilde{N}_i = x_i \xi \qquad \text{and} \qquad N_i = x_i \xi + v_i$$

where N_i is the local area population size, and x_i the known list enumeration, and ξ the known global adjustment factor $\xi = N/x$ for $N = \sum_{i=1} m N_i$ and $x = \sum_{i=1}^{m} x_i$. Then, $\tilde{N}_i$ is the best estimator for an out-of-sample area, and $E(\tilde{N}_i - N_i | N_i) = -v_i$ is its finite-population (FP) bias for fixed $U_N = \{N_1, \ldots, N_m\}$, or $U_v = \{v_1, \ldots, v_m\}$.

Zhang (2007) uses an area-level mixed model to assess the squared FP-bias of a synthetic estimator. An application to the register-based census employment rate is given by Fosen and Zhang (2011). For population size estimation, consider the following adaption. Let $\hat{N}_i = N_i + e_i = x_i \xi + v_i + e_i$ be the direct estimator for an in-sample area, where e_i is its FP sampling error such that $E(\hat{N}_i | v_i) = N_i$ and $V(\hat{N}_i | v_i) = \psi_i$. Put

$$z_i = \hat{N}_i - \tilde{N}_i = v_i + e_i \tag{4.3}$$

and assume independence between v_i and e_i. Next, since $N = x\xi$, assume $E(v_i) = 0$. Finally, for the variance, assume $V(v_i) = x_i^{\alpha} \sigma_v^2$ for some fixed constant α. The mixed model (4.3) is then a special case of the area-level models (Rao 2003). Having fitted the model to the in-sample areas, we obtain $\hat{\sigma}_v^2$ as the estimate of σ_v^2 and, for an out-of-sample area,

$$\hat{E}(v_i^2) = x_i^{\alpha} \hat{\sigma}_v^2$$

One may interpret $E(v_i^2)$ as the anticipated squared FP-bias under model (4.3). Other measures are possible, for example $E(|v_i|)$ under an additional normality assumption of v_i. Or, it may be more suitable to model v_i on a different scale.

In practice, the global factor ξ needs to be estimated. Let $\hat{\xi} = \xi + \epsilon$ with sampling error ϵ. Then, instead of model (4.3), put

$$z_i = \hat{N}_i - x_i \hat{\xi} = v_i + u_i \qquad \text{and} \qquad u_i = e_i - x_i \epsilon$$

where e_i and ϵ are correlated. Re-sampling methods may be used to estimate the variance of u_i directly, instead of those of e_i and ϵ and their covariance separately.

More importantly, the FP-bias v_i is uncontrollable for an out-of-sample area. This suggests that, subjected to the total sample size affordable, one may consider adopting a sampling design which ensures that all the local areas are represented in the sample. A necessary consequence is the reduction of the sample size in certain (if not all) local areas, and the direct estimation may no longer have an acceptable accuracy in a number of areas.

Notice that indirect SAE via the random effects is also FP-biased. To fix the idea, assume known ξ, σ_v^2 and $\psi_i(> 0)$ for $i = 1, \dots, m$. The best predictor (BP) of v_i is then

$$\tilde{v}_i = \gamma_i(\hat{N}_i - x_i\xi) \qquad \text{and} \qquad \gamma_i = x_i^\alpha \sigma_v^2 / (x_i^\alpha \sigma_v^2 + \psi_i)$$

For fixed U_N, or U_v, the FP-bias of the BP is then $E(\tilde{v}_i - v_i|v_i) = -(1 - \gamma_i)v_i$. Thus, the unavailability of an acceptable direct estimator for the in-sample areas calls for a careful evaluation of the FP-bias under alternative sampling designs.

Future research in this area is important because a viable methodology for population size estimation that is able to address non-negligible over- and under-coverage errors in the input registers can potentially be useful in many situations beyond population census, including the production of poverty indicators as indicated above.

4.3.2 Relevance Error and Benchmarked Synthetic Small Area Estimation

Benchmarking the aggregates of mixed-effects model-based SAEs to accepted estimates or known totals can yield some protection against model misspecification and achieve output consistency that is important in Official Statistics. See Pfeffermann (2013, section 6.3) for a review. But there is a limit to which sample survey data can support the various mixed-effects models. As one descends the hierarchy of aggregation, sooner or later, a level will be reached where many (or most) areas are not represented in the sample, and many areas will have only very few sample observations. Synthetic methods are necessary from then on. However, the stringent assumptions required for the synthetic estimates to be unbiased are often plainly unattainable. So the issue becomes more than protection against the misspecification of certain aspects of the model. It is about actively reducing the model bias.

Proxy measures from census or administrative sources can provide substitutions to the 'random effects' whose estimation is poorly supported by the sample data available. A well known technique in the SAE literature is the Structure PREserving Estimation (SPREE) following Purcell and Kish (1980). See also Noble *et al.* (2002) for a generalized linear model (GLM) framework which includes the log-linear model underpinning SPREE. Identical or similar applications exist for example in demography and population geography, where the approach is known under different names such as iterative proportional fitting (IPF) or models with spatial-interaction offsets, and so on. See for example Simpson and Tranmer (2005) and Raymer *et al.* (2011). Benchmarking by IPF in all these applications of proxy data can as well be motivated as a means for reducing the definition bias (or relevance error) of the proxy measure, so as to achieve statistical relevance at which level the estimates are benchmarked.

The SPREE and related methods mentioned above can easily become relevant to the production of multidimensional poverty indicators, where the dimensions of poverty classification may involve economic, social and other factors. For instance, suppose that a complete cross-classification is feasible in the previous census. However, based on the current separate surveys and/or registers, only the updated marginal distributions are available, but not the joint distribution of the cross-classifications.

Below we outline some techniques for reducing the model bias, or the relevance error, of the synthetic estimation methods for SAE, from one-way benchmarking to multivariate generalized structure preserving estimation. A unifying formulation is optimal adjustment of the initial synthetic estimates subjected to the benchmarking constraints.

4.3.2.1 One-way Optimal Benchmarking

Take first the one-way case. Let μ_i be the best synthetic estimate of area (or domain) mean θ_i, which is given when all the involved parameters are known. Let μ_i^B be the *one-way* benchmarked best synthetic estimates that satisfy the constraint

$$\sum_{i=1}^{m} W_i \mu_i^B - \sum_{i=1}^{m} W_i \theta_i = 0 \tag{4.4}$$

where $\boldsymbol{\mu}^B = (\mu_1^B, \ldots, \mu_m^B)^T$ and $W_i = N_i/N$, and N_i is the area population size and $N = \sum_{i=1}^m N_i$. To obtain $\boldsymbol{\mu}^B$ as the solutions of optimal adjustment of $\boldsymbol{\mu}$ subjected to the benchmark constraint (4.4), one needs to specify: (i) the form of adjustment; and (ii) the loss function. Consider the following.

- Global additive adjustment δ and loss function Δ given by, respectively,

$$\mu_i^B = \mu_i + \delta$$
$$\Delta = \frac{1}{2}\sum_{i=1}^{m} W_i(\mu_i^B - \theta_i)^2 = \frac{1}{2}\sum_{i=1}^{m} W_i(\mu_i + \delta - \theta_i)^2$$
$$\Rightarrow \quad \mu_i^B = \mu_i + \delta \quad \text{and} \quad \delta = \bar{\theta}_w - \bar{\mu}_w$$

where $\bar{\theta}_w = \sum_{i=1}^m W_i\theta_i$ and $\bar{\mu}_w = \sum_{i=1}^m W_i\mu_i$.

- Global multiplicative adjustment δ and loss function Δ given by

$$\mu_i^B = \delta\mu_i$$
$$\Delta = \frac{1}{2}\sum_{i=1}^{m} W_i(\mu_i^B - \theta_i)^2/\mu_i = \frac{1}{2}\sum_{i=1}^{m} W_i(\delta\mu_i - \theta_i)^2/\mu_i$$
$$\Rightarrow \quad \mu_i^B = \delta\mu_i \quad \text{and} \quad \delta = \bar{\theta}_w/\bar{\mu}_w$$

- Area-specific additive adjustment δ and loss function Δ given by

$$\mu_i^B = \mu_i + \delta_i$$
$$\Delta = \frac{1}{2}\sum_{i=1}^{m} W_i(\mu_i^B - \theta_i)^2 = \frac{1}{2}\sum_{i=1}^{m} W_i(\mu_i + \delta_i - \theta_i)^2$$
$$\Rightarrow \quad \mu_i^B = \mu_i + \delta_i \quad \text{and} \quad \delta_i = \theta_i - \mu_i$$

In practice, θ_i is replaced by an unbiased direct estimate $\hat{\theta}_i$, and μ_i by a synthetic estimate $\hat{\mu}_i$ depending on some global parameter estimates. The benchmarked synthetic estimate $\hat{\mu}_w^B = \sum_{i=1}^m W_i\hat{\mu}_i^B$ is unbiased for $\bar{\theta}_w$, because $\hat{\mu}_w^B = \hat{\bar{\theta}}_w$. Moreover, it is clear that area-specific optimal adjustment is difficult because it essentially recast the SAE problem one had from the beginning, that is to obtain a good estimate of θ_i.

Wang *et al.* (2008) derive benchmarked best linear unbiased predictor (BBLUP) under the Fay–Herriot model. Replacing the empirical best linear unbiased predictor (EBLUP) $\hat{\theta}_i^{EBLUP}$

by $\hat{\mu}_i$, one obtains a benchmarked synthetic estimate in analogy to the BBLUP, which is given by

$$\hat{\theta}_i^B = \hat{\mu}_i + (a_i/N_i)N(\hat{\hat{\theta}}_w - \hat{\hat{\mu}}_w)$$

where $\sum_{i=1}^m a_i = 1$. First, under the additive adjustment, apportioning the overall difference $N\hat{\hat{\theta}}_w - N\hat{\hat{\mu}}_w$ to all the areas according to the area population sizes leads to a difference estimator for θ_i, that is

$$a_i = N_i/N \quad \Rightarrow \quad \hat{\theta}_i^B = \hat{\mu}_i + (\hat{\hat{\theta}}_w - \hat{\hat{\mu}}_w) = \hat{\hat{\theta}}_w + (\hat{\mu}_i - \hat{\hat{\mu}}_w)$$

which is identical to the optimal additive adjustment above. In particular, $\approxeq_i = \hat{\mu}_i - \hat{\hat{\mu}}_w$ can be considered as a substitution estimate of the random effect $u_i = \theta_i - \bar{\theta}_w$. Secondly, the commonly used proportional or pro-rata adjustment can be given by

$$a_i = (N_i\hat{\mu}_i)/(N\hat{\hat{\mu}}_w) \quad \Rightarrow \quad \hat{\theta}_i^B = (\hat{\mu}_i/\hat{\hat{\mu}}_w)\hat{\hat{\theta}}_w$$

which is identical to the optimal multiplicative adjustment above. Moreover, $\hat{\mu}_i = \hat{\mu}_i/\hat{\hat{\mu}}_w$ can be regarded as a substitution estimate of the multiplicative random effect $u_i = \theta_i/\bar{\theta}_w$.

As a property of the additive adjustment, we notice that it leads to the familiar post-stratification estimator as follows. Let the categorical target (Y) and proxy (X) variables take value $j = 1, \ldots, J$. Let Y_{ij} and $Y_{.j}$ be the area and overall count of units with $y = j$, and θ_{ij} and $\theta_{.j}$ the corresponding area and overall proportions. Similarly for X_{ij} and $X_{.j}$. Let $Y_{ij;k} = X_{ik}p_{i,kj}$ be the area count of the units with $(x, y) = (k, j)$. Let the initial proxy substitution estimate be $\hat{Y}^S_{ij;k} = X_{ij} = X_{ik}p^S_{i,kj}$ where $p^S_{i,kj} = 1$ if $j = k$ and $p^S_{i,kj} = 0$ if $j \neq k$. Overall we have $\hat{Y}^S_{j;k} = X_k p^S_{kj}$, where $p^S_{kj} = 1$ if $j = k$, and $p^S_{kj} = 0$ if $j \neq k$. It is straightforward to verify that, whether $j = k$ or not, additive adjustment of $p^S_{i,kj}$ yields $\hat{p}_{i,kj} = \hat{p}_{kj} = \hat{Y}_{j;k}/X_{.k}$, where $\hat{Y}_{j;k}$ is an unbiased estimate of the population count of the units with $(x, y) = (k, j)$. Thus, the additive adjustment yields

$$\hat{Y}^B_{ij} = \sum_{k=1}^{J} X_{ik}(\hat{Y}_{j;k}/X_{.k})$$

which is the synthetic post-stratification estimate of Y_{ij} for $i = 1, \ldots, m$ and $j = 1, \ldots, J$.

4.3.2.2 Two- or Multi-way Optimal Benchmarking

Consider now multi-way benchmarking constraints. We explain the approaches for the two-way case; the multi-way cases follow similarly. Let Y_{ij} denote the target two-way totals of interest, where $i = 1, \ldots, m$ denotes the areas and $j = 1, \ldots, J$ denotes the categories. The underlying variable can either be categorical such as the number of people by activity status, or continuous such as the VAT turnover by product type. Let $Y_{i.} = \sum_{j=1}^J Y_{ij}$ and $Y_{.j} = \sum_{i=1}^m Y_{ij}$. Let $(\hat{Y}_{i.})_{i=1}^m$ and $(\hat{Y}_{.j})_{j=1}^J$ be the benchmark totals for the two-way table $\{Y_{ij}\}$. In particular, due to the constraints $(\hat{Y}_{i.})_{i=1}^m$, the data are referred to as compositions, and it is equivalent to estimate the totals or proportions of the categories.

Let X_{ij} be a substitution estimate of Y_{ij} based on a proxy measure. Consider the ANOVA decomposition

$$X_{ij} = \mu_0^X + \mu_i^X + \mu_j^X + \mu_{ij}^X$$

where $\mu_0^X = \bar{X} = \sum_{i,j} X_{ij}/(mJ)$, and $\mu_i^X = \bar{X}_i - \bar{X} = \sum_j X_{ij}/J - \bar{X}$, and $\mu_j^X = \bar{X}_j - \bar{X} = \sum_i X_{ij}/m - \bar{X}$, and $\mu_{ij}^X = X_{ij} - \bar{X}_i - \bar{X}_j + \bar{X}$. Notice that we have $\sum_i \mu_i^X = \sum_j \mu_j^X = \sum_i \mu_{ij}^X = \sum_j \mu_{ij}^X = 0$ by construction. Put

$$\hat{Y}_{ij}^B = \hat{\mu}_0^Y + \hat{\mu}_i^Y + \hat{\mu}_j^Y + \mu_{ij}^X \tag{4.5}$$

where $\hat{\mu}_0^Y = \hat{\bar{Y}} = \sum_i \hat{Y}_{i.}/m = \sum_j \hat{Y}_{.j}/J$, and $\hat{\mu}_i^Y = \hat{\bar{Y}}_{i.} - \hat{\mu}_0^Y = \hat{Y}_{i.}/J - \hat{\mu}_0^Y$, and $\hat{\mu}_j^Y = \hat{\bar{Y}}_{.j} - \hat{\mu}_0^Y = \hat{Y}_{.j}/m - \hat{\mu}_0^Y$. Notice that $\sum_i \hat{\mu}_i^Y = \sum_j \hat{\mu}_j^Y = 0$. We observe the following.

1. Now that $\sum_j \mu_{ij}^X = \sum_i \mu_{ij}^X = 0$, $\sum_i \hat{Y}_{ij}^B = \hat{Y}_{.j}$ and $\sum_j \hat{Y}_{ij}^B = \hat{Y}_{i.}$ by construction.
2. The estimate (4.5) can be obtained from X_{ij} via the following additive adjustments

$$\hat{\mu}_0^Y - \mu_0^X \quad \text{and} \quad (\hat{\mu}_i^Y - \mu_i^X)_{i=1}^m \quad \text{and} \quad (\hat{\mu}_j^Y - \mu_j^X)_{j=1}^J$$

   The adjustments are uniquely determined by the benchmark constraints $(\hat{Y}_{i.})_{i=1}^m$ and $(\hat{Y}_{.j})_{j=1}^J$, together with the definitional constraints $\sum_i \hat{\mu}_i^Y = \sum_j \hat{\mu}_j^Y = 0$. Optimal adjustment is trivial in this case, irrespective of the choice of the loss function.
3. The interaction offsets μ_{ij}^X can be regarded as a substitution estimate for the ANOVA random effect μ_{ij}^Y. The approach is straightforward given a proxy measure. It is not so effective based only on auxiliaries. The key challenge then is how to construct a synthetic estimate of the interaction μ_{ij}^Y that is non-trivial.

When the underlying y-variable is non-negative, or when the Y_{ij}'s are the counts of a categorical variable, it is more common to apply the two-way proportional adjustment to the proxy table via IPF, or raking. At each iteration, the rows and the columns are adjusted successively by a multiplicative factor to satisfy the corresponding benchmark totals one at a time. One obtains the benchmarked estimates $\hat{Y}_{ij}^B$ on convergence of the IPF. In SAE this is known as the SPREE (Purcell and Kish 1980), where the log-linear interactions of the proxy table are preserved by the IPF.

The SPREE is an approach based on the log-linear ANOVA decomposition. Put

$$\log(X_{ij}) = \alpha_0^X + \alpha_i^X + \alpha_j^X + \alpha_{ij}^X$$

where

$$\alpha_0^X = \sum_{i,j} \log(X_{ij})/(mJ) \quad \text{and} \quad \alpha_i^X = \sum_j \log(X_{ij})/J - \alpha_0^X$$
$$\alpha_j^X = \sum_i \log(X_{ij})/m - \alpha_0^X \quad \text{and} \quad \alpha_{ij}^X = \log(X_{ij}) - \alpha_i^X - \alpha_j^X - \alpha_0^X$$

The IPF updates all the other α-terms except the α_{ij}^X's.

Let $\log\ (Y_{ij}) = \alpha_0^Y + \alpha_i^Y + \alpha_j^Y + \alpha_{ij}^Y$ be the log-linear ANOVA decomposition of Y_{ij}, defined in the same way as for X_{ij}. Under the SPREE, α_{ij}^X can be considered a proxy substitution estimate of the random effect α_{ij}^Y. Moreover, the IPF yields the maximum likelihood estimate of the log-linear model parameters, given the sufficient marginals $(\hat{Y}_{i.})_{i=1}^m$ and $(\hat{Y}_{.j})_{j=1}^J$ and with the interactions constrained at the α_{ij}^X's. One may therefore interpret the multiplicative adjustments from X_{ij} to Y_{ij} as the result of optimal adjustment subjected to these constraints, where the loss function is the Kullback–Leibler divergence between the constrained log-linear model and the saturated model for the two-way table $\{Y_{ij}\}$.

4.3.2.3 Modelling under Benchmark Constraints

The two-way benchmarking adjustments of the proxy table above amounts to substitution estimation of the random effects. By definition, however, the proxy measure suffers relevance error and entails definition bias. Flexible fixed-effects modelling of the relationship between the target interactions and the proxies can help to further reduce the bias. Conceptually, this points to a modelling approach that respects the benchmark constraints inherent to the problem at hand, rather than treating the necessary benchmark adjustments as an alien element to be administered ad hoc to the model-based estimates.

For instance, take the target two-way table $\{Y_{ij}\}$. Given the two sets of marginal benchmark totals $(\hat{Y}_{i.})_{i=1}^m$ and $(\hat{Y}_{.j})_{j=1}^J$, the only unknowns that remain are the linear or log-linear interactions. it is therefore natural and appealing that a modelling approach under the benchmark constraints should be developed in terms of these interactions. Zhang and Chambers (2004) propose a generalized SPREE (GPREE) model:

$$\alpha_{ij}^Y = \beta\alpha_{ij}^X \tag{4.6}$$

It is shown that the model (4.6) of the log-linear interactions can be fitted using the associated GLM of the within-area proportions $\theta_{ij}^Y = Y_{ij}/Y_{i.}$, that is

$$\eta_{ij}^Y = \lambda_j + \beta\eta_{ij}^X$$

where $\eta_{ij}^Y = \log\ \theta_{ij}^Y - \sum_{j=1}^J \log\ \theta_{ij}^Y$, and likewise for η_{ij}^X. The λ_j's are nuisance parameters subject to the constraint $\sum_{j=1}^J \lambda_j = 0$. On fitting the GLM, one obtains $\{\exp\ (\hat{\beta}\alpha_{ij}^X)\}$ as the starting table for the IPF, which then yields the final benchmarked synthetic estimates of $\hat{Y}_{ij}$. The SPREE is the special case of constraining β at 1.

More recently, Luna-Hernandez and Zhang (2013) have proposed an extension given by

$$\alpha_{ij}^Y = \sum_{k=1}^J \beta_{jk}\alpha_{ik}^X \tag{4.7}$$

where $\sum_{k=1}^J \beta_{jk} = \sum_{j=1}^J \beta_{jk} = 0$. This may be referred to as the multivariate GSPREE, in analogy to multivariate linear regression. Again, the model (4.7) can be fitted using the associated GLM

$$\eta_{ij}^Y = \lambda_j + \sum_{k=1}^J \beta_{jk}\eta_{ik}^X$$

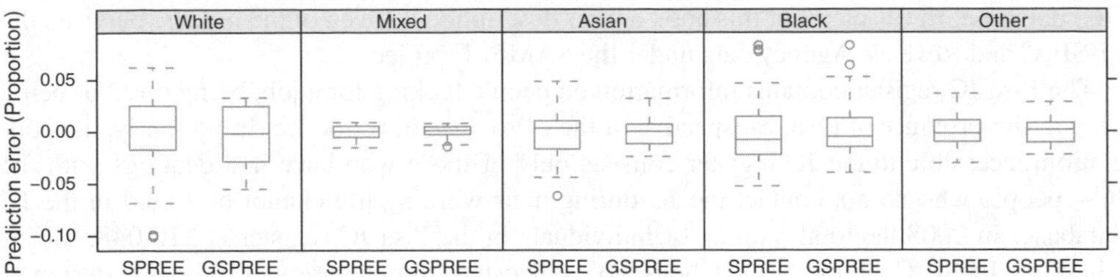

Figure 4.1 Boxplot of prediction error for ethnicity composition, Hackney Borough of London

with nuisance parameters λ_j's. On fitting the model, one obtains $\{\exp\,(\sum_{k=1}^{J}\hat{\beta}_{jk}\alpha_{ik}^{X})\}$ as the starting table of the IPF, and so on. Notice that the multivariate GSPREE model (4.7) includes as a special case the logistic multinomial model

$$\log\,(\theta_{ij}^{Y}/\theta_{iJ}^{Y}) = \gamma_j + \phi_j \log\,(\theta_{ij}^{X}/\theta_{iJ}^{X})$$

where category J is the arbitrarily chosen reference category, so that the model has $(J-1)$ free intercepts γ_j and slopes ϕ_j for $j \neq J$.

For an illustration, consider the proportions of ethnicity groups (White, Mixed, Asian, Black, Other) in the Hackney Borough of London. The UK Census 2001 and 2011 compositions are treated as X_{ij} and Y_{ij}, respectively. A boxplot of the prediction error of the SPREE and the multivariate GSPREE model (4.7) is given in Figure 4.1. It can be seen that the extra flexibility of the model (4.7) leads to further reduction of the bias of the SPREE. Notice that in practice the multivariate GSPREE modelling approach does not require any additional data compared with the SPREE. Notice also that the ability to produce less biased synthetic estimates is critical at a low aggregation level, where the random-effects modelling approach is not supported due to the sparseness of the sample survey data.

4.3.3 Probability Linkage Error

The EU-SILC survey is a national survey. It is a major source for poverty and living conditions estimates in Italy. Data on income and on working and living conditions are collected at both the household and individual levels. The target population of the Italian SILC are all the Italian households (and the individuals living in these households). However, the sample is not large enough to allow reliable estimates to be calculated at the local level, that is below the regional level (NUTS 2 level in the European Nomenclature for Territorial Units).

Probability linkage of EU-SILC data with several administrative data from the Province of Pisa (PI-SILC) was carried out under the FP7 project SAMPLE–Small Area Methods for Poverty and Living condition Estimates (www.sample-project.eu). The sample size for Pisa was boosted as part of the SAMPLE project. Indicators relevant to the measurement of poverty and living conditions were transferred from the administrative sources to the linked dataset. One of the primary aims was to explore the use of the EU-SILC sampling weights for the administrative indicators and to enhance the administrative data with the household characteristics collected in the survey. We explore here the linked data of the PI-SILC and the Job Centre

(JC) database. In Chapter 3 of this book a brief description is given of the linkage between the PI-SILC and Revenue Agency data under the SAMPLE project.

The Pisa JC register contains information on people looking for a job, being hired or being fired in the Province of Pisa, irrespective of their Province of residence. Importantly, the population accessible to the JC register consists only of those who have had dealings with the JC—people who do not contact the JC during their working life cannot be found in the JC database. In 2008 the total number of individuals in the Pisa JC register is 216 048. Meanwhile, the PI-SILC sample has 818 households, boosted from the original EU-SILC design of 162 households. The number of responding households and persons are 675 and 1685, respectively. There are 1476 persons with age above 15, which is the maximum number of persons that can be linked to the JC register.

Due to privacy restrictions, anonymized probability linkage between PI-SILC and JC data is based on the following key variables: municipality of residence, gender, birth month and birth year. A multiple pass procedure was applied. In the first pass, a pair of records is identified as a match provided exact match on all the key variables, referred to as level-1 linkage. In the next two passes, an increasing degree of mismatch between the key variables is allowed, resulting in level-2 and level-3 linkage.

The results, obtained using the package 'RecordLinkage' of the R software, are summarized in Table 4.4. It can be seen that multiple matches are found for the same PI-SILC unit at all three levels of linkage. The number of PI-SILC units with at least one matched JC record is 1113 out of the 1476 available units. The average number of matches is 7.9, irrespective of the level of linkage. The epiWeight shown in the table is a weight calculated for each matched pair of records based on the approach used by Contiero *et al.* (2005) in the EpiLink record linkage software: it is equal to 1 only in the case of exact match on all the key variables. Generally speaking, the weight reflects how likely a match identifies a true pair of records. For each of the 1113 PI-SILC units with at least one match, the match with the highest epiWeight is accepted as the linkage for that unit. In the case of ties, one of them is chosen at random. A linked dataset of 1113 units is obtained in this way.

Clearly, linkage errors exist in the linked dataset, which is the case if the accepted pair of records does not correspond to a true pair of records. From the SAE perspective, it is of interest to examine the heterogeneity of this linkage error. Of the 39 Municipalities in Pisa, 25 are represented in the linked dataset. The Municipalities are a partition of the provincial administrative area. These are treated as the small areas here.

Table 4.4 Results of the record linkage procedure between PI-SILC and JC data. Total number of matched pairs, number of unique PI-SILC units and average value of the epiWeights, by level of linkage

Level of linkage	Matched pairs	PI-SILC records	Average epiWeights
1	353	321	1
2	3097	715	0.76
3	5348	910	0.54
Total	8798	1946	0.64

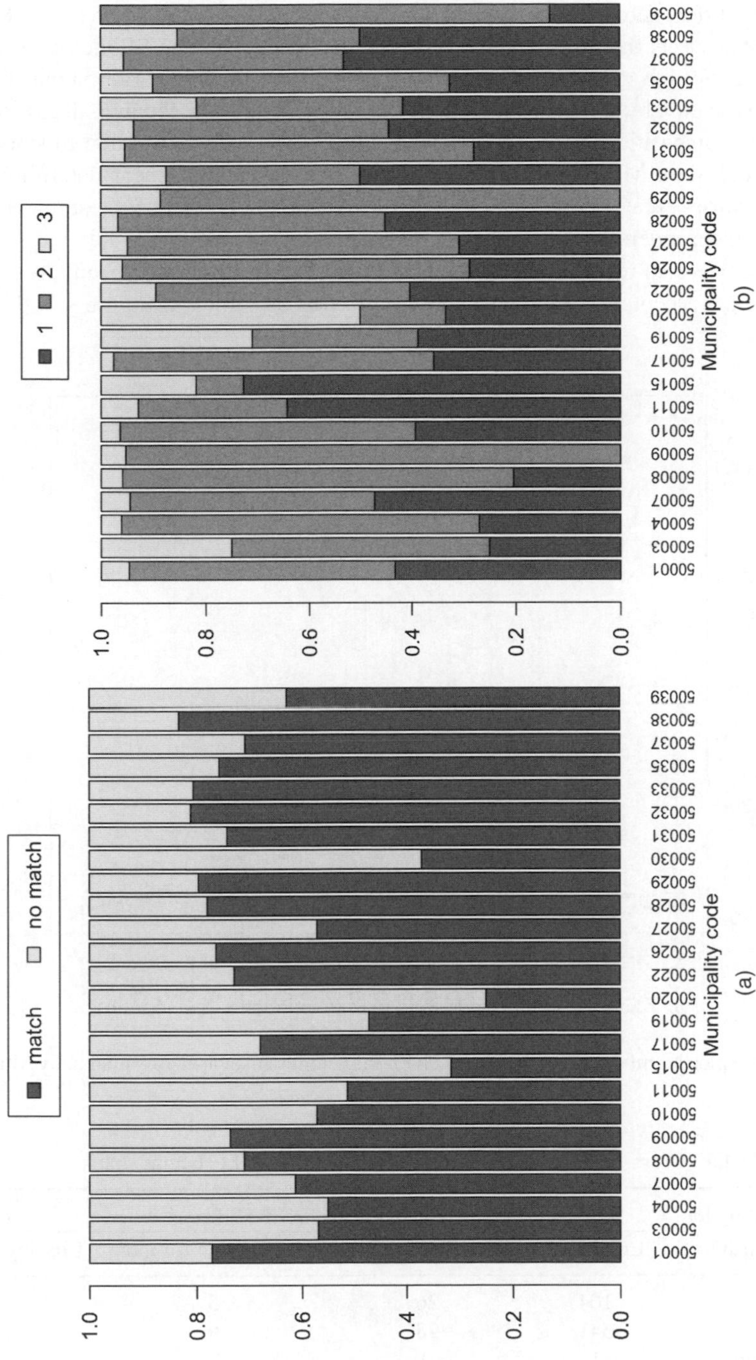

Figure 4.2 Percentage of linked units (a) and composition of linkage level (b) by Municipality

Simple exploratory analysis provides clear evidence for heterogeneous linkage error across the areas. For instance, it can be seen in Figure 4.2a that the overall percentage of linked PI-SILC units differs across the 25 Municipalities, with values between 25% and 83%. The Pearson χ^2-test rejects the null hypothesis of independence between Municipality and linkage percentage (p-value < 0.0001). Taking into consideration that the variation may have been caused, at least partly, by the varying sampling fraction across the Municipalities, we examine further the composition of the levels of linkage. Figure 4.2b shows that this also differs considerably across the Municipalities. For example, the percentage of near-deterministic linkages (level 1) varies greatly between 0% and 67%. Again, the Pearson χ^2-test rejects the null hypothesis of independence between Municipality and linkage level.

Heterogeneity is also found in the number of matches for each PI-SILC unit, regardless of whether these are accepted as linkage or not in the end. As shown in Figure 4.3, the average

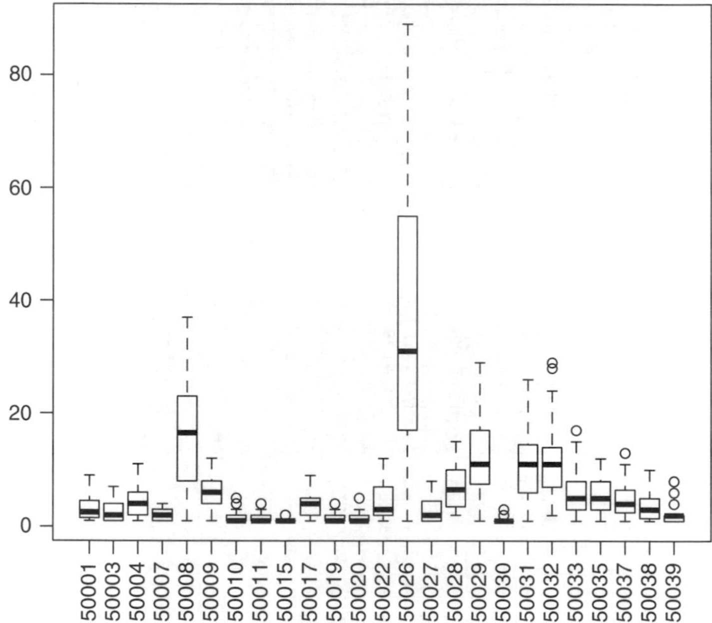

Figure 4.3 Boxplot: Number of matches for each PI-SILC unit with at least one match, by Municipality

Table 4.5 Cross-classification of JC job-seeker and PI-SILC labour status

Looking for job in JC	Labour status in PI-SILC			
	Employed	Looking for job	Not looking for job	Missing
Yes	104	26	55	5
No	541	28	336	16
Missing	1	0	1	0

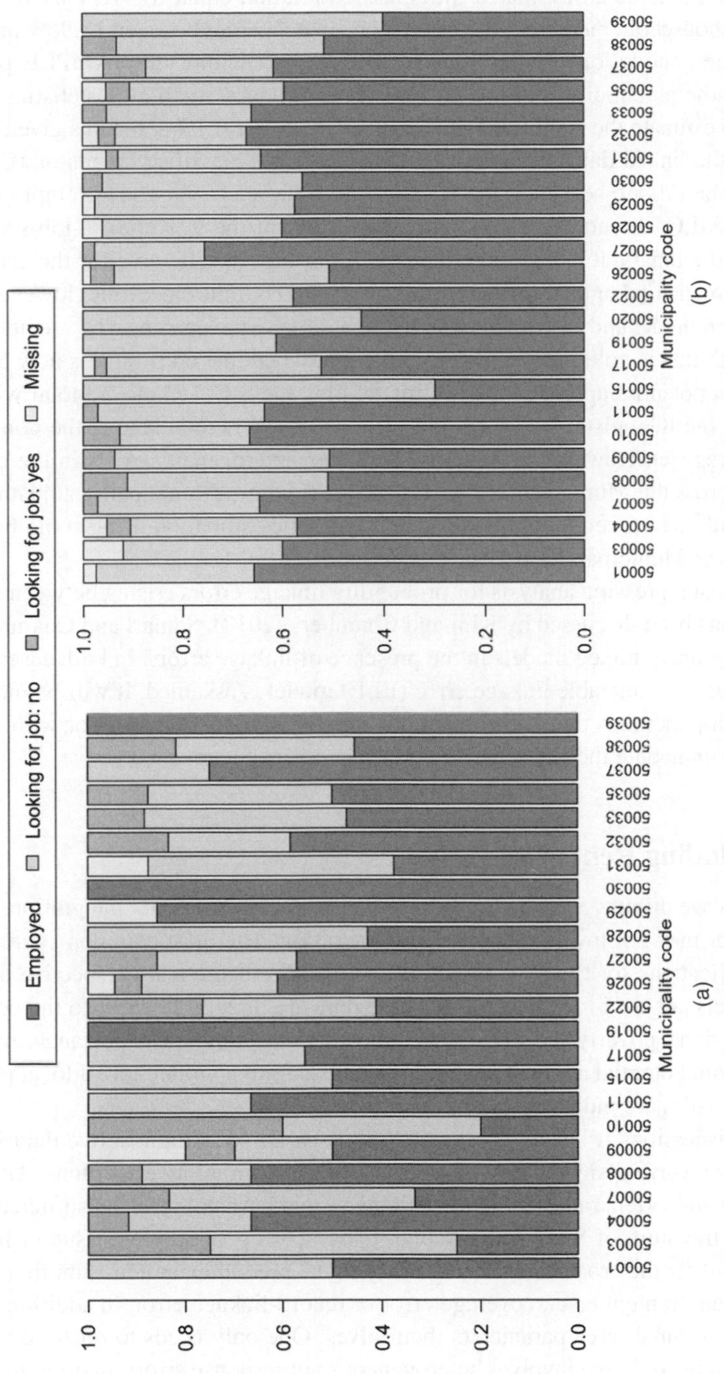

Figure 4.4 Distribution of PI-SILC status by Municipality conditional on JC status: looking for a job (a) and not looking for a job (b)

number of matches varies between 1.15 and 35.43. Fitting an 'empty' mixed model with only random intercept yields an estimated intraclass correlation equal to 41.77%. Moreover, the percentage of household members with at least one match varies between 18.96% and 87.70%.

As mentioned above, one of the aims of the research within the SAMPLE project was to investigate the potential use of the JC register to produce small area statistics. A natural approach is to estimate the conditional distribution of the SILC labour-status given the JC status based on the linked dataset. Table 4.5 shows the cross-classified counts of JC and SILC status. Out of the 190 job-seekers in the JC, there are 104 (or 54.7%) who are employed according to the PI-SILC; the number is 541 (or 58.7%) out of the 921 non JC-job-seekers. This suggests that the binary JC job-seeker status will not be very effective for the estimation of SILC-employment total or rate, since the odds ratio 0.849 is quite close to 1. However, the association between the JC and PI-SILC is much higher among the persons who are not employed in the PI-SILC, that is columns 3 and 4 in Table 4.5, where the odds ratio is now 5.673. This suggests that a potential approach is to estimate, first, the SILC-employed total without necessarily using the JC register and, then, the SILC job-seekers total among the non-employed using the JC register. However, as Figure 4.4 shows, heterogeneity exists in the conditional distribution across the Municipalities, such that direct within-Municipality adjustment of the JC counts would have been too unstable, while synthetic estimation using some fixed effects model across the Municipalities may be too biased.

Adjustment of regression analysis for probability linkage errors arising between survey and register data has been discussed by Kim and Chambers (2012). Samart and Chambers (2010) consider fitting linear mixed models in the presence of linkage errors. In both cases, however, a homogeneous exchangeable linkage error (ELE) model is assumed. It will be interesting in future to develop methods that allow the application of SAE models, with or without random effects, to accommodate the heterogeneous linkage errors documented above.

4.4 Concluding Remarks

In this chapter we discuss SAE methods based on administrative data integration, where the settings may or may not involve survey data in addition. The most common sources of error and their implications for the SAE are described from a theoretical perspective and several real-life datasets are used for illustration. These data are directly relevant to the construction of multidimensional poverty and well-being indicators, including life expectancy, employment status, and so on. Potential applications of the methods for the production of local poverty and well-being statistics are indicated.

From the discussions it becomes clear that SAE based on administrative data integration can potentially involve many other types of errors than the sampling error alone. This presents some different and, often, rather difficult challenges, that have not received sufficient attention in the current literature of SAE. In particular, there arises generally the issue of how potentially to account for the cross-area heterogeneity in the predominant non-sampling error, be it progressive measurement error, coverage error or record-linkage error, in addition to the heterogeneity of the small area parameters themselves. One only needs to reflect on how scant the literature is, on SAE that involves heterogeneous nonresponse errors, in order to appreciate the challenges and open issues for future research.

References

Chiang CL 1984 *The Life Table and its Applications.* Malabar, FL: Robert E. Krieger Publishing Co.

Contiero P, Tittarelli A, Tagliabue G, Maghini A, Fabiano S, Crosignani P, Tessandori R 2005 The EpiLink Record Linkage Software Presentation and Results of Linkage Test on Cancer Registry Files. *Methods of Information in Medicine*, **4**, 66–71.

ESSnet SAE 2011 *WP5 Final Report on the Case Studies*. September 1, 2015. Available at http://www.cros-portal.eu/sites/default/files//ESSnet%20SAE%20WP5%20Report-final-rev2.pdf.

Fay R and Herriot R 1979 Estimation of income from small places: An application of James-Stein procedures to census data. *Journal of the American Statistical Association* **74**, 269–277.

Fosen J and Zhang L-C 2011 Quality evaluation of employment status in register-based census. *Bulletin of the ISI 58th World Statistics Congress of the International Statistical Institute, Dublin*.

Hedlin D, Fenton T, McDonald JW, Pont M and Wang S 2006 Estimating the undercoverage of a sampling frame due to reporting delays. *Journal of Official Statistics*, **22**, 53–70.

Hogan H 1993 The Post-Enumeration Survey: Operations and results. *Journal of the American Statistical Association*, **88**, 1047–1060.

Kim G and Chambers RL 2012 Regression analysis under incomplete linkage. *Computational Statistics & Data Analysis*, **56**, 2756–2770.

Linkletter CD and Sitter RR 2007 Predicting natural gas production in Texas: An application of nonparametric reporting lad distribution estimation. *Journal of Official Statistics*, **23**, 239–251.

Luna-Hernandez A and Zhang L-C 2013 On models for small area compositions. *Proceedings of Small Area Conference 2013,* Bangkok.

Nirel R and Clickman H 2009 Sample surveys and censuses. In *Sample Surveys: Design, Methods and Applications*, Vol 29A (eds. D. Pfeffermann and C.R. Rao), North-Holland, Elsevier, Amsterdam, The Netherland. Chapter 21, pp. 539–565.

Noble A, Haslett S and Arnold G 2002 Small area estimation via generalized linear models. *Journal of Official Statistics*, **18**, 45–60.

Pfeffermann D 2013 New important developments in small area estimation. *Statistical Science*, **28**, 40–68.

Purcell NJ and Kish L 1980 Postcensal estimates for local areas (or domains). *International Statistical Review*, **48**, 3–18.

Rao JNK 2003 *Small Area Estimation*. Hoboken: John Wiley & Sons, Inc, Hoboken, NJ.

Raymer J, Smith PWF and Giuletti C 2011 Combining census and registration data to analyse ethnic migration patterns in England from 1991 to 2007. *Population, Space and Place*, **17**, 73–88.

Samart K and Chambers RL 2010. *Fitting linear mixed models using linked data.* Centre for Statistical and Survey Methodology, University of Wollongong, September 1, 2015. Working Paper 18-10, 25 pp. Available at http://ro.uow.edu.au/cssmwp/68.

Simpson L and Tranmer M 2005 Combining sample and census data in small area estimates: Iterative Proportional Fitting with standard software. *The Professional Geographer*, **57**, 222–234.

Wallgren A and Wallgren B 2006 *Register-based Statistics - Administrative Data for Statistical Purposes*. Chichester: John Wiley & Sons, Ltd.

Wang J, Fuller WA and Qu Y 2008 Small area estimation under a restriction. *Survey Methodology*, **34**, 29–36.

Wolter K 1986 Some coverage error models for census data. *Journal of the American Statistical Association*, **81**, 338–346.

Zhang L-C 2007 Finite population small area interval estimation. *Journal of Official Statistics*, **23**, 223–237.

Zhang L-C 2011 A unit-error theory for register-based household statistics. *Journal of Official Statistics*, **27**, 415–432.

Zhang L-C 2012 Topics of statistical theory for register-based statistics and data integration. *Statistica Neerlandica*, **66**, 41–63.

Zhang L-C 2015 On modelling register coverage errors. *Journal of Official Statistics*, **31**, 381–396.

Zhang L-C and Chambers RL 2004 Small area estimates for cross-classifications. *Journal of the Royal Statistical Society: Series B*, **66**, 479–496.

Zhang L-C and Fosen J 2012 A modelling approach for uncertainty assessment of register-based small area statistics. *Journal of the Indian Society of Agricultural Statistics*, **66**, 91–104.

References

Part II

Impact of Sampling Design, Weighting and Variance Estimation

5

Impact of Sampling Designs in Small Area Estimation with Applications to Poverty Measurement

Jan Pablo Burgard, Ralf Münnich and Thomas Zimmermann
Department of Economics and Social Statistics, University of Trier, Trier, Germany

5.1 Introduction

In 2000, the Lisbon Council pronounced *social cohesion* as one of the challenging responsibilities of the EU. The aim was to reduce poverty and social exclusion in Europe. In order to adequately monitor the objectives, a first set of indicators had been developed and been agreed at the European Council in December 2000—the indicators on poverty and social exclusion. Prepared by the Indicators Sub-Group (ISG) of the Social Protection Committee (SPC), they were finally decided on at the Laeken Council in December 2011 (cf. Atkinson *et al.*, 2005, as well as Kolb *et al.*, 2011, and the references cited therein). After the 2010 European Year of combating poverty and social exclusion (http://ec.europa.eu/employment_social/2010againstpoverty/index_en.htm), the European Commission stipulated further goals to reduce poverty as part of the Europe2020 strategy (http://ec.europa.eu/europe2020/index_en.htm).

According to the European needs to monitor the progress against poverty and social exclusion, the EU Statistics on Income and Living Conditions (EU-SILC; http://epp.eurostat.ec.europa.eu/portal/page/portal/microdata/eu_silc) was set up. This European survey was installed to provide the necessary information to measure progress in terms of the indicators

Analysis of Poverty Data by Small Area Estimation, First Edition. Edited by Monica Pratesi.

Companion Website: www.wiley.com/go/pratesi/poverty

on poverty and social exclusion. A comprehensive overview on EU-SILC, its aims and recent results from the Network for the Analysis of EU-SILC (Net-SILC) can be drawn from Atkinson and Marlier (2010).

Additionally to the Net-SILC group, the European Commission has launched two research projects within the 7th Research Framework Programme to support further research on poverty indicators and the SILC survey, namely SAMPLE (Small Area Methods for Poverty and Living Condition Estimates; http://www.sample-project.eu/) and AMELI (Advanced Methodology for European Laeken Indicators; http://ameli.surveystatistics.net).

In contrast to the classical view of measuring progress on country level, recent developments urge the needs on developing poverty indicators on regional level. This is essentially important when considering regional policies for supporting underdeveloped areas. Revisiting the earlier introduced EU statistics on poverty and living conditions, the sample sizes are not built to furnish estimating regional poverty estimates with the necessary reliability. Thus, model-based methods, such as small area estimators, have to be considered for improving the accuracy of the statistics of interest.

The aim of this study is to investigate small area methods for estimating the at-risk-of-poverty rate (ARPR), as one of the most widely used poverty measures. The focus is held on a European view considering the SILC data and different administrative backgrounds of the member countries, mainly resulting in the availability of auxiliary information. In contrast to the Scandinavian countries, which have a rich source of register data at their disposal, many other countries lack such information and, hence, may have difficulties to provide adequate auxiliary data which allow the building of strong models for income or poverty data. This stresses the robustness of the different approaches against the strength of the models.

Since the methodology is meant to be applied by National Statistical Institutes, special attention is put on the applicability in terms of SILC surveys and their peculiarities with respect to the corresponding sampling designs.

We summarize the research tasks as follows:

1. How precise are small area poverty estimates using survey data which are designed to deliver reliable national estimates?
2. What effect do the different sampling designs have on the model-based estimates in comparison with classical design-based methods?
3. Which estimation method is preferable in what context? How robust is this preference with regards to the model requirement?
4. How reliably can we estimate the precision of the local poverty estimates?

This chapter is organized as follows. In the next Section, the concurring sampling designs are presented. The choice can be considered as a set of SILC sampling plans. Next, the different estimators of interest will be depicted. The main focus is on the ARPR. In a Monte Carlo study, the performance of the different estimators is investigated using a variety of sampling designs. The study shall allow recommendations to be given for an adequate use of the methods in the context of official applications not using a rich set of register data. Finally, the chapter closes with a summary of the findings and an outlook on further research topics.

5.2 Sampling Designs in our Study

Household surveys generally use multi-stage designs mainly due to practical reasons. In the European SILC surveys, mainly one- or two-stage designs are applied, which are conducted separately within the participating countries (cf. Graf *et al.*, 2011 for an overview of SILC sampling designs).

The main emphasis of the present chapter is to elaborate the impact of sampling designs on poverty estimates. Hence, we focus on the main possible peculiarities of the surveys of interest. Similarly to 2012, we use two-stage stratified designs. The first stage represents $D = 78$ small areas and at the second stage within the areas, households are stratified according to the total disposable household income. These designs avoid introducing unplanned domains, since the domains of interest are in accordance with the first stage of our stratified random sampling design. Thus, in view of small area estimates, our designs can be viewed as one-stage stratified designs within areas. Because our focus is on small area estimation, we shall refer to a two-stage design with the stratified random sampling within areas as the second stage simply as stratified random sampling.

The stratification schemes were chosen with regards to the total disposable household income within areas. This is related but not identical to the equivalized disposable household income, that is used to identify the poor households. This setting is chosen against real-life application, since the total disposable household income is among the variables collected in the SILC survey (HY020). Nonetheless, we use it to stratify within areas since it allows to consider optimization via stratification. Thus, it reflects a situation in which the stratification variable could be very useful for improving the estimation of the variables of interest, but is not a perfect proxy. These settings are a compromise between mimicking real-life situations and allowing for improving survey designs in the design-based context.

As already pointed out, our study aims at identifying survey designs which are suitable for the small area estimation of the ARPR. Since the relationship between the target variable and the strata is of crucial importance for the efficiency of a design-based method in stratified sampling, we consider different scenarios for this relationship. What is common in those scenarios is that we group the households within areas in five strata, related to their total disposable income.

In the first scenario, which we call rand, the households within one area are randomly assigned to the different strata. This situation should not allow for efficiency gains compared with the benchmark case of simple random sampling (SRS) within areas using a design-based approach. This is due to the fact that the sum of squares within the strata will be large compared with the sum of squares between strata, which is a less efficient scenario for stratified random sampling.

In the second scenario, referred to as sort, we assign the households to a particular stratum based on the quantiles of total disposable income. With five strata per area, this implies that the households with the 20% lowest total disposable incomes belong to the first stratum, and so on. Hence, we expect that the sum of squares within should be small relative to the sum of squares between, such that a substantial variance reduction can be achieved. This scenario may yield an informative design for poverty measures based on the disposable household income.

Table 5.1 Stratification for the poverty analysis

Method	Description
rand	The households are randomly assigned to a stratum
sort	The households are assigned to a stratum based on the quantiles of their disposable income
raso	A mixture of the two above patterns

In the third scenario, called raso, we mix the above two strategies in the following way. The first stratum comprises every fifth household from the sorted disposable income and therefore contains the whole spectrum of households. The second and the third strata are composed of households from the first and second quartile of the sorted disposable income vector, respectively, which do not belong to the first stratum. Hence, both strata should exhibit relatively small within variation. Finally, the remaining households are randomly assigned to either the fourth or the fifth stratum. We briefly summarize these stratification strategies in Table 5.1 for convenience.

Besides the stratification effect discussed above, the allocation effect may play an important role for the efficiency under stratified random sampling (StrRS) designs. This amounts to the impact of different strategies to divide the sample size among the strata within the areas on the quality of the estimates.

A particularly easy approach is to spread the total sample size evenly among the strata, such that all strata have the same sample size. This procedure is known as equal allocation and frequently yields good results for domain estimation since it guarantees minimal sample sizes in all areas. On the other hand, it is usually less recommended for the estimation of aggregate statistics (cf. Burgard *et al.*, 2014) which is mainly due to possible efficiency losses.

Another common strategy is proportional allocation, where the stratum sample size is proportional to the number of units within each stratum. This procedure yields equal sampling fractions in the different strata and leads to ignorable design weights. In many cases, proportional allocation will lead to better estimation of national statistics at the expense of the reliability of domain estimates due to possibly very small sample sizes in small areas or strata. This issue is explored in further detail in Costa *et al.* (2004) who propose a convex combination of equal and proportional allocation to account for the trade-off between domain and national estimates.

Moreover, in some applications, we might be willing to minimize the variance of the national estimate of a single survey variable such as number of inhabitants in a census. In this case, optimal allocation can be employed, provided that the stratum-specific standard deviations or some proxy information is available. In this study, we used the standard deviations of total disposable income within strata to compute the sample sizes under optimal allocation. Note, that this allocation may lead to very small sample sizes in some domains, which may produce very inefficient estimates using design-based methods. Furthermore, highly different design weights may result, which could be an issue for model-based procedures. To address these issues, Gabler *et al.* (2012) proposed an extension of the optimal allocation introducing upper and lower bounds on the sampling fractions or sizes. For complexity reasons, we decided to focus the simulation study on the three classical allocation schemes.

The combination of these allocations with the different stratification patterns yield the sampling designs used in our study. Besides these StrRS design within areas, we also included SRS within areas as the benchmark case. The resulting sample sizes are given in Table 5.2.

Table 5.2 Sample sizes for the areas in the study

	Proportional	Equal	Optimal		
			rand	raso	sort
Min.	10.00	75.00	10.00	10.00	11.00
1st Quartile	56.25	75.00	51.00	52.00	47.00
Median	75.00	75.00	69.50	70.50	65.00
Mean	76.92	76.92	76.92	76.92	76.92
3rd Quartile	93.75	80.00	99.75	97.00	96.25
Max.	385.00	80.00	462.00	427.00	468.00

Table 5.3 Design effects

	Equal			Optimal			Proportional		
	rand	raso	sort	rand	raso	sort	rand	raso	sort
Min.	0.943	0.478	0.347	0.590	0.483	0.689	0.587	0.419	0.334
1st Quartile	0.945	0.497	0.369	0.929	0.929	1.162	0.927	0.490	0.363
Median	0.946	0.527	0.397	0.956	1.185	1.341	0.946	0.527	0.393
Mean	0.947	0.520	0.395	0.990	1.196	1.567	0.923	0.513	0.392
3rd Quartile	0.949	0.539	0.417	1.112	1.438	2.012	0.956	0.537	0.414
Max.	0.950	0.560	0.451	1.608	2.441	3.947	0.990	0.556	0.444

For the SRS case we set the area-specific sample sizes to the corresponding ones for proportional allocation. It can be seen from Table 5.2 that only the equal allocation guarantees considerably larger minimal sample sizes for all areas. In the case of the other allocation schemes a minimal sample size of two was introduced in all strata. Already the proportional allocation displays a large variation of the sample sizes between areas. While the largest area has a sample size of 385 units, the smallest area has only 10 units. This could be a major issue for design-based and model-assisted estimation strategies. Using the optimal allocation amounts to an even higher degree of variation in the sample sizes. Now the sample sizes differ by a factor of more than 40 in all stratification patterns. Note that the sample sizes under equal and proportional allocation do not vary with the stratification pattern used.

To investigate for the effectiveness of stratification and allocations we may look at the distribution of design effects for area-specific total Horvitz–Thompson estimates given in Table 5.3.

It can be seen from Table 5.3 that under equal and proportional allocations, the area-specific design effects decrease when moving from the random assignment to the raso and to the sorted pattern. This is in line with our reasoning that a stronger stratification should decrease the variance. In the case of the optimal allocation, however, a stronger stratification leads to increasing design effects at the area level. Thus, design optimization may backfire, which is due to the fact that the correlation between stratum-specific standard deviations of the dependent variable and the variable used to determine the optimal allocation is negative in the case of the sort pattern. Hence, under the sort scheme with optimal allocation, the sample size is higher, if the standard deviation of the dependent variable is lower. This means, we gather more information in homogeneous strata.

5.3 Estimation of Poverty Indicators

Our aim in the following is to estimate the ARPR, which is defined as the share of households whose equivalized disposable household income is less than a certain poverty threshold. Since our focus is on small area estimation, we consider the ARPR for the different domains d. It is given by

$$\text{ARPR}_d = \frac{1}{N_d} \sum_{j=1}^{N_d} \mathbb{I}(y_{jd} < z), \quad d = 1, \ldots, D, \tag{5.1}$$

where N_d is the number of households in domain d, y_{jd} refers to the equivalized disposable income of household j in domain d and z denotes the poverty threshold, usually taken as 60% of the median of the equivalized disposable household income. In equation (5.1), $\mathbb{I}(A)$ is taken to be the indicator function which takes the value 1 if condition A is met and 0 otherwise. In the literature on small area estimation, two methods have been predominantly discussed to estimate the ARPR. The first approach due to Molina and Rao (2010) is to model the y_{jd} or some transformation of it and obtain the best predictor of $\mathbb{I}(y_{jd} < z)$ for the non-sampled units by means of Monte Carlo integration. The second group of methods exploits the fact that $\mathbb{I}(y_{jd} < z)$ is a Bernoulli-distributed random variable and models these probabilities directly under a logistic regression model (cf. Lehtonen and Veijanen, 1998).

5.3.1 *Design-based Approaches*

Traditionally, design-based procedures have been used in finite population sampling. One example of a design-based estimator is the Horvitz–Thompson type estimator defined as

$$\widehat{\text{ARPR}}_d^{\text{HT}} = \frac{1}{N_d} \sum_{j=1}^{n_d} w_{jd} \mathbb{I}(y_{jd} < z), \quad d = 1, \ldots, D, \tag{5.2}$$

with $w_{jd} = \pi_{jd}^{-1}$ and π_{jd} being the inclusion probabilities. Estimator (5.2) is design-unbiased (Horvitz and Thompson, 1952) but its variance may be too large to produce reliable estimates for small sample sized n_d. A ratio estimator where N_d is replaced by the sum of the design-weights in area d, $\widehat{N}_d = \sum_{j=1}^{n_d} w_{jd}$, may yield more precise estimates in sampling designs with variable sample size, at the expense of a design bias for small n_d (cf. Lehtonen and Veijanen, 2009). In our study, however, with SRS and StrRS within planned domains, $\widehat{N}_d = N_d$, and hence there is no difference between the Horvitz–Thompson type estimator (5.2) and the corresponding ratio estimator. To derive an expression for the variance of (5.2) is slightly more complicated, since the ARPR is a non-linear statistic and z is typically estimated from the sample itself. Hence, following Osier (2009) both sources of randomness should be considered: the estimation of the proportion and the estimation of the threshold. For the explicit expression of the linearized ARPR and the details of deriving it, we refer the reader to Osier (2009) or Osier *et al.* (2013) and the references therein. After linearizing the ARPR we may simply approximate the variance of (5.2) by

$$\text{Var}\left(\widehat{\text{ARPR}}_d^{\text{HT}}\right) \approx \text{Var}\left(\sum_{j=1}^{n_d} w_{jd} \zeta_{jd}\right), \quad d = 1, \ldots, D, \tag{5.3}$$

with ζ_{jd} as the linearized value of household j in area d (cf. Deville, 1999; Osier, 2009). As can be seen from the formula for ζ (Graf and Tillé, 2014)

$$\zeta_{jd} = \frac{1}{\widehat{N}}(\mathbb{I}(y_{jd} < \widehat{z}) - \widehat{\text{ARPR}}_d^{\text{HT}}) + F'(\widehat{z})\zeta_{jd}^z, \quad \text{with} \tag{5.4}$$

$$\zeta_{jd}^z = -\frac{0.6}{F'(\widehat{z})\widehat{N}}\mathbb{I}(y_{jd} < 0.5), \tag{5.5}$$

where F is the cumulative distribution function of y and ζ_{jd}^z the linearized values for the estimation of the at-risk-of-poverty threshold z. In the present case of a fixed at-risk-of-poverty threshold the second part of (5.4) is constant, and thus does not play a role in the variance estimation. If auxiliary information is available, it may not only be used at the design stage but also at the estimation stage in a model-assisted framework.

The most common estimator in this framework is the generalized regression estimator (GREG) as discussed in (Särndal *et al.*, 2003, chapter 6). For this estimator a linear regression is used as assisting model producing predicted values $\widehat{y}_{jd}$. Suppose the model is such that these $\widehat{y}_{jd}$ may be interpreted as the probability that household j in area d is poor. Now, the usual expression for the GREG of a mean (cf. Lehtonen and Veijanen, 2009) yields an estimator of the ARPR, given by

$$\widehat{\text{ARPR}}_d^{\text{GREG}} = \frac{1}{N_d}\left[\sum_{j\in \text{U}_d} \widehat{y}_{jd} + \sum_{j\in \text{s}_d} w_{jd}(\mathbb{I}(y_{jd} < z) - \widehat{y}_{jd})\right], \quad d = 1, \ldots, D. \tag{5.6}$$

It can be seen from (5.6) that the GREG is composed of two parts: the sum of the predictions for all units in area d, $\sum_{j\in \text{U}_d} \widehat{y}_{jd}$ and the sum of the weighted residuals from the sampled units in area d, $\sum_{j\in \text{s}_d} w_{jd}(\mathbb{I}(y_{jd} < z) - \widehat{y}_{jd})$. Since the predictions from the model are adjusted by the weighted residuals, the estimator is asymptotically design-unbiased. If the model is reasonable, we expect that the residuals and thus their weighted sum is small in absolute size, such that (5.6) leans towards the model-based sum of the predictions. If, however, the model does not yield a good fit for the sample, the weighted sum of the residuals will be an important contribution to the GREG and prevent it from suffering biases due to the effects of a poorly specified model.

Up to now, we only assumed that the $\widehat{y}_{jd}$ should be interpreted as probabilities, but did not address the issue of specifying the model used to compute these probabilities. A simple solution is to use a linear probability model with $\mathbb{I}(y_{jd} < z)$ as the dependent variable, where $\widehat{y}_{jd} = \mathbf{x}_{jd}\widehat{\boldsymbol{\beta}}$ under a fixed-effects model. A drawback of the linear probability model is that its predictions are not restricted to the interval $[0, 1]$ and may thus lead to values that are invalid for probabilities. This may turn out to be crucial for many applications, though it does not necessarily impose a big problem in small area estimation, since we are not interested in predicting individual outcomes but rather in the ARPR for a finite population. Hence, if the sum of the predictions, $\sum_{j\in \text{U}_d} \widehat{y}_{jd}$, is close to the sum of the poor households in the population, $\sum_{j=1}^{N_d} \mathbb{I}(y_{jd} < z)$, the average residual within area d is close to zero and the correlation between residuals and weights within area d is small, then a linear probability model may yield ARPR predictions (5.6) with a small variance.

Alternatively, we can fit a logistic regression model, such that $\widehat{y}_{jd} = (1 + \exp(-\mathbf{x}_{jd}\widehat{\beta}))^{-1}$. Inserting these probabilities into (5.6) leads to a GREG under a logistic fixed-effects model

(LGREG). The use of the LGREG was proposed by Lehtonen and Veijanen (1998) for the prediction of national statistics. Myrskylä (2007) gives a detailed account for using the LGREG in small area estimation. Besides yielding predictions that can easily be interpreted as probabilities, a logistic regression model may yield a better fit to the sample data and thus lead to a smaller variance of the domain predictions in many cases. Note that the LGREG requires knowledge about $\mathbf{x}_{jd}$ for all units in the population, unless all auxiliary variables are categorical (cf. Lehtonen and Veijanen, 1998).

Equation (5.6) corresponds to a direct estimator, if a model (linear or logistic) is fitted separately in each domain. But it can be an indirect estimator as well and borrow strength from other areas, when the model is fitted for groups of areas or over the whole sample. The latter approaches may be preferred if there are areas with small sample sizes or if an elaborate model with many covariates is used. Moreover, the unobserved heterogeneity can be accounted for by modeling the areas as random effects. Despite their intuitive appeal, in applications with realistic data the gains from including a random effects specification may be negligible (cf. Lehtonen *et al.*, 2003). To estimate the variance of (5.6) residual variance estimators are commonly employed (cf. Särndal *et al.*, 2003, chapter 6). This is straightforward if the model is fitted separately within each domain, but may be problematic when using a regression model incorporating information from other areas.

5.3.2 *Model-based Estimators*

As pointed out by Pfeffermann (2013), model-assisted procedures such as LGREG or GREG (5.6) are reasonable if the assisting model works well, but they may suffer from large variances if the sample sizes are too small. To overcome this issue, we may break with the requirement of (approximate) design-unbiasedness if this enables us to obtain estimates with smaller mean squared errors (mses). This leads to a trade-off between bias and variance, as discussed by Lehtonen and Veijanen (2009). Note that since design-unbiasedness is no longer presumed, we have to use the mse instead of the variance, as the former includes the squared bias. One approach which leads to estimators which have smaller variances at the expense of a possible design bias are obtained by model-based procedures. These procedures assume that the sample data are one particular realization from a super-population model and obtain small area estimates as optimal predictions given the data and the assumed model (cf. Pfeffermann, 2013). Hence, any properties of the small area estimates are obtained with respect to the model and depend crucially on its validity. Detailed accounts on model-based estimators for general small area prediction problems are given in Pfeffermann (2013) and Jiang and Lahiri (2006).

A prominent approach to obtain small area estimates for non-linear statistics such as the ARPR (5.1) is to use a micro-simulation approach. This idea dates back to Elbers *et al.* (2003) and was extended by Molina and Rao (2010). It should be noted that this approach is not restricted to the estimation of the ARPR but can be used for the whole class of poverty measures due to Foster *et al.* (1984) and other indicators. Molina and Rao (2010) assume that a nested-error regression model holds for the transformed income variable $\eta_{jd} = g(y_{jd})$ in both the population and the sample, such that the sampling design is non-informative. This model is given by

$$\begin{aligned} \eta_{jd} &= \mathbf{x}_{jd}\boldsymbol{\beta} + u_d + e_{jd}, \quad j = 1, \ldots, N_d, \quad d = 1, \ldots, D \\ u_d &\sim N(0, \sigma_u^2), \quad d = 1, \ldots, D, \\ e_{jd} &\overset{iid}{\sim} N(0, \sigma_e^2), \quad j = 1, \ldots, N_d, \quad d = 1, \ldots, D. \end{aligned} \tag{5.7}$$

Under model (5.7), units from different areas will be uncorrelated whereas units from the same area will be correlated. The empirical best predictor (EBP) for the ARPR in domain d can be approximated by

$$\widehat{\text{ARPR}}_d^{\text{MR}} = \frac{1}{N_d}\left\{\sum_{j\in s_d} \mathbb{I}(y_{jd} < z) + \sum_{j\in r_d} \frac{1}{K}\sum_{k=1}^{K} \mathbb{I}\left(g^{-1}\left(\hat{\eta}_{jd}^{k}\right) < z\right)\right\}, \quad d = 1, \ldots, D, \tag{5.8}$$

where $r_d = u_d / s_d$ is the non-sampled part of the population in domain d, $g^{-1}(\cdot)$ is the back-transformation and K is the number of Monte Carlo replications (cf. Molina and Rao, 2010). In (5.8), $\hat{\eta}_{jd}^{k}$ is the prediction for unit j in area d in the kth Monte Carlo replication of model (5.7). It is obtained by sampling from the conditional distribution of the non-sampled units in domain d given the sampled units in domain d. Under model (5.7) this conditional distribution is normal (see Molina and Rao, 2010 for details). It can be seen from (5.8) that for the non-sampled part $j \in r_d$ the EBP for whether a unit is poor or not is obtained by micro-simulation. Furthermore, it may be noted under a nested-error regression model with block-diagonal variance covariance structure, the Monte Carlo integration for the non-sampled units can be easily replaced by a numerical integration. To estimate the mse, Molina and Rao (2010) propose a parametric bootstrap. The predictor (5.8) is approximately mse-optimal if the assumed model (5.7) holds. Since this model is a unit-level model, strong predictive covariates $\mathbf{x}_{jd}$ are required.

In some cases, however, covariates with explanatory power may be available at some aggregated level only. If this level coincides with the level at which small area estimates are required, area-level models can be taken into consideration. Their use in small area estimation was pioneered by Fay and Herriot (1979), who used a two-level model. At the first stage, referred to as a sampling model, it is assumed that unbiased direct design-based estimators with known sampling variances are available. At the second stage, these estimators are linked to known auxiliary information, which is taken for example from registers. Typically, normality assumptions are used at both levels. Together this model reads (cf. Jiang and Lahiri, 2006)

$$\begin{aligned} &\text{sampling model} \quad \hat{\mu}_d^{Dir} | \mu_d \overset{ind}{\sim} N(\mu_d, \psi_d), \quad d = 1, \ldots, D, \\ &\text{linking model} \quad \mu_d \overset{ind}{\sim} N(\overline{\mathbf{X}}_d\beta, \sigma_u^2), \quad d = 1, \ldots, D, \end{aligned} \tag{5.9}$$

with $\hat{\mu}_d^{Dir}$ as the direct estimator, ψ_d as its known variance, μ_d as the small area mean, $\overline{\mathbf{X}}_d$ as the auxiliary information and σ_u^2 as the error term in the linking model. The assumption of known sampling variances is rarely fulfilled in applications. It is a common practice to replace ψ_d by sample estimates. In our study the direct estimator is given by (5.2), and hence the mse-optimal predictor assuming a fixed poverty threshold under model (5.9) follows as:

$$\widehat{\mathrm{ARPR}}_d^{\mathrm{FH}} = \widehat{\gamma}_d \widehat{\mathrm{ARPR}}_d^{\mathrm{HT}} + (1 - \widehat{\gamma}_d)\overline{\mathbf{X}}_d \widehat{\beta}, \quad d = 1, \ldots, D \quad \text{where}$$

$$\widehat{\gamma}_d = \frac{\widehat{\sigma}_u^2}{\psi_d + \widehat{\sigma}_u^2}. \tag{5.10}$$

Predictor (5.10) is a convex combination of the direct estimator $\widehat{\mathrm{ARPR}}_d^{\mathrm{HT}}$ and the regression-synthetic component $\overline{\mathbf{X}}_d \widehat{\beta}$ with weights $\widehat{\gamma}_d$ and $(1 - \widehat{\gamma}_d)$. The assumptions regarding the area-level model defined by (5.9) are less likely to be violated in practice than those present in the unit-level model (5.7). This is due to the fact that in (5.9) the assumptions are made on the behavior of aggregate statistics not on individual observations. Even if the latter are not normally distributed, the $\widehat{\mu}_d^{Dir}$ may be approximately normal, provided the sampled y_{jd} are independently and identically distributed. To ensure normality of the sampling model, variance-stabilizing transformations of the direct estimators may be employed. In the case where the direct estimator is a proportion, the arc-sine square root transformation $\theta_d = \sin^{-1}(\sqrt{\mathrm{ARPR}_d})$ is commonly applied, see for example Jiang *et al.* (2001). As detailed by those authors, the variance of the transformed direct estimator is approximately equal to the design effect divided by the area sample size times four. A simple estimator of the ARPR is then obtained by applying the back-transformation to the EBP of θ_d as

$$\widehat{\mathrm{ARPR}}_d^{\mathrm{FHtrans}} = \sin\left(\widehat{\theta}_d^{\mathrm{FH}}\right)^2, \quad d = 1, \ldots, D, \tag{5.11}$$

where $\widehat{\theta}_d^{\mathrm{FH}}$ is given by employing (5.10) on the transformed variable θ_d. As noted by Jiang *et al.* (2001), this simple estimator is not optimal under the model, since the back-transformation is a non-linear function, but provided the area-specific sample sizes are not to small, it could work reasonably well. It is tempting to rely on a Taylor linearization to estimate the mse of the transformed predictor (5.11) and use

$$\widehat{\mathrm{mse}}(\widehat{\mathrm{ARPR}}_d^{\mathrm{FHtrans}}) = 2\sin\left(\widehat{\theta}_d^{\mathrm{FH}}\right)\cos\left(\widehat{\theta}_d^{\mathrm{FH}}\right)\widehat{\mathrm{mse}}\left(\widehat{\theta}_d^{\mathrm{FH}}\right), \quad d = 1, \ldots, D. \tag{5.12}$$

It has been noted, however, that the naive mse estimator (5.12) is not accurate to the desired order, and hence not reliable (cf. Rao, 2003, chapter 7.1.8). Alternatively, the mse of (5.12) can be estimated using the jackknife method due to Jiang *et al.* (2002). Moreover, we can compute the EBP

$$\widehat{\mathrm{ARPR}}_d^{\mathrm{AEBP}} = \mathrm{E}\left(\mathrm{ARPR}_d | \widehat{\theta}_d^{\mathrm{FH}}, \widehat{\beta}, \widehat{\sigma}_u^2\right)$$

using numerical integration, as the estimated posterior density of the transformed variable is given by $N(\widehat{\theta}_d^{\mathrm{FH}}, \widehat{\gamma}_d \psi_d)$ (Rao 2003, chapter 9.2.1). Since the EBP is simply the expectation of a transformation of $\widehat{\theta}_d^{\mathrm{FH}}$, it can be computed as

$$\widehat{\mathrm{ARPR}}_d^{\mathrm{AEBP}} = \int_{-\infty}^{\infty} \sin(\zeta)^2 \frac{1}{\sqrt{2\pi\widehat{\gamma}_d\psi_d}} \exp\left(-\frac{1}{2}\left(\frac{\left(\zeta - \widehat{\theta}_d^{\mathrm{FH}}\right)^2}{\widehat{\gamma}_d\psi_d}\right)\right) d\zeta, \quad d = 1, \ldots, D. \tag{5.13}$$

While the use of variance stabilizing transformations such as the arc-sine square root transformation may be reasonable to achieve normality of the transformed direct estimators, we can alternatively account for their non-normality directly. In order to pursue this idea, note that the direct sample total $n_d\widehat{\text{ARPR}}_d^{\text{HT}}$ is a count which may be modeled by a binomial distribution. Hence, we modify model (5.9) in the following manner:

$$\begin{aligned}&\text{sampling model} \quad n_d\widehat{\text{ARPR}}_d^{\text{HT}}\,|\text{ARPR}_d \overset{ind}{\sim} Bin(n_d, p_d)\\ &\text{linking model} \quad \text{logit}(\text{ARPR}_d) \overset{iid}{\sim} N(\overline{\mathbf{X}}_d\beta, \sigma_u^2), \quad d = 1, \ldots, D. \qquad (5.14)\end{aligned}$$

Model (5.14) is an extension of the model developed by Jiang and Lahiri (2001) for binary data. In (5.14), we assume that conditional on the true ARPR, the sample total is independently binomially distributed within each domain. For the linking model, the commonly used logit-transformation is employed, to transform the probabilities ARPR_d to the real line. The EBP under model (5.14) does not have a closed-form solution, but emerges as ratio of two one-dimensional integrals (Jiang and Lahiri, 2001). Hence, it can be readily estimated using numerical integration. One issue with this procedure is that the values of $\widehat{\sigma}_u^2$ are typically very small. This may lead to problems for the integration routine, as the integrals in the numerator and the denominator of the EBP may be estimated to zero and thus the EBP may collapse. Moreover, even if $\widehat{\sigma}_u^2 > 0$ but takes small values, the gain from incorporating a random effect in the linking model may be overcompensated by the difficulties in estimating the model. In our simulations, we could not estimate model (5.14) due to very small $\widehat{\sigma}_u^2$. Thus, we changed the linking model in (5.14) to a deterministic link, that is

$$\text{logit}(\text{ARPR}_d) = \overline{\mathbf{X}}_d\beta, \quad d = 1, \ldots, D,$$

which is a synthetic model, since the unobserved heterogeneity in ARPR_d not explained by the covariates $\overline{\mathbf{X}}_d$ is not accounted for. Instead of modeling the direct sample total $n_d\widehat{\text{ARPR}}_d$, we decided to consider modeling the estimated population total. This leads to the following binomial synthetic area-level model:

$$\begin{aligned}&\text{sampling model} \quad \widehat{N}_d\widehat{\text{ARPR}}_d^{\text{HT}}\,|\text{ARPR}_d \overset{ind}{\sim} Bin(\widehat{N}_d, \text{ARPR}_d)\\ &\text{linking model} \quad \text{logit}(\text{ARPR}_d) = \overline{\mathbf{X}}_d\beta, \quad d = 1, \ldots, D, \qquad (5.15)\end{aligned}$$

where $\widehat{N}_d = \sum_{j=1}^{n_d} w_{jd}$ is the estimated population size in domain d. Note that for the designs in Chapter 2, $\widehat{N}_d = N_d$. We consider a synthetic estimator under model (5.15), which is given by

$$\widehat{\text{ARPR}}_d^{\text{SynAL}} = \frac{1}{1 + \exp\,(-\overline{\mathbf{X}}_d\widehat{\beta})}, \quad d = 1, \ldots, D, \qquad (5.16)$$

where $\widehat{\beta}$ is obtained from the linking model in (5.15). Since predictor (5.16) is synthetic, its variance can be easily estimated, for example using the jackknife, but estimating its prediction mse is not straightforward. We follow the approach of Marker (1995), which is discussed in Rao (2003, chapter 4.2.4). This approach assumes that squared design bias in each domain is close to the average squared bias, which in turn can be estimated from the average prediction mse minus the average variance. The prediction mse of (5.16) follows from applying the mse identity, that is mse = Var + Bias^2.

5.4 Monte Carlo Comparison of Estimation Methods and Designs

For comparing the performance of the estimators at hand under different sampling designs, a Monte Carlo study was conducted. The data set used for the simulation study is a publicly available data set generated within the Ameli Project (cf. Alfons *et al.*, 2011). For each combination of stratification and allocation, 10 000 independent samples are drawn. And for every sample, the estimators and their corresponding precision estimates are computed. $\widehat{\theta}^{*d} = (\widehat{\theta}^*_{d\,1}, \widehat{\theta}^*_{d\,2}, \ldots, \widehat{\theta}^*_{d\,M})$ is the vector of all the estimates obtaind by an combination $*$ of an estimator and a sampling design. θ_d denotes the true parameter which is to be estimated in area d.

The dependent variable is a dichotomous variable, taking the value 1 if the equivalized disposable household income is below the at-risk-of-poverty threshold and 0 if this is not the case. For the selection of the independent variables, a stepwise backward elimination departing from the model including all available variables is performed on the complete data set. Due to the dichotomous dependent variable, a logit model is used. The resulting set of variables is: number of retired person in household (RB210old), number of employed persons in household (RB210work), any unemployed person in household (RB210unem), household type (HHT), taxes on income (HY140C), indicator of urbanization (DOU), and region (REG). A χ^2 test against the null model results in a p-value numerically zero. According to (Faraway 2005, p. 158) the significance of the random effect can be tested conservatively by performing a likelihood-ratio test using the model with and without the random effect. The null hypothesis that the random effect on the areas is zero results in a p-value smaller than 0.0001 and thus can be rejected. The model for the Molina-Rao estimator (MR) estimator has the equivalized disposable household income as dependent variable. The underlying mixed model does not fit well on the data. For comparison the same covariates are used as above.

For the evaluation of the performance of the point estimates, we focus on calculating the observed relative bias and relative root mean squared error in the simulation.

The relative bias is given by

$$\text{rbias}(\widehat{\theta}^*_d) := \frac{\frac{1}{M}\sum_{m=1}^{M} \widehat{\theta}^*_{d\,m} - \theta_d}{\theta_d} \quad . \tag{5.17}$$

The resulting values for the relative bias are in the real line and values closer to zero indicate a lower bias. For accessing the precision besides the bias also the variability of the estimates is of importance. As the model-based small area estimates are not necessarily unbiased, a compensatory measure including both, bias and variability is chosen—the relative root mean squared error. It is computed as the root of the mean squared deviation from the estimates to the true value in relation to the true value:

$$\text{rrmse}(\widehat{\theta}^*_d) := \frac{\sqrt{\frac{1}{M}\sum_{m=1}^{M} (\widehat{\theta}^*_{d\,m} - \theta_d)^2}}{\theta_d} \quad . \tag{5.18}$$

As stated before, model-based estimators may have a considerable design bias. Therefore, as precision estimate the mse is used instead of the variance. The mse is defined as the variance plus the squared bias. Therefore, for unbiased estimators the variance estimator is at the same time a mse estimator. For simplicity, we will refer to both, mse and variance estimators, as mse estimators. One way of validating an mse estimator is to study its relative bias. That is, the bias of the estimated mse of the point estimate. The resulting value lies in the interval $[-1, \infty)$, with values above zero indicating overestimation and values under zero indicating underestimation of the mse. The nearer the value is to zero the lower is the bias of the mse estimate. The relative bias of the combination $*$ of mse estimator and sampling design is

$$\text{rbiasmse}_d^* := \frac{\frac{1}{R}\sum_{m=1}^{M} \widehat{\text{mse}}(\widehat{\theta}_{d\,m}^*) - \text{mse}_{\text{MC}}(\widehat{\theta}_d^*)}{\text{mse}_{\text{MC}}(\widehat{\theta}_d^*)}, \tag{5.19}$$

$$\text{with} \qquad \text{mse}_{\text{MC}}(\widehat{\theta}_d^*) := \frac{1}{M}\sum_{m=1}^{M} (\widehat{\theta}_{d\,m}^* - \theta_d)^2$$

being the empirically observed mse in the simulation study.

Confidence intervals are helpful in evaluate simulation studies, as they combine two important and related components, the *confidence interval coverage rate* (cicr) and the confidence interval length (cil). The cicr is defined as follows:

$$\text{cicr}_d^* := \frac{1}{M}\sum_{m=1}^{M} \mathbb{I}\left(\theta_d \in \widehat{\text{ci}}(\widehat{\theta}_{d\,m}^*)\right), \tag{5.20}$$

where $\widehat{\text{ci}}(\widehat{\theta}_{d\,m}^*)$ is the estimated confidence interval for the estimator/sampling design combination $*$ in area d in simulation run m.

For a deliberately high mse estimate, the cicr will always be 1. Therefore, the comparision of the cicr with the magnitude of the mse is of interest. This can be done by incorporating the mean confidence interval length (mcil) over the simulation runs. This is defined as

$$\text{mcil}_d^* := \frac{1}{M}\sum_{m=1}^{M} \text{cil}(\widehat{\theta}_{d\,m}^*). \tag{5.21}$$

Large values of mcil indicate high mse estimates.

In Figure 5.1 the relative bias of the point estimates are plotted in a lattice plot. In the horizontal direction the different allocations and on the vertical direction the three different stratifications are depicted. Within each of the combinations of the allocations and stratifications, boxplots of the relative bias of the estimators over all areas plotted.

As expected from theory the design-based estimators do not show to have any bias in any design. In contrast to that, the MR estimator has a considerable positive bias. This bias is clearly dependent on the design and rises with both, the optimality of the stratification and the optimality of the allocation with respect to a design-based national estimate. The big issue with the MR in our data set is that we could not validate the model in use. Therefore the comparision is not fair. On the other hand, it seems important to us to show how cumbersome a misspecified model for the MR can be and emphasize to use it only with a validated model. In many countries, the available information for model building is rather scarce, and thus it can be difficult to find reasonably good covariates for the MR. For the Fay and Herriot (FH) estimator,

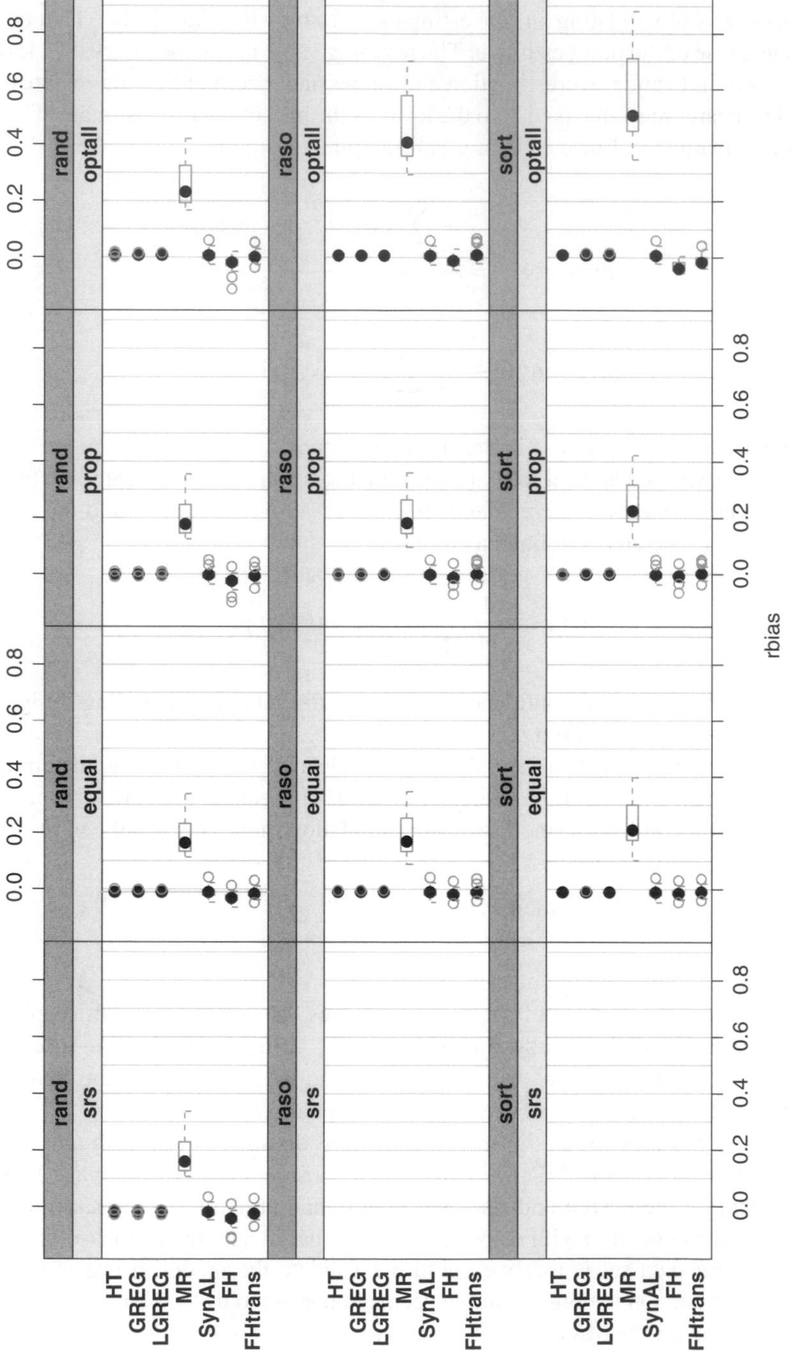

Figure 5.1 Relative bias (rbias) of the ARPR estimators

a negative bias can be observed, which rises again with the optimality of the allocation and the stratification for a national design-based estimate. Neither for the FHTrans nor the SynAL the designs under study seem to affect the bias considerably.

Figure 5.2, depicting the results for the rrmse, is structured as Figure 5.1 for the rbias. As the mse is the sum of the variance and the squared bias, the high bias of the MR increases strongly its rrmse. When comparing the design-based estimators, the improvement of using the GREG over the HT is visible in all situations but rather small. Further, using the LGREG improves slightly over using the GREG with one exception. In the case of the stratification raso with the allocation optall the LGREG is even underperforming the HT.

Now the question arises, why the general relationship breaks down under this particular setting. To explain why this is the case, it will be helpful to think about the circumstances in which a GREG-type estimator is going to perform well. As mentioned earlier, the GREG can be written in the following manner:

$$\widehat{\text{ARPR}}_d^{\text{GREG}} = N_d^{-1}\left(\sum_{i\in \text{U}_d} \hat{y}_{dj} + \sum_{i\in \text{S}_d} w_{id}(\mathbb{I}(y_{id} < z) - \hat{y}_{id})\right). \tag{5.22}$$

The expression within brackets in (5.22) comprises two parts: the sum of the model prediction for all units in the population in a particular domain and the sum of weighted residuals in this domain from the sample. If the second term in brackets is equal to zero, this implies that a bias correction is not necessary. Hence a fully model-based approach might be considered, since this would lead to a lower variance. To achieve a low bias of the model-based part $\sum_{i\in \text{U}_d} \hat{y}_{id}$, one might want to include many covariates. However, this is generally accompanied by an increase in the variance of predictions, so that this does not solve the bias-variance trade-off. As a measure of the bias of the model-based predictions we can consider the bias-adjustment ratio

$$\text{bar}_d = \frac{\sum_{i\in \text{S}_d} w_{id}(\mathbb{I}(y_{id} < z) - \hat{y}_{id})}{\sum_{i\in \text{U}_d} \hat{y}_{id}}. \tag{5.23}$$

The larger the absolute value of (5.23), the larger the necessary adjustment to the model-based predictions. Thus, we may want to study whether the differences between the linear GREG and the logistic GREG observed in the previous section are somehow related to the bias-adjustment ratio (5.23).

In Figure 5.3 the Monte Carlo mean and standard deviation of the bar for the LGREG and GREG are plotted against each other. The diagonal line indicates the angle bisection. As can be seen in all designs but one, the mean and the standard deviation of the bar for the GREG are higher than the ones for the LGREG. Only in the combination optall–raso these values favor the GREG over the LGREG, which is in line with the results shown in Figure 5.2.

Comparing the design-based to the model-based estimators, the clear domination of the model-based estimators in terms of rrmse is apparent (see Figure 5.2). In all situations, the model-based estimators FH, FHTrans and SynAL outperform the other estimators by a considerable amount. Even though, the SynAL is not always the best estimator, it shows to have a constantly low rrmse over all designs under study. The arc-sine transformation in the FH seems to improve the estimation fo the FH in all situations but the combinations of the stratifications

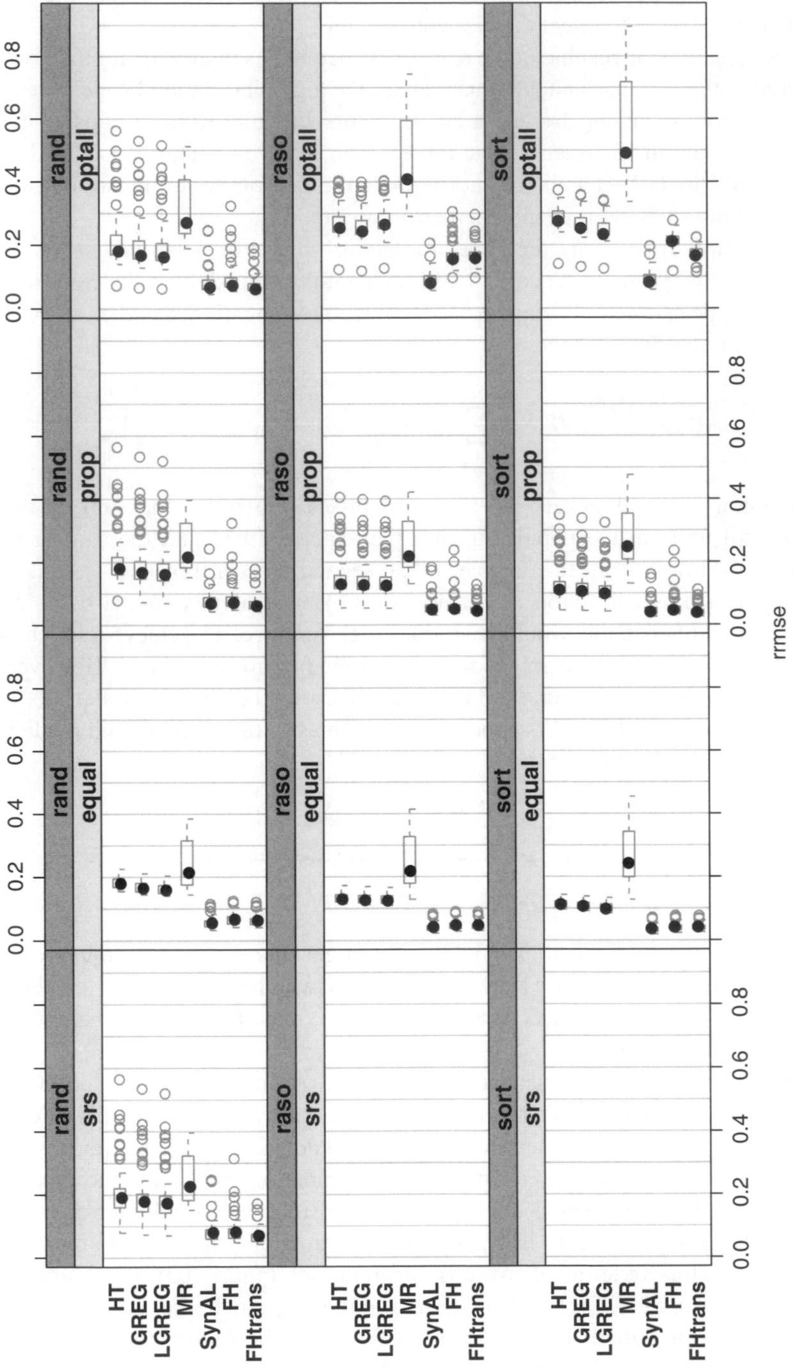

Figure 5.2 Relative root mean squared error (rrmse) of the ARPR estimators

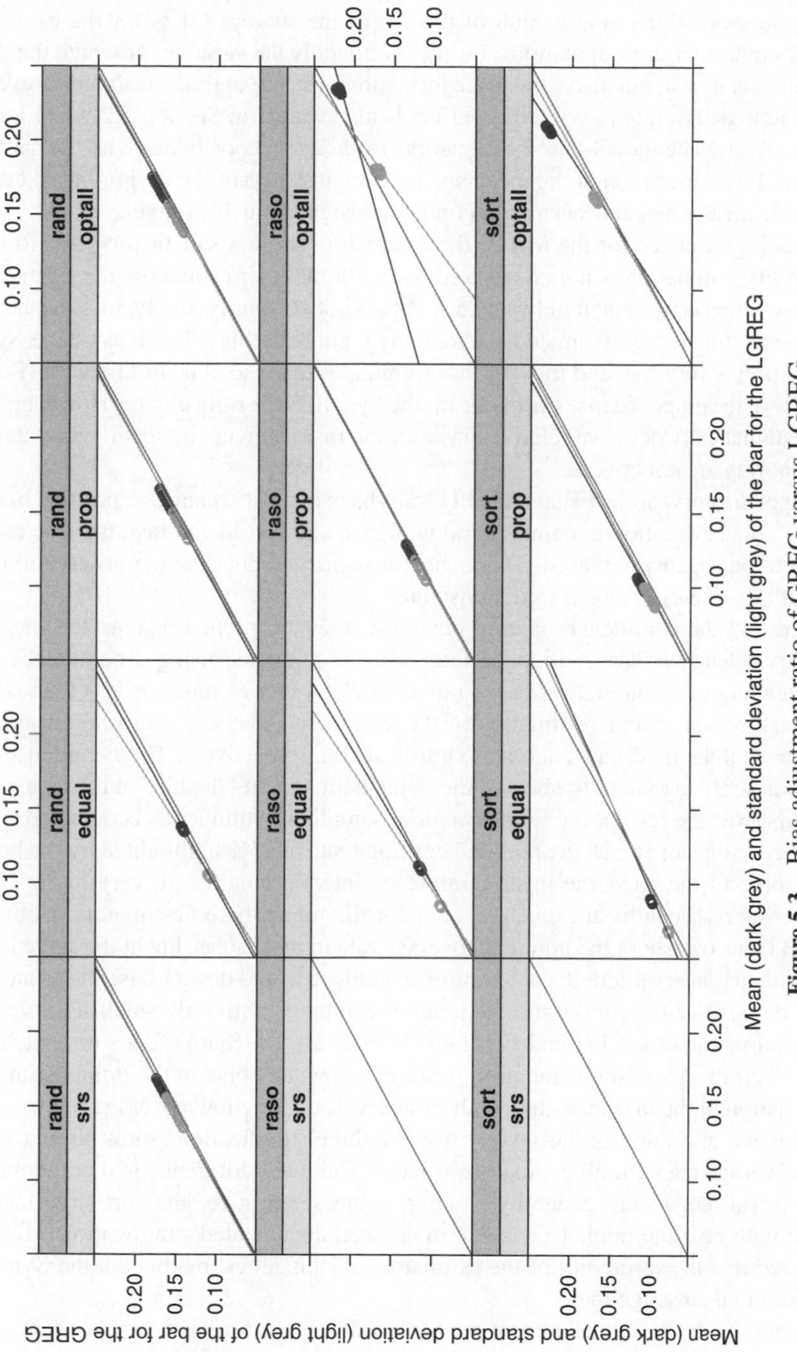

Figure 5.3 Bias adjustment ratio of GREG versus LGREG

raso with the allocation optall. With this finding, in the small area context the idea of optimality of design has to be reflected. Optimal designs, assuring a precise estimation of national estimates, do show severe implication of the use of the surveyed data for the estimation of regional or contextual sub-populations. Further, if not only the variable for which the design is optimized is of interest, but also some transformation of it, the optimal design may even cause the outcome to deteriorate heavily. This effect is also visible in Section 5.2 when looking at the design effects. The model-based estimators are shown to cope better with the sample size restriction. The deterioration of the mean and median precision of the design-based estimators for the single areas when approaching an optimal design is much stronger.

As the point estimates for the MR in these scenarios are not satisfactory, due to the poor model, the mse estimation is not considered here. For all other estimators the relative bias of the mse estimators is presented in Figure 5.4. Again not very surprisingly, the variance for the design-based estimators is estimated without any significant bias. The bias of the SynAL in our application is very low and the variance dominates the mse, thus in line with (Rao 2003, chapter 4.2.4), the proposed mse estimator for the SynAL is the right choice. However, it shows to have systematically negative relative biases of the mse estimate, indicating that major parts of the variability are not covered.

The mse estimators for the FH and the FHTrans have in most scenarios a positive bias. In the case of the optall allocation in combination with raso and sort in contrast, the mse estimators are shown to be negatively biased. This behavior seems to underline the observation by Bell (2001) that this mse estimator is rather unstable.

In Figure 5.5 the confidence interval coverage rates are plotted against the mean confidence interval lengths. The confidence intervals are computed using a confidence level of $\alpha = 0.05$ leading to a nominal coverage rate of 0.95. A cicr of more or less than 0.95 indicates overestimation or underestimation of the mcil, respectively. The crosses, triangles, and circles represent the rand, raso, and sort stratifications, respectively. The confidence interval estimation is sensible to both the bias of the point estimator and the bias of the mse estimator.

In the first row the results for simple random sampling within areas is depicted. All three design-based estimators reach the nominal coverage rate of 95% highlighted by the horizontal line. For some of the areas the mean confidence interval lengths are very high. The mean confidence interval lengths are much smaller for the model-based estimators. Both, the FH and the FHTrans overshoot the nominal coverage rate in most areas, but at the same time their mean confidence interval lengths are shorter than those of the design-based estimators. This is due to the positive bias in the mse estimation combined with only small observed bias in the point estimation of the FH and FHTrans. In contrast, the SynAL does not reach 95% in any area. Even though in the simulation it did not show any bias in the point estimation the underestimation of the mse leads to too short intervals causing low coverage rates.

For the equal allocation in the second row the three stratifications show similar results in terms of coverage rates for all estimators at hand. The main difference is that the mean confidence interval lengths are generally smaller, being smallest for the sort stratification and increasing with raso and rand. This is due to the well documented stratification effect, which generally reduces the variability of the estimators. Again, all estimators but the SynAL meet the 95% nominal coverage rate.

An advantage of the equal allocation for small area estimation can be deduced from the variance of the HT leading to $\sigma^2/n_{d,h}$ in each stratum and $\sum_{h=1}^{H} \sigma^2_{d,h}/n_{d,h}$ for the area. Therefore, if the $\sigma^2_{d,h}/n_{d,h}$ do not differ strongly between the areas, the obtained precision

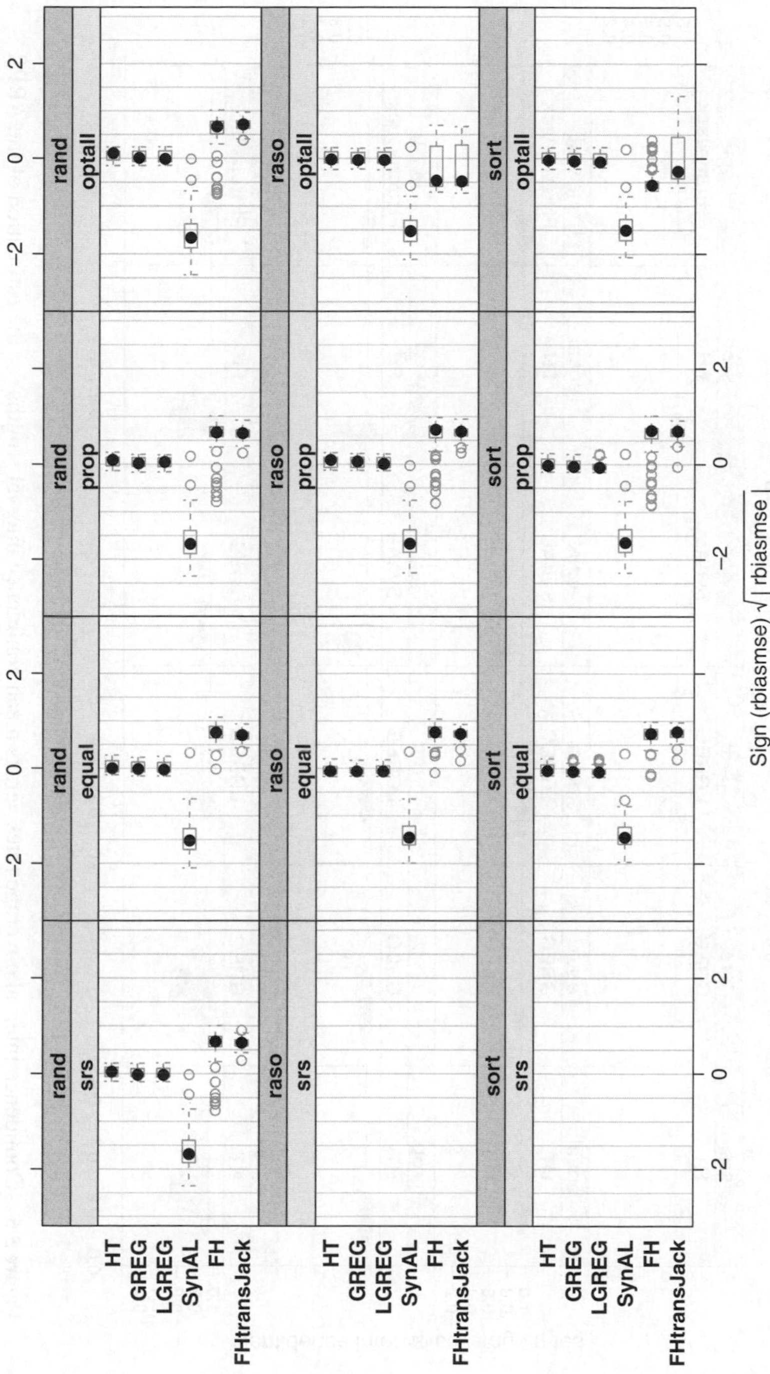

Figure 5.4 Relative bias of the mse and variance estimators for the ARPR

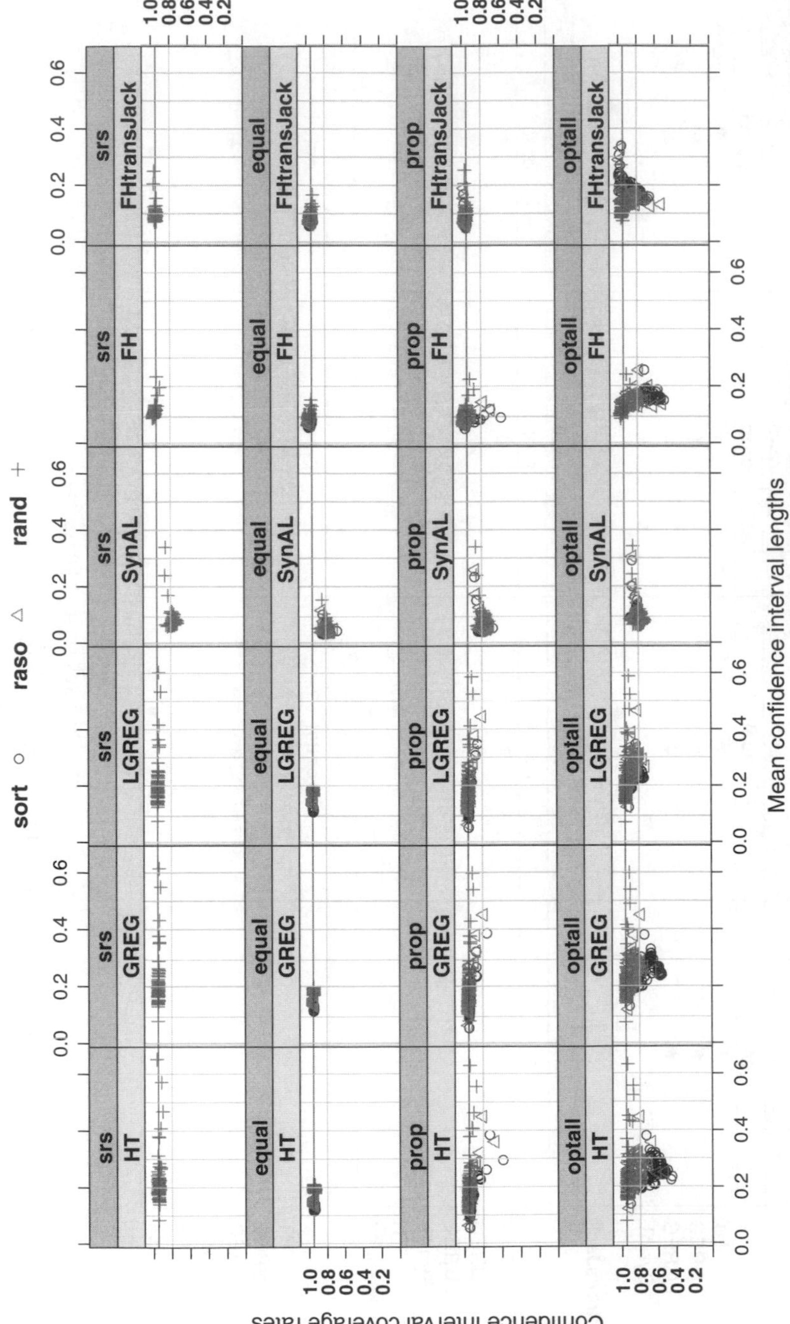

Figure 5.5 Confidence interval coverage rates versus mean confidence interval lengths for the estimation of the ARPR

between the areas is at the same level, independently of the area size. If the sample size is allocated proportionally to the area sizes, however, the $n_{d,h}$ also differ from area to area leading to highly varying variances for the design-based estimators. This fact is also observable in the third row showing the results for the proportional allocation. The mean confidence interval lengths, which are proportional to the estimated variance or mse of the estimators, rae shown to be highly variable for the design-based estimators along the different areas. Also for the model-based estimators, the variability of the mean confidence interval length rises compared with equal allocation, but to a far lesser extent. The confidence interval rates for the design-based estimates reach in most areas the nominal coverage rate, but in some areas they break down notably. This is mainly due to the effect of stratification in combination with the binary variable of interest. As both stratifications sort and raso contain strata which either have a very large or very low equivalized disposable household income, in many samples there will be no observed variability.

In the small sample sizes setting, arguing for the GREG with asymptotic design-unbiasedness is problematic as shown by Lepik (2007), when the assisting model is poor. In line with this, also the asymptotically unbiased design-based variance estimators seem to suffer from small sample sizes and poor assisting models.

When using an optimal allocation, the $n_{d,h}$ vary even stronger. As the optimality is computed on the total disposable household income, strata with high variable total disposable household incomes have a high $n_{d,h}$ and vice versa. However, the variable of interest in this simulation is dichotomous, being one if the equivalized disposable household income lies under the poverty threshold or else zero. This leads to an interesting effect. In the case of the rand stratification, where the optall allocation will lead to a proportional allocation, there is no visible difference to the results for the proportional allocation. In contrast, if either the raso or sort stratification is used, then the coverage rates heavily break down.

5.5 Summary and Outlook

The present study has shown that optimizing sampling designs most likely will have to be re-investigated in the future. This is mainly due to the policy needs to rely on area-specific information, for example poverty indicators. The traditional design-based view will have to consider the use of model-based methods on small sub-populations which are constructed either on area or domain level.

Given the example of the ARPR, we have seen that reliable estimates can be produced even in cases which are not too convenient, that is where highly correlated auxiliary variables are less available. In these cases, however, the robustness of the method plays a very important role. Hence, it is crucial to know the environment where the estimators of interest shall be applied. One good example is the Molina–Rao estimator which can be extremely efficient once the model assumptions hold but lacks robustness against model misspecification.

In the future, one may expect increasing use of model-based methods in official statistics. To further support model-based methods, register variables shall further be made available which may happen by merging different sources. This may also furnish a wider range of estimators to be applied for regional poverty measurement.

Acknowledgements

The research was conducted within the research infrastructure InGRID (Inclusive Growth Research Infrastructure Diffusion; https://inclusivegrowth.be/), which is financially supported by the European Commission within the 7th Framework Programme (FP7) under grant agreement no. 312691. The dataset AMELIA was developed within the AMELI FP7 project and will be made available via the project homepage (http://ameli.surveystatistics.net).

References

Alfons A, Filzmoser P, Hulliger B, Kolb JP, Kraft S, Münnich R and Templ M 2011 Synthetic data generation of silc data. Research Project Report WP6—D6.2, FP7-SSH-2007-217322 AMELI.

Atkinson AB, Cantillon B, Marlier E and Nolan B 2005 Taking forward the EU social inclusion process. An Independent Report Commissioned by the Luxembourg Presidency of the Council of the European Union.

Atkinson AB and Marlier E (eds) 2010 *Income and Living Conditions in Europe*. European Union.

Bell WR 2001 Discussion on jackknifing in the Fay–Herriot model with an example. *Proceedings of the Seminar on Funding Opportunity in Survey Research*. Bureau of Labor Statistics 2 Massachusetts Avenue, NE Washington, DC 20212 June 11, 2001

Burgard JP, Münnich R and Zimmermann T 2014 The impact of sampling designs on small area estimates for business data. *Journal of Official Statistics, accepted*. Volume 30, Issue 4 (Dec 2014)

Costa A, Satorra A and Ventura E 2004 Improving both domain and total area estimation by composition. *Statistics and Operations Research Transactions* **28**(1), 69–86.

Deville JC 1999 Variance estimation for complex statistics and estimators: linearization and residual techniques. In Income and living conditions in Europe Statistical books. Edited by Anthony B. Atkinson and Eric Marlier. Luxembourg: Publications Office of the European Union, 2010.

Elbers C, Lanjouw JO and Lanjouw P 2003 Micro-level estimation of poverty and inequality. *Econometrica* **71**(1), 355–364.

Faraway JJ 2005 *Extending the Linear Model with R: Generalized Linear, Mixed Effects and Nonparametric Regression Models*. CRC Press.

Fay RE and Herriot RA 1979 Estimates of income for small places: An application of James–Stein procedures to census data. *Journal of the American Statistical Association* **74**(366), 269–277.

Foster J, Greer J and Thorbecke E 1984 A class of decomposable poverty measures. *Econometrica: Journal of the Econometric Society*, Vol. 52, No. 3. (May, 1984), 761–766.

Gabler S, Ganninger M and Münnich R 2012 Optimal allocation of the sample size to strata under box constraints. *Metrika* **75**(2), 151–161.

Graf E and Tillé Y 2014 Variance estimation using linearization for poverty and social exclusion indicators. *Catalogue no. 12-001-XWE* in Survey Methodology, Volume 40, Number 1, June 2014, pp 61–80.

Graf M, Wenger A and Nedyalkova D 2011 Description and quality of the user data base. Research Project Report WP5—D5.1, FP7-SSH-2007-217322 AMELI.

Horvitz DG and Thompson DJ 1952 A generalization of sampling without replacement from a finite universe. *Journal of the American Statistical Association* **47**(260), 663–685.

Jiang J and Lahiri P 2001 Empirical best prediction for small area inference with binary data. *Annals of the Institute of Statistical Mathematics* **53**, 217–243.

Jiang J and Lahiri P 2006 Mixed model prediction and small area estimation. *TEST: An Official Journal of the Spanish Society of Statistics and Operations Research* **15**(1), 1–96.

Jiang J, Lahiri P and Wan SM 2002 A unified jackknife theory for empirical best prediction with m-estimation. *The Annals of Statistics* **30**(6), 1782–1810.

Jiang J, Lahiri P, Wan SM and Wu CH 2001 Jackknifing in the Fay–Herriot model with an example. *Proceedings of the Seminar on Funding Opportunity in Survey Research*. Bureau of Labor Statistics 2 Massachusetts Avenue, NE Washington, DC 20212 June 11, 2001, pp 75–97.

Kolb JP, Münnich R, Beil S, Chatziparadeisis A and Seger J 2011 Policy use of indicators on poverty and social exclusion. Research Project Report WP9—D9.1, FP7-SSH-2007-217322 AMELI.

Lehtonen R, Särndal CE and Veijanen A 2003 The effect of model choice in estimation for domains, including small domains. *Survey Methodology* **29**(1), 33–44.

Lehtonen R and Veijanen A 1998 Logistic generalized regression estimators. *Survey Methodology* **24**(1), 51–55.

Lehtonen R and Veijanen A 2009 Design-based methods of estimation for domains and small areas. In *Handbook of Statistics—Sample Surveys: Inference and Analysis* (ed. Rao C) vol. 29, Part 2 of *Handbook of Statistics*. Elsevier, pp. 219–249.

Lepik N 2007 On the bias of the generalized regression estimator in survey sampling. *Acta Applicandae Mathematicae* **97**(1–3), 41–52.

Marker DA 1995 *Small Area Estimation: A Bayesian Perspective*. University of Michigan.

Molina I and Rao JNK 2010 Small area estimation of poverty indicators. *Canadian Journal of Statistics* **38**(3), 369–385.

Münnich R and Burgard JP 2012 On the influence of sampling design on small area estimates. *Journal of the Indian Society of Agricultural Statistics* **66**(1), 145–156.

Myrskylä M 2007 Generalized regression estimation for domain class frequencies. Research Report no. 247. University of Helsinki.

Osier G 2009 Variance estimation for complex indicators of poverty and inequality using linearization techniques. *Survey Research Methods*, **3**, 167–195.

Osier G, Berger YG and Goedeme T 2013 Standard error estimation for the eu-silc indicators of poverty and social exclusion. *Eurostat Methodologies and Working Papers Series*. Populations and social conditions Collection: Methodologies & Working papers, Luxembourg: Publications Office of the European Union, 2013.

Pfeffermann D 2013 New important developments in small area estimation. *Statistical Science* **28**(1), 40–68.

Rao JNK 2003 *Small Area Estimation*. John Wiley & Sons, Inc.

Särndal CE, Swensson B and Wretman J 2003 *Model Assisted Survey Sampling*. Springer.

6

Model-assisted Methods for Small Area Estimation of Poverty Indicators

Risto Lehtonen[1] and Ari Veijanen[2]

[1]*Department of Social Research, University of Helsinki, Helsinki, Finland*

[2]*Statistics Finland, Finland*

6.1 Introduction

6.1.1 General

In this chapter we discuss design-based model-assisted methods for the estimation of indicators on inequality and poverty for population subgroups or domains and small areas. We apply the design-based framework for inference. The design-based approach is widely used in official statistics production. A typical national statistical agency aims at delivering reliable statistics with good accuracy for important finite population characteristics, such as average level of income in the country and sub-regions. Small (or negligible) design bias of design-based estimates is often considered as a benefit, even at the cost of increased variance, when compared with the corresponding model-based estimates. The inferential framework is discussed in Section 6.4.

The Gini index and at-risk-of poverty rate (poverty rate for short) have been chosen to represent examples of important indicators of income inequality and poverty. Poverty rate is one of the components of the so-called combined AROPE indicator (at risk of poverty or social exclusion). The AROPE rate is the key indicator in monitoring the poverty target in the EU 2020 Strategy [1]. The Gini index measures income inequality and is widely used in assessing inequality differences between regions within a country and in country comparisons in international statistics of the United Nations, OECD and World Bank.

Analysis of Poverty Data by Small Area Estimation, First Edition. Edited by Monica Pratesi.

Companion Website: www.wiley.com/go/pratesi/poverty

We use design-based model-assisted methods for the estimation of the Gini index and poverty rate for domains (small or large). In the construction of the estimators we assume access to register-based information covering the target population and an option to link the sample survey data to the register data at the unit level. The motivation of this type of statistical infrastructure is discussed in Section 6.4. The auxiliary information is incorporated in the estimation procedure by using carefully chosen assisting models, such as linear and logistic mixed models. Mixed models offer a flexible tool in modelling regional differences.

For the Gini index we examine the percentile-adjusted predictor method of Veijanen and Lehtonen [2]. In this method, unit-level auxiliary data from registers are effectively incorporated in the estimation procedure by a mixed model, and the predictions from the model are successfully transformed. The method is robust in the sense that distributional assumptions on income such as log-normal are not imposed. For comparison we used a parametric method based on log-normal assumption. A Horvitz–Thompson (HT) estimator is used as a reference.

For poverty rate we consider generalized regression (GREG; [3, 4]) and model calibration (MC; [5, 6]) estimators. In these methods, unit-level auxiliary data are incorporated in the estimation procedure by assisting models chosen from the family of generalized linear mixed models. Both GREG and MC share the property of being robust against model misspecification; this property is independent of the choice of the assisting model. Lehtonen and Veijanen [7] discuss GREG in domain estimation. MC has been developed for domain estimation in Lehtonen, Särndal and Veijanen [8] and Lehtonen and Veijanen [9]. In GREG estimation and MC for poverty rate for domains, our study variable is binary and we employ logistic fixed-effects and mixed models involving domain-level random intercept effects in addition to the fixed effects. Also here, a HT estimator acts as reference.

Statistical properties (design bias and accuracy) of the methods for the Gini index and poverty rate are assessed by design-based simulation experiments. In the construction of the population we use register data maintained by Statistics Finland. The possibility of unit-level linking of the sample data sets with the register data is at our disposal.

The chapter is organized as follows. The estimation of the Gini index is examined in Section 6.2. Section 6.3 discusses the estimation of poverty rate. Discussion is in Section 6.4.

6.1.2 Concepts and Notation

The type of domain structure (planned or unplanned) underlying the data and the type of the estimator (direct or indirect) of domain parameters are two concepts relevant in a typical small area estimation situation. The domain structure is of *planned* type if the domains of interest are defined as strata in the sampling design [10]. It is often favourable to define the important domains as strata and apply suitable allocation techniques to control the domain sample sizes. Singh, Gambino and Mantel [11] introduced allocation strategies in order to attain reasonable accuracy for small domains, still retaining good accuracy for large domains. Falorsi, Orsini and Righi [12] proposed sample balancing and coordination techniques for situations where the number of strata is very large.

In a more common situation in small area estimation practice, the domain structure cuts across the stratum structure. In this case, the domains are called *unplanned*. For unplanned domains, a single sample is drawn from the entire population. The domain sample sizes are not under control and thus are random. Variance estimators that account for the extra variation due to randomness are suggested in Särndal, Swensson and Wretman [3] and Lehtonen and Veijanen [7]. In this chapter we deal with unplanned domain structures, where a single equal or unequal probability sample is drawn from the population.

A *direct* estimator of a domain parameter uses values of the variable of interest only from the time period of interest and only from units in the domain of interest. An *indirect* estimator uses values of the variable of interest from a domain and/or time period other than the domain and time period of interest [13]. A HT estimator and a GREG estimator (whose assisting model is fitted separately for each domain) are typical examples of direct estimators. A direct domain estimator can still incorporate auxiliary data outside the domain of interest; this is the case for example for a model-assisted estimator if the auxiliary data only are available at a higher aggregate level. In small area estimation literature, indirect estimators are said to "borrow strength" from other domains and/or from earlier time points.

Direct estimators are often used for planned domain structures because too small domain sample sizes can be avoided. Indirect estimators are important in small area estimation practice, where unplanned domains are common. Indirect model-assisted estimators are discussed widely in the literature (e.g. [7, 14–19]). In this chapter we concentrate in most cases on indirect estimators. Direct estimators are used as references.

Our basic notation is as follows: the fixed and finite population of interest is denoted $U = \{1, \ldots, k, \ldots, N\}$, where k refers to the label of population element and N is the size of the population. A *domain* U_d, $d = 1, \ldots, D$, is a subset of the population U, and D is the number of domains in the population. In practice, regional sub-populations and socio-demographic breakdown within regional areas constitute typical domains. Poverty rate estimates, for example, are required not only for regions but also for classes within regions, defined by age group and gender. For this case, consider a region r and a class c. They define a domain d: in population U, a subset $U_d = U_r \cap U_c$ contains people belonging to class c (U_c) in region r (U_r). In typical small area estimation situations, the number D of domains can be fairly large. The number of population units in a domain is denoted by N_d. In sample $s \subset U$, corresponding subsets for unplanned domain structure are defined as $s_d = s \cap U_d$ with n_d observations, and $\sum_{d=1}^{D} n_d = n$, the sample size. A small area is a domain whose realized sample size n_d is small. Inclusion probabilities are denoted π_k, $k = 1, \ldots, N$; their inverses $a_k = 1/\pi_k$ are the design weights. Second-order inclusion probabilities are denoted π_{kl}, and $a_{kl} = 1/\pi_{kl}$.

Many poverty indicators are composed of domain totals, frequencies and medians. The domain total of the study variable y (e.g. equivalized incomes) is defined as $t_d = \sum_{k \in U_d} y_k$, where y_k denotes the value of the study variable for element k. The frequency f_d of a class C, such as the frequency of low-income persons with income smaller than a threshold, is written as a sum of class indicators $v_k = I\{y_k \in C\}$ as $f_d = \sum_{k \in U_d} v_k$.

6.2 Design-based Estimation of Gini Index for Domains

6.2.1 Estimators

6.2.1.1 Direct Estimator

To define the Gini index we first define the *Lorenz curve* in a population ordered by equivalized income from the poorest to the richest. In this population, the curve plots the cumulative share of total income of people against their numerical proportion. We present a definition that is equivalent with equation (4) in Langel and Tillé [20]. Consider a population domain U_d of size N_d where the equivalized incomes are ordered: $y_{(1)} \leq y_{(2)} \leq \dots \leq y_{(N_d)}$. The Lorenz curve $L_d(.)$ in domain d is defined at the following points for persons $k \in U_d$:

$$L_d\left(\frac{k}{N_d}\right) = \frac{\sum\limits_{i \leq k; i \in U_d} y_{(i)}}{\sum\limits_{t \in U_d} y_t}. \tag{6.1}$$

The x-coordinate represents the first k persons' numerical proportion of the population and the y-coordinate represents their share of the total income. For practical purposes, we define the Lorenz curve as a piecewise linear function, approximated by a line between consecutive points. If all incomes were equal, the curve would be a line from $(0, 0)$ to $(1, 1)$. In real data, the Lorenz curve lies below this line.

The *Gini index* G_d in domain d is defined as:

$$G_d = 1 - 2\int_0^1 L_d(x)dx.$$

With equal incomes, $G_d = 0$. Typical values for a country range from 0.2 to 0.4.

To estimate the Lorenz curve in a sample domain s_d, we first establish points corresponding to points (6.1) in the population. Their x-coordinates are estimated numerical proportions, calculated using the design weights a_k, as each sample unit 'represents' a_k units in the population. The y-coordinates, shares of the total income, are estimated as ratios of weighted sums of equivalized incomes. The persons in the sample domain are first ordered by equivalized income: $y_{(1)} \leq y_{(2)} \leq \dots \leq y_{(n_d)}$. The weights are correspondingly ordered by the income; the design weight of the observation at the ith position is denoted by a_i^s. For the k first persons in the ordered sample, HT-type estimates define a point through which the Lorenz curve passes:

$$L_{HT;d}\left(\frac{\sum\limits_{i \leq k; i \in s_d} a_i^s}{\sum\limits_{t \in s_d} a_t}\right) = \frac{\sum\limits_{i \leq k; i \in s_d} a_i^s y_{(i)}}{\sum\limits_{t \in s_d} a_t y_t}.$$

Consecutive points are joined by a line. The estimator of the Gini index in the domain d is then defined as:

$$\widehat{G}_{HT;d} = 1 - 2\int_0^1 L_{HT;d}(x)dx, \quad d = 1, \dots, D. \tag{6.2}$$

It can be proved that this definition agrees with other definitions, such as equations (6)–(8) in Langel and Tillé [20]. The estimator (6.2) is of direct type, as each domain estimate is determined by observations in the domain only.

In order to estimate the variance of an estimator, Langel and Tillé [20] approximate the true Gini index by a sum of linearized variables z_k and the estimator by a weighted sum of variables $\widehat{z}_k$, derived from the influence function. For estimation in domain d,

$$\widehat{z}_k = \frac{1}{\widehat{N}_d\widehat{Y}_d}\left[2\widehat{N}_k\left(y_k - \widehat{\bar{Y}}_k\right) + \widehat{Y}_d - \widehat{N}_d y_k - G_{HT;d}\left(\widehat{Y}_d + \widehat{N}_d y_k\right)\right] \qquad (k \in s_d),$$

where $\widehat{N}_d = \sum_{k\in s_d} a_k$ is the domain sum of design weights, $\widehat{Y}_d = \sum_{k\in s_d} a_k y_k$ is the HT estimator of domain total $Y_d = \sum_{k\in U_d} y_k$ of income,

$$\hat{\widehat{Y}}_k = \frac{\sum\limits_{i\leq k; i\in s_d} a_i^s y_{(i)}}{\widehat{N}_d}, \text{ and } \widehat{N}_k = \sum_{i\leq k; i\in s_d} a_i^s.$$

The difference between the estimator and the true value is approximated by:

$$\widehat{G}_{HT,d} - G_d \approx \sum_{k\in s_d} a_k \widehat{z}_k - \sum_{k\in U_d} z_k,$$

which shows that the variance of $\widehat{G}_{HT,d}$ can be approximated by estimating the variance of the domain total estimator $\sum_{k\in s_d} a_k \widehat{z}_k$. This is treated as an HT estimator, for which simple variance estimators are available [7]. When the domains are of planned type, the variance is estimated by:

$$\widehat{V}_1 = \sum_{k\in s_d} \sum_{l\in s_d} (a_k a_l - a_{kl})\, \widehat{z}_k \widehat{z}_l, \tag{6.3}$$

where the a_{kl} are second-order inclusion probabilities. These are difficult to calculate under some sampling designs. This is avoided for example by an approximation presented in Berger [21, 22]:

$$\widehat{V}_2 = \sum_{k\in s_d} (1-\pi_k)\, e_k^2; \quad e_k = \widehat{z}_k - \frac{\sum\limits_{i\in s_d} (1-\pi_i)\widehat{z}_i}{\sum\limits_{i\in s_d} (1-\pi_i)}. \tag{6.4}$$

In the case of unplanned domains, the randomness of the domain size introduces more variability, which is taken into account by so-called extended variables ([7], p. 223). From $\widehat{z}_k$ we derive domain-specific extended variables $\widehat{z}_{dk} = I\{k \in s_d\}\widehat{z}_k$ $(d = 1, \ldots, D)$. In an unplanned type SRSWOR (simple random sampling without replacement) sample of size n, the variance of the Gini index estimator is estimated by:

$$\widehat{V}_3 = \frac{n}{n-1} \sum_{k\in s} \left(a_k \widehat{z}_{dk} - \frac{1}{n} \sum_{i\in s_d} a_i \widehat{z}_i \right)^2. \tag{6.5}$$

6.2.1.2 Percentile-adjusted Predictor

The distribution of equivalized income is sometimes approximated by log-normal distribution. It is known [23] that the Gini index of log-normal distribution in $N(\mu, \sigma^2)$ is:

$$G_{\ln N} = 2\Phi\left(\frac{\sigma}{\sqrt{2}}\right) - 1. \tag{6.6}$$

It is seemingly straightforward to estimate the Gini index by plugging the estimated parameter σ into (6.6). The parameter σ^2 is estimated by the design-based sample variance of logarithms of equivalized incomes $\delta_k = \log(y_k + 1)$ in domain d:

$$\widehat{G}_{d;\ln N} = 2\Phi\left(\frac{\widehat{\sigma}_d}{\sqrt{2}}\right) - 1; \quad \widehat{\sigma}_d^2 = \frac{\sum_{k\in s_d} a_k(\delta_k - \overline{\delta}_d)^2}{\sum_{k\in s_d} a_k}; \quad \overline{\delta}_d = \frac{\sum_{k\in s_d} a_k\delta_k}{\sum_{k\in s_d} a_k}. \tag{6.7}$$

The estimator (6.7) is of direct type. However, deviations from the assumed distribution may result in biased estimates of the Gini index. It is necessary to apply more flexible distributions, such as the generalized beta distribution of the second kind (e.g. [24]).

Strict assumptions on the income distribution are relaxed in the percentile-adjusted predictor method of Veijanen and Lehtonen [2]. In order to use auxiliary information, we use predictions from a model. However, as the predictions typically vary much less than the true values, the income distribution estimated from the predictions is unrealistically uniform, resulting in too small Gini estimates. This problem is addressed in Veijanen and Lehtonen [2] by transforming the predictions to have distribution that is closer to the distribution of sample observations y_k. The difference between the distributions is assessed by the average squared difference between certain percentiles of log-transformed predictions and observations. The adjusted predictions $\widetilde{y}_k$ in domain d are derived from the original predictions $\widehat{y}_k$ as $\log(\widetilde{y}_k + 1) = a_d + b\log(\widehat{y}_k + 1)$ $(k \in U_d)$, where a_d is specific to the domain and b is common to all domains. The adjusted predictor of the Gini index is obtained by substituting the adjusted predictions $\widetilde{y}_k$ for the y_k in (6.1):

$$\widehat{G}_{pred;d} = 1 - 2\int_0^1 \widetilde{L}_d(x)\,dx; \quad \widetilde{L}_d\left(\frac{k}{N_d}\right) = \frac{\sum_{i\leq k; i\in U_d} \widetilde{y}_{(i)}}{\sum_{t\in U_d} \widetilde{y}_t}. \tag{6.8}$$

For income data, it is sensible to fit a model to logarithms of observations, but the distribution does not have to be exactly specified. As all observations contribute to the model of percentiles, the percentile-adjusted predictor for a domain is of indirect type and borrows strength from other domains. Its mean squared error (MSE) is probably best estimated by resampling methods such as bootstrap.

The percentile-adjusted predictor provides essentially a practical and robust method for estimating a poverty indicator. From a theoretical point of view, a more rigorous approach is offered by M-quantile small area modelling presented in Chambers and Tzavidis [25]. Design weights can be incorporated in the fitting of an M-quantile model. An application of the M-quantile method has been proposed in Marchetti, Tzavidis and Pratesi [26] for the estimation of poverty rate and poverty gap and in Marchetti [27] for the Gini index. The estimation of the Gini index by M-quantile-based small area methodology is discussed in Chapter 9.

6.2.2 *Simulation Experiments*

Statistical properties (design bias and accuracy) were examined by simulation experiments. We constructed a unit-level population based on register data from Statistics Finland. The population consisted of 1 million persons in 36 NUTS 4 regions in Western Finland.

The equivalized income, age class (0–15, 16–24, 25–49, 50–64, or at least 65 years) and gender were obtained from registers, whereas the labour force status and the socio-economic status were obtained from a household survey for the household head and imputed for the other members of each household. In the simulations, $K = 1000$ unequal probability samples of $n = 2000$ were drawn from the population. For unequal probability sampling, an artificial size variable depended on the socio-economic status: in classes 1–6, the value of size variable was inversely proportional to 2, 3, 5, 1, 1 and 1, respectively. We calculated domain estimates for the $D = 36$ NUTS 4 regions. They were classified by expected domain sample size into three classes (11–25, 25–50, or at least 50 units).

The quality of an estimator $\widehat{\theta}_d$ of θ_d ($d = 1, \ldots, D$) over samples s_k ($k = 1, 2, \ldots, 1000$) was assessed by absolute relative bias (ARB) and relative root mean squared error (RRMSE):

$$\text{ARB}(\widehat{\theta}_d) = \frac{\left| \frac{1}{1000} \sum_{k=1}^{1000} \widehat{\theta}_d(s_k) - \theta_d \right|}{\theta_d}; \quad \text{RRMSE}(\widehat{\theta}_d) = \frac{\sqrt{\sum_{k=1}^{1000} (\widehat{\theta}_d(s_k) - \theta_d)^2}}{\theta_d}.$$

The HT-type estimator of the Gini index (6.2) performed relatively well (Table 6.1). As expected, it had small bias but in small domains, especially, its RRMSE was large. The true distribution of equivalized income differed slightly from the log-normal distribution, having fewer rich people than could be expected. Therefore, the Gini indices calculated under the assumption of log-normality by (6.7) were, on average, about 20% too large. For the adjusted predictor, the predictions were obtained from a mixed model fitted to log-transformed equivalized income. The model incorporated domain (NUTS 4) level random effects and class indicators corresponding to all the auxiliary variables listed above and, moreover, interactions of age class with gender. The adjusted predictions were constructed by calculating n_d, or at most 100, percentiles at points evenly distributed over (0,100) in each domain.

The ordinary predictor was inferior to the HT-type estimator, having negative bias of about 45%. The percentile-adjusted predictor yielded much better results. It appeared to be robust against model misspecification, as mixed models with or without logarithm transformation of the response yielded similar results.

Table 6.1 Mean absolute relative bias (ARB) and mean relative root mean squared error (RRMSE) of estimators of the Gini index by domain size class in 1000 unequal probability samples

Estimator	Mean ARB (%)				Mean RRMSE (%)			
	Expected domain sample size				Expected domain sample size			
	11–25	25–50	50–	All	11–25	25–50	50–	All
Direct estimators								
HT-type (6.2)	8.9	4.5	1.8	5.0	22.8	18.1	11.6	17.6
Log-normal (6.7)	10.1	14.1	35.3	18.9	59.8	52.4	64.0	57.7
Indirect estimator								
Adjusted predictor (6.8)	3.9	2.8	1.7	2.8	6.5	5.7	4.7	5.7

6.2.3 Empirical Application

In a numerical example, we used the population data set described in Section 6.2.2. A SRSWOR sample of $n = 2000$ was drawn. The domains ($D = 36$ NUTS 4 regions) were of unplanned type, that is, the domains were not included in the sampling design. Therefore, the domain sample sizes were random. The smallest regions had 14 persons in the sample, the largest had 323 persons.

Consider first the HT-type direct estimator $\widehat{G}_{HT;d}$ (6.2). Due to the small domain sizes, the estimated Gini indices ranged from 0.17 to 0.42, whereas the true values varied less (from 0.215 to 0.255). The estimated Gini indices were not strongly correlated with the true values ($r = 0.33$). Nevertheless, the estimates were accurate enough to be useful when looking for correlations with known properties of regions. Both the estimates and the true values were positively correlated with the region's population size ($r = 0.43$ and $r = 0.45$, respectively). Equation (6.3) was applied for variance estimation. The estimated standard errors of the estimators ranged from 0.03 to 0.23, with typical values close to 0.04, that is, about 20% of the Gini index. Approximate 95% confidence intervals involving the standard errors covered the true value in 86% of the regions. This implies that the estimated standard errors are too small. Moreover, in a simulation experiment incorporating 1000 SRSWOR samples from the population, the estimated standard errors were, on average, 15% smaller than the Monte Carlo standard deviations of the point estimates.

The absolute relative error (ARE) of $\widehat{G}_{HT;d}$ in domain d is defined as:

$$\text{ARE}(\widehat{G}_{HT;d}) = \frac{|\widehat{G}_{HT;d} - G_d|}{G_d}$$

and their mean over a size class of domains is called the mean absolute relative error (MARE). The mean coefficient of variation (MCV) is the mean of coefficients of variation (CVs) over a domain size class. Table 6.2 shows the MARE and MCV of $\widehat{G}_{HT;d}$ over three size classes of domains. As expected, small sample size resulted in large errors and large standard error.

The adjusted predictor had stronger correlations with the true value ($r = 0.55$) and the domain population size ($r = 0.64$) than the direct estimator. In contrast with the direct estimator, the MARE of the predictor increased with domain size (Table 6.2). However, in the

Table 6.2 Mean absolute relative error (MARE) and mean coefficient of variation (MCV) of estimators of the Gini index over domain classes defined by sample size

Estimator	Quality indicators	Domain sample size		
		14–24	25–49	50–323
Direct estimator				
HT-type (6.2)	MARE (%)	19.6	13.5	11.2
	MCV (%)	18.6	13.2	10.0
Indirect estimator				
Adjusted predictor (6.8)	MARE (%)	11.9	13.7	15.1
	MCV (%)	4.8	4.6	4.5

simulation experiment (Table 6.1), the RRMSE of the adjusted predictor was smallest in the largest domains.

The standard error of the adjusted predictor was estimated by bootstrap as follows. One hundred bootstrap populations were constructed from the mixed model fitted to the sample: the random effect of a region was chosen randomly among the random effects in the fitted model and each error term was selected by simple random sampling, with replacement, from the residuals of the model. From each bootstrap population, a single bootstrap sample was drawn with SRSWOR. The standard error of the adjusted predictor in a domain was estimated by the standard deviation of domain estimates in the bootstrap samples. The MCV of the predictor was smaller than that of the direct estimator. However, a small simulation experiment showed that the standard errors have substantial negative bias, about 30% in all domain size classes.

6.3 Model-assisted Estimation of At-risk-of Poverty Rate

6.3.1 Assisting Models in GREG and Model Calibration

6.3.1.1 Logistic Models

In the estimation of the Gini index, it is difficult to benefit from a statistical model, as the statistics describes the shape of the income distribution. In the case of poverty rate, in contrast, the estimation of the proportions of poor people can be interpreted as a problem of estimating the domain means of an indicator variable. Therefore, model-assisted methods such as ordinary GREG, logistic GREG (LGREG; [28]) and calibration are readily applicable.

The available auxiliary information consists of an auxiliary **x**-vector and a domain membership indicator $I_{dk} = 1$ if $k \in U_d$, and $I_{dk} = 0$ otherwise, $d = 1, 2, \ldots, D$, for every population unit $k \in U$. Letting $\mathbf{x}_k = (1, x_{1k}, \ldots, x_{Jk})'$ denote the value of the auxiliary vector for unit k, we thus assume that both $\mathbf{x}_k$ and domain membership I_{dk} is known for every $k \in U$.

The poverty indicator shows when a person's equivalized income is smaller than or equal to the poverty threshold, 60% of the median equivalized income M in the population, according to the EU definition. To estimate the median income, we first derive the HT estimator of the distribution function of equivalized income in the population. The distribution function of y is:

$$F_U(t) = \frac{1}{N} \sum_{k \in U} I\{y_k \leq t\}.$$

This is estimated by HT:

$$\hat{F}_U(t) = \frac{1}{\hat{N}} \sum_{k \in s} a_k I\{y_k \leq t\},$$

where $\hat{N} = \sum_{k \in s} a_k$. Median equivalized income $\hat{M}$ is obtained from $\hat{F}_U$ as the smallest y_k ($k \in s$) for which $\hat{F}_U(y_k) > 0.5$. In the special case of $\hat{F}_U(y_{(k)}) = 0.5$ for the kth observation in sorted data, the median is the average of $y_{(k)}$ and $y_{(k+1)}$.

For the 0–1-valued poverty indicator $v_k = I\{y_k \leq 0.6\hat{M}\}$, a natural model describes how the probability of value 1 ('in poverty') depends on auxiliary variables. This setting excludes ordinary linear fixed-effects regression models. A logistic fixed-effects regression model is defined by:

$$p_k = P\{v_k = 1\} = \frac{\exp(\mathbf{x}_k'\boldsymbol{\beta})}{1+\exp(\mathbf{x}_k'\boldsymbol{\beta})},$$

where $\boldsymbol{\beta} = (\beta_0, \beta_1, \beta_2, \ldots, \beta_J)'$ is the vector of regression parameters and $\mathbf{x}_k = (1, x_{1k}, x_{2k}, \ldots, x_{Jk})'$. In order to have a well fitting model, we may be tempted to define domain-specific parameters $\boldsymbol{\beta}_d$ for each domain: $\boldsymbol{\beta} = (\boldsymbol{\beta}_1, \boldsymbol{\beta}_2, \ldots, \boldsymbol{\beta}_D)$. A corresponding change is made in the vectors of auxiliary variables: if the vectors of original auxiliary variables are $\mathbf{x}_{0k} = (x_{1k}, x_{2k}, \ldots, x_{Jk})'$, the new model incorporates vectors $\mathbf{x}_k = (I_{1k}x_{1k}, I_{1k}x_{2k}, \ldots, I_{1k}x_{Jk}, I_{2k}x_{1k}, I_{2k}x_{2k}, \ldots, I_{Dk}x_{Jk})'$, which contain domain membership indicators. Then the model is essentially fitted separately in every domain. However, in small domains lack of sufficient sample data may yield unstable estimates of the parameters. On the other hand, a domain-specific model might be realistic for domains with a large sample size. To describe differences between domains, it often suffices to define domain-specific fixed-effects intercepts by including the domain membership indicators I_{dk} in the model. For this model, $\mathbf{x}_k = (I_{1k}, I_{2k}, \ldots, I_{Dk}, x_{1k}, x_{2k}, \ldots, x_{Jk})'$ and $\boldsymbol{\beta} = (\beta_{01}, \beta_{02}, \ldots, \beta_{0D}, \beta_1, \beta_2, \ldots, \beta_J)'$. In model-based methods, notably empirical best linear unbiased prediction (EBLUP; e.g. [29]), a more common approach is to treat the domain intercepts as random variables, 'random effects' u_d, generated by the superpopulation model from a common distribution $N(0, \sigma_u^2)$. A logistic mixed model incorporates these random effects:

$$p_k^{(m)} = P\{v_k = 1 | u_d\} = \frac{\exp(\mathbf{x}_k'\boldsymbol{\beta} + u_d)}{1+\exp(\mathbf{x}_k'\boldsymbol{\beta} + u_d)} \qquad (k \in U_d,\ d = 1, \ldots, D),$$

where $\mathbf{x}_k = (1, x_{1k}, x_{2k}, \ldots, x_{Jk})'$ and $\boldsymbol{\beta} = (\beta_0, \beta_1, \beta_2, \ldots, \beta_J)'$. The parameters $\boldsymbol{\beta}$ and σ_u^2 are first estimated from the sample using ML, and the random effects are predicted for each domain, as they are not interpreted as parameters of the model. For logistic regression we used R functions `glm` and `glmer`. A mixed model is more parsimonious than the fixed-effects model. Due to the imposed common distribution of the random effects, the differences between random effects are often smaller than the differences between fixed domain-specific intercepts, which may vary erratically in small domains. Mixed models are discussed for example in Demidenko [30].

6.3.1.2 Accounting for Sampling Complexities

In an ideal case, design-based model fitting incorporates design weights in functions fundamental to estimation. For example, a pseudo maximum likelihood estimator maximizes the pseudo likelihood function

$$\log\widehat{f}(\boldsymbol{\beta}) = \sum_{k\in s} a_k \log f(y_k; \boldsymbol{\beta}),$$

which is interpreted as an HT estimator of the population-level likelihood function that would be maximized if complete information about the population units was available. Another example is weighted least squares, where the squared errors are weighted by design weights. Again, this is an HT-type estimator of a function that would be minimized by least squares in population data. In practice, design weights can be included in this fashion in certain R functions, such as `lm` (linear fixed-effects models) and `glm` (logistic fixed-effects model). In fitting a mixed model, it is more difficult to include design weights. If the model is fitted without the design weights, a problem known as informative sampling [31] may arise: the study variable correlates with the sample inclusion indicator even given the auxiliary variables. Then the sample observations do not follow the same model as units in the population, leading to bias at least in the estimated model parameters. This problem may be avoided or alleviated by introducing new auxiliary variables associated with the sampling design. The design variables that determine the sampling design, such as the size variable in PPS or the stratum variable indicators in stratified sampling, can be included in the model as auxiliary variables. If this is not feasible, it has been suggested in Pfeffermann and Sverchkov ([31], p. 463) that the design weights are included in the model, or preferably the inclusion probabilities. On the other hand, GREG estimators are somewhat robust in the case of model misspecification.

6.3.2 Generalized Regression Estimation

GREG estimation (linear GREG) usually incorporates auxiliary information by linear fixed-effects models. The GREG approach has been generalized to include logistic fixed-effects models (LGREG; [28, 32]) and logistic mixed models (MLGREG; [15, 16]). The LGREG estimator of the number of poor people in domain d consists of the population domain sum of estimated probabilities $\widehat{p}_k$ and a weighted sample domain sum of residuals:

$$\widehat{t}_{d;LGREG} = \sum_{k \in U_d} \widehat{p}_k + \sum_{k \in s_d} a_k (v_k - \widehat{p}_k), \quad d = 1, \dots, D. \tag{6.9}$$

The poverty rate is then estimated as a mean whose denominator is either the known domain size N_d or the estimated domain size, the domain sum of design weights, $\widehat{N}_d = \sum_{k \in s_d} a_k$:

$$\widehat{r}_d = \frac{\widehat{t}_{d;LGREG}}{N_d}, \quad d = 1, \dots, D \tag{6.10}$$

or

$$\widetilde{r}_d = \frac{\widehat{t}_{d;LGREG}}{\widehat{N}_d}, \quad d = 1, \dots, D. \tag{6.11}$$

Variance estimation of GREG assisted by a logistic model is based on similar equations as in the case of a linear assisting model. The variance of (6.9) is estimated by:

$$\widehat{V}(\widehat{t}_{d;LGREG}) = \sum_{k \in s_d} \sum_{l \in s_d} (a_k a_l - a_{kl}) e_k e_l; \quad (e_k = v_k - \widehat{p}_k). \tag{6.12}$$

This variance estimator is for planned-type domains. In the unplanned domains case, we use the variance formula (5) of Lehtonen and Veijanen [7], applied to extended residuals $e_{dk} = I\{k \in s_d\} e_k$ $(d = 1, \dots, D)$. The variance estimator is given by:

$$\widehat{V}(\widehat{t}_{d;LGREG}) = \frac{n}{n-1} \sum_{k \in s} \left(a_k e_{dk} - \sum_{k \in s} a_k e_{dk} / n \right)^2$$

The variance estimator of the domain mean estimator (6.10) is simply:

$$\widehat{V}(\widehat{r}_d) = \frac{\widehat{V}(\widehat{t}_{d;LGREG})}{N_d^2}. \tag{6.13}$$

The variance estimator of (6.11) takes into account that the denominator $\widehat{N}_d$ is a random variable. The mean is interpreted as a ratio of two GREG estimators, and the following equation from section 4.3.1 of Lehtonen and Veijanen [7] is applied:

$$\widehat{V}(\widehat{R}_{dGREG}) = \frac{1}{\widehat{t}^2_{dzGREG}} \sum_{k\in s}\sum_{l\in s}(a_k a_l - a_{kl}) g_{dk}(e_{yk} - \widehat{R}_{dGREG} e_{zk}) g_{dl}(e_{yl} - \widehat{R}_{dGREG} e_{zl}),$$

where $\widehat{R}_d = \widehat{t}_{dy}/\widehat{t}_{dz}$ is the ratio, $\widehat{t}^2_{dzGREG}$ is the square of the denominator, g_{dk} are the g-weights, e_{yk} is the residual from the model for the nominator and e_{zk} is the corresponding residual for the denominator. In (6.11), the denominator is a domain sum of design weights, which is interpreted as a GREG estimator without auxiliary information, and we apply $e_{zk} = I_{dk}$. The g-weights are also rewritten as domain indicators, so the variance estimator of (6.11) has the form:

$$V(\widetilde{r}_d) = \frac{1}{\widehat{N}_d^2} \sum_{k\in s_d}\sum_{l\in s_d}(a_k a_l - a_{kl})(e_k - \widetilde{r}_d)(e_l - \widetilde{r}_d). \tag{6.14}$$

When the samples are drawn by SRSWOR, the equations can be simplified, as the second-order inclusion probabilities are tractable. If the domains are planned and SRSWOR is applied separately in each domain, the inclusion probabilities may depend on the domain. For other planned-type designs with varying inclusion probabilities, an approximation is constructed using Berger's equation (6.4). The terms $\widehat{z}_k$ therein are replaced by residuals e_k when approximating (6.12) and by $e_k - \widetilde{r}_d$ when approximating (6.14).

6.3.3 Model Calibration Estimation

Ordinary (model-free) calibration involves construction of weights satisfying *calibration equations*: the weighted sum of each auxiliary variable coincides with the known population sum of the variable. Design bias is reduced by demanding that the calibrated weights are as close to the design weights as possible. The calibrated weights are then applied to study variables. The reasoning behind this idea is that if the weights reproduce the known totals of auxiliary variables, they might yield estimates close to the true totals of the study variable as well. This can happen if the relationship of the study variable and the auxiliary variables is strong. Furthermore, calibration equations are often important in official statistics production for the sake of consistency with published statistics. Calibration described in this way is equivalent with GREG estimation assisted by a linear fixed-effects model. Therefore, ordinary calibration is not necessarily suitable for estimation of a total of an indicator variable, such as the number of poor people.

MC was introduced by Wu and Sitter [5] to combine flexible modelling with calibration. In MC, predictions are used instead of auxiliary variables: the weighted sum of fitted values must equal the known population sum of predictions. With non-linear models, this requires access to unit-level information about the population. Moreover, the sum of weights must agree with the known population size. Thus, the calibration equations are defined using vectors $\mathbf{u}_k = (1, \widehat{p}_k)$ under the logistic fixed-effects model or $\mathbf{u}_k = (1, \widehat{p}_k^{(m)})$ under the logistic mixed model.

For the logistic fixed-effects model, the calibration equation is:

$$\sum_{i \in s} w_i \mathbf{u}_i = \sum_{i \in U} \mathbf{u}_i = \left(N, \sum_{i \in U} \widehat{p}_i \right).$$

The calibrated weights minimize a function

$$\sum_{i \in s} \frac{(w_i - a_i)^2}{a_i} - \boldsymbol{\lambda}' \left(\sum_{i \in s} w_i \mathbf{u}_i - \sum_{i \in U} \mathbf{u}_i \right),$$

where $\boldsymbol{\lambda}$ is the Lagrange coefficient. The solution is:

$$w_k = a_k (1 + \boldsymbol{\lambda}' \mathbf{u}_k); \quad \boldsymbol{\lambda}' = \left(\sum_{i \in U} \mathbf{u}_i - \sum_{i \in s} a_i \mathbf{u}_i \right)' \left(\sum_{i \in s} a_i \mathbf{u}_i \mathbf{u}_i' \right)^{-1}. \tag{6.15}$$

These weights are applied in each domain to calculate the weighted domain totals of the poverty indicator, and domain means are calculated using the known domain size.

MC is generalized in Lehtonen and Veijanen [9] by defining calibration equations at different regional levels. We have called (6.15) population-level calibration. In domain-level calibration, the weighted domain sum of fitted values in the sample has to agree with the known domain sum of predictions in the population:

$$\sum_{i \in s_d} w_i \mathbf{u}_i = \sum_{i \in U_d} \mathbf{u}_i = \left(N_d, \sum_{i \in U_d} \widehat{p}_i \right).$$

In a 'semi-indirect' spatial method, calibration takes into account the neighbouring regions of a domain: the weighted sample sum of fitted values over a set h_d of units in the neighbourhood, including the domain and its neighbours, must equal the population domain sum of predictions. Simultaneously, the calibrated weights must be close to zero outside the domain and close to the design weights within the domain. The calibration weights ω_i minimize

$$\sum_{i \in h_d} \frac{(\omega_i - I_{di} a_i)^2}{a_i} - \boldsymbol{\lambda}' \left(\sum_{i \in h_d} \omega_i \mathbf{u}_i - \sum_{i \in U_d} \mathbf{u}_i \right).$$

The calibrated weights are defined in h_d:

$$\omega_k = a_k I_{dk} + a_k \boldsymbol{\lambda}' \mathbf{u}_k; \quad \boldsymbol{\lambda}' = \left(\sum_{i \in U_d} \mathbf{u}_i - \sum_{i \in s_d} a_i \mathbf{u}_i \right)' \left(\sum_{i \in h_d} a_i \mathbf{u}_i \mathbf{u}_i' \right)^{-1}.$$

The inclusion of neighbouring regions may improve accuracy, as spatial correlations of poverty rates are possibly significant. Similar derivations hold for the $\widehat{p}_k^{(m)}$ under the logistic mixed model.

The poverty rate for domain d is estimated by:

$$\widehat{r}_d^{MC} = \frac{\sum_{i \in h_d} \omega_i v_i}{N_d}, \quad d = 1, \ldots, D. \tag{6.16}$$

6.3.4 Simulation Experiments

In simulation experiments, we used the population described in Section 6.2.2. A thousand samples of size $n = 1000$ were drawn by SRSWOR. We compared GREG and MC estimators with a direct HT-type method that does not incorporate auxiliary information. The HT-CDF (cumulative distribution function) estimator of poverty rate is based on estimating the distribution function in a domain,

$$F_d(t) = \frac{1}{N_d} \sum_{k \in U_d} I\{y_k \leq t\}.$$

Distribution function is estimated by an estimator

$$\widehat{F}_d(t) = \frac{1}{\widehat{N}_d} \sum_{k \in s_d} a_k I\{y_k \leq t\}.$$

The poverty rate is then estimated by:

$$\widehat{r}_{d;HT} = \widehat{F}_d(0.6\widehat{M}), \ d = 1, \ldots, D. \tag{6.17}$$

To study the effect of model type, we fitted three types of models: (a) fixed-effects logistic common model without domain-specific terms; (b) fixed-effects logistic model with NUTS 4 intercepts; and (c) logistic mixed model with random intercepts associated with the NUTS 4 domains. As in the experiments with the Gini index, these models included class indicators corresponding to main effects and interactions of age class with gender, labour force status and the socio-economic status. GREG estimators (6.10) and MC estimators (6.16) were used with models (a), (b) and (c). The MC estimator was of semi-indirect type. The direct HT-type estimator (6.17) serves as a reference.

All estimators were nearly design unbiased as expected (Table 6.3). This property also holds for the model-assisted estimators, irrespective of model choice. Model choice had larger effect

Table 6.3 Absolute relative bias (ARB) and relative root men squared error (RRMSE) of estimators of poverty rate in an experiment of 1000 SRSWOR samples

	Estimator	ARB (%)			RRMSE (%)		
		Expected domain sample size			Expected domain sample size		
		5–12	12–25	25–151	5–12	12–25	25–151
Direct estimator	HT (6.17)	1.7	2.2	0.9	83.7	60.1	38.9
Indirect estimators							
Assisting models							
(a) Fixed-effects model with common	LGREG (6.10)	2.1	1.7	0.9	72.8	55.5	37.1
intercept for all domains	MC (6.16)	2.0	1.9	1.1	72.5	55.3	37.0
(b) Fixed-effects model with	LGREG (6.10)	1.8	1.9	0.9	83.7	59.9	38.5
domain-specific intercepts	MC (6.16)	1.6	2.1	0.9	83.9	59.6	38.3
(c) Mixed model with domain-specific	MLGREG (6.10)	2.0	1.8	0.9	72.4	55.0	36.8
random intercepts	MC (6.16)	1.9	1.8	0.8	72.1	54.8	36.9

on RRMSE. Fixed-effects model with domain-specific intercepts did not yield good results with the model-assisted LGREG estimator and MC (6.16). The best results were obtained with the mixed model, and MC yielded slightly more accurate estimates than the GREG estimator assisted by MLGREG.

6.3.5 *Empirical Example*

We calculated HT-CDF, MLGREG and MC estimates from the same SRSWOR sample ($n = 2000$) as in Section 6.2.3. The poverty threshold was 6482 euros, and the overall poverty rate in the sample was 14.5%. In the smallest domains, the true poverty rate varied more, from 10 to 22%, than in the five largest domains, where the proportion of poor people ranged from 11.4 to 13.7%. In the population, the poverty rate was negatively correlated with the degree of urbanization in a region ($r = -0.53$), that is, the average poverty rate was larger in rural areas than in cities. The logistic mixed model fitted to the data did not contain strong explanatory variables, as the largest estimated probability of being poor was about 0.48. This is relevant in MLGREG assisted by the mixed model and in the MC methods incorporating the fitted values. HT-CDF and population-level calibration did not yield good results (Table 6.4). The domain-level calibration performed well only in the largest domains. Even the best estimates, obtained by MLGREG and semi-indirect MC, were not strongly correlated with the true values ($r \leq 0.19$). Therefore, the estimates did not provide much evidence on correlation with the degree of urbanization: the correlation coefficient calculated from the estimates was at most $r = -0.2$. The MLGREG and the semi-indirect MC were strongly correlated ($r = 0.997$). This implies that by increasing the size of the neighbourhoods in the semi-indirect MC we probably obtain similar results as with MLGREG.

The estimated standard errors of HT-CDF estimates were large, and the median CV was 60% in regions with sample size smaller than 25, 45% in regions with sample size between 25 and 49, and 26% for regions with sample size larger than 49. The corresponding median CVs for MLGREG were slightly smaller: 44, 42 and 25%, respectively. If a statistic is considered accurate enough for publication only if its CV does not exceed 30%, poverty rate estimation by methods discussed here seems to require sample sizes over 50.

Table 6.4 Mean absolute relative error (%) of estimators of poverty rate in a SRSWOR sample ($n = 2000$) over NUTS 4 regions

Estimator	Domain sample size		
	14–24	25–49	50–323
Direct estimator			
HT-CDF	48.7	39.2	24.1
Indirect estimators			
Population-level MC	48.7	38.9	23.7
Domain-level MC	56.7	36.7	18.4
Semi-indirect MC	44.5	31.9	18.2
MLGREG	43.5	31.8	18.0

6.4 Discussion

6.4.1 Empirical Results

The population that we used in simulation experiments comprised 36 NUTS 4 regions (domains) in Western Finland. Rural areas always had small population sizes but there were also urban regions with small populations. The correlation between urbanization and population size was $r = 0.52$. In the population, the Gini index was larger in the largest cities than in regions with small populations. The poverty rate was larger in rural than in urban areas, including small cities. These correlations could be recognized in samples, especially with methods involving auxiliary information. We did not have strong explanatory variables, as the coefficient of determination was only $R^2 = 0.1$ in a linear fixed-effects model fitted to log-transformed equivalized income. Nevertheless, our estimation results improved when using the auxiliary data in the estimation procedures.

The Gini index and poverty rate measure different aspects of income distribution. Notably, the indicators were not substantially correlated in our population ($r = -0.06$). The Gini index describes the overall inequality of income distribution, whereas the poverty rate is related to occurrence of poor standard of living.

Our simulation results for the Gini index for domains showed that the accuracy of the HT estimator can be fairly good for large regions (with large domain sample size). However, in small regions, the variance of the HT estimator can become too large. Simulation experiments under unequal probability sampling, and the empirical example, showed that the indirect percentile-adjusted predictor technique can significantly improve the accuracy over the direct HT-based method. The parametric method assuming log-normal income distribution was the worst of the methods. Simulation results for poverty rate for domains showed that the HT estimator and the model-assisted generalized regression and MC methods were nearly design unbiased. In using model-assisted methods, our aim was to improve the accuracy over the HT estimator. GREG and MC clearly outperformed HT in most cases. MC tended to be slightly more accurate than GREG.

GREG can be assisted by mixed models as well as fixed-effects models. A mixed model with random effects associated with domains yielded much better results than a fixed-effects model with domain-specific intercept terms. The semi-indirect MC method involving neighbouring regions of each domain yielded slightly better estimates of poverty rate than other MC methods and GREG.

In small regions, the sample size was often too small for accurate estimation of poverty rate. Estimates of poverty rate appeared to be sufficiently accurate, with index of variation smaller than 30%, only in regions containing at least 50 units. A sample size as small as 14 seemed to suffice for the estimation of the Gini index.

Variance estimation of design-based domain estimators discussed here has not been studied thoroughly in the literature. We presented empirical examples of variance estimation. We show how the linearization method of Langel and Tillé [20] can be applied in estimating the standard errors of domain estimators of the Gini index. The variance of the percentile-adjusted predictor of the Gini index was estimated by bootstrap, but the method yielded too small estimates. For a HT-type estimator of the poverty rate and the GREG estimators for domains, approximate estimators of standard error are available. Similar estimators for MC have not yet been published.

6.4.2 Inferential Framework

The estimation of the Gini index and at-risk-of poverty rate for domains was based on the design-based methodology. The design bias of design-based estimators is, under mild conditions, an asymptotically insignificant contribution to the estimator's MSE ([4], p. 99). Our design-based simulation results showed that the design bias of the domain estimator for the Gini index was small. Design bias was negligible for GREG estimator of regional poverty rate. In official statistics production, small bias is often considered as a benefit and is appreciated by clients. Societal importance of regional estimates is increasing for example in monitoring territorial poverty at regional level and allocating funds in combating territorial poverty (e.g. [33–35]). Small anticipated bias of published sub-national figures may contribute to credible monitoring and fair justification of decision-making. Such arguments are not necessarily valid for model-based estimates.

A drawback of design-based estimators is that variances for small domains can be large when compared with MSEs of model-based estimators. On the other hand, accuracy of a model-based estimator can be poor if the bias dominates the MSE. This can happen for example if the model does not hold for a given domain or if the sampling design is informative (see Chapter 5).

We used models as assisting tools in successfully improving accuracy when compared with direct design-based estimators. The property of near design unbiasedness of indirect GREG and model-calibration estimators for the domains of interest is preserved even under model failure. In this sense, model-assisted estimators can provide a safe choice, when compared with model-based estimators, whose design bias for a given domain is difficult to assess in survey practice.

6.4.3 Data Infrastructure

The data infrastructure for our estimation purposes consisted of sample survey data (typically based on a national income survey) and administrative register data on variables that are related to the income variable. We assumed that the sample survey data and the register data can be linked uniquely at the micro-level. In this framework, predicted values from the fitted model can be calculated for every population element and successively incorporated in the construction of the estimators of the desired indicators for the domains of interest. This kind of estimation strategy is applicable in statistical infrastructures of the so-called register countries in particular.

There are an increasing number of register-based statistical infrastructures for official statistics production, notably in Europe (Denmark, Finland, Norway and Sweden as forerunners; see [36, 37]). The combination of different administrative data sources into integrated statistical registers at the unit level is based on unique identifiers such as personal identification numbers and business identification codes.

Much of the recent methodological research and development work in small area estimation is assuming a statistical infrastructure described here. An obvious advantage is flexibility in the construction of assisting models and estimators of the indicators, when compared with statistical infrastructures where area-level or higher aggregated auxiliary data only are available. Recent trends in small area estimation are reviewed in Pfeffermann [38].

References

[1] European Commission (2013) *Smarter, greener, more inclusive? Indicators to support the Europe 2020 strategy*. Publications Office of the European Union, Luxembourg.

[2] Veijanen, A. and Lehtonen, R. (2011) Percentile-adjusted estimation of poverty indicators for domains under outlier contamination. *Statistics in Transition*, **12**, 345–356.

[3] Särndal, C.-E., Swensson, B. and Wretman, J. (1992) *Model Assisted Survey Sampling*. Springer-Verlag, New York.

[4] Särndal, C.-E. (2007) The calibration approach in survey theory and practice. *Survey Methodology,* **33**, 99–119.

[5] Wu, C. and Sitter, R.R. (2001) A model-calibration approach to using complete auxiliary information from survey data. *Journal of the American Statistical Association,* **96**, 185–193.

[6] Montanari, G.E. and Ranalli, M.G. (2005) Nonparametric model calibration estimation in survey sampling. *Journal of the American Statistical Association*, **100**, 1429–1442.

[7] Lehtonen, R. and Veijanen, A. (2009) Design-based methods of estimation for domains and small areas. Chapter 31 in Rao, C.R. and Pfeffermann, D. (eds) *Handbook of Statistics Vol. 29B. Sample Surveys. Inference and Analysis*. Elsevier, Amsterdam, 219–249.

[8] Lehtonen, R., Särndal, C.-E. and Veijanen, A. (2009) Model calibration and generalized regression estimation for domains and small areas. SAE2009 Conference on Small Area Estimation, June 2009, Elche, Spain.

[9] Lehtonen, R. and Veijanen, A. (2012) Small area poverty estimation by model calibration. *Journal of the Indian Society of Agricultural Statistics*, **66**, 125–133.

[10] Drew, J.D., Singh, M.P. and Choudhry, G.H. (1982) Evaluation of small area estimation techniques for the Canadian Labour Force Survey. *Proceedings of the Survey Research Methods Section*, ASA, 545–550. August 16–19, 1982, Cincinnati, Ohio. http://www.amstat.org/sections/srms/Proceedings/.

[11] Singh, A.C., Gambino, J. and Mantel, H.J. (1994) Issues and strategies for small area data. *Survey Methodology*, **20**, 3–14.

[12] Falorsi, P.D., Orsini, D. and Righi, P. (2006) Balanced and coordinated sampling designs for small domain estimation. *Statistics in Transition*, **7**, 805–829.

[13] U.S. Office of Management and Budget (1993) Indirect Estimators in Federal Programs. *Statistical Policy Working Paper 21,* National Technical Information Service, Springfield, VA.

[14] Estevao, V.M. and Särndal, C.-E. (1999) The use of auxiliary information in design-based estimation for domains. *Survey Methodology*, **25**, 213–221.

[15] Lehtonen, R., Särndal, C.-E. and Veijanen, A. (2003) The effect of model choice in estimation for domains, including small domains. *Survey Methodology*, **29**, 33–44.

[16] Lehtonen, R., Särndal, C.-E. and Veijanen, A. (2005) Does the model matter? Comparing model-assisted and model-dependent estimators of class frequencies for domains. *Statistics in Transition*, **7**, 649–673.

[17] Hidiroglou, M.A. and Patak, Z. (2004) Domain estimation using linear regression. *Survey Methodology*, **30**, 67–78.

[18] Torabi, M. and Rao, J.N.K. (2008) Small area estimation under a two-level model. *Survey Methodology*, **34**, 11–17.

[19] Lehtonen, R., Veijanen, A., Myrskylä, M. and Valaste, M. (2011) Small area estimation of indicators on poverty and social exclusion. European Union, 7th Framework Programme Research Project AMELI, Research Report 2.2 (FP7-SSH-2007-217322 AMELI).

[20] Langel, M. and Tillé, Y. (2013) Variance estimation of the Gini index: revisiting a result several times published. *Journal of the Royal Statistical Society, Series A*, **176**, 521–540.

[21] Berger, Y.G. (2004) A simple variance estimator for unequal probability sampling without replacement. *Journal of Applied Statistics*, **31**, 305–315.

[22] Berger, Y.G. (2005) Variance estimation with highly stratified sampling designs with unequal probabilities. *Australian & New Zealand Journal of Statistics*, **47**, 365–373.

[23] Aitchinson, J. and Brown, J.A.C. (1963) *The Lognormal Distribution*. Cambridge University Press, New York.

[24] Graf, M., Nedyalkova, D., Münnich, R., Seger, J. and Zins, S. (2011) Parametric estimation of income distributions and indicators of poverty and social exclusion. European Union, 7th Framework Programme Research Project AMELI, Research Report 2.1 (FP7-SSH-2007-217322 AMELI).

[25] Chambers, R. and Tzavidis, N. (2006) M-quantile models for small area estimation. *Biometrika*, **93**, 255–268.

[26] Marchetti, S., Tzavidis, N. and Pratesi, M. (2012) Non-parametric bootstrap mean squared error estimation for M-quantile estimators of small area averages, quantiles and poverty indicators. *Computational Statistics & Data Analysis*, **56**, 2889–2902.

[27] Marchetti, S. (2013) Estimating inequalities at local level in Italy. *Proceedings of the ISI World Statistics Congress*, August 2013, Hong Kong.

[28] Lehtonen, R. and Veijanen, A. (1998) Logistic generalized regression estimators. *Survey Methodology*, **24**, 51–55.

[29] Rao, J.N.K. (2003) *Small Area Estimation*. John Wiley & Sons, Inc., Hoboken.

[30] Demidenko, E. (2013) *Mixed Models: Theory and Applications with R*, 2nd Edition. John Wiley & Sons, Inc., Hoboken.

[31] Pfeffermann, D. and Sverchkov, M. (2009) Inference under informative sampling. Chapter 39 in Rao, C.R. and Pfeffermann, D. (eds) *Handbook of Statistics Vol. 29B. Sample Surveys. Inference and Analysis*. Elsevier, Amsterdam, 455–487.

[32] Lehtonen, R. and Veijanen, A. (1999) Domain estimation with logistic generalized regression and related estimators. *Proceedings of the IASS Satellite Conference on Small Area Estimation*, August 1999, Riga, 121–128.

[33] Böhme, K., Doucet, P., Komornicki, T., Zaucha, J., Świątek, D. (2011) *How to strengthen the territorial dimension of 'Europe 2020' and the EU cohesion policy*. Report based on the Territorial Agenda 2020, European Union, Warsaw.

[34] European Commission (2010) *Investing in Europe's future. Fifth report on economic, social and territorial cohesion*. Publications Office of the European Union, Luxembourg.

[35] World Bank (2014) *European Union (EU) accession countries – Poverty mapping of new members in EU: completion memo*. World Bank Group, Washington, DC.

[36] United Nations Economic Commission for Europe (2007) *Register-based statistics in the Nordic countries. Review of best practices with focus on population and social statistics*. United Nations, New York.

[37] Wallgren, A. and Wallgren, B. (2014) *Register-based Statistics: Statistical Methods for Administrative Data*, 2nd Edition. John Wiley & Sons, Ltd, Chichester.

[38] Pfeffermann, D. (2013) New important developments in small area estimation. *Statistical Science*, **28**, 40–68.

7

Variance Estimation for Cumulative and Longitudinal Poverty Indicators from Panel Data at Regional Level

Gianni Betti, Francesca Gagliardi and Vijay Verma
Department of Economics and Statistics, University of Siena, Siena, Italy

7.1 Introduction

The extent to which income inequality and poverty vary within countries across different regions is very relevant for policy decisions and monitoring. However, sub-national measures are scarce, given the complexity of producing indicators at the regional level from the available data and the methodological issues related to comparability of the results.

Moreover, the effect of poverty on a person or a household is directly related to the duration of their poverty. If people's experience of marginality and want is only temporary, their life-chances will be less seriously impaired. A persistent risk of poverty, on the other hand, is more likely to be associated with the erosion of resources and a qualitatively different experience of deprivation.

The scope of this chapter is to address the issue of estimating sampling error of cumulative and longitudinal poverty indicators from panel data. It also concerns the robustness of a few EU Statistics on Income and Living Conditions (EU-SILC)-based Laeken indicators at regional level and aims to assess their statistical reliability at one point in time, primarily in terms of sampling reliability, that is, in terms of the magnitude of the sampling error of estimates based on EU-SILC. In 'direct' estimation of such indicators based on a single wave of cross-sectional survey data, the primary concern is the increased sampling error when the results are broken down by region. Starting with direct estimates from the survey, the chapter explores what

Analysis of Poverty Data by Small Area Estimation, First Edition. Edited by Monica Pratesi.

Companion Website: www.wiley.com/go/pratesi/poverty

can be done for calculating more reliable regional estimates by, for example, cumulating over waves.

The chapter comprises nine sections; after this brief introduction, Section 7.2 addresses the differences between cumulative and longitudinal measures of poverty and inequality. Section 7.3 introduces the most commonly used methods for cross-sectional variance estimation, with particular focus on the Jackknife Repeated Replication (JRR) method, and Section 7.4 provides an extension of JRR to cumulative and longitudinal poverty and inequality measures. Sections 7.5 and 7.6 present some practical aspects of variance estimation: specification of sample structure variables, and design effects and correlation, respectively. Section 7.7 describes differences between cumulative measures and measures of net change, while the special case of cumulating 3-years' averages, with an empirical application, is presented in Section 7.8. Section 7.9 concludes the chapter. The numerical illustrations later in the chapter are based on EU-SILC data from two countries – Austria and Spain. Spain has been selected since the available sample size is large and, more importantly, full information on the sample structure is available to us. Austria has a particularly simple sample structure – essentially a weighted simple random sample.

7.2 Cumulative vs. Longitudinal Measures of Poverty

The aim of this section is to address some statistical aspects relating to improving the sampling precision of indicators, in particular through the cumulation of data over rounds of regularly repeated national surveys. This issue has been discussed in Verma *et al.* (2013).

7.2.1 Cumulative Measures

Two types of measures can be constructed by aggregating information on individual elementary units:

- Average measures, that is ordinary measures such as totals, means, rates and proportions constructed by aggregating or averaging individual values (such as population proportions in an area having certain characteristics relating to welfare).
- Distributional measures, such as measures of variation or dispersion among households and persons in the region. These measures may depend on the distribution of characteristics in each region, or on the overall distribution in the whole national (or even EU-level) population.

The patterns of variation and relationship for the two types of measures can differ from each other, and hence involve separate statistical considerations. Average measures are often more easily constructed or are available from alternative sources. Distributional measures tend to be more complex and are less readily available from sources other than complex surveys; at the same time, such measures are more pertinent to the analysis of poverty, social exclusion and other aspects of well-being.

Both types of measures can be constructed using pooled data, for instance over three consecutive years.

An important point to note is that, more than at the national level, many measures of averages can also serve as indicators of disparity and deprivation when seen in the regional context:

the dispersion of regional means is of direct relevance in the identification of geographical disparity.

We are concerned with the pooling of different data sources pertaining to the same population or largely overlapping populations. In particular, the interest is in pooling over survey waves in a national survey in order to increase the precision of regional estimates.

A difficulty in pooling samples is that, in the presence of complex sampling designs, the structure of the resulting pooled sample can become too complex to permit proper variance estimation. Also, different waves of a survey such as EU-SILC do not correspond to exactly the same population. Pooling of wave-specific estimates rather than of micro-datasets is generally the appropriate approach to aggregation over time from surveys such as EU-SILC.

Consider that for each wave, a person's poverty status (poor or non-poor) is determined based on the income distribution of that wave separately, and the proportion poor at each wave is computed. These proportions are then averaged over a number of consecutive waves. The issue is to quantify the gain in sampling precision from such pooling, compared with results based on a single wave, given that data from different waves of a rotational panel are highly correlated.

A large proportion of the individuals are common in the different cross-sections of a panel. However, a certain proportion of individuals are different from one wave to the other. The cross-sectional samples are not independent, resulting in correlation between measures from different waves. Apart from correlations at the individual level, we have to deal also with additional correlation that arises because of the common structure (stratification and clustering) of the waves of a panel. Such correlation would exist in, for instance, samples coming from the same clusters even if there is no overlap in terms of individual households.

7.2.2 Longitudinal Measures

The effect of poverty on a person or a household is directly related to the duration for which they are poor. In the longitudinal dimension, indicators may be designed to capture the experience of poverty and deprivation at any time during a period, or persistently or continuously over the period. One can expect that simpler indicators will be more robust and less demanding on the data available. The main simplification we propose is to focus on *longitudinal indicators defined over a short time period*. Where the available statistical data cover a longer time period for constructing longitudinal indicators, those indicators can themselves be averaged over time to obtain more robust measures.

In specific terms, we can define indicators based on the persistence of poverty over *pairs of adjacent years*:

- Persons are *persistently* poor over two consecutive years if, in relation to the poverty line specific to each of the years, they are classified as poor in *both* of the years.
- Persons are in *any-time* poverty over two consecutive years if, in relation to the poverty line specific to each of the years, they are classified as poor in *either or both* of the years.

7.3 Principle Methods for Cross-sectional Variance Estimation

A diversity of variance estimation approaches have been developed for computation of sampling errors for complex statistics arising from complex samples of households and persons with complex designs, both general and some very specific for particular applications. Among the former, for the 'typical' social surveys based on reasonably large samples but with complex designs, the applicability of at least two broad approaches is generally well-established in the literature. These are the approaches based on (a) Taylor linearisation and (b) on resampling such as the Bootstrap, Balanced repeated replication (BRR) and JRR.

We briefly summarise the differences among the two most used methodologies, namely Taylor linearisation and JRR. For details on the two methodologies, see Verma and Betti (2011).

Concerning JRR, consider a design in which two or more primary sampling units (PSUs) have been selected independently from each stratum. Within each PSU, subsampling of any complexity may be involved, including weighting of the ultimate units.

Let u be the full-sample estimate of a statistic of any complexity, and $u_{(hi)}$ be the estimate produced using the same procedure, but after eliminating PSU i in stratum h and increasing the weight of the remaining (a_h-1) PSUs in the stratum by an appropriate factor (see below). Let $u_{(h)}$ be the simple average of the $u_{(hi)}$ over the a_h values of i in h. The variance of u is then estimated as:

$$\mathrm{var}(u) = \Sigma_h \left[(1\text{-}f_h) \cdot \frac{a_h - 1}{a_h} \cdot \Sigma_i (u_{(hi)} - u_{(h)})^2 \right], \tag{7.1}$$

where (1-f_h) is the finite population correction, usually ~ 1.

The second approach, namely linearisation, is based on the use of Taylor approximation to reduce non-linear statistics to a linear form, justified on the basis of asymptotic properties of large populations and samples. For each ultimate unit (e.g. household or person) in the sample, it seeks a linearised 'indicative' variable, in such a way that the simple expression estimating the variance of the total of this linearised variable, under the given sampling design, approximates the required variance of the original complex statistic. The basis of Taylor linearisation is the following simple variance estimation formula for aggregates in multistage stratified samples of large size.

Let

$$(\text{Kish\_taylor})^2 = \left(\frac{\text{se\_wtd}}{\text{se\_srs}}\right)^2_{\text{Taylor}} = \left(\frac{n}{\sum w_j}\right) \times \frac{\sum w_j^2 u_j^2}{\sum w_j u_j^2}$$

be a sample aggregate or a linear function of sample aggregates such as of $y = \Sigma_h y_h; y_h = \Sigma_i y_{hi}; y_{hi} = \Sigma_j (w_{hij} \cdot y_{hij})$. Then its variance is estimated as:

$$\mathrm{var}(u) = \Sigma_h \left[(1\text{-}f_h) \cdot \frac{a_h}{a_h\text{-}1} \cdot \Sigma_i \left(u_{hi} - \frac{u_h}{a_h} \right)^2 \right], \tag{7.2}$$

with the quantity u_{hi} defined at the level of primary selection (h,i) as the weighted sum of values of the ultimate units j in the PSU. Here j refers to ultimate sampling unit, i to PSU, and h to stratum; $a_h > 1$ is the number of sample PSUs in stratum h; and (1-f_h) is the finite population correction, usually $\sim$1.

As noted in Verma and Betti (2011), the two methods tend to give similar results and often the choice among them is dictated by practical considerations. In situations where either of the two approaches – linearisation or JRR – may be used, their relative merits and limitations are briefly as in the following.

Linearisation tends to be lighter in terms of the computational work involved, especially if the sample contains many PSUs. The main drawback of the linearisation procedure is that different computational formulae have to be used for different types of statistics, the development of which can be complex and – what is more critical – may be intractable in some cases. Another complexity is the need for numerical evaluation of the slope at various points of the income distribution function. This can be problematic because of irregularities in the empirical income distribution based on sample survey data.

The JRR procedure, by comparison, is considerably heavier in terms of the computational work involved, especially if the sample contains many PSUs. This is becoming less important given the rapid increase in computing power. In addition, care is needed in the application of the JRR method to non-smooth statistics, such as the median or other quantiles of the income distribution, as it may not always provide a consistent variance estimator for such statistics. The same applies, but to a lesser degree, with regard to measures based on quantiles, such as the poverty rate defined with reference to the median income. For this reason, as well as to reduce computational work, the grouping of existing PSUs and strata is often needed to define new (fewer and more uniform) computational units.

The relative advantages of the JRR method include the following. The same variance estimation formula applies to different types of statistics. This also permits the development of highly standardised software for its application. The method can be extended to take into account the effect on variance of aspects such as imputation and weighting in the estimation process, insofar as those aspects can be repeated for each replication. Most importantly, the JRR methodology can be easily extended to longitudinal samples and to measures based on several cross-sections. Once a common structure (PSUs and strata) is defined for all waves of a dataset, the JRR can be used to estimate variance of any measure that incorporates measures from different waves.

By contrast, incorporating such additional aspects into the variance estimation is generally much more complex with the linearisation technique. In fact, extensions of the linearisation method to deal with the added complexity involved in dealing with multiple samples from different waves of a panel are not available, but a replication-based approach such as the JRR may be extended more readily to deal with longitudinal and other measures involving correlated samples.

7.4 Extension to Cumulation and Longitudinal Measures

JRR provides a versatile and straightforward technique for variance estimation of comparisons and cumulation over correlated cross-sections. As described in the previous section, it is one of the classes of variance estimation methods based on comparisons among replications generated through repeated re-sampling of the same parent sample. Once the set of replications has been appropriately defined for any complex design, the same variance estimation algorithm can be applied to a statistic of any complexity. We have extended and applied this method for estimating variances for subpopulations including regions and other geographical domains, longitudinal measures such as persistent poverty rates, and measures of net changes and averages over cross-sections in the rotational panel design of EU-SILC using JRR (Verma and Betti, 2007, 2010). Application of Taylor linearisation methodology has not been extended to cumulation or longitudinal measure; Osier *et al.* (2013) have tried an application, but with rather unsatisfactory results.

7.5 Practical Aspects: Specification of Sample Structure Variables

Sampling error computations need to take into account variations in the sampling design. This is done through the definition of the sample structure. For the type of sample designs involved in EU-SILC, and in the practical procedures for variance estimation used, generally *all the necessary information about the sample structure can be provided in the form of two variables defined for each unit*: the 'computational stratum' and 'computational PSU' to which the unit belongs. Normally, we may expect the new variable 'computational stratum' to be related (and sometimes identical) to the original UDB (User Data Base) variable DB050; similarly for 'computational PSU' and DB060. *However, very often the UDB variables require some redefinition before they can be used for the purpose of variance estimation.* To do this in a statistically valid way requires sampling expertise. In order to correctly define the computational strata and PSUs, information concerning the following three aspects must be available:

1. Codes of the sample structure in the micro-data files.
2. Detailed description of the sample design, for instance identifying features such as the presence of self-representing units, systematic selection, and so on.
3. Information connecting the sample structure codes in the micro-data with descriptions of the particular sample design features, so as to be able to identify the design features applicable to particular units.

Practical variance estimation methods need to make some basic assumptions about the sample design. These include:

1. The sample selection is independent between strata.
2. Two or more primary selections are drawn from each stratum.
3. These primary selections are drawn at random, independently and with replacement.
4. The number of primary selections is large enough for valid use of the variance estimation equations.

The computational strata and PSUs have to be defined to meet, or at least approximate, these requirements.

In many practical situations some aspects of sample structure need to be redefined to make variance computation possible, efficient and stable:

1. It may be necessary to regroup ('collapse') strata so as to ensure that each stratum has at least two sample PSUs – the minimum number required for the computation of variance.
2. Units which are included into the sample automatically ('self-representing units') are in fact strata rather than PSUs, and computational PSUs have to be defined at a lower stage within each such unit.
3. In samples selected systematically, the implied implicit stratification is often used to define explicit strata, from each of which an independent sample is supposed to have been selected. Such strata have to be formed by pairing or otherwise grouping of PSUs in the order of their selection from the systematic list, ensuring that each resulting computational stratum has at least two primary selections.
4. Sometimes non-response can result in the disappearance from the sample of whole PSUs. This can disturb the structure of the sample, such as leaving fewer than two PSUs in some

strata. Variance computation requires some redefinition of the computational units to meet the basic requirement of having at least two PSUs per stratum.

5. The above-mentioned problem arises more frequently and seriously when computing sampling errors for subclasses (subpopulations). The risk can be reduced by aggregating PSUs and strata to create fewer, larger computational units.
6. There are restrictions on the detail with which information identifying individual sampling units, such as PSUs, and strata, can be included in the public-release micro-data. Grouping of units and strata can help in preserving the confidential nature of the data, and make the suppression of the sample structural information unnecessary.

   Note that such considerations apply equally irrespective of the particular technique adopted for variance computation – whether the JRR or linearisation, for example.
7. In a procedure like the JRR, the number of replications is equal or at least similar to the number of PSUs in the sample. In a large sample where elements (households, persons) have been selected directly, the number of replications which can be formed will be of the order of the sample size, normally running into thousands. This necessitates forming much fewer computational units, such as creating 'pseudo-cluster' from random grouping of sample elements, and then random pairing of these 'clusters' to construct computational strata.
8. The above issue in fact arises in the case of any sample irrespective of its structure when we want to estimate not only variances but also design effects. The denominator of the design effect is variance under simple random sampling (SRS). That variance can be normally estimated by assuming the sample structure to be SRS.
9. At a minimum, the replication approach requires re-computation of the statistic of interest at each replication. For complex statistics such as poverty rates, this may require a considerable amount of computational time, and it can be desirable to reduce the number of times the process has to be repeated. The same also applies to many other forms of complex analysis, such as estimation involving multivariate analysis and complex parameters, especially if they require iterative procedures.
10. Variance estimation with replications captures the effect on variance of those features of the data treatment and estimation process used in the actual survey which are applied repeatedly to each replication, in the same way that they were applied to the full sample. For instance, in order to fully capture the effect of calibration on variance, it is necessary to recalibrate the sample of each replication using the same procedure as used in the actual sample. The same applies to other aspects of sample weighting, such as adjustment for non-response. Another even more demanding example is imputation for missing data. The need to repeat such heavy procedures at each replication can greatly increase the computational task. Means are required to reduce the number of replications involved.

Appropriate coding of the sample structure, in the survey micro-data and accompanying documentation, is an essential requirement in order to compute sampling errors taking into account the actual sample design. Lack of information on the sample structure in survey data files is a long-standing and persistent problem in survey work, and unfortunately affects EU-SILC as well. Indeed, the major problem in computing sampling errors for EU-SILC is the lack of sufficient information for this purpose in the micro-data available to researchers: currently this information *is not readily available* at the central level for all countries in EU-SILC data; presumably (and hopefully) it is available within each country for its own national survey. We have developed approximate procedures in order to overcome these limitations at least partially, and used them to produce useful estimates of sampling errors (Verma *et al.*, 2010).

7.6 Practical Aspects: Design Effects and Correlation

Design effect (Kish, 1965) is the ratio of the variance (v) under the given sample design, to the variance (v_0) under a simple random sample of the same size:

$$d^2 = v/v_0, \quad d = se/se_0. \tag{7.3}$$

Computing design effects requires the additional step of estimating the error under SRS (se_0), apart from its estimate under the actual design (se).

Proceeding from standard errors to design effects is essential for understanding the patterns of variation and determinants of the magnitude of the error, for smoothing and extrapolating the results for diverse statistics and population subclasses, and for evaluating the performance of the sampling design. Analysing design effects into components also helps to better understand from where inefficiencies of the sample arise, to identify patterns of variation, and through that, to improve 'portability' of the results to other statistics, designs, situations. In applications to EU-SILC, there is in addition a most important and special reason for having procedures for appropriate decomposition of the total design effect into its components. Because of the limited information on sample structure included in the micro-data available to researchers, direct and complete computation of variances cannot be done in many cases. Decomposition of variances and design effects identifies more 'portable' components, which may be more easily imputed (carried over) from a situation where they can be computed with the given information, to another situation where such direct computations are not possible. On this basis, valid estimates of variances can be produced for a wider range of statistics, thus overcoming at least partly the problem due to the lack of information on sample structure in EU-SILC micro-data.

All the above reasons apply even more strongly for statistics at the regional level than they do at the country level. Smaller sample sizes and less information at the regional level make the computation of sampling errors more difficult, sometimes impossible. The results of individual computations also tend to be less stable, and therefore there is a greater need for averaging over them. Disaggregation to the regional level also increases greatly the amount of computations involved, unless the results from a limited set of computations can be extrapolated to other statistics and situations. All these operations require portable measures such as individual components of the design effect.

7.6.1 Components of the Design Effect

We may decompose the design effect into components as follows:

$$v = v_0 \cdot d^2 = v_0 \cdot (d_W \cdot d_H \cdot d_D \cdot d_X)^2. \tag{7.4}$$

Here v_0 is the variance (for the statistic concerned and given sample size) in a simple random sample of individual persons; d_W is the effect of sample weights; if relevant, d_H is the effect of clustering of individual persons into households and d_D is the effect of clustering of households into dwellings; and finally, d_X is the effect of other complexities of the design, mainly clustering and stratification.

All factors other than d_X do not involve clusters or strata, but essentially depend only on the number elements (households, persons, etc.), and the sample weight associated with each such element in the sample. Hence normally they are well estimated, even for quite small regional samples. Procedures for estimating components of the design effect are summarised in the following.

7.6.1.1 Effect of Weights (d_W)

The effect of weights d_W does not depend on the sample structure, other than the presence of unequal sample weights for the elementary units of analysis. Weighting generally inflates variance (weighting is primarily introduced to reduce bias). With the complex weighting procedures of EU-SILC, variation in weights can become large, inflating the design effect. This effect needs to be evaluated and controlled. In principle (but rarely in practice) the factor can be <1, for example with particularly efficacious calibration.

7.6.1.2 Clustering of Persons within Households (d_H)

Factor d_H applies if v_0 refers to variance in a simple random sample of individuals, while v refers to a variable measured at the household level. For example, this factor equals the square-root of household size for variables relating to household income when v_0 is defined to refer to a SRS of individual persons. This applies equally to register and survey countries in EU-SILC, since in both cases income is defined and measured at the household level.[1]

The factor equals 1 for personal interview variables in register countries, since there is only one such interview per household.

For variables constructed to the household level on the basis of separate but correlated observations on individual household members, d_H will be lower than the square-root of household size but generally >1, depending on the strength of the correlation. In principle, d_H *is* <1 for variables which are negatively correlated among members of the same household, but this situation is rare.

7.6.1.3 Clustering of Persons and Households within Dwellings (d_D)

The effect of clustering of households within dwellings or addresses is absent ($d_D = 1$) when we have a direct sample of households or persons, or when such units are selected directly within sample areas – as is the case in most of the EU-SILC surveys. This effect is present when the ultimate units are dwellings, some of which may contain multiple households, but it is small in so far as there is generally a one-to-one correspondence between addresses and households.

Factor d_D cannot be estimated separately unless we have unit identifiers linking households to the dwellings from which they come. Such information has not been recorded in EU-SILC for the few countries using samples of dwellings or addresses.

[1] Actually, such a design involving a SRS of individual persons is never used in EU-SILC, because income of any individuals is defined and measured only in terms of income of all its household members.

Note that when the sample has multiple stages, with one or more area stages preceding the selection of dwellings, d_D can be incorporated into the estimation of d_X – the effect of clustering, stratification and other complexities – and hence into the estimation of the overall design effect; the separation of d_D requires unit identifiers linking households to the dwellings from which they come. By contrast, in a direct sample of dwellings, d_D cannot be estimated at all in the absence of linking information, and therefore is neglected in the estimated overall design effect.

7.6.1.4 Multi-stage Sampling, Stratification and Other Design Complexities (d_X)

Factor d_X represents the effect on sampling error of various complexities of the design such as multiple stages and stratification. Normally this effect exceeds 1 because the loss in efficiency of the sample due to clustering tends to be larger than the gain from stratification. We can expect it to be <1 in stratified random samples of element.

Components of the design effect other than d_X can be estimated without reference to the information on sample structure, except for weighting and identifiers linking different types of units (e.g. persons with their households). By contrast, computation of d_X requires information on the *sample* structure linking elementary units to their strata and higher stage units.

7.6.2 Estimating the Components of Design Effect

7.6.2.1 Effect of weights (d_W)

A very simple expression for estimating d_W is the following from Kish (1965):

$$d_W^2 = \left[\frac{n \cdot \sum (w_i^2)}{\left(\sum w_i\right)^2} \right] = 1 + cv^2(w_i). \tag{7.5}$$

This provides a very good approximation when the sample weights are 'external', not correlated with survey variables. Generally it over-estimates the effect.

In situations for which the 'linearisation method' of variance estimation can be formulated, the effect can be estimated more precisely as:

$$d_W^2 = \left[\frac{n}{\sum w_i} \right] \cdot \frac{\sum (w_i^2 \cdot z_i^2)}{\sum (w_i \cdot z_i^2)}.$$

Here z_i is the 'linearised variate' corresponding to a complex statistic, which is used in the linearisation method to estimate the variance of the complex statistic.

In the linearisation variance estimation method, the above-mentioned linearised variate is always in the form $z_i = z_{1i} + t_i$, where: z_{1i} is defined as $z_{1i} = (y_i - r \cdot x_i)$, with the complex statistic concerned written as if it were a simple ratio of the form $r = \Sigma w_i \cdot y_i / \Sigma w_i \cdot x_i$; and t_i are additional, generally complex, terms (Verma and Betti, 2011). Now, empirically we have found that the following simpler expression yields values indistinguishable to the above for all the complex statistics encountered in analysis of income inequality and poverty:

$$d_W^2 = \left[\frac{n}{\sum w_i}\right] \cdot \frac{\sum(w_i^2 \cdot z_{1i}^2)}{\sum(w_i \cdot z_{1i}^2)}. \tag{7.6}$$

The complex variate z_i is available only when the linearisation procedure for variance estimation can be developed, but the simpler z_{1i} is available in most cases.[2] Hence (7.6) can be used with other procedures such as JRR to estimate the effect of weighting without having to refer to the full linearised variable.

7.6.3 *Estimating other Components using Random Grouping of Elements*

The estimation of design effect due to design complexity with a replication method such as JRR requires an indirect approach. Consider variance computed under the following two assumptions about structure of the design:

1. Variance (v) under the actual design.
2. Using the same procedure, variance (say, v_R) computed by assuming the design to be (weighted) SRS of elements. This can be estimated from a 'randomised sample' created from the actual sample by completely disregarding its structure other than the weight attached to individual elements.

For computation (2), the JRR replications are constructed as in the normal application of the JRR, but in place of the actual strata and primary selections, random grouping of the sample elements are used for this purpose. This provides a variance estimate corresponding to a sample of elements, that is without the effect of stratification, clustering or other complexities, but which still differs from the SRS estimate due to the effect of sample weights on variance. Actually, the result v_R depends on which type of elements are used for constructing random groupings. When we have random groupings of persons (i.e. without regard to whether they come from the same or from different households), the variance estimate obtained is:

$$v_R = v_0 \cdot d_W^2. \tag{7.7}$$

In a sample of households, we may use random groupings of households instead (i.e. keeping all members of a household together in the same group), the variance estimate obtained is:

$$v_R = v_0 \cdot \ (d_W \cdot \ d_H)^2. \tag{7.8}$$

In a sample with dwellings as the ultimate units, provided that the identifiers linking households to their dwellings are available in the micro-data, we may use random groupings of dwelling (i.e. keeping all persons and households of a dwelling together in the same group), and obtain the variance estimate:

$$v_R = v_0 \cdot (d_W \cdot d_H \cdot d_D)^2. \tag{7.9}$$

[2] However, the above expressions cannot be applied to the median and other quantiles of the distribution for which the linearised variable z_i does not contain the term z_{1i}. For these we either have to borrow the result from 'similar' variables where the above procedure is applicable (see, e.g. Verma and Betti, 2011), or use the simpler expression (7.5) based on Kish (1965).

With v computed from the standard application of JRR to the actual design, and d_W estimated from (7.6), the application of (7.7) gives v_0 – and hence overall design effect $d^2 = v/v_0$ without the need to separate out other components.

If applicable and necessary, the separate components can be obtained from (7.8) and (7.9): (7.8) gives d_H and (7.9) gives d_D. The ratio (v/v_R) gives $(d_X)^2$ if we use (7.9), gives $(d_X \cdot d_D)^2$ using (7.8), and gives $(d_X \cdot d_D \cdot d_H)^2$ using (7.7).

7.7 Cumulative Measures and Measures of Net Change

The aim of this section is to address some statistical aspects relating to improving the sampling precision of indicators for subnational regions, in particular through the pooling of data over rounds of regularly repeated national surveys.

7.7.1 *Estimation of the Measures*

Survey data such as from EU-SILC can be used in different forms or manners to construct regional indicators.

When two or more data sources contain – for the same type of units such as households or persons – a set of variables measured in a comparable way, then the information may be pooled either (a) by combining estimates from the different sources, or (b) by pooling data at the micro level. Technical details and relative efficiencies of the procedures depend on the situation. The two approaches may give numerically identical results, or the one or the other may provide more accurate estimates; in certain cases, only one of the two approaches may be appropriate or feasible in any case.

As noted earlier, here we are concerned with the pooling of different sources pertaining to the same population or largely overlapping populations. In particular, the interest is in pooling over survey waves in a national survey in order to increase the precision of regional estimates. Estimates from samples from the same population are most efficiently pooled with weights inversely proportional to their variances (meaning, with similar designs, in direct proportion to their sample sizes). Alternatively, the samples may be pooled at the micro level, with unit weights inversely proportion to their probabilities of appearing in any of the samples. This latter procedure may give more efficient results (e.g. O'Muircheataigh and Pedlow, 2002), but may be impossible to apply as it requires information, for every unit in the pooled sample, on its probability of selection into each of the samples irrespective of whether or not the unit appears in the particular sample (Wells, 1998). Another serious difficulty in pooling samples is that, in the presence of complex sampling designs, the structure of the resulting pooled sample can become too complex or even unknown to permit proper variance estimation. In any case, different waves of a survey such as EU-SILC do not correspond to exactly the same population. The problem is akin to that of combining samples selected from multiple frames, for which it has been noted that micro level pooling is generally not the most efficient method (Lohr and Rao, 2000). For the above reasons, pooling of wave-specific estimates rather than of micro-datasets is generally the appropriate approach to aggregation over time from surveys such as EU-SILC.

In any case, measure of net change, such as net change in the poverty rate from one year to the next, can be constructed only on the basis of pooling of wave-specific estimates at the macro level. The procedure of estimation (both of the measures and of their variances) for measures of net change is essentially the same as that for cumulative measures based on pooling of wave-specific measures.

7.7.2 *Variance Estimation*

Consider that for each wave, a person's poverty status (poor or non-poor) is determined based on the income distribution of that wave separately, and the proportion of poor at each wave is computed. These proportions are then averaged over a number of consecutive waves or are differenced to compute measures of net change. The issue is to quantify the effect on sampling precision of such pooling, given that data from different waves of a rotational panel are highly correlated. The positive correlation between waves reduces the precision of averaged measures and increases the precision of measures of net change, compared with what would be the case in the absence of such correlation. The JRR variance estimation methodology can be easily extended to deal with the sample structure of the pooled sample as follows:

1. The total sample of interest is formed by the union of all the cross-sectional samples being compared or aggregated. Using as basis the common structure of this total sample, a common set of JRR replications is defined for it in the usual way. Constructing a 'common set of replications' essentially requires that when an element is to be excluded in the construction of a particular replication, it must be excluded simultaneously from every cross-sectional sample (included in the above-mentioned total sample) where it appears.
2. For each replication, the required measure is constructed for each of the cross-sectional samples involved. These replication-specific cross-sectional measures are differenced or aggregated to obtain the required net change and average measures for the replication.
3. Variance is then estimated from the resulting replicated measures in the usual way.

7.8 An Application to Three Years' Averages

An example of the application of cumulation as described in Section 7.7.1 is the following.

If we have a dataset for three consecutive years, we proceed as follows to estimate the average of the three years.

7.8.1 *Computation Given Limited Information on Sample Structure in EU-SILC*

Appropriate coding of the sample structure in the survey micro-data and accompanying documentation is an essential requirement in order to compute sampling errors taking into account the actual sample design. As noted, the major problem in computing sampling errors for EU-SILC is the lack of sufficient information for this purpose in the micro-data available to researchers.

We have developed approximate procedures in order to overcome these limitations at least partially, and used them to produce useful estimates of sampling errors (Verma *et al.*, 2010). EU-SILC provides two versions of the dataset for each year: the full cross-sectional data for the year, and the longitudinal data covering panel for the current year and up to three preceding years. In order to construct averaged measures (and also measures of net change) we actually need the cross-sectional datasets. Unfortunately these datasets *do not* provide linkage between household and persons across waves. Therefore, we need to pass through the *longitudinal dataset* to get the measure of the correlation between the datasets. So, as an example, if we want to produce the estimates for the average of three years, we take the longitudinal dataset covering three years, say years 1, 2 and 3. We work separately with each pair of years, 1–2, 2–3 and 1–3. In the longitudinal dataset we can construct the common structure (that we are not able to construct in the cross-sectional one) since households and individuals are linkable over different years in the former. For each pair of years the dataset can be divided into three parts depending on whether the two sample are composed of: (a) the same persons; (b) the same area, but different persons; or (c) different persons and different area. We have to work separately with each of these parts.

Part (*a*). For this part, common sample of persons, we apply exactly the same procedure that we describe later in Section 7.8.2 for the cross-sectional dataset and estimate the variance of the first year $V(Y_t)$, the variance of the second year $V(Y_{t+1})$ and the variance of their average

$$V\left(\frac{Y_t + Y_{t+1}}{2}\right).$$

From the relationship

$$V\left(\frac{Y_t + Y_{t+1}}{2}\right) = \frac{1}{4}(V(Y_t) + V(Y_{t+1}) \ + \ 2\rho\sqrt{V(Y_t)V(Y_{t+1})})$$

we can derive the value of correlation ρ, estimated for the three pairs of years being, respectively, ρ_{12}, ρ_{23} and ρ_{13}.

For the correlation between adjacent years, a more stable estimate is provided by taking the average of the two estimates: $\rho^L_{adj} = \frac{\rho_{12}+\rho_{23}}{2}$ where L stands for 'longitudinal'. For the wave separated by a year, we have $\rho^L_{13} = \rho_{13}$.

Part (*b*). For this part of the sample, same area but different persons, again we apply the same procedure for each pair of years and get ρ^A_{12}, ρ^A_{23}, ρ^A_{13}, where A stands for 'area level'. We use their average

$$\rho^A = \frac{\rho^A_{12} + \rho^A_{23} + \rho^A_{13}}{3}$$

as a more stable estimate of the correlation at area level.

Part (*c*). For this part, with no overlap even at area level, we can take correlation between years to be zero.

Now we have all the information we need from the longitudinal dataset: ρ^L_{adj}, ρ^L_{13} and ρ^A.

We can return to the cross-sectional datasets (1, 2 and 3 in our case), and estimate at national level the three variances $V_1(Y)$, $V_2(Y)$, $V_3(Y)$.

The general expression for variance of measures of average and net change taking into account the correlation between waves is:

$$V\left(\sum_i a_i Y_i\right) = \sum_i a_i^2 V(Y_i) \ + \ 2\sum_{j>i} a_i a_j \rho_{ij}\sqrt{V(Y_i)V(Y_j)} \tag{7.10}$$

With parameters $a_i = 1/3$ for the average measure.[3]

[3] For net change from year i to year $(i+1)$, the corresponding coefficient would be $a_i = 1$, $a_i + 1 = -1$.

The main problem is to estimate ρ_{ij}^{CS} between the correlated cross-sectional samples from the correlations for the three types of sample overlaps, (a), (b) and (c), defined above. The former can be taken as a weighted sum of the latter since it is a mixture of the three types of overlaps. With reference to a particular pair of waves (i, j), we take

$$\rho^{CS} = w^L \rho_{\text{adj}}^L + w^A \rho^A + w_0 0$$

(for simplicity, subscript *i,j* has been dropped in the above).

The weights are determined by the following considerations: $w^L + w^A + w_0 = 1$, since they define a weighted average $w_0 = \frac{1}{4}$ from the structure of EU-SILC $w_{ij}^L = \frac{n_L}{n_C}$ where n_L is the sample size of the panel considered and n_C is the larger of the samples of the two cross-sectional datasets considered.

With the above-defined set of correlations and weights, the set of ρ_{ij} in equation (7.10) are defined as follows:

$$\rho_{12} = (\rho_{\text{adj}}^L w_{12}^L + \rho^A w_{12}^A),\ \rho_{23} = (\rho_{\text{adj}}^L w_{23}^L + \rho^A w_{23}^A),\ \rho_{13} = (\rho_{\text{adj}}^L w_{13}^L + \rho^A w_{13}^A).$$

Now we have to get the estimates at regional level.

In order to simplify the calculations, we estimate them from the decomposition of the design effect, as described in Section 7.6.

At country level (C) the total design effect can be decomposed as:

$$d^{2(C)} = d_W^{2(C)} \times d_H^{2(C)} \times d_D^{2(C)} \times d_X^{2(C)} \times d_R^{2(C)} \tag{7.11}$$

d_W^2, d_H^2, d_D^2, d_X^2 have been defined in Section 7.6. Remember, d_W is the effect of unequal weights, d_H and d_D are the effects of clustering of persons within households and dwellings, and d_X is the effect of other deviations from SRS.

The new factor ${d_R}^2$ is the effect of correlation between dependent samples of waves in a panel. It is given by:

$$d_R^2 = \frac{V\left(\sum a_i Y_i\right)}{\sum a_i^2 \times V_R(Y_i)} \tag{7.12}$$

With quantity $V_R(Y_i)$ defined above in equation (7.9), and taking d_X^2 as an averaged value over the three waves, that is writing

$$V(Y_i) = d_X^2 \times V_R(Y_i)$$

we get

$$\frac{V\left(\sum a_i Y_i\right)}{\sum a_i^2 V_R(Y_i)} = d_X^2 \times d_R^2 \quad \text{or} \quad d_X^2 = \frac{\sum a_i^2 V(Y_i)}{\sum a_i^2 V_R(Y_i)} \tag{7.13}$$

The main difficulty in moving from national to regional level is the reduced sample size, which makes it difficult (and more laborious) to estimate quantities d_R and d_X directly and separately for each region. Our practical strategy in estimating design effects at the regional level

$$d^{2(G)} = \mathrm{d}_W^{2(G)} \times \mathrm{d}_H^{2(G)} \times \mathrm{d}_D^{2(G)} \times \mathrm{d}_X^{2(G)} \times \mathrm{d}_R^{2(G)} \tag{7.14}$$

is as follows (Verma *et al.*, 2010):

1. Quantities d_W, d_H and d_D do not depend on structure (especially clustering) of the sample, and can be easily estimated from samples of elements at the regional level.
2. With the common structure of the panel in all regions of a country, it is reasonable to take $\mathrm{d}_R^{(G)} = \mathrm{d}_R^{(C)}$, already estimated at the country level.
3. Depending on the relation between national and regional sample design, d_X^G can be reasonably defined as a function of already computed d_X^C, thus avoiding its re-estimation at the regional level. For instance, if the sample design in the region is the same or very similar to that for the country as a whole – which is quite often the case – we can take $\mathrm{d}_\mathrm{X}^{2(\mathrm{G})} = \mathrm{d}_\mathrm{X}^{2(\mathrm{C})}$. For models for other cases, see Verma *et al.* (2010).
4. In moving from design effects to actual variances, we need regional estimates of simple random sample variance, $V_{SRS}^{(G)}$. It can be inferred from $V_{SRS}^{(C)}$ on the basis of the very reasonable assumption that the coefficient of variation

$$CV^2(Y) = n \cdot \frac{V_{SRS}(Y)}{Y^2}$$

at the regional level are the same as the country level, so that

$$V_{SRS}^{(G)} = \left[\frac{Y^{(G)}}{Y^{(C)}}\right]^2 \cdot \left(\frac{n^{(C)}}{n^{(G)}}\right) \cdot V_{SRS}^{(C)}$$

7.8.2 *Direct Computation*

Direct computation is possible when fuller information of sample structure is available.

We refer in particular to the situation when identifiers are provided to link strata and PSUs in the EU-SILC *cross-sectional* datasets; so that correlation between waves can be estimated directly.

We first construct a common structure of strata and PSUs from the union of the three cross-sectional datasets (Figure 7.1) and assign to this common structure new weights equal to the average of the weights of the three years:

$$w_t^{(\mathrm{Common})} = (w_t)^{\mathrm{Average}} = (w_1 + w_2 + w_3)/3$$

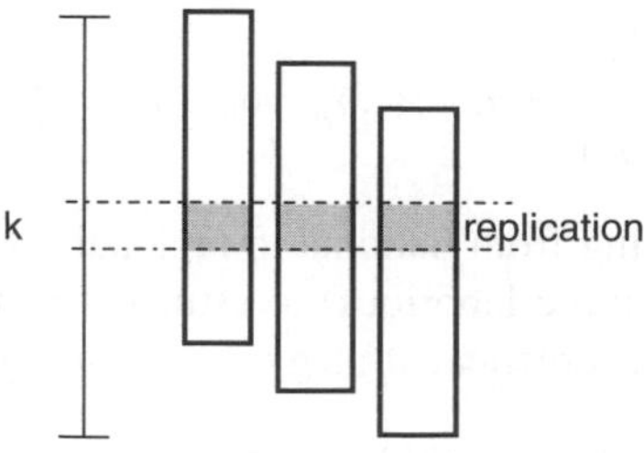

Figure 7.1 Union of the cross-sectional datasets

For each year (t) and for each replication (k), we can estimate $y_k^{(t)}$ where $t = 1, 2, 3$ and from this, the required statistic $y_k^{\text{Average}} = \sum_t a_t y_k^{(t)}$; in our case this is just $y_k^{\text{Average}} = (y_k^1 + y_k^2 + y_k^3)/3$.

Variance estimate of this measure $V = \sum_t a_t V^{(t)}$ can be easily estimated applying the usual JRR for variance estimation procedure, as if the statistic was a common cross-sectional measure.

7.8.3 Empirical Results

We have applied the methodologies described above to calculate the average measures for three years (2009, 2010 and 2011) to EU-SILC data for Austria and Spain. For Austria, all necessary information (linkage of the structure for the three cross-sectional datasets) was not available to us, therefore we used the indirect procedure, described above. For Spain we had all the information, and could use the direct methodology.

Results at the national level for Austria and Spain are shown in Table 7.1, and results at regional NUTS 2[4] level in Austria are shown in Table 7.2.

In general the two methodologies (direct and indirect) for the estimation of standard errors of averaged measures over three years perform well and give similar results both at national and regional level, as we have already shown in our past work.

The results are very stable across regions in Austria.

Furthermore, the results for mean and median measures are nearly the same, showing a reduction in variance of about 25–30% with pooling over 3 years.

Table 7.1 Average over three years, Austria and Spain

	Estimate 2011	standard error (s.e.) 2011	s.e. three years' average	Ratio of s.e. three years' average over s.e. single year	Ratio of s.e. three years' average over s.e. three years' average for independent samples
Austria					
HCR 60% national poverty line (p.l.)	*13.8*	0.61	0.43	0.70	1.22
S80/S20	*4.0*	0.08	0.07	0.88	1.52
Gini	*26.9*	0.38	0.31	0.82	1.41
Spain					
HCR 60% national p.l.	*22.0*	0.48	0.31	0.65	1.12
S80/S20	*6.5*	0.15	0.11	0.73	1.27
Gini	*33.8*	0.30	0.24	0.80	1.39

[4] NUTS is an abbreviation for Nomenclature of Statistical Territorial Units. This is Eurostat's hierarchical classification of regions, from Member States (NUTS 0) down to smaller area.

Table 7.2 Average over three years, Austria (AT) regional NUTS 2 level

	s.e. 2011	s.e. three years' average	Ratio of s.e. three years' average over s.e. single year	Ratio of s.e. three years' average over s.e. three years' average for independent samples
HCR 60%, national p.l. AT11	3.72	2.44	0.66	1.14
HCR 60%, national p.l. AT12	1.07	0.78	0.74	1.28
HCR 60%, national p.l. AT13	1.86	1.35	0.73	1.26
HCR 60%, national p.l. AT21	3.35	2.24	0.67	1.16
HCR 60%, national p.l. AT22	1.30	1.10	0.85	1.47
HCR 60%, national p.l. AT31	1.12	0.72	0.64	1.11
HCR 60%, national p.l. AT32	1.88	1.32	0.70	1.21
HCR 60%, national p.l. AT33	1.91	1.15	0.60	1.04
HCR 60%, national p.l. AT34	1.99	1.59	0.80	1.39
Mean			*0.71*	*1.23*
Median			*0.7*	*1.21*
S80/S20 AT11	0.48	0.32	0.68	1.18
S80/S20 AT12	0.18	0.14	0.78	1.35
S80/S20 AT13	0.23	0.17	0.75	1.30
S80/S20 AT21	0.31	0.24	0.75	1.30
S80/S20 AT22	0.22	0.16	0.75	1.30
S80/S20 AT31	0.18	0.14	0.77	1.33
S80/S20 AT32	0.36	0.26	0.71	1.23
S80/S20 AT33	0.30	0.22	0.71	1.23
S80/S20 AT34	0.43	0.40	0.95	1.65
Mean			*0.76*	*1.32*
Median			*0.75*	*1.30*
Gini AT11	2.10	1.48	0.70	1.21
Gini AT12	0.90	0.70	0.78	1.35
Gini AT13	0.95	0.73	0.77	1.33
Gini AT21	1.49	1.11	0.74	1.28
Gini AT22	1.04	0.77	0.74	1.28
Gini AT31	0.93	0.74	0.80	1.39
Gini AT32	1.82	1.31	0.72	1.25
Gini AT33	1.50	1.10	0.73	1.26
Gini AT34	1.99	1.79	0.90	1.56
Mean			*0.77*	*1.32*
Median			*0.74*	*1.28*

7.9 Concluding Remarks

In this chapter we have described practical procedures for estimation of variance of complex statistics such as poverty rates under complex sample designs. We have also developed and applied procedures to deal with the additional difficulties caused by the lack of full information on sample structure in datasets such as EU-SILC.

Our specific concern is with estimation at subnational (regional) level, where additional numerical difficulties can arise as a consequence of the reduced sample sizes available. In conclusion, we would like to point out that the production of more adequate statistics at the regional level requires a much broader strategy beyond simply pooling of data over waves of the panel. The most important elements of such a strategy may be briefly stated as:

1. Treating each region as a separate sample design domain is highly desirable in order to improve estimation at the regional level.
2. Significantly increasing the overall sample size to meet regional reporting requirements is not a practical option in many, perhaps most, situations. A more practical option is to reallocate the affordable sample to meet regional requirements.
3. In this context, explicit consideration should be given to the determination of sample size and allocation for individual regions in the country. As far as possible, each region, irrespective of its size, should receive a certain minimum sample size.
4. The use of small area estimation procedures for producing regional estimates is limited in the context of international surveys such as EU-SILC, because these estimates may lack comparability across countries, and also because estimates cannot be replicated without access to the auxiliary information used in their original construction.

References

Kish, L. (1965) *Survey Sampling*, New York: John Wiley & Sons, Inc.

Lohr, S.L. and Rao, J.N.K. (2000) Inference from dual frame surveys. *Journal of American Statistical Association*, 95, pp. 271–280.

O'Muircheataigh, C. and Pedlow, S. (2002) Combining samples vs. cumulating cases: a comparison of two weighting strategies in NLS97. *American Statistical Association Proceedings of the Joint Statistical Meetings,* August 11–15, 2002, New York. pp. 2557–2562.

Osier, G., Berger, Y. and Goedemè, T. (2013) Standard error estimation for the EU-SILC indicators of poverty and social exclusion, Eurostat Methodologies and Working Papers, Eurostat, Luxembourg.

Verma, V. and Betti, G. (2007) Cross-sectional and Longitudinal Measures of Poverty and Inequality: Variance Estimation using Jackknife Repeated Replication. In: Conference 2007 'Statistics under one Umbrella', Bielefeld University.

Verma, V. and Betti, G. (2011) Taylor linearization sampling errors and design effects for poverty measures and other complex statistics. *Journal of Applied Statistics*, 38(8), pp. 1549–1576.

Verma, V., Betti, G. and Gagliardi, F. (2010) An assessment of survey errors in EU-SILC, Eurostat Methodologies and Working Papers, Eurostat, Luxembourg.

Verma, V., Gagliardi, F. and Ferretti, C. (2013) Cumulation of poverty measures to meet new policy needs, in *Advances in Theoretical and Applied Statistics*. Torelli, N., Pesarin, F. and Bar-Hen, A. (Eds), Springer-verlag, pp. 511–522.

Wells, J.E. (1998), Oversampling through households or other clusters: comparison of methods for weighting the oversample elements. *Australian and New Zealand Journal of Statistics*, 40, Springer-verlag, Berlin-Heidelberg 2013. pp. 269–277.

Part III

Small Area Estimation Modeling and Robustness

8

Models in Small Area Estimation when Covariates are Measured with Error

Serena Arima[1], Gauri S. Datta[2] and Brunero Liseo[1]

[1]*Department of Methods and Models for Economics Territory and Finance, University of Rome La Sapienza, Rome, Italy*

[2]*Department of Statistics, University of Georgia, Athens, USA*

8.1 Introduction

In this chapter we will discuss some issues related to the use of covariates measured with error in a small area statistical model. We will present a review of the existing literature, both from a frequentist and a Bayesian perspective. We will also discuss some new methodological issues and illustrate the methods through a real data application.

Measurement error issues are ubiquitous in statistical practice. As masterly illustrated in Carroll *et al.* (2006), independently of the complexity of the model, the use of covariates which are affected by measurement error (ME) has three main effects:

- it causes bias in parameter estimation for statistical models;
- it leads to a loss of power, sometimes profound, for detecting interesting relationships among variables;
- it masks the features of the data, making graphical model analysis difficult.

There is insufficient space for solutions to the above problems except for the first one. Several different methods have been proposed to correct for biases of estimation caused by ME.

The classical ME model assumes that for each single unit, the value of the covariate X, say X_i $(i = 1, \dots, n)$, is not available. Instead we have k $(k \geq 1)$ replications of measurements of X_i, namely

$$W_{ij} = X_i + U_{ij}, \quad j = 1, \cdots, k \tag{8.1}$$

Analysis of Poverty Data by Small Area Estimation, First Edition. Edited by Monica Pratesi.

Companion Website: www.wiley.com/go/pratesi/poverty

where the U_{ij}'s are independent and identically distributed (i.i.d.) with zero mean. The values of the X_i's might be either unknown and fixed quantities or random variables with given mean μ_x and variance σ_x^2.

In practice, in a simple regression example and with $k = 1$, instead of having the standard

$$Y_i = \beta_0 + \beta_1 X_i + \epsilon_i, \quad i = 1, \ldots, n,$$

one has

$$Y_i = \beta_0 + \beta_1 W_i + \epsilon_i - \beta_1 U_i, \quad i = 1, \ldots, n.$$

If we assume, for example, that $X_i \overset{i.i.d.}{\sim} N(\mu_x, \sigma_x^2)$, $i = 1, \ldots, n$, this is still a standard linear regression model, but the variance of the "error" component is now $\sigma_\epsilon^2 + \beta_1^2 \sigma_U^2 > \sigma_\epsilon^2$ (unless $\beta_1 = 0 \ldots$), with obvious meaning of the symbols. Also, since the covariate is now W, the ordinary least squares estimator, say $\tilde{\beta}_1$ is not a consistent estimator of β_1, but rather of

$$\frac{\sigma_x^2}{\sigma_x^2 + \sigma_U^2} \beta_1 \neq \beta_1.$$

So one obtains bias in estimation and more variability in the data.

The classical ME model contrasts with the so-called Berkson ME model where one assumes that

$$X_i = W_i + U_i, \tag{8.2}$$

that is, there is only one measurement on the covariate, and the error is added to the measurement rather than to the "true" value. We will not go into the details of this model: in general, the Berkson approach leads to an over-optimistic assessment of the power of the procedure.

Small area models are generally classified into two broad types: (i) area-level models where for the study of response variable only summary statistics, such as sample mean or the Horvitz–Thompson estimate, are available for each small area; and (ii) unit-level models which are appropriate when data for the study variable is at the unit-level. Details of model-based small area estimation under area-level and unit-level models are given in the previous chapters of this book and in Rao (2003). Other general overviews are Jiang and Lahiri (2006), Datta (2009) and Pfeffermann (2013).

Although ME models have been mainly developed for the analysis of experimental data, their role in social studies and, in particular, in official statistics is crucial. For example, most of the unit-level small area models make use of covariates which are based on administrative data, or survey data, in which the respondents provide information about their habits and the reported data are often imprecise, due to lack of memory, rounding, and other obvious reasons. In other situations, ME is artificially included for privacy reasons. National Statistical Offices release data and often, to prevent identification of a unit or to protect its privacy, a noise generated from an appropriate distribution is independently added to the true values. This is an example of classical ME where the error distribution is known exactly.

Within the classical ME framework, there are two major classes of models one can consider, the classical functional model and the classical structural one. In the former case, the "true" values of the covariate X_i's are assumed to be unknown but fixed constants; in the latter case, the X_i's are considered as random variables. Carroll *et al.* (2006) introduce a slightly different distinction: since the orthodox functional approach is known to be impractical in many situations (e.g., it might be hard to find sensible maximum likelihood estimates of the unknown and unobserved quantities), researchers today distinguish between functional modeling, where the X_i's are considered fixed or random, but in the latter case no, or minimal, assumptions are made about the distribution of the X_i's, and structural modeling, where the X_i's are explicitly modeled, usually through reasonably simple parametric models.

In a broader sense, the distinction between functional and structural models is something related to the more general problem of how to model the latent structure of a statistical model: within a frequentist or empirical Bayes approach one can either use functional or structural models, while a Bayesian model is, by definition, structural.

As a consequence, functional approaches are useful when a sort of robustness of inference – to misspecification of the distribution of the true values – is needed. Structural modeling plays a predominant role in practical applications, although concerns about the robustness of inference to assumptions on the X_i's may remain an issue. We defer the discussion of these problems to Section 8.4.1.

8.2 Functional Measurement Error Approach for Area-level Models

The most popular model in small area estimation for area-level data is the model due to Fay and Herriot (1979). A direct estimator Y_i, which is an area-level summary of the response variable for area i, estimates the small area mean θ_i. Along with Y_i, auxiliary data z_i and x_i, in partitioned form, are available for the covariates z and x for the ith area, where z_i is $q \times 1$ and x_i is $p \times 1$ vectors. To develop improved prediction of θ_i, Fay and Herriot (1979) suggested the following model:

$$Y_i = \theta_i + \epsilon_i, \qquad \theta_i = z_i'\delta + x_i'\beta + u_i, \qquad i = 1, \dots, m, \tag{8.3}$$

where the sampling errors ϵ_i's and the model errors u_i's are all independently distributed. It is assumed that $\epsilon_i \sim N(0, \psi_i)$ and $u_i \sim N(0, \sigma_u^2)$, $i = 1, \dots, m$. The sampling variances ψ_i's are assumed known, and the unknown model parameters δ, β and σ_u^2 are collectively denoted by ϕ.

First uses of ME ideas in area-level models appeared in Fuller (1990) where predictors and small area estimators were derived in the context of the Fay–Herriot model. In that case, the measurement error came from "random effect" u, and not from an incorrect measure of x.

Ybarra and Lohr (2008) have considered a Fay–Herriot area-level model with the auxiliary information used in the covariates as measured with error. As an example, suppose it is required to estimate the number of individuals with cardiac problems. We may have area-level estimates of these numbers. Cardiac problems are often related to obesity or being overweight; however, it is very common that overweight persons tend to underestimate their own weight. In their functional approach to the problem, Ybarra and Lohr assumed that the true values x_i for the covariate x in the ith area are measured with error as W_i. They assumed that the mean squared error (MSE) of W_i is a known positive definite (p.d.) matrix C_i.

8.2.1 Frequentist Method for Functional Measurement Error Models

When the model parameters ϕ and the auxiliary variables are *known*, the best unbiased predictor of θ_i, that minimizes the MSE of prediction is given by

$$\tilde{\theta}_i = \gamma_{iu} Y_i + (1 - \gamma_{iu})\{z_i'\delta + x_i'\beta\}, \tag{8.4}$$

where $\gamma_{iu} = \sigma_u^2/(\psi_i + \sigma_u^2)$, and the MSE of $\tilde{\theta}_i$ is given by $\psi_i\gamma_{iu}$.

Ybarra and Lohr (2008) showed that if the unknown x_i in the best predictor $\tilde{\theta}_i$ is replaced by its estimator W_i, which is subject to error, the resulting naive predictor

$$\tilde{\theta}_{i,na} = \gamma_{iu} Y_i + (1 - \gamma_{iu})\{z_i'\delta + W_i'\beta\}, \tag{8.5}$$

has MSE that is larger than $\psi_i\gamma_{iu}$. In fact, it may be unacceptably large, even larger than ψ_i, the MSE of the direct estimator Y_i. Indeed,

$$MSE(\tilde{\theta}_{i,na}) = \psi_i\gamma_{iu} + (1 - \gamma_{iu})^2\beta' C_i\beta = \frac{\psi_i\sigma_u^2}{\psi_i + \sigma_u^2} + \frac{\psi_i^2}{(\psi_i + \sigma_u^2)^2}\beta' C_i\beta, \tag{8.6}$$

which exceeds ψ_i if and only if $\beta' C_i\beta > \sigma_u^2 + \psi_i$. Thus a large variability in W_i may make $\tilde{\theta}_{i,na}$ less accurate than Y_i. It is intuitively clear that due to error in the covariate W_i, a synthetic "estimator" $z_i'\delta + W_i'\beta$ for θ_i should be given a smaller weight in a composite estimator of the form $a_iY_i + (1 - a_i)\{z_i'\delta + W_i'\beta\}$ for θ_i. Ybarra and Lohr (2008) [see also Datta *et al.* (2010)] showed that the estimator in this class that minimizes the prediction MSE corresponds to choosing $a_i = \gamma_i$ with

$$\gamma_i = \frac{\sigma_u^2 + \beta' C_i\beta}{\psi_i + \sigma_u^2 + \beta' C_i\beta}.$$

The resulting best predictor of θ_i is given by

$$\tilde{\theta}_{i,BP} = \gamma_i Y_i + (1 - \gamma_i)\{z_i'\delta + W_i'\beta\}, \tag{8.7}$$

and its associated MSE is given by $\psi_i\gamma_i$. The new predictor is a convex combination, that is, a weighted average of the direct estimator Y_i and the synthetic estimator $z_i'\delta + W_i'\beta$. The weight to the synthetic component is $\psi_i/(\psi_i + \sigma_u^2 + \beta' C_i\beta)$, which is less than the weight to the synthetic component in (8.5). The coefficient, γ_i, however, depends on the variability of the estimator W_i as well as on σ_u^2, ψ_i, and the magnitude of the regression coefficient.

Remark 1. Interestingly, it can be checked that if $x_i'\beta$ is "estimated" by a "composite estimator" of the form $h_i\{Y_i - z_i'\delta\} + (1 - h_i)W_i'\beta$ by minimizing the MSE, the "optimal composite estimator" turns out to be $\frac{\beta' C_i\beta}{\psi_i+\sigma_u^2+\beta' C_i\beta}(Y_i - z_i'\delta) + \frac{\psi_i+\sigma_u^2}{\psi_i+\sigma_u^2+\beta' C_i\beta}W_i'\beta$. This estimator extracts information about $x_i'\beta$ from $Y_i - z_i'\delta$ as well as from $W_i'\beta$. If this estimator is substituted in to the right-hand side of (8.4) in place of $x_i'\beta$, then the right-hand side simplifies to $\gamma_iY_i + (1 - \gamma_i)\{z_i'\delta + W_i'\beta\}$.

Remark 2. The predictor $\tilde{\theta}_{i,BP}$ may be derived as the best predictor of θ_i from a newly formulated conditional model, by conditioning on W_i. Consider the model

$$Y_i = \theta_i + \epsilon_i, \qquad \theta_i = z_i'\delta + W_i'\beta + u_i^*, \quad i = 1, \ldots, m, \tag{8.8}$$

where $\epsilon_1, \ldots, \epsilon_m, u_1^*, \ldots, u_m^*$ are mutually independent; $\epsilon_i \sim N(0, \psi_i)$ and $u_i^* \sim N(0, \sigma_u^2 + \beta' C_i \beta), i = 1, \ldots, m$. We refer to the model in (8.8) as the *extended Fay–Herriot* model. The difference between the standard Fay–Herriot model in (8.3) and the extended Fay–Herriot model in (8.8) is that unlike in (8.3), the model error variance in the latter model is not constant over areas. However, as we obtained the best predictor $\tilde{\theta}_i$ of θ_i in (8.4) from the model in (8.3), if we proceed along the same line to derive the best predictor of θ_i under the model in (8.8), the resulting predictor will be $\tilde{\theta}_{i,BP}$ in (8.7).

In the more practical case when the model parameters ϕ are unknown, Ybarra and Lohr (2008) modified Prasad–Rao's method of moments estimators for their case based on modified least squares to estimate the regression coefficients using the results of Cheng and van Ness (1999) to account for sampling variability in the auxiliary information. In particular, to estimate the regression coefficients we follow these authors to define an estimating equation below. Let $R_i' = (z_i', W_i'), G_i = \text{Diag}(0, C_i)$. We denote the regression coefficients vectors (δ', β') by α'. Then for a finite set of weights $w_1, \ldots, w_m$ bounded away from 0,

$$\sum_{i=1}^{m} w_i (R_i R_i' - G_i)\alpha = \sum_{i=1}^{m} w_i R_i Y_i \tag{8.9}$$

is an unbiased estimating equation for α. If $\sum_{i=1}^m w_i(R_i R_i' - G_i)$ is non-singular, $\hat{\alpha}_w = \left\{\sum_{i=1}^m w_i(R_i R_i' - G_i)\right\}^{-1} \sum_{i=1}^m w_i R_i Y_i$ will be the estimator of α.

To estimate σ_u^2, Ybarra and Lohr (2008) suggested

$$\hat{\sigma}_u^2(w) = (m - p - q)^{-1} \sum_{i=1}^{m} \left\{ \left(Y_i - R_i'\hat{\alpha}_w\right)^2 - \psi_i - \hat{\beta}_w' C_i \hat{\beta}_w \right\}. \tag{8.10}$$

If the weights $w_1, \ldots, w_m$ depend on the parameters β and σ_u^2, then these equations need to be solved iteratively, starting with some reasonable initial values for α and σ_u^2. One of the suggestions of Ybarra and Lohr (2008) is $w_i = 1/(\psi_i + \sigma_u^2 + \beta' C_i \beta), i = 1, \ldots, m$. For details on parameter estimation, we refer to Ybarra and Lohr (2008).

By replacing the unknown parameters α and σ_u^2 in the best predictor $\tilde{\theta}_{i,BP}$ by their estimators, we get the empirical best predictor $\hat{\theta}_{i,EBP}$ of θ_i. Ybarra and Lohr (2008) showed that the new predictor is almost unbiased, provided that W_i estimates the true values of the x_i without any bias. They also showed that the leading term of the approximation to the prediction MSE of $\hat{\theta}_{i,EBP}$ is $M_{1i}(\sigma_u^2, \alpha) = \psi_i \gamma_i$, where the neglected terms are of the order $O(m^{-1})$. They estimated the MSE by using the jackknife method, proposed by Jiang *et al.* (2002). Their method reduces the bias of the estimator of the leading term of the MSE, and captures the variability in the Empirical Best Predictor (EBP) due to uncertainty associated with estimation of the model parameters. An approximately unbiased estimator of the MSE, obtained by Ybarra and Lohr (2008), is

$$\widehat{MSE}(\hat{\theta}_{i,EBP}) = \hat{M}_{1i} + \hat{M}_{2i}, \tag{8.11}$$

with

$$\hat{M}_{1i} = \psi_i \hat{\gamma}_i + \frac{m-1}{m} \sum_{j=1}^{m} \left(\psi_i \hat{\gamma}_i - \psi_i \hat{\gamma}_i(-j)\right),$$

and

$$\hat{M}_{2i} = \frac{m-1}{m} \sum_{j=1}^{m} \left(\hat{\theta}_{i,EBP}(-j) - \hat{\theta}_{i,EBP}\right)^2,$$

where $\hat{\gamma}_i(-j)$ denotes the estimator of γ_i based on all data except for the jth area; similarly, $\hat{\theta}_{i,EBP}(-j) = \tilde{\theta}_{i,EBP}(\hat{\phi}(-j))$.

Remark 3. The Empirical Best (EB) predictor $\hat{\theta}_{i,EBP}$, based on $\hat{\sigma}_u^2(w)$ defined above, has some drawbacks. Estimator $\hat{\sigma}_u^2(w)$, which is a modification of Prasad–Rao's estimator for the measurement error setup, is sometimes subject to considerable variability, just like the original estimator for the non-measurement error case proposed by Prasad and Rao (1990). This large variability results in less reliable estimators of the MSE.

8.2.2 *Bayesian Method for Functional Measurement Error Models*

In a recent article, Arima *et al.* (2014) provided a Bayesian solution for the ME problem for the area-level small area setup. They reformulated the frequentist model given above into a hierarchical Bayesian model where

1. $Y_i|W_i, \theta_i, \phi, x_i \overset{ind}{\sim} N(\theta_i, \psi_i)$, $i = 1, \ldots, m$;
2. $W_i|\theta_i, x_i \overset{ind}{\sim} N(x_i, C_i)$, $\quad i = 1, \ldots, m$;
3. $\theta_i|x_i, \phi \overset{ind}{\sim} N(x_i'\beta + z_i'\delta, \sigma_u^2)$, $i = 1, \ldots, m$;
4. Finally, a uniform prior on $x_i, i = 1, \ldots, m$, and ϕ, that is, $\pi(x_1, \ldots, x_m, \beta, \delta, \sigma_u^2) \propto 1$.

After setting $Z = (z_1, \ldots, z_m)'$ as the $m \times q$ matrix of the q precisely observed covariates for the m small areas, $X = (x_1, \ldots, x_m)'$ and $W = (W_1, W_2, \ldots, W_m)'$, and assuming that $[X : Z]$ is a full column rank matrix, the resulting posterior distribution is

$$\pi(\theta_1, \ldots, \theta_m, x_1, \ldots, x_m, \beta, \delta, \sigma_u^2 | Y_1, \ldots, Y_m, W_1, \ldots, W_m)$$
$$\propto \frac{1}{\sigma_u^m} \prod_{i=1}^{m} \exp\left(- \left[\frac{(y_i - \theta_i)^2}{2\psi_i} + \frac{(\theta_i - x_i'\beta - z_i'\delta)^2}{2\sigma_u^2} + \frac{(W_i - x_i)' C_i^{-1} (W_i - x_i)}{2} \right] \right) \tag{8.12}$$

Arima *et al.* (2014) showed that the above posterior density is proper under the very mild condition that $m > p + q + 2$. Also, when $m > p + q + 6$, the model parameters $\beta, \delta, \sigma_u^2$ have finite posterior variances. Computation with this model is very simple, since all full conditional distributions are standard distributions and a straightforward Gibbs sampling can be used.

8.3 Small Area Prediction with a Unit-level Model when an Auxiliary Variable is Measured with Error

In standard small area contexts, with data available at unit level, we suppose there are m areas and let N_i denote the known population size of area i. We denote Y_{ij} the response of the jth unit in the ith area ($i = 1, \ldots, m; j = 1, \ldots, N_i$). A random sample of size n_i is drawn from the ith area. The goal is to predict the finite population means $\theta_i = N_i^{-1} \sum_{j=1}^{N_i} Y_{ij}$ based on an observed sample. To develop reliable estimates of the small area means it is important that we have useful auxiliary information, often in the form of covariates, which may be available at different levels, sometimes at unit level, in other cases at a more aggregate level.

Battese *et al.* (1988), Prasad and Rao (1990), and Datta and Ghosh (1991) used a unit-level nested error linear regression model where the covariates are measured without errors. However, often it is not possible to measure the covariates at their precise value. For example, to quote a historical and famous example in Ghosh *et al.* (2006), in order to predict the yield of corn for counties (small areas), an effective covariate is the nitrogen in the soil. To measure the nitrogen level, it is necessary to sample the soil of experimental plot and perform lab analysis. As a result, we do not observe the true level of nitrogen, but only its estimate. In the following sections, we will review for unit-level data both functional and structural approaches to account for ME in covariates from frequentist and Bayesian perspectives.

Ghosh *et al.* (2006) and Ghosh and Sinha (2007) were the first to consider the problem of ME in small area models for unit-level data. They considered both the functional and the structural approaches and used a nested error linear regression population model with a single area-level covariate, with value x_i, at the ith area, $i = 1, \ldots, m$. True covariate x_i is measured with error for a sampled unit when observed at unit level. For example, we may be interested in estimating average crop yield for different small areas. One important covariate in predicting the yield is soil nitrogen level for that small area. True soil nitrogen level x_i is measured by taking several soil samples from each small area, which are subject to ME. As another example, the response variable may be lung capacity of a person with respiratory problems and a useful area-level covariate may be air quality in the area the individual lives in. The air quality readings, obtained from several monitoring stations in the area, may be subject to ME. In our presentation below, we consider a useful extension of the model in Ghosh *et al.* (2006).

8.3.1 Functional Measurement Error Approach for Unit-level Models

In the superpopulation approach to finite population sampling, for the response variable Y we propose the model

$$Y_{ij} = \delta + \beta x_i + u_i + \epsilon_{ij}, \quad i = 1, \ldots, m, \quad j = 1, \ldots, N_i, \tag{8.13}$$

where N_i is the population size in the ith area on which measurements are available for the response variable. It is assumed that $\epsilon_{ij}, j = 1, \ldots, N_i, i = 1, \ldots, m$ are i.i.d. $N(0, \sigma_\epsilon^2)$, and are independent of $u_i, i = 1, \ldots, m$, where $u_i \sim N(0, \sigma_u^2)$. We assume that a random sample $Y_{ij}, j = 1, \ldots, n_i$ of size n_i is selected from the ith small area, $i = 1, \ldots, m$. Sampling is assumed to be noninformative so that the population model remains valid for the sampled data. To measure the true area-level covariate, suppose there are T_i units in the ith small area, and a random sample of size t_i is taken from the ith small area, resulting in the data $X_{il}, l = 1, \ldots, t_i, i = 1, \ldots, m$. The sample $\{X_{il} : l = 1, \ldots, t_i, i = 1, \ldots, m\}$ may be obtained from a survey or may be repeated measurements of the true covariate x_i. For the sample we assume the model

$$X_{il} = x_i + \eta_{il}, \quad l = 1, \ldots, t_i, \quad i = 1, \ldots, m, \tag{8.14}$$

where η_{il}'s are i.i.d. $N(0, \sigma_\eta^2)$, $l = 1, \ldots, t_i$, $i = 1, \ldots, m$. Furthermore, $\epsilon_{ij}, j = 1, \ldots, N_i$, $i = 1, \ldots, m$, $u_i, i = 1, \ldots, m$ and η_{il}, $l = 1, \ldots, t_i, i = 1, \ldots, m$ are assumed to be mutually independent. The model described above by (8.13) and (8.14) will reduce to the model described by Ghosh and Sinha (2007) [also considered by Datta *et al.* (2010)] if we take $T_i = N_i, t_i = n_i$ and the sampled data for $\{X_{ij}, Y_{ij}\}$ are for the same units in the population.

However, this needs not be the case and area-level information on the covariate might also be gathered from different samples.

8.3.1.1 Frequentist Method for Functional Measurement Error Models

Based on the data $\{Y_{ij}, j = 1, \ldots, n_i, X_{il}, l = 1, \ldots, t_i, i = 1, \ldots, m\}$ we like to predict $\theta_i = N_i^{-1}\sum_{j=1}^{N_i} Y_{ij}, i = 1, \ldots, m$. Let $\bar{Y}_i = n_i^{-1}\sum_{j=1}^{n_i} Y_{ij}$ denote the ith small area sample mean for the response variable and $\bar{X}_i = t_i^{-1}\sum_{l=1}^{t_i} X_{il}$ denote the sample mean for the covariate. We denote the model parameters $\delta, \beta, \sigma_u^2, \sigma_\epsilon^2$ and σ_η^2 by ϕ. For known ϕ and x_i, the best unbiased predictor of θ_i that minimizes the prediction MSE is given by

$$\tilde{\theta}_i = f_i\bar{Y}_i + (1 - f_i)\left[\delta + \beta x_i + \frac{\sigma_u^2}{\sigma_u^2 + n_i^{-1}\sigma_\epsilon^2}(\bar{Y}_i - \delta - \beta x_i)\right], \tag{8.15}$$

where $f_i = n_i/N_i$ is the sampling fraction for the ith area.

In the "predictor" $\tilde{\theta}_i$ above, x_i is unknown, which needs to be estimated. To motivate an optimal "estimator" of x_i, we consider the simplest situation, when the model parameters are known. It can be checked that the likelihood function $L(x_i)$ for x_i is given by

$$L(x_i) = f(\{Y_{ij}\}_{j=1}^{n_i}, \{X_{il}\}_{l=1}^{t_i}|x_i) \propto N(\bar{Y}_i|\delta + \beta x_i, \psi_i + \sigma_u^2) \times N(\bar{X}_i|x_i, C_i), \tag{8.16}$$

where $\psi_i = \sigma_\epsilon^2/n_i$ and $C_i = \sigma_\eta^2/t_i$. This representation connects this unit-level model to the area-level model discussed earlier, where $\bar{Y}_i$ is identified to Y_i, and $\bar{X}_i$ is identified to W_i in the last section. Then as in Remark 1, an "optimal estimator" of βx_i is given by

$$\begin{aligned}&\frac{\beta^2 C_i}{\psi_i + \sigma_u^2 + \beta^2 C_i}(\bar{Y}_i - \delta) + \frac{\psi_i + \sigma_u^2}{\psi_i + \sigma_u^2 + \beta^2 C_i}\beta\bar{X}_i\\ &= \beta\bar{X}_i + \frac{\beta^2 C_i}{\psi_i + \sigma_u^2 + \beta^2 C_i}(\bar{Y}_i - \delta - \beta\bar{X}_i).\end{aligned}$$

An estimator of this type has been derived earlier by Datta *et al.* (2010) (see Theorem 2.1). Substituting this estimator of βx_i in (8.15), we get the best predictor of θ_i, which we again denote by $\tilde{\theta}_{i,BP}$, is given by

$$\begin{aligned}\tilde{\theta}_{i,BP} &= f_i\bar{Y}_i + (1 - f_i)\left[\bar{Y}_i - \frac{\psi_i}{\psi_i + \sigma_u^2 + \beta^2 C_i}(\bar{Y}_i - \delta - \beta\bar{X}_i)\right]\\ &= \bar{Y}_i - \frac{(1 - f_i)\psi_i}{\psi_i + \sigma_u^2 + \beta^2 C_i}(\bar{Y}_i - \delta - \beta\bar{X}_i).\end{aligned} \tag{8.17}$$

Note that as $f_i = n_i/N_i \to 0$, the best predictor in (8.17) approaches the best predictor of θ_i given by (8.7) for the area-level model.

We need to estimate the model parameters from the data. Sampling variances ψ_i and ME variances C_i depend on model parameters. We estimate them by estimating σ_ϵ^2 and σ_η^2 by $\hat{\sigma}_\epsilon^2 = (n - m)^{-1}\sum_{i=1}^m\sum_{j=1}^{n_i}(Y_{ij} - \bar{Y}_i)^2$ and $\hat{\sigma}_\eta^2 = (T - m)^{-1}\sum_{i=1}^m\sum_{l=1}^{t_i}(X_{il} - \bar{X}_i)^2$, respectively. Here $T = \sum_{i=1}^m t_i$. Using $\hat{\sigma}_\epsilon^2/n_i$ as ψ_i and $\hat{\sigma}_\eta^2/t_i$ as C_i, we can use the estimation method outlined in (8.9) and (8.10) to estimate β, δ and σ_u^2. Finally, we substitute these estimators in to the right-hand side of (8.17) to obtain the empirical best predictor of θ_i.

For their model, which is a special case of the model given by (8.13) and (8.14), an EBP of θ_i derived by Ghosh and Sinha (2007) is different from the one derived by Datta *et al.* (2010). While the predictor of Ghosh and Sinha (2007) is similar to the naive predictor in (8.5), outlined for the area-level model, the predictor of Datta *et al.* (2010) is an efficient version of $\tilde{\theta}_{i,BP}$ given in (8.17). Ghosh and Sinha (2007) developed consistent estimators of the model parameters and both sets of authors used these estimators to derive their respective final predictors. While Ghosh and Sinha (2007) emphasized asymptotic optimality of their final predictors, Datta *et al.* (2010), in addition to establishing the asymptotic optimality of their predictors, also provided estimators of the MSE of these predictors based on jackknife. Datta *et al.* (2010) also conducted a simulation study that showed the superiority of their final predictors in terms of empirical prediction MSE over the final predictors of Ghosh and Sinha (2007). A reason for this dominance is that unlike Ghosh and Sinha (2007) who used the sample mean $\bar{X}_i$ to estimate x_i, Datta *et al.* (2010) used an optimal estimator for x_i.

8.3.1.2 Bayesian Method for Functional Measurement Error Models

We will now consider a hierarchical Bayes (HB) analysis of the functional ME model presented by (8.13) and (8.14). Ghosh and Sinha 2007 and Datta *et al.* 2010 considered only an Empirical Bayesian (EB) analysis of the functional measurement error for unit-level data with an area-level covariate. We complete our HB specification of the model by specifying a prior distribution for the x_i, $i = 1, \dots, m$, and the model parameters $\phi = (\beta, \delta, \sigma_\epsilon^2, \sigma_u^2, \sigma_\eta^2)^T$ which is, as in the area-level model, a uniform prior, given by

$$\pi\left(x_1, \dots, x_m, \beta, \delta, \sigma_\epsilon^2, \sigma_u^2, \sigma_\eta^2\right) \propto 1.$$

Then an HB predictor, $\hat{\theta}_{i,HB}$ of θ_i will be given by

$$\hat{\theta}_{i,HB} = E\left[\tilde{\theta}_{i,BP}(\phi)|data\right], \tag{8.18}$$

where the expectation is taken with respect to the posterior distribution of ϕ, and the data are given by $Y_{ij}, j = 1, \dots, n_i, X_{il}, l = 1, \dots, t_i, i = 1, \dots, m$.

We compute this expectation by Monte Carlo integration by generating samples for $x_1, \dots, x_m$ and ϕ from the posterior distribution via Gibbs sampling. It can be shown that the set of full conditional distributions of $x_1, \dots, x_m, \beta$ and δ are all normal distribution, and the distributions of $\sigma_\epsilon^2, \sigma_u^2$ and σ_η^2 are all inverse gamma. We use the standard notation $\gamma|\cdots$ to denote the distribution of the random variable γ conditionally on all the remaining quantities. Recall that $\psi_i = n_i^{-1}\sigma_e^2$ and $C_i = t_i^{-1}\sigma_\eta^2$.

1. $x_i|\cdots \sim N\left(\frac{\beta C_i(\bar{Y}_i-\delta-u_i)+\psi_i\bar{X}_i}{\beta^2 C_i+\psi_i}, \frac{\psi_i C_i}{\psi_i+\beta^2 C_i}\right)$, independently, $i = 1, \cdots m$;
2. $u_i|\cdots \sim N\left(\frac{\sigma_u^2}{\psi_i+\sigma_u^2}(\bar{Y}_i - \delta - \beta x_i), \frac{\psi_i\sigma_u^2}{\psi_i+\sigma_u^2}\right)$, independently, $i = 1, \cdots m$;
3. $\delta|\cdots \sim N\left(\frac{\sum_{i=1}^m \psi_i^{-1}(\bar{Y}_i-\beta x_i-u_i)}{\sum_{i=1}^m \psi_i^{-1}}, \left(\sum_{i=1}^m \psi_i^{-1}\right)^{-1}\right)$;

4. $\beta|\cdots \sim N\left(\frac{\sum_{i=1}^m \psi_i^{-1}(\bar{y}_i-\delta-u_i)x_i}{\sum_{i=1}^m \psi_i^{-1}x_i^2}, \left(\sum_{i=1}^m \psi_i^{-1}x_i^2\right)^{-1}\right)$;
5. $\sigma_u^2|\cdots \sim IG\left(\frac{m-2}{2}, \frac{1}{2}\sum_{i=1}^m u_i^2\right)$;
6. $\sigma_\epsilon^2|\cdots \sim IG\left(\frac{\sum_{i=1}^m n_i-2}{2}, \frac{1}{2}\sum_{i=1}^m \sum_{j=1}^{n_i}(Y_{ij}-\delta-\beta x_i-u_i)^2\right)$
7. $\sigma_\eta^2|\cdots \sim IG\left(\frac{\sum_{i=1}^m t_i-2}{2}, \frac{1}{2}\sum_{i=1}^m \sum_{l=1}^{t_i}(X_{il}-x_i)^2\right)$.

In the above expressions $IG(a, b)$ denotes the inverse gamma distribution with density proportional to $x^{-a-1}\exp\left(-\frac{b}{x}\right)$ for $a > 0, b > 0$.

Finally, the posterior variance of θ_i will be computed from

$$V_{i,HB} = (1-f_i)^2 E\left[\frac{\psi_i^2(\sigma_u^2+\beta^2 C_i)}{\psi_i+\sigma_u^2+\beta^2 C_i} \mid data\right] + (1-f_i)N_i^{-1}E(\sigma_\epsilon^2 \mid data)$$
$$+(1-f_i)^2 V\left[\frac{\psi_i}{\psi_i+\sigma_u^2+\beta^2 C_i}(\bar{Y}_i-\delta-\beta\bar{X}_i) \mid data\right]. \qquad (8.19)$$

It can be shown that the posterior density of the above HB model will be proper if $m > 4$, $\sum_{i=1}^m n_i > 6$ and $\sum_{i=1}^m t_i > m+1$.

8.3.2 *Structural Measurement Error Approach for Unit-level Models*

In their first article dealing with measurement error in covariates for the unit-level nested error linear regression model, Ghosh *et al.* (2006) focus on the structural ME model. They considered a super-population model, given by

$$Y_{ij} = \delta + \beta x_i + u_i + \epsilon_{ij}, \quad X_{ij} = x_i + \eta_{ij}, \quad j = 1, \ldots, N_i, \quad i = 1, \ldots, m, \qquad (8.20)$$

where, as before, N_i is the population size in the ith area, and m is the number of small areas. The model given by (8.20), is a special case of the model presented in Section 8.3.1. Data consist of pairs (Y_{ij}, X_{ij}), with $j = 1, \ldots, n_i$, $i = 1, \ldots, m$, and n_i represents the sample size for the ith area. In the structural approach, in addition to the random errors ϵ_{ij}, η_{ij} and random effects u_i, unknown covariates x_i's are also considered random. We assume independence among all these components, with corresponding distributions specified by

$$\epsilon_{ij} \overset{i.i.d.}{\sim} N(0, \sigma_\epsilon^2), \quad \eta_{ij} \overset{i.i.d.}{\sim} N(0, \sigma_\eta^2), j = 1, \ldots, N_i, i = 1, \ldots, m,$$
$$x_i \overset{i.i.d.}{\sim} N(\mu_x, \sigma_x^2), \quad u_i \overset{i.i.d.}{\sim} N(0, \sigma_u^2), i = 1, \ldots, m.$$

Define the model parameter vector as $\phi = (\delta, \beta, \mu_x, \sigma_u^2, \sigma_\epsilon^2, \sigma_\eta^2, \sigma_x^2)$. While Ghosh *et al.* (2006) considered both EB and HB approaches to derive predictors of small area means θ_i, Torabi *et al.* (2009) considered only an EB approach and Arima *et al.* (2012) considered only an HB approach for the model in (8.20). We restrict our presentations in this section to the setup in (8.20), although a generalization to the model given by (8.13) and (8.14) is possible and appears more realistic.

8.3.2.1 Empirical Bayes Method for Structural Measurement Error Models

In their EB approach, Ghosh *et al.* (2006) first derived a predictor for the unsampled response vector of $N_i - n_i$ units, conditional on ϕ, and the observed sample, which they denoted by $Y^{(1)}$. In particular, for any unsampled $Y_{ij}, j = n_i + 1, \ldots, N_i$, they obtained

$$E(Y_{ij} \mid Y^{(1)}, \phi) = (1 - B_i)\bar{Y}_i + B_i(\delta + \beta\mu_x), \tag{8.21}$$

where $B_i = \sigma_e^2/[\sigma_e^2 + n_i(\sigma_u^2 + \beta^2\sigma_x^2)]$. Statistical methods of course differ in the way in which the vector ϕ is dealt with. To pursue the EB approach, Ghosh *et al.* (2006) replaced the unknown model parameters in (8.21) by their estimators to derive an EB predictor of an unsampled unit and finally of θ_i. For details on the method of moments estimation of the parameters, one can refer to Ghosh *et al.* (2006).

Torabi *et al.* (2009) considered the same model and focused on the EB estimation. They noticed that the above Bayes predictor for an unsampled Y_{ij}, proposed by Ghosh *et al.* (2006) does not involve sample information on the covariate values, which is present in the actual data and it should be adequately used in the predictor, by conditioning on the observed values, say $X^{(1)}$'s, as well. The expression of the new Bayes predictor for each of the $N_i - n_i$ not sampled units is then obtained as

$$E\left(Y_{ij} \mid Y^{(1)}, X^{(1)}, \phi\right) = \left(1 - A_i\right)\bar{Y}_i + A_i\left(\delta + \beta\mu_x\right) + A_i\frac{n_i\sigma_x^2}{\sigma_\eta^2 + n_i\sigma_x^2}\,\beta(\bar{X}_i - \mu_x),$$

with

$$A_i = \frac{\sigma_\epsilon^2(\sigma_\eta^2 + n_i\sigma_x^2)}{n_i\beta^2\sigma_x^2\sigma_\eta^2 + (n_i\sigma_u^2 + \sigma_\epsilon^2)(\sigma_\eta^2 + n_i\sigma_x^2)}.$$

Notice that both the A_i's and the B_i's depend on the area index only through the corresponding sample sizes. Using the method-of-moments estimators of the model parameters, suggested in Ghosh *et al.* (2006), the latter authors also provided their EB predictor of θ_i. Both groups of authors established asymptotic optimality of their respective EB predictors by showing that, as the number of areas $m \to \infty$, the empirical mean of the mean squared of differences of the EB predictors and the Bayes predictors goes to zero.

While Ghosh *et al.* (2006) did not provide any estimators of the MSE of their EB predictors, Torabi *et al.* (2009), using the jackknife method as in the functional case, provided an approximately unbiased estimator of the MSE of their predictors. Furthermore, both analytically and by simulations, Torabi *et al.* (2009) demonstrated the superiority of their EB predictors over the EB predictors proposed by Ghosh *et al.* (2006) [cf. Tables 1 and 2 of Torabi *et al.* (2009)].

8.3.2.2 Hierarchical Bayes Method for Structural Measurement Error Models

For the structural ME model described at the beginning of Section 8.3.2, Ghosh *et al.* (2006) proposed an HB model by specifying a prior distribution for the model parameters ϕ. While they used uniform prior distributions on the location/regression parameters μ_x, δ, β, they used conjugate inverse gamma distributions on the model variance parameters, namely, $\sigma_\epsilon^2, \sigma_u^2, \sigma_\eta^2, \sigma_x^2$. Bayesian computations with these prior distributions are straightforward via Gibbs sampling. Ghosh *et al.* (2006) proved that the posterior distribution based on their improper prior (on the regression parameters) is proper under some mild conditions.

Since Bayesian methods in small area models are often used by many National Institutes of Statistics, a certain degree of objectivity in the procedure seems natural to be required. While the use of conjugate priors by Ghosh *et al.* (2006) simplifies the computation, the shape and the scale parameters of the inverse gamma distributions were selected by them arbitrarily in a subjective manner. In an attempt to produce objective Bayesian methods for the model (8.20), Arima *et al.* (2012) computed HB predictors using the Jeffreys' prior on the model parameters derived from the marginal distribution of the data.

Standard but lengthy calculations show that the Jeffreys' prior cannot be written in a closed, convenient and readable form – the interested reader can find the details in Arima *et al.* (2012). The authors also proved that, if the number of areas is at least seven, the corresponding posterior distribution of the parameter vector is proper. For this prior, while the computing is less than straightforward, samples from the posterior distribution can be obtained via a Markov chain Monte Carlo algorithm based on Adaptive Rejection Metropolis Sampling (cf. Gilks *et al.* 1995).

Noninformative priors on hyperparameters are routinely used in hierarchical linear mixed models. The usual situation is that, while the HB and the EB predictors are in agreement, the corresponding frequentist and Bayesian measures of uncertainty are not necessarily close. This conclusion is confirmed by simulation studies in Arima *et al.* (2012).

8.4 Data Analysis

We now illustrate the Bayesian approach and the empirical best linear unbiased prediction (EBLUP) method proposed in Section 8.2 to two applications of interest to the U.S. Census Bureau. In our first illustration, we use 4-person and 3-person household state median incomes data from the 1979 U.S. Current Population Survey (CPS) to estimate the population 4-person median incomes for the 50 states and Washington, DC. In the second illustration we consider estimation of poverty ratios for the 0–4 age group again for the same domains by using data from the CPS and the American Community Survey (ACS). For more details on these methods, a reader may refer to Arima *et al.* (2014) and Ybarra and Lohr (2008).

8.4.1 Example 1: Median Income Data

We develop estimates of 4-person family median incomes for all the states; we use the 4-person household state median incomes from the CPS response, and the 3-person household state median incomes from the same survey as an auxiliary variable which is subject to ME due to sampling.

Values of the 4-person family median incomes in 1979 are also available from the 1980 Census long form. These values, being based on a much larger sample, are treated to be the true values of the median income. From the census, the median household income nationally in 1979 was \$22 014. The corresponding value from the CPS data was \$22 159. Census household income varied from state to state, ranging from a median of \$17 978 for Alabama to \$30 788 for Alaska.

The map in Figure 8.1 shows the 4-person household median incomes of the states based on the 1979 Census data. Median incomes in 17 states were below \$21 000, while 10 were above \$24 000. For the remaining states, the median income ranged from \$21 000 to \$24 000.

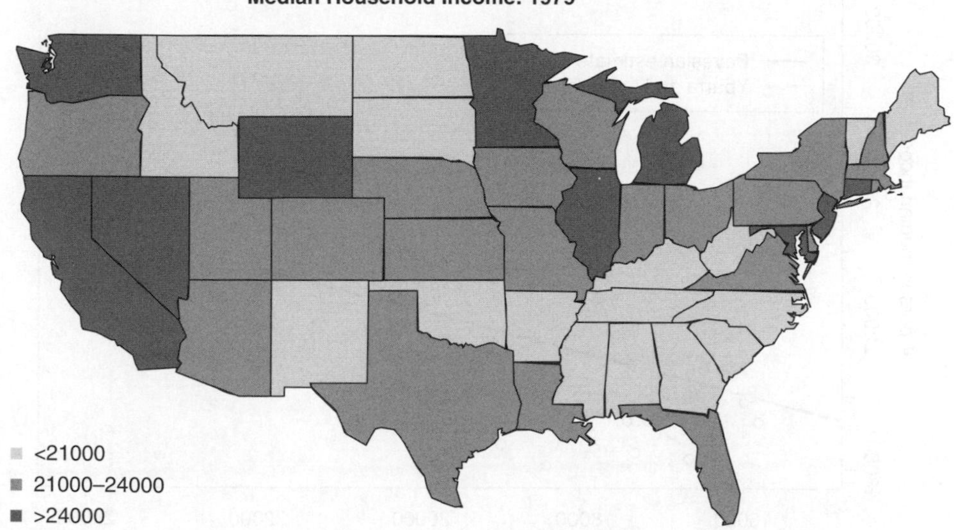

Figure 8.1 Four-person median income in 1979 (Census data)

Figure 8.2 shows the CPS 4-person family median income Y_i against the CPS 3-person family median income W_i for 51 states. The error variance C_i's have been provided in the 1979 CPS data. We have also added the regression lines estimated with the method in Ybarra and Lohr (2008) (dotted line) and with the Bayesian method (solid line). The two regression lines are quite different from each other: with the Bayesian method posterior means of the intercept and of the slope are, respectively, equal to 10775 and 0.620. Their posterior variances are, respectively, equal to 5 253 776 and 0.015. The posterior mean of σ_u^2 is equal to 3 294 921 and its posterior variance is larger than 10^{12}. On the other hand, the frequentist approach in Ybarra and Lohr (2008) yields the following estimates: −1265 for the intercept and 1.191 for the slope.

In order to evaluate the reliability of the estimates, we compare the Bayesian estimates $\hat{\theta}_B$ and the frequentist estimates $\hat{\theta}_{EBP}$ with the values of the 4-person family median income θ from the 1979 Census, which are based on the Census long form distributed to about 15% of all US households. Being based on a large sample size, the census state median incomes have negligible sampling errors. Table 8.1 shows the relative bias (in percentage) defined as

$$RB = \left(\frac{\hat{\theta} - \theta}{\theta} \right) \cdot 100.$$

The Bayesian approach yields more precise estimates, since the RB_B's are smaller than RB_{EBP}'s in more than 70% of the states.

Figure 8.3 shows a comparison of the relative accuracies of the Bayes and the frequestist predictors of 4-person family median income. Based on the 1979 Census median income for 4-person families for each state, we calculated the percent deviations of the estimates from the Census values, which are mapped in Figure 8.3. In this map, the result for the Bayesian estimates of Arima *et al.* (2012) are given in Figure 8.3a and the frequentist estimates of Ybarra

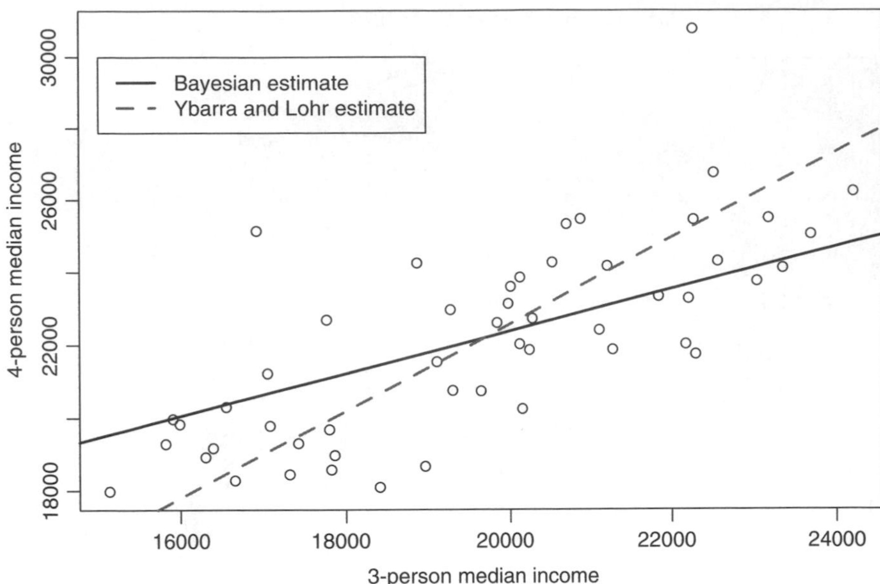

Figure 8.2 Regression lines estimated with the Bayesian method and the approach in Ybarra and Lohr (2008)

and Lohr (2008) are given in Figure 8.3b. For an estimator a state is shaded light gray if the percent relative deviation of that estimator for that state is less than −6%; it is shaded gray if the percent relative deviation is between −6% and 6%, and it is shaded black if the percent relative deviation is bigger than 6%. From the two maps we find that overall for a vast majority of states the Bayesian estimates are closer to the Census median income values than the frequentist estimates are. While for only one state the Bayesian method underestimates the "true" income by more than 6%, the income for eight states are underestimated by more than 6% by the frequentist estimates. On the other hand, the Census median income is overestimated by more than 6% for five states by the Bayesian method and for four states by the frequentist method.

We notice that the Gibbs sampling produced estimates with posterior variances of the model variance and the intercept are so large that the corresponding credible intervals are very large. It is likely that the Gibbs sampler can be trapped in a local mode indefinitely, resulting in inefficient and even unreliable samples. Moreover, since the estimates obtained with the Gibbs sampling and those obtained with the method in Ybarra and Lohr (2008) are very different, further investigations are necessary. We analytically derive the marginal posterior distribution of (β, σ_u^2) and then we approximate numerically the moments of this distribution through the Gaussian quadrature method. The posterior mean of β is equal to 0.796, with variance equal to 0.009 and the posterior mean of σ_u^2 is 3 762 684, with variance equal to 4597.102. Then we derive the posterior mean and variance of δ, via iterated expectation. The posterior mean is 6534 with variance 138,218.5. The three computational methods yield different results with a partial agreement between the estimates obtained with the Gibbs sampling and the Gaussian quadrature. Indeed, none of the estimation methods seem to outperform the others: since no method can be considered as an intrinsically infallible benchmark for estimating the model,

Table 8.1 Relative bias of the small area means obtained with the Bayesian method (RB_B) and the Ybarra and Lohr method (RB_{EBP})

State ID	RB_B	RB_{EBP}	State ID	RB_B	$RBEBP$
ME	5.72	7.58	WV	−5.96	−8.68
NH	3.53	3.53	NC	2.77	−8.87
VT	−0.18	−2.67	SC	−1.79	−10.02
MA	2.63	−2.23	GA	7.54	−0.29
RI	3.64	3.88	FL	−4.97	−3.71
CT	−3.32	1.32	KY	−0.43	3.72
NY	−3.13	−15.08	TN	1.93	−2.96
NJ	−3.59	0.57	AL	−7.30	−13.95
PA	4.84	47.94	MS	8.53	1.44
OH	−3.33	16.22	AR	8.47	3.17
IN	−1.41	−0.41	LA	−7.74	−8.10
IL	−4.76	2.33	OK	−3.35	−6.10
MI	−3.88	7.18	TX	1.11	−21.32
WI	0.30	4.40	MT	1.16	−4.34
MN	−2.90	−13.46	ID	8.47	7.78
IA	4.18	3.58	WY	−8.23	−3.41
MO	−3.52	−1.94	CO	7.30	8.55
ND	−4.83	−5.23	NM	3.86	−3.76
SD	6.65	−1.03	AZ	3.03	7.98
NE	4.72	4.79	UT	−1.59	0.62
KS	0.29	1.85	NV	0.55	3.70
DE	−5.47	−4.43	WA	0.61	−1.63
MD	−2.20	3.24	OR	2.84	0.53
DC	−0.55	−1.31	CA	−2.97	3.82
VA	4.41	5.63	AK	−10.85	−9.04
			HI	−4.28	−0.58

a consensus of multiple methods may well result in a safer estimation and better-grounded conclusions.

8.4.2 *Example 2: SAIPE Data*

The Census Bureau's Small Area Income and Poverty Estimates (SAIPE) program produces poverty estimates for various age groups for states, counties, and school districts. The SAIPE program uses the Fay–Herriot model with dependent variables obtained from direct survey poverty estimates, currently obtained from the ACS and predictor variables coming from previous census income and poverty estimates and administrative record data. Through 2004 the SAIPE program used CPS direct poverty estimates in Fay–Herriot models to produce poverty estimates for age groups 0–4, 5–17, 18–64, and 65+ by fitting the Fay–Herriot model for the CPS direct estimates with suitable covariates. During 2000–2004 both ACS demonstration surveys and the CPS provided poverty estimates for these age groups at different geographic level.

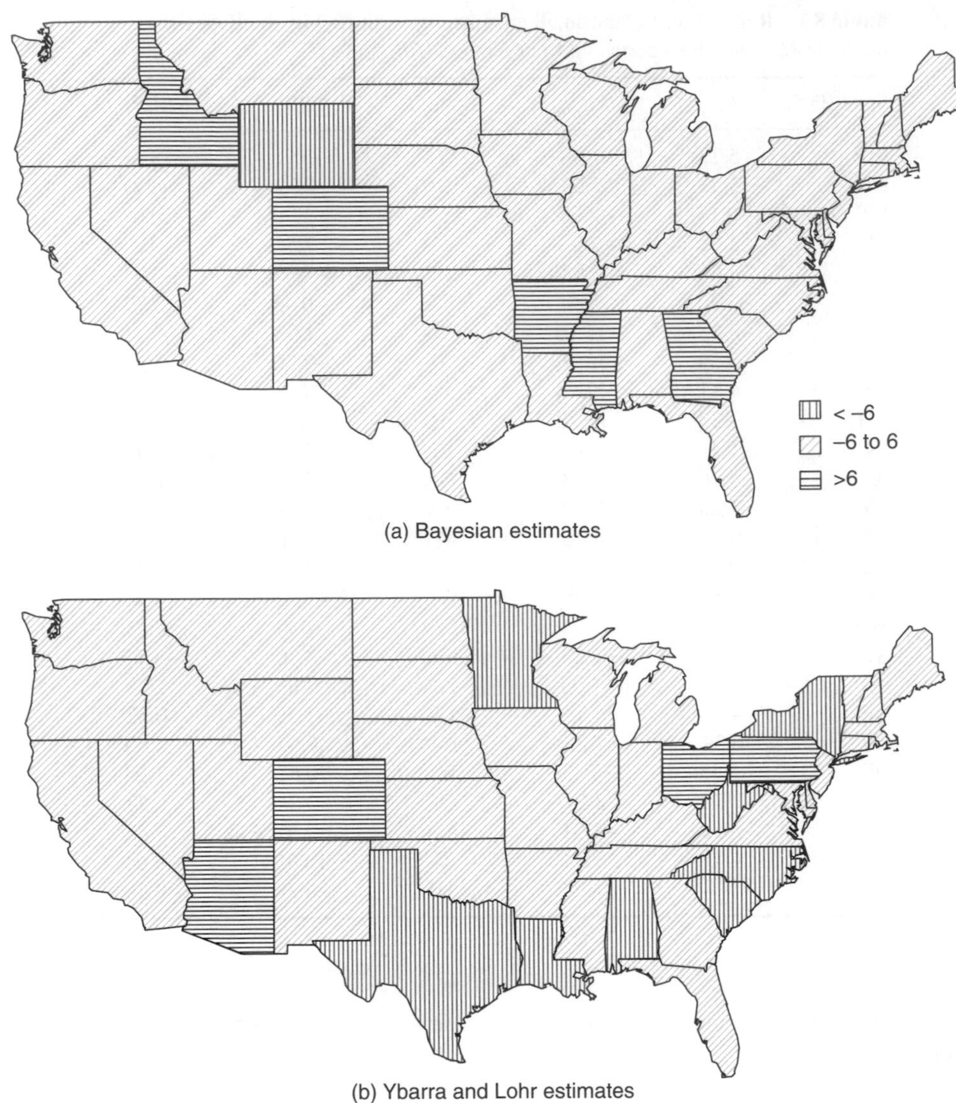

Figure 8.3 Percent difference from 1979 Census median income: Bayesian estimates (a) and frequentist estimates (b)

In a technical report, Huang and Bell (2007) investigated potential benefits to state poverty models of using data from both the surveys via a bivariate model. In the absence of any covariates, a bivariate model is equivalent to a structural ME model with one of the variables as the response and the other variable as the covariate measured with error. In this report, the authors focus mostly on the potential of the bivariate models to reduce prediction error variances for the CPS model. Their empirical results "suggest that use of the ACS data has some potential to reduce prediction error variances below those of the CPS univariate state poverty ratio models, but there are two qualifications" (p. 2). These qualifications were that the results varied over states in each case (with the bivariate model doing worse for some states), and that using restrictive assumptions in the bivariate model (like common regression coefficients in the two equations) produced apparently greater improvements over the univariate Fay–Herriot model.

Huang and Bell (2007, 2012) argue that additional precision for Fay–Herriot estimates can be obtained by using data from two different surveys ostensibly measuring the same quantity. Suppose $\{Y_{1i}, i = 1, \ldots, m\}$ and $\{Y_{2i}, i = 1, \ldots, m\}$ represent data collected from two independent surveys. Then one built a bivariate regression model with potential additional covariates. Their interesting analysis shows strong connections with our approach based on a ME model. In our set-up one assumes that the former survey – the Y_{1i}'s – represents the response variable while the latter – the Y_{2i}'s – acts as the covariate, measured with error. In the Huang and Bell scenario, the two vectors are both used as response variables, and the ME component gets, in some sense, translated into a random component. While Huang and Bell (2007) included covariates in their bivariate models, we use 2004 state level poverty ratio data for the 0–4 age group from the CPS and the ACS alone to illustrate our ME model [i.e., without using the additional covariates used by Huang and Bell (2007)].

We fit a Fay–Herriot model for the CPS poverty ratio with the ACS estimate as the lone covariate, first ignoring the sampling error in the ACS estimate and later using the ME approach to account for the sampling error. Figure 8.4 shows the regression lines obtained with the

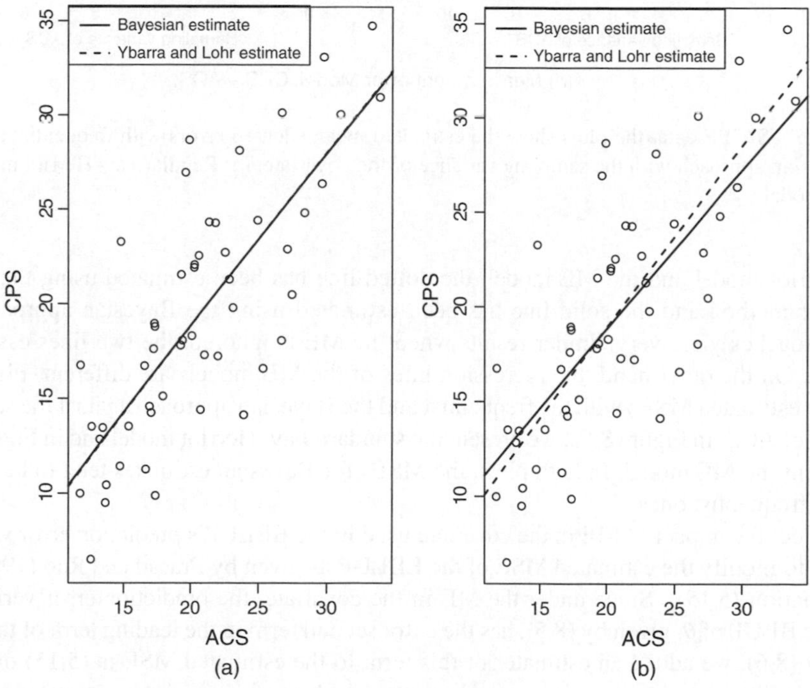

Figure 8.4 SAIPE data: regression lines estimated with the frequentist (broken line) and the Bayesian approach (solid line). (a) Regular Fay–Herriot model and (b) ME model

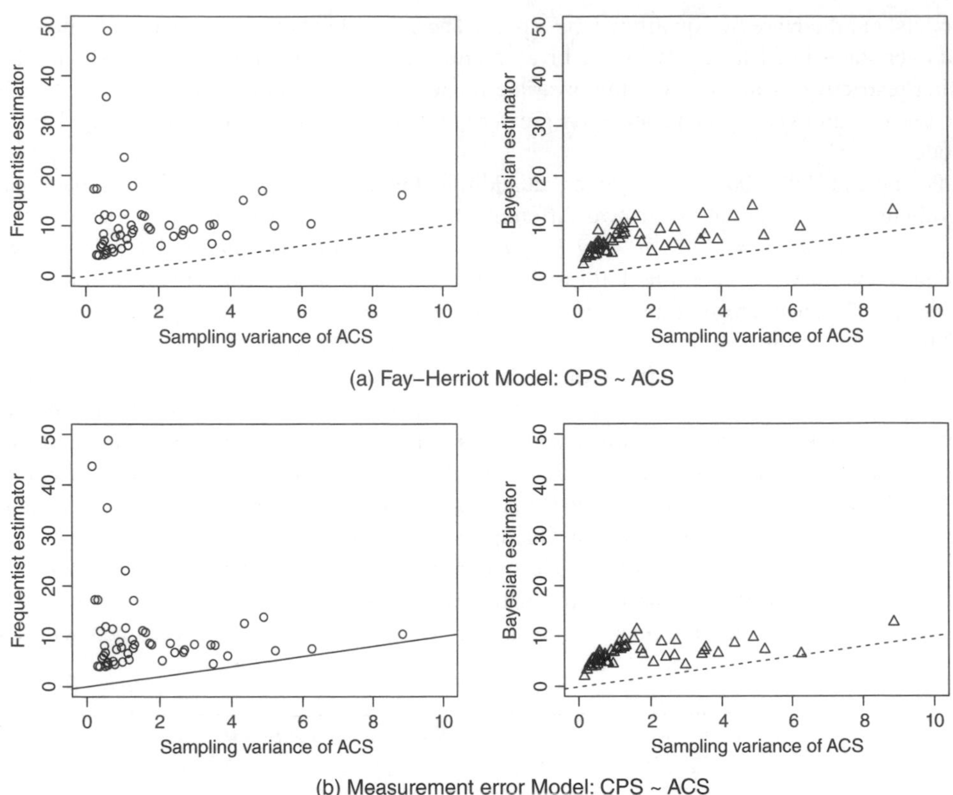

Figure 8.5 SAIPE data: the plots show the estimated mean squared errors with frequentist approach and Bayesian approach with the sampling variance of the covariate. (a) Regular Fay–Herriot model and (b) ME model

Fay–Herriot model and the ME model: the dotted line has been estimated using the Ybarra and Lohr method and the solid line has been estimated using the Bayesian approach. The two approaches yield very similar results when the ME is ignored: the two lines essentially coincide. On the other hand, the regression lines of the ME models are different. Figure 8.5 show the estimated MSEs with the frequentist and the Bayesian approach against the sampling variance of ACS. In Figure 8.5a, we present the standard Fay–Herriot model and in Figure 8.5b we present the ME model. In both plots, the MSEs for Bayesian estimates tend to be smaller than the frequentist ones.

To reflect the impact of ME in the covariate used in the EBLUP's prediction error variance, we need to modify the estimated MSE of the EBLUP as given by Prasad and Rao (1990) [see their equation (5.15)]. Since under the ME in the covariate, the prediction error variance of the naive BLUP of θ_i given by (8.5), has the extra second term in the leading term of the MSE, given by (8.6), we added an estimator of this term to the estimated MSE in (5.15) of Prasad and Rao (1990). Admittedly, this is an ad hoc attempt. A similar argument, presented below, is needed for correctly measuring the Bayesian prediction MSE when the covariate is measured with error. Let $\hat{\theta}_i$ denote an estimator of θ_i. Then, under the ME version of the Fay–Herriot–HB model, the uncertainty associated with $\hat{\theta}_i$ is measured by the posterior MSE, denoted by $pm(\hat{\theta}_i)$,

$$pm(\hat{\theta}_i) = V_i + \{\hat{\theta}_i - \hat{\theta}_{i,ME}\}^2, \tag{8.22}$$

where $\hat{\theta}_{i,ME}$ is the HB estimate of θ_i based on the posterior density (8.12) and V_i is the posterior variance of θ_i under the same posterior density.

Now, if we take $\hat{\theta}_i = \hat{\theta}_{i,FH-NM}$, the HB estimate of θ_i under the standard Fay–Herriot model *without* ME, in (8.22), then we get the posterior MSE under *no ME* model of the usual Fay–Herriot–HB estimate of θ_i.

8.5 Discussion and Possible Extensions

Small area estimation methods typically combine direct estimates from a survey with prediction from a model in order to obtain accurate estimates of population quantities. Covariates used in the model strongly affect the accuracy of these estimates. Often, however, it is not possible to measure the covariates without any error. In this chapter we have presented a review of the existing literature, considering unit-level and area-level models from a frequentist and a Bayesian perspective. We have illustrated some of the reviewed methods using data from the 1979 U.S. CPS. We developed a state level estimate of the 4-person family median income; we used the 3-person family median income from the CPS as auxiliary variable which is subject to ME. The ACS poverty ratio for 2004 was used as a covariate when modeling the 2004 CPS poverty ratios.

Although the topic of ME has been deeply investigated in the small area literature, some aspects may be improved. All the approaches reviewed in this chapter consider the auxiliary variables as independent, that is the p covariates $X_{1i}, \ldots, X_{pi}$, recorded at the area i are assumed to be independent. This assumption may be largely unrealistic in those situations where the auxiliary information comes from the same source and a reasonable degree of correlation among the X_i's is reasonable. We are currently working on extending the model in Arima *et al.* (2014) as a multivariate Fay–Herriot model with structural ME and correlated auxiliary variables.

Small area model selection and variable selection are other interesting issues recently addressed in a frequentist context in Datta *et al.* (2011). These problems have not been considered in the context of ME in the covariates, and the general methods proposed in Bayarri *et al.* (2012) cannot be used without some adjustment. This is a promising topic of research which we will consider in the future.

Acknowledgements

The authors express their gratitude to Dr William R. Bell for his very careful reading and many useful suggestions on an earlier version of the manuscript. His extensive comments and the data he provided for the SAIPE example led to a significant improvement of the initial draft. The authors are also grateful to Dr Jerry Maples for his comments on the original chapter.

Disclaimer

This chapter is released to inform interested parties of research and to encourage discussion of work in progress. The views expressed are those of the authors and not necessarily those of the U.S. Census Bureau.

References

Arima S, Datta G and Liseo B 2012 Objective Bayesian analysis of a measurement error small area model. *Bayesian Analysis* **7**, 363–384.

Arima S, Datta G and Liseo B 2014 Bayesian estimators for small area models when auxiliary information is measured with error. *Scandinavian Journal of Statistics*. DOI: 10.1111/sjos.12120

Battese G, Harter R and Fuller W 1988 An error components model for prediction of county crop areas using survey and satellite data. *Journal of the American Statistical Association* **83**, 23–36.

Bayarri M, Berger J, Forte A and García-Donato G 2012 Criteria for Bayesian model choice with application to variable selection. *Annals of Statistics* **40**, 1550–1577.

Carroll RJ, Ruppert D, StefanskiL and CrainiceanuC 2006 *Measurement Error in Nonlinear Models: a Modern Perspective* 2nd edn. Chapman & Hall/CRC, London.

Cheng CL and van Ness JW 1999 *Statistical Regression with Measurement Error vol. 6 of Kendall's Library of Statistics*. Arnold, London; Oxford University Press, New York.

Datta G 2009 Model-based approach to small area estimation. In *Handbook of Statistics: Sample Surveys: Inference and Analysis*, Eds D. Pfeffermann and C.R. Rao, vol. 29B. North-Holland, Amsterdam, pp. 251–288.

Datta G and Ghosh M 1991 Bayesian prediction in linear models: applications to small area estimation.. *Annals of Statistics* **19**, 1748–1770.

Datta G, Hall P and Mandal A 2011 Model selection by testing for the presence of small-area effects and application to area-level data. *Journal of the American Statistical Association* **106**, 362–374.

Datta G, Rao J and Torabi M 2010 Pseudo-empirical Bayes estimation of small area means under a nested error linear regression model with functional measurement errors. *Journal of Statistical Planning Inference* **250**, 2952–2962.

Fay RE and Herriot RA 1979 Estimates of income for small places: an application of James-Stein procedures to census data. *Journal of the American Statistical Association* **74**, 269–277.

Fuller WA 1990 Prediction of true values for the measurement error models *Statistical Analysis of Measurement Error Models and Applications* vol. 112 of *Contemporary Mathematics*. American Mathematical Society, Amer. Math. Soc., Providence, RI, pp. 41–57.

Ghosh M and Sinha K 2007 Empirical Bayes estimation in finite population sampling under functional measurement error models. *Journal of Statistical Planning Inference* **137**, 2759–2773.

Ghosh M, Sinha K and Kim D 2006 Empirical and hierarchical Bayesian estimation in finite population sampling under structural measurement error models. *Scandinavian Journal of Statistics* **33**, 591–608.

Gilks W, Best N and Tan K 1995 Adaptive rejection metropolis sampling. *Applied Statistics* **44**, 455–472.

Huang E and Bell W 2007 An Empirical Study on Using CPS and ACS Survey Data in Bivariate State Poverty Models. *Rearch Report Series – Center for Statistical Research & Methodology, U.S. Bureau of Census*.

Huang E and Bell W 2012 An Empirical Study on Using Previous American Community Survey Data versus Census 2000 Data in SAIPE Models for Poverty Estimates. *Research Report Series – Center for Statistical Research & Methodology, U.S. Bureau of Census*.

Jiang J and Lahiri P 2006 Mixed model prediction and small area estimation. *Test* **15**, 1–96.

Jiang J, Lahiri P and Wan S 2002 A unified jackknife theory for empirical best prediction with m-estimation. *Annals of Statistics* **30**, 1782–1810.

Pfeffermann D 2013 New important developments in small area estimation. *Statistical Science* **28**, 40–68.

Prasad N and Rao J 1990 The estimation of mean squared error of small area estimators. *Journal of the American Statistical Association* **85**, 163–171.

Rao J 2003 *Small Area Estimation*. John Wiley & Sons, Inc., Hoboken, NJ.

Torabi M, Datta G and Rao J 2009 Empirical Bayes estimation of small area means under nested error linear regression model with measurement errors in the covariates. *Scandinavian Journal of Statistics* **36**, 355–368.

Ybarra L and Lohr S 2008 Small area estimation when auxiliary information is measured with error. *Biometrika* **95**, 919–931.

9

Robust Domain Estimation of Income-based Inequality Indicators

Nikos Tzavidis[1] and Stefano Marchetti[2]
[1]*Department of Social Statistics and Demography, University of Southampton, Southampton, UK*
[2]*Department of Economics and Management, University of Pisa, Pisa, Italy*

9.1 Introduction

Social deprivation is the reduction or prevention of culturally normal interaction between an individual and the rest of society. Social deprivation is included in a broad network of correlated factors that contribute to social exclusion and include mental illness, poverty, poor education, and low socioeconomic status. The EU has proposed a core of statistical indicators to quantify deprivation and social exclusion in each country commonly known as Laeken indicators (see Chapter 2 for more information).

The EU Survey of Income and Living Conditions (EU-SILC) represents the most important source of information for estimating Laeken indicators at national or regional levels (NUTS 1 and 2 levels). Nevertheless, policy makers require estimates of social exclusion and living conditions at finer levels of geographical/domain disaggregation. Direct estimation at the requested levels of disagregation by using the EU-SILC data will in most cases lead to inaccurate estimates. This is mainly due to the small sample sizes for the domains of interest. Given that oversampling is not feasible due to budget constraints, there is need to resort to small area estimation methods. Note that in this chapter the terms domain and area will be used interchangeably.

During the last two decades there has been substantial growth in the development and application of model-based small area methods. Although the majority of this literature focuses on the estimation of domain averages, recent methodology has also focused on the estimation of non-linear statistics for example, the estimation of the incidence of income poverty, the

Analysis of Poverty Data by Small Area Estimation, First Edition. Edited by Monica Pratesi.

Companion Website: www.wiley.com/go/pratesi/poverty

poverty gap and percentiles of the income distribution function. More specifically, until very recently the industry standard for estimating poverty indicators was based on the World Bank (WB) method (Elbers *et al.*, 2003). Molina and Rao (2010) proposed Empirical Best Predictors (EBP) for poverty indicators. Provided that the normality assumptions of the nested error regression model hold, the Molina and Rao approach offers the best estimation method for in-sample domains. Finally, Marchetti *et al.* (2012) and Tzavidis *et al.* (2013) proposed robust small area methods of poverty indicators when the parametric assumptions of the nested error regression model do not hold.

In this chapter we will exclusively focus on extending the methodology in Marchetti *et al.* (2012) and Tzavidis *et al.* (2013) for estimating two Laeken indicators namely, the income quintile share ratio and the Gini coefficient. The remainder of the chapter is structured as follows. In Section 9.2 we define the two income-based inequality measures of interest for this chapter. In Section 9.3 we present the M-quantile-based small area methodology for estimating the two inequality measures and in Section 9.4 we present a Mean Squared Error (MSE) estimator for estimating the uncertainty of the small area estimates. Section 9.5 presents results from Monte Carlo empirical evaluations and Section 9.6 presents results from the application of the methodology to data from the Italian EU-SILC 2008 and the Italian Census 2001. The aim of this application is to obtain estimates of the Gini coefficient and the income quintile share ratio for unplanned domains in Tuscany. The term unplanned is used to denote domains that do not feature in the design and allocation of the EU-SILC sample. For the application in this chapter these domains are defined by provinces in Tuscany cross-classified by the gender of the head of the household. We conclude the chapter with some final remarks and open areas for research.

9.2 Definition of Income-based Inequality Measures

In this section we provide a general definition of the two income-based inequality measures of interest. Denote by n the sample size, by $q_{0.2}$ and $q_{0.8}$ the 20th and 80th percentiles of the income distribution and by y_j the corresponding income for household j. The income quintile share ratio ($S80/S20$) is the ratio of the total income received by the 20% of the country's population with the highest income (top quintile) to that received by the 20% of the country's population with the lowest income (lowest quintile). Income, here, is defined as the equivalized total net income. For each individual the equivalized total net income is calculated as its household total net income divided by the equivalized household size according to the modified OECD scale. In the OECD scale the head of the household has weight equal to 1, other household members aged 14 or more have a weight of 0.5, and members aged 13 or less have a weight of 0.3. The quintile share ratio is defined as follows:

$$S80/S20 = \frac{\sum_{j=1}^{N} [y_j I(y_j > q_{0.8})]}{\sum_{j=1}^{N} [(y_j I(y_j < q_{0.2})]}. \tag{9.1}$$

Denote further by (j) the order of the income of household j. We now assume that y_j denotes the ordered income values. The Gini coefficient is defined as follows:

$$G = \frac{n+1}{n} - \frac{-2\sum_{j=1}^{n}(n+1-(j))y_j}{n\sum_{j=1}^{n} y_j}. \tag{9.2}$$

9.3 Robust Small Area Estimation of Inequality Measures with M-quantile Regression

In this section we describe the methodology for estimating the two Laeken inequality measures for small area (domains) of interest. The methodology we employ is based on the use of M-quantile models for small area estimation (Chambers and Tzavidis, 2006) and their extension for estimating deprivation indicators (Marchetti *et al.*, 2012; Tzavidis *et al.*, 2013).

In what follows we assume that a vector of p auxiliary variables $\mathbf{x}_{jd}$ is known for each population unit j in small area d and that values of the income variable of interest y_{jd} are available from a random sample, s, that includes units from all target domains. We denote the population size, sample size, sampled part of the population and non-sampled part of the population in each domain, respectively, by N_d, n_d, s_d and r_d. We further assume that conditional on the covariates available, the sampling design is ignorable.

The small area estimation methodology we describe in this section assumes the availability of survey data on an income variable and explanatory variables that can be used for modeling the outcome variable. In addition, the methods assume the availability of Census/administrative data on the same set of explanatory variables. Some methods further assume that the Census and survey data are linked. However, this assumption is fairly unrealistic as in most cases the link between the survey and the Census data is unknown. Having said this, the estimation methods can be modified so that the linkage assumption is not necessary. Finally, the methods that are based on the nested error regression model (WB and EBP) conventionally use a logarithmic transformation of the welfare variable. In contrast, the M-quantile approach uses the raw values of the income variable. Nevertheless, before proceeding to small area estimation, it is always advisable to use model diagnostics. Depending on the results of the model diagnostics, all methods can be implemented either by using the raw or the transformed values of the income variable.

The classical regression model summarizes the behavior of the mean of a random variable at each point in a set of covariates. Instead, quantile regression summarizes the behavior of different parts of the conditional distribution $f(y|\mathbf{x})$ at each point in the set of the $\mathbf{x}$'s. Let us for the moment and for notational simplicity drop the area-specific subscript d. Suppose that $(\mathbf{x}_j, y_j)$, $j = 1, \ldots, n$, denotes the observed values of a random sample consisting of n units, where $\mathbf{x}_j$ are row p-vectors of a known design matrix $\mathbf{x}$ and y_j is a scalar response variable corresponding to a realization of a continuous random variable with unknown continuous cumulative distribution function. A linear regression model for the q conditional quantile of $f(y|\mathbf{x})$ is

$$Q_y(q|\mathbf{x}) = \mathbf{x}^T \boldsymbol{\beta}_q.$$

Estimates of the quantile regression parameter $\boldsymbol{\beta}_q$ are obtained by minimizing

$$\sum_{j=1}^{n} |y_j - \mathbf{x}_j^T \boldsymbol{\beta}_q| \{(1-q)I(y_j - \mathbf{x}_j^T \boldsymbol{\beta}_q \le 0) + qI(y_j - \mathbf{x}_j^T \boldsymbol{\beta}_q > 0)\}.$$

M-quantile regression (Breckling and Chambers, 1988) is a "quantile-like" generalization of regression based on influence functions (M-regression). The M-quantile q of the conditional density $f(y|\mathbf{x})$, m, is defined as the solution to the estimating equation

$$\int \psi_q(y - m)f(y|\mathbf{x})\, dy = 0,$$

where ψ_q denotes an asymmetric influence function, which is the derivative of an asymmetric loss function ρ_q. A linear M-quantile regression model is defined by

$$m_y(q|\mathbf{x}) = \mathbf{x}^T \boldsymbol{\beta}_{\psi,q}.$$

Estimates of $\boldsymbol{\beta}_{\psi,q}$ are obtained by minimizing

$$\sum_{j=1}^{n} \rho_q(y_j - \mathbf{x}_j^T \boldsymbol{\beta}_{\psi,q}). \tag{9.3}$$

Throughout this chapter we will take the linear M-quantile regression model to be defined by using as ρ_q the Huber loss function (Breckling and Chambers, 1988). The estimating equation defined by (9.3) is

$$\sum_{j=1}^{n} 2\psi(r_{jq})\{(1-q)I(r_{jq} \leq 0) + qI(r_{jq} > 0\} = 0,$$

where $r_{jq} = s^{-1}(y_j - \mathbf{x}_j^T \boldsymbol{\beta}_{\psi,q})$, and s is an estimate of scale such as the Mean Absolute Deviation. Provided that the tuning constant of the influence function is strictly greater than zero, estimates of $\boldsymbol{\beta}_{\psi,q}$ are obtained by using iterative weighted least squares (IWLS).

Chambers and Tzavidis (2006) extended the use of M-quantile regression to small area estimation. Following their development, these authors characterize the conditional variability across the population of interest by the M-quantile coefficients of the population units. For unit j with values y_j and $\mathbf{x}_j$, this coefficient is the value θ_j such that $m_y(\theta_j|\mathbf{x}_j) = y_j$. If a hierarchical structure does explain part of the variability in the population data, we expect units within domains to have similar M-quantile coefficients. An area specific semi-parametric (empirical) pseudo-random effect, θ_d, is then computed by taking the expected value of the M-quantile coefficients θ_j in area d.

Having presented the M-quantile small area model, we now focus on the estimation of the Laeken indicators of interest namely, the Gini coefficient and the quintile share ratio. Let us define the *S80/S20* for area d by

$$S80/S20_d = \frac{\sum_{j\in s_d} y_j I(y_j > q_{d,0.8}) + \sum_{k\in r_d} y_k I(y_k > q_{d,0.8})}{\sum_{j\in s_d} y_j I(y_j < q_{d,0.2}) + \sum_{k\in r_d} y_k I(y_k < q_{d,0.2})}. \tag{9.4}$$

Notice that for the purposes of this chapter we will be using definition (9.4) which assumes area-specific income quantiles. Alternatively, one may decide to use income quantiles defined at an aggregate level leading to an alternative definition of *S80/S20*. In (9.4) y_j, $j \in s_d$, are the realized household income values in the sample and y_k, $k \in r_d$, are the unobserved out of sample income values. In order to estimate the area-specific *S80/S20*, the y_k's are replaced by their expectation under the model. Moreover, since linked Census and survey data hardly ever exist, we further replace the sample y values also by their expectation under the model. This leads to the following definition of *S80/S20* at the population level

$$S80/S20_d = \frac{\sum_{j \in U_d} E(y_j I(y_j > q_{d,0.8}))}{\sum_{j \in U_d} E(y_j I(y_j < q_{d,0.2}))}, \tag{9.5}$$

where U_d is the set of population units in area d.

A non-parametric approach of estimating $E(y_j I(y_j > q_d))$ is offered by using a smearing-type estimator motivated by the work of Duan (1983) leading to

$$E(y_j I(y_j > q_d)) = \int (\mathbf{x}_j^T \boldsymbol{\beta}_{\psi,\theta_d} + \epsilon) I((\mathbf{x}_j^T \boldsymbol{\beta}_{\psi,\theta_d} + \epsilon) > q_d)\, dF(\epsilon). \tag{9.6}$$

Assuming that the distribution of the error term $F(\epsilon)$ is unknown, we estimate $F(\epsilon)$ by using the empirical distribution function of the estimated model residuals. Further, substituting $\boldsymbol{\beta}_{\psi,\theta_d}$ by their estimates under the M-quantile small area model $\hat{\boldsymbol{\beta}}_{\psi,\hat{\theta}_d}$ leads to

$$\hat{E}(y_j I(y_j > q_d)) = \int (\mathbf{x}_j^T \hat{\boldsymbol{\beta}}_{\psi,\hat{\theta}_d} + \hat{\epsilon}) I((\mathbf{x}_j^T \hat{\boldsymbol{\beta}}_{\psi,\hat{\theta}_d} + \hat{\epsilon}) > q_d)\, d\hat{F}(\epsilon). \tag{9.7}$$

Evaluating $\hat{E}(y_j I(y_j > q_d))$ is achieved by means of Monte Carlo simulation. In particular, having estimated the M-quantile model residuals, we draw a replacement sample of size N_d from the empirical distribution of these residuals. This allows us to microsimulate a population of synthetic income values for each small area from which we estimate $q_{0.8,d}$, $q_{0.2,d}$ and $\hat{E}(y_j I(y_j > q_d))$. The Monte Carlo simulation is repeated for $h = 1, \ldots, H$ times each time estimating $\widehat{S80/S20}_d^{\,h}$. The *S80/S20* estimate in area d is obtained by taking the average of the *S80/S20* values over the Monte Carlo replications.

A similar approach to the one described above is used for estimating the area-specific Gini coefficient G_d. Let us define the Gini coefficient in area d by

$$G_d = \frac{N_d + 1}{N_d} - \frac{-2 \sum_{j \in U_d} (N_d + 1 - (j)) y_j}{N_d \sum_{j \in U_d} y_j}. \tag{9.8}$$

Similarly to the case of *S80/S20*, y_j in (9.8) is replaced by its expectation under the model

$$\hat{E}(y_j) = \int (\mathbf{x}_j^T \hat{\boldsymbol{\beta}}_{\psi,\hat{\theta}_d} + \hat{\epsilon})\, d\hat{F}(\epsilon).$$

The expectation is evaluated also by using Monte Carlo simulation.

9.4 Mean Squared Error Estimation

MSE estimation for small area estimators under the M-quantile model is discussed in detail in Marchetti *et al.* (2012) and is based on the use of a non-parametric bootstrap scheme. Here we recall the main steps of this bootstrap scheme. Starting from sample s, selected from a finite population U without replacement, we fit the M-quantile small area model and obtain estimates $\hat{\theta}$ and $\hat{\boldsymbol{\beta}}_{\psi,\hat{\theta}_d}$ which are used to compute the model residuals. We then generate B bootstrap populations, U^{*b}. From each bootstrap population we select L bootstrap samples using simple random sampling within the small areas and without replacement such that $n_d^* = n_d$.

Using the bootstrap samples we obtain estimates of the inequality indicators of interest by using the methodology described in Section 9.2. Bootstrap populations are generated by sampling from the empirical distribution of the residuals or from a smoothed version of this distribution conditionally or unconditionally on the small areas. Denoting by $\hat{\tau}_d$ the estimated small area parameter, bootstrap estimators of the bias and variance are defined, respectively, by

$$\hat{B}(\hat{\tau}_d) = B^{-1}L^{-1}\sum_{b=1}^{B}\sum_{l=1}^{L}(\hat{\tau}_d^{*bl} - \tau_d^{*b}),$$

$$\hat{V}(\hat{\tau}_d) = B^{-1}L^{-1}\sum_{b=1}^{B}\sum_{l=1}^{L}(\hat{\tau}_d^{*bl} - \hat{\bar{\tau}}_d^{*bl})^2.$$

In the expressions for the bias and the variance τ_d^{*b} is the small area parameter of the b bootstrap population, $\hat{\tau}_d^{*bl}$ is the small area parameter estimated by using the l sample from the b bootstrap population, and $\hat{\bar{\tau}}_d^{*bl} = L^{-1}\sum_{l=1}^{L}\hat{\tau}_d^{*bl}$. The bootstrap MSE estimator of the estimated small area target parameter is then defined as

$$\widehat{MSE}(\hat{\tau}_d) = \hat{V}(\hat{\tau}_d) + \hat{B}(\hat{\tau}_d)^2.$$

9.5 Empirical Evaluations

In this section we present results from a simulation study that was used to compare the performances of the various small area estimators defined in the preceding sections. The study was conducted by a model-based simulation in which small area population and sample data were simulated based on a two-level linear mixed model with different parametric assumptions for the area- and unit-level random effects. The sample design used in this simulation was a stratified random sampling, where the strata correspond to the small areas and the allocation is proportional to the population size in the small area.

We generated data for $D = 30$ small areas. The total population size is $N = 6000$ with $100 \leq N_d \leq 300, d = 1, \ldots, D$. For each area, we selected a simple random sample (without replacement) of size $5 \leq n_d \leq 8$, leading to an overall sample size of $n = 175$. The sample values of the target variable y and the population values of the auxiliary variable x were then used to estimate the small area quintile share ratio and the Gini coefficient. This process was repeated 1000 times. Values of y are generated using a random intercept model $y_{jd} = 5000 + \beta x_{jd} + u_d + e_{jd}$, with x_{jd} generated as independently and identically distributed realizations from a mixture model $(1-\gamma)N(\mu_d, 1) + \gamma N(\eta_d, 1)$. The small area x-means μ_d and η_d were themselves drawn at random from the uniform distribution on the interval $[1, 10]$ and

[81, 90], respectively, and held fixed over the simulations. The weight of the mixture, γ, was set to 0.2 and held fixed over the simulations.

In the first simulation experiment (Scenario 1) the random effects u_d and e_{dj} were independently and identically generated as $N(0, 40\,000)$ and $N(0, 640\,000)$ realizations respectively. In this scenario β was set equal to 250. In the second simulation experiment (Scenario 2) the random effects u_d and e_{jd} were independently and identically generated as mean-corrected Singh–Maddala distribution realizations, both with parameters $(a = 2.8, b = 100^{-5/14}, q = 1.7)$. The Singh–Maddala distribution has density function equal to

$$f(y) = \frac{aqy^{a-1}}{b^a\left(1 + \frac{y^a}{b^a}\right)^{1+q}}.$$

The mean correction was set as $15\,000u_d - D^{-1}\sum_{d=1}^{D} 15\,000u_d$ and $15\,000e_{jd} - D^{-1}n_d^{-1}\sum_{d=1}^{D}\sum_{j=1}^{n_j} 15\,000e_{jd}$. The purpose of Scenario 2 was to examine the effect of misspecification of the Gaussian assumptions of a mixed model using a density function able to mimic an income distribution.

The use of a mixture distribution for generating the values of the auxiliary variable is to ensure realistic values of the Gini coefficient and the quintile share ratio in the simulated population. In Scenario 2 the Gini coefficient varies between 0.12 and 0.5 and the quintile share ratio between 1.9 and 11.8 over areas and simulations. In Scenario 1 the Gini coefficient and the quintile share ratio range, respectively, between 0.29 and 0.39, and 3.9 and 6.2. These values are considered realistic in the sense that in the Euro Area in 2012 Eurostat reported a Gini coefficient between 0.23 (Norway) and 0.36 (Latvia) and a quintile share ratio between 3.5 (Czech Republic) and 7.2 (Spain) (http://epp.eurostat.ec.europa.eu/portal/page/portal/income_social_inclusion_living_conditions/data/main_tables).

Using the simulated data, point estimation and MSE estimation is performed using the methodology in Sections 9.3 and 9.4. All computations were performed by using R (R Development Core Team, 2013). Biases and root MSEs over these simulations, summarized over the 30 areas, are shown in Table 9.1, Table 9.2, and Table 9.3. Moreover, in Table 9.4 and Table 9.5 we also show the bias of the root MSE estimator we presented in Section 9.4.

Starting with the results in Table 9.1 we observe a tendency for both estimators to have negative bias. However, the bias is fairly small with the maximum absolute relative bias being 3.7% for the Gini coefficient and 7.3% for the quintile share ratio.

Results for Scenario 2 are shown in Table 9.2. Using a heavy tailed error distribution results in an increase, albeit small compared with Scenario 1, in the bias for both estimators.

For the Gini coefficient the mean of the relative bias is now equal to −3% and the mean of the absolute relative bias equal to 5.6%. For the quintile share ratio the mean of the relative bias and the absolute relative bias, respectively, are equal to −3.5 and 8%.

Table 9.3 shows the empirical MSE of the Gini coefficient and quintile share ratio estimators for both scenarios summarized over areas. The empirical MSE for an estimator $\hat{\theta}_d$ of θ_d is computed as

$$MSE(\hat{\theta}_d) = N^{-1}\sum_{h=1}^{H}(\hat{\theta}_{dh} - \theta_{dh})^2,$$

where θ_{dh} and $\hat{\theta}_{dh}$ are, respectively, the true and the estimated values of the target statistics (Gini coefficient or quintile share ratio) for area d in the iteration h of the Monte Carlo

Table 9.1 Scenario 1 (normally distributed errors). Distribution over areas and Monte Carlo simulations of the bias, absolute bias, relative bias and relative absolute bias of the Gini coefficient (*G*) and the quintile share ratio (*S80/S20*)

	Min.	1st Qu.	Median	Mean	3rd Qu.	Max.
bias($\hat{G}$)	−0.005	−0.004	−0.004	−0.003	−0.003	−0.002
Abs. bias($\hat{G}$)	0.009	0.010	0.011	0.011	0.012	0.013
Rel. bias($\hat{G}$) %	−1.340	−1.130	−1.050	−1.000	−0.890	−0.630
Rel. abs. bias($\hat{G}$) %	2.770	3.080	3.410	3.310	3.580	3.690
bias($\widehat{S80/S20}$)	−0.133	−0.104	−0.090	−0.091	−0.079	−0.058
Abs. bias($\widehat{S80/S20}$)	0.228	0.279	0.312	0.313	0.350	0.391
Rel. bias($\widehat{S80/S20}$) %	−2.430	−1.980	−1.800	−1.770	−1.510	−1.140
Rel. abs. bias($\widehat{S80/S20}$) %	5.120	5.540	6.240	6.170	6.700	7.320

Table 9.2 Scenario 2 (Singh–Maddala distributed errors). Distribution over areas and Monte Carlo simulations of the bias, absolute bias, relative bias and relative absolute bias of the Gini coefficient (*G*) and the quintile share ratio (*S80/S20*)

	Min.	1st Qu.	Median	Mean	3rd Qu.	Max.
bias($\hat{G}$)	−0.012	−0.010	−0.009	−0.009	−0.008	−0.006
Abs. bias($\hat{G}$)	0.014	0.016	0.017	0.017	0.018	0.020
Rel. bias($\hat{G}$) %	−3.800	−3.400	−3.070	−2.990	−2.660	−1.970
Rel. abs. bias($\hat{G}$) %	4.690	5.230	5.840	5.640	6.070	6.360
bias($\widehat{S80/S20}$)	−0.247	−0.188	−0.162	−0.163	−0.137	−0.096
Abs. bias($\widehat{S80/S20}$)	0.281	0.338	0.389	0.380	0.417	0.484
Rel. bias($\widehat{S80/S20}$) %	−4.920	−4.020	−3.560	−3.520	−2.950	−2.120
Rel. abs. bias($\widehat{S80/S20}$) %	6.540	7.320	8.040	8.040	8.580	9.620

simulation, and H is the total number of Monte Carlo runs. The true values are computed from the corresponding Monte Carlo population in each small area. The empirical MSEs are treated as true MSEs of the proposed estimators and are used in Table 9.4 and Table 9.5 to compute the bias, absolute bias and relative bias of the root MSE bootstrap estimator we proposed in Section 9.4.

Table 9.4 and Table 9.5 show an overall reasonable performance of the root MSE bootstrap estimator under both scenarios and for both the target parameters. There is a known tendency to underestimate the true root MSE as confirmed in Marchetti *et al.* (2012). Improvements of the bootstrap scheme proposed in Marchetti *et al.* (2012) are currently under study. There are also no remarkable differences in the behavior of the root MSE estimators under both scenarios. This implies that the non-parametric root MSE bootstrap estimator is robust to the assumed unit- and area-level error distributions.

Table 9.3 Distribution over areas of the empirical root MSE (RMSE) of the Gini coefficient (*G*) and the quintile share ratio (*S80/S20*)

	Scenario 1					
	Min.	1st Qu.	Median	Mean	3rd Qu.	Max.
$RMSE(\hat{G})$	0.012	0.013	0.014	0.014	0.015	0.016
$RMSE(\widehat{S80/S20})$	0.288	0.347	0.388	0.393	0.438	0.492
	Scenario 2					
	Min.	1st Qu.	Median	Mean	3rd Qu.	Max.
$RMSE(\hat{G})$	0.018	0.021	0.022	0.022	0.023	0.026
$RMSE(\widehat{S80/S20})$	0.386	0.460	0.532	0.530	0.593	0.676

Table 9.4 Scenario 1 (normally distributed errors). Distribution over areas and Monte Carlo simulations of the bias, absolute bias and relative bias of the root MSE (RMSE) bootstrap estimator of the Gini coefficient (*G*) and the quintile share ratio (*S80/S20*)

	Min.	1st Qu.	Median	Mean	3rd Qu.	Max.
Bias $RMSE(\hat{G})$	−0.003	−0.002	−0.001	−0.001	0.000	0.001
Abs. bias $RMSE(\hat{G})$	0.001	0.001	0.001	0.002	0.002	0.003
Rel. bias $RMSE(\hat{G})$ %	−17.070	−11.620	−5.860	−6.000	−2.270	12.680
Bias $RMSE(\widehat{S80/S20})$	−0.108	−0.046	−0.033	−0.032	−0.011	0.047
Abs. bias $RMSE(\widehat{S80/S20})$	0.043	0.058	0.068	0.071	0.077	0.125
Rel. bias $RMSE(\widehat{S80/S20})$ %	−26.410	−12.160	−8.040	−7.290	−2.710	19.260

Table 9.5 Scenario 2 (Singh–Maddala distributed errors). Distribution over areas and Monte Carlo simulations of the bias, absolute bias and relative bias of the root MSE (RMSE) bootstrap estimator of the Gini coefficient (*G*) and the quintile share ratio (*S80/S20*)

	Min.	1st Qu.	Median	Mean	3rd Qu.	Max.
Bias $RMSE(\hat{G})$	−0.006	−0.002	−0.001	−0.001	0.001	0.003
Abs. bias $RMSE(\hat{G})$	0.003	0.004	0.004	0.004	0.005	0.007
Rel. bias $RMSE(\hat{G})$ %	−20.210	−9.800	−5.790	−4.840	2.560	15.390
Bias $RMSE(\widehat{S80/S20})$	−0.132	−0.049	−0.008	−0.009	0.04	0.119
Abs. bias $RMSE(\widehat{S80/S20})$	0.141	0.194	0.24	0.245	0.284	0.366
Rel. bias $RMSE(\widehat{S80/S20})$ %	−19.410	−9.940	−2.160	−0.440	7.860	28.240

9.6 Estimating the Gini Coefficient and the Quintile Share Ratio for Unplanned Domains in Tuscany

In this section we present an application of the proposed methodology to EU-SILC data from Italy. The aim is to estimate the Gini coefficient and the quintile share ratio for unplanned domains in the region of Tuscany in Italy. The domains are defined by the cross-classification of provinces in Tuscany by the gender of the head of the household leading to a total of 20 domains (10 provinces × 2 gender categories).

The outcome we model is the household equivalized income which is available for each sampled household from the EU-SILC survey 2008. The explanatory variables are the marital status of the head of the family (four categories: single, married, divorced and widow), the employment status of the head of the household (working/not working), the years of education of the head of the household, the mean house surface (in square meters) at municipality level (LAU 2 level), and the number of household members. These covariates are available both from the EU-SILC and from the Population Census of Italy in 2001. It is important to underline that EU-SILC and Census datasets are confidential. The datasets were provided by ISTAT, the Italian National Institute of Statistics, to the researchers of the SAMPLE project (http://www.sample-project.eu) and were analyzed by respecting the confidentiality restrictions.

Figure 9.1 shows box plots of the household equivalized income in each of the 20 domains. The box plots highlight the asymmetry of the income distribution. Moreover, for the majority of provinces we observe differences in the income distribution between households whose head is female and those whose head is male. Income quantiles in the former case tend to be lower. Figure 9.2 shows normal probability plots of level one and level two residuals obtained by fitting a two-level random effects model to the EU-SILC data. Households are the level one units and the 20 domains define the level two units. The plot suggests departures from the normality assumptions of the random effects model. The use of the Shapiro test statistic confirms that the hypothesis of normally distributed level one residuals, both when using the raw and log-transformed income variable, is rejected. It may be appropriate in this case to use a small area estimation approach that imposes less strict parametric assumptions.

Table 9.6 reports model-based estimates of the Gini coefficient and of the quintile share ratio for each of the 20 domains derived by applying the M-quantile-based small area methodology. Estimated root MSEs are reported in parentheses. Table 9.6 also reports the sample and population sizes in each domain.

By examining the results in Table 9.6 we observe that the inequality in those domains where the head of the household is female is in the majority of cases higher than in those domains where the head of the household is male. The results for the Gini coefficient provide some evidence of higher inequality in the Provinces of Grosseto (GR), Pistoia (PT), and Florence (FI) for households with a female as the head of the household. However, the uncertainty around these estimates does not allow more detailed comparisons. According to estimates by Eurostat (http://epp.eurostat.ec.europa.eu/tgm/table.do?tab=table&language=en&pcode=tessi190) the Gini coefficient in Italy in 2008 was 31% a number consistent with the estimates we present in Table 9.6. The 95% normal confidence intervals for the point estimates indicate that for some domains the upper bound of the interval is less than 31%. This is not surprising since Tuscany is an affluent region of Italy. The analysis uses 2008 EU-SILC data that refer to income in

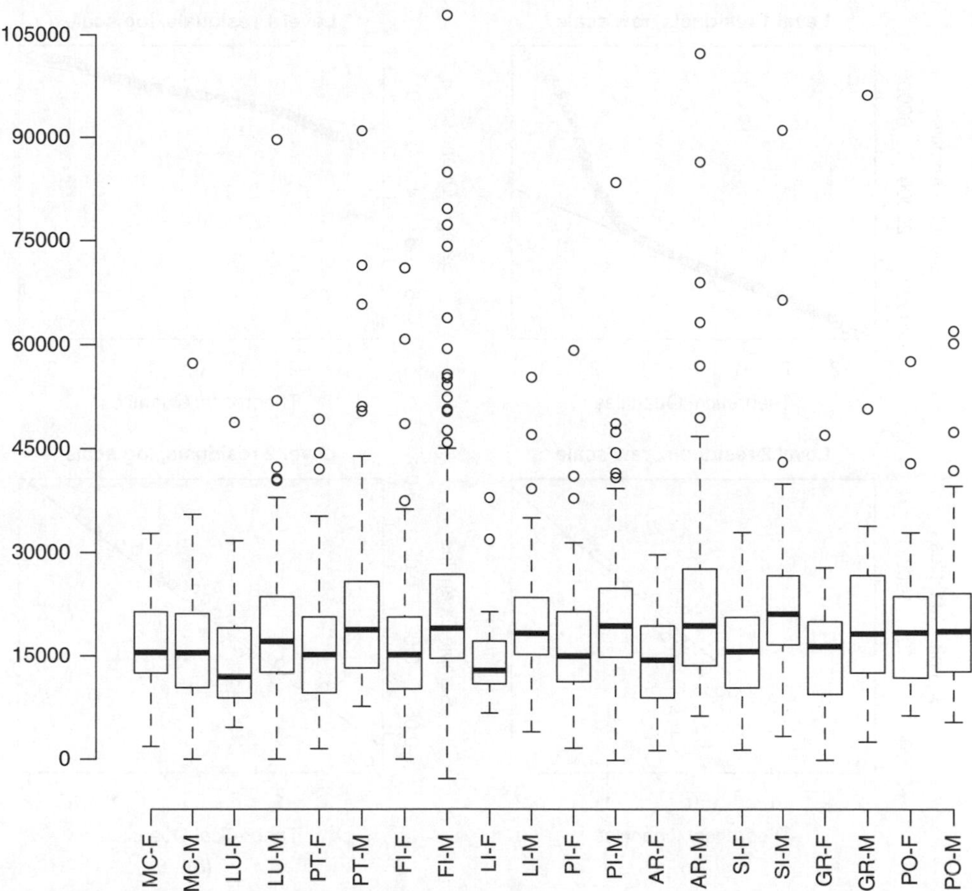

Figure 9.1 Boxplots of the equivalized household income for Tuscany Provinces by gender of the head of the household (F=female, M=male)

2007. It will be interesting to replicate the same analysis with the data after 2008 that marked the start of the financial crisis.

With respect to the quintile share ratio, the results in Table 9.6 indicate again that inequality in those domains where the head of the household is a female is very often higher than in those domains where the head of the household is a male. The estimates indicate higher inequality in the Province of Pistoia (PT), Livorno (LI), and Florence (FI) when the head of the household is a female. According to estimates by Eurostat (http://epp.eurostat.ec.europa.eu/tgm/table.do?tab=table&init=1&language=en&pcode=tessi180&plugin=1) the quintile share ratio (*S80*/*S20*) in Italy in 2008 is estimated to be 5.1, which is consistent with the estimates in Table 9.6.

A more effective representation of the model-based estimates of the Gini coefficient and of the quintile share ratio is offered in Figure 9.3 and Figure 9.4. In each figure the map in (a) refers to the provinces with a female as the head of the household, while (b) refers to

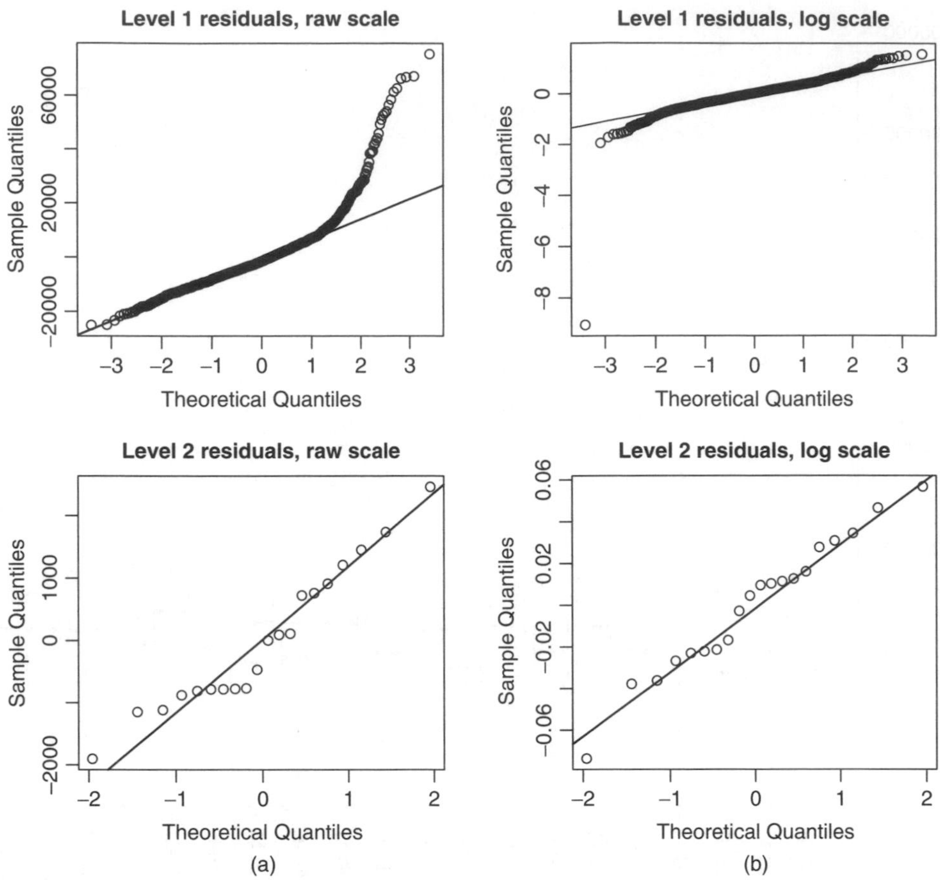

Figure 9.2 Normal probability plots of level one and level two residuals derived by fitting a two-level linear mixed model to the EU-SILC data using the original income variable (a) and a log-transformed income variable (b)

households with a male as the head of the household. In each map the provinces are grouped into four different shades. The darker the shade, the lower the Gini coefficient and the quintile share ratio. To facilitate the comparison between males and females we used a common scale for the different shades.

The inequality in the provinces of Tuscany, measured by the small area estimates of the Gini coefficient and quintile share ratio, is in line with national estimates. What emerges from this analysis is some evidence for differences in inequality between male and female heads of household. This offers useful information because in Italy most households where the head is a female are households with either a widow living alone or a widow living with dependents. These results allow us to identify domains with potentially higher inequality, which alongside other poverty indicators can assist policy makers and stakeholders to plan and implement appropriate social policies.

Table 9.6 Model-based (M-quantile) estimates of the Gini coefficient (G) and the quintile share ratio ($S80/S20$) with corresponding estimated root MSEs (in parentheses). The sample of households and population in each domain (n and N, respectively) are also reported

Province	Head of the household gender	n	N	$\hat{G}(\%)$	$\widehat{S80/S20}$
MC	Female	34	24608	30.07 (3.66)	6.57 (3.96)
	Male	71	56202	28.13 (2.07)	4.56 (0.50)
LU	Female	38	41622	29.63 (3.68)	5.90 (2.14)
	Male	112	104495	28.98 (2.84)	4.88 (0.93)
PT	Female	51	27684	36.85 (3.96)	10.41 (6.35)
	Male	85	76782	28.89 (2.93)	4.24 (0.86)
FI	Female	140	110484	32.63 (2.36)	6.92 (1.45)
	Male	275	265771	28.64 (1.70)	4.45 (0.60)
LI	Female	31	37646	31.79 (3.49)	6.32 (2.28)
	Male	74	96083	24.33 (2.00)	3.53 (0.44)
PI	Female	44	37673	30.81 (3.89)	5.74 (1.90)
	Male	105	112586	26.55 (2.53)	4.32 (0.87)
AR	Female	34	30589	26.86 (2.15)	4.73 (0.76)
	Male	109	93291	31.47 (2.79)	4.97 (1.11)
SI	Female	29	25699	28.13 (3.04)	5.47 (1.54)
	Male	75	75700	28.07 (3.34)	4.40 (1.08)
GR	Female	30	24531	32.89 (3.62)	6.98 (2.41)
	Male	35	63189	33.54 (6.32)	6.77 (8.57)
PO	Female	37	19130	27.27 (3.40)	4.52 (1.24)
	Male	86	64487	25.92 (2.47)	3.89 (0.67)

9.7 Conclusions

In recent years there has been a rapid development in model-based small area methodologies as a response to the increased demand for producing estimates of deprivation and inequality indicators at the level of unplanned domains. Applications of small area estimation have mainly focused on income deprivation in particular, estimating the incidence of poverty. In this chapter we presented a small area methodology that is based on the use of M-quantile modeling and is used for estimating inequality indicators in particular, the Gini coefficient and the quintile share ratio. The proposed methodology is semi-parametric and hence avoids the use of strong parametric assumptions in estimation. Point and MSE estimation is facilitated by the availability of open source software that has been written in the statistical language *R* (R Development Core Team, 2013) and can be easily adapted for estimating other Laeken indicators.

Despite the availability of a wide range of small area methodologies for estimating income-based deprivation and inequality, there are practical problems that require the development of new methodology. A frequent demand by policy makers is the estimation

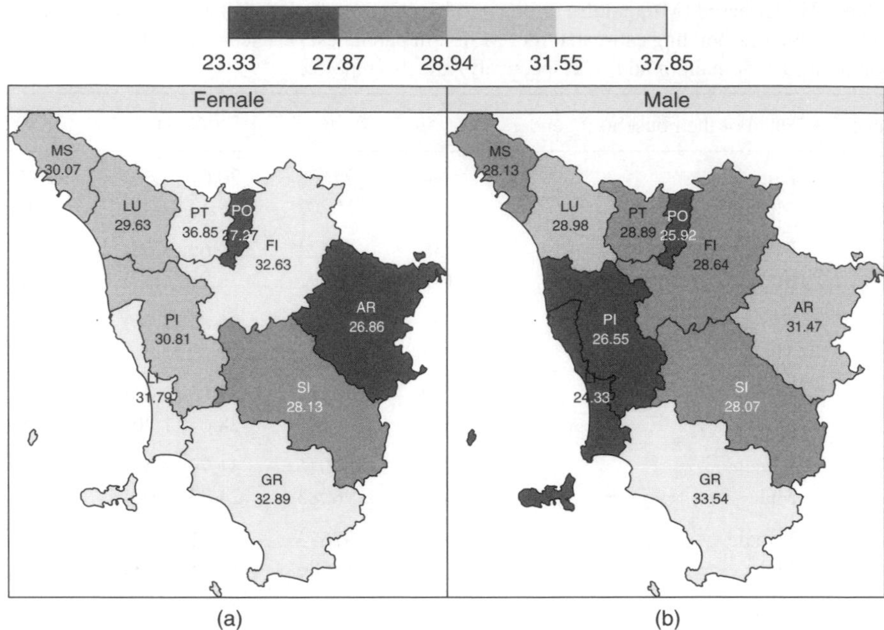

Figure 9.3 Model-based estimates of Gini coefficient for Provinces in Tuscany by gender of the head of the household: female (a) and male (b).

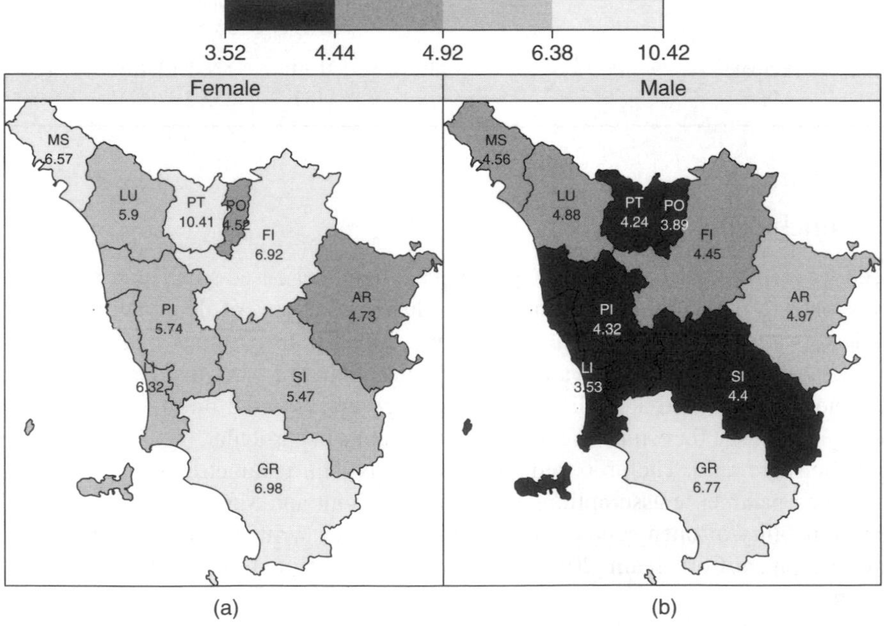

Figure 9.4 Estimates of quintile share ratio for Provinces in Tuscany by gender of the head of the household: female (a) and male (b).

of multidimensional indicators that extend beyond income deprivation and incorporate additional dimensions such as educational, health and social security inequality. Research for developing small area methodologies appropriate to tackle this problem is currently under way.

References

Breckling J and Chambers R 1988 M-quantiles. *Biometrika* **75** (4), 761–771.

Chambers R and Tzavidis N 2006 M-quantile models for small area estimation. *Biometrika* **93** (2), 255–268.

Duan N 1983 Smearing estimate: A nonparametric retransformation method. *Journal of the American Statistical Association* **78** (383), 605–610.

Elbers C, Lanjouw JO, and Lanjouw P 2003 Micro-level estimation of poverty and inequality. *Econometrica* **71** (1), 355–364.

Marchetti S, Tzavidis N, and Pratesi M 2012 Non-parametric bootstrap mean squared error estimation for m-quantile estimators of small area averages, quantiles and poverty indicators. *Computational Statistics and Data Analysis* **56** (10), 2889–2902.

Molina I and Rao J 2010 Small area estimation of poverty indicators. *Canadian Journal of Statistics* **38** (3), 369–385.

R Development Core Team 2013 *R: A Language and Environment for Statistical Computing.* R Foundation for Statistical Computing, Vienna.

Tzavidis N, Marchetti S, and Donbavand S 2013 *Outlier robust semi-parametric small area methods for poverty estimation.* In *Poverty and Social Exclusion: New Methods of Analysis*. Routledge, New York.

10

Nonparametric Regression Methods for Small Area Estimation

M. Giovanna Ranalli[1], F. Jay Breidt[2] and Jean D. Opsomer[2]
[1]*Dipartimento di Scienze Politiche, Università degli Studi di Perugia, Perugia, Italy*
[2]*Department of Statistics, Colorado State University, Fort Collins, USA*

10.1 Introduction

Nonparametric regression methods have significant advantages compared with parametric approaches when the functional form of the relationship between the variable of interest and available covariates cannot be specified a priori. In such situations, erroneous specification of a model can result in biased estimators, and nonparametric regression models are good robustification tools. Even when a specific functional form appears reasonable, the nonparametric model can be useful in the process of model checking and validation. Because of technical difficulties in incorporating smoothing techniques into small area estimation (SAE) tools, nonparametric approaches have only recently gained attention in this field (see Pfeffermann 2013, section 6.6). This chapter is devoted to a review of nonparametric methods in SAE, with a particular focus on issues that are relevant in the estimation of poverty indices.

As elsewhere in the book, the finite population of interest U has size N and is partitioned into D sub-populations, U_d, $d = 1, \ldots, D$, each of size N_d. The sample s of size n is partitioned accordingly, so that s_d and $r_d = U_d \backslash s_d$ denote the sampled and unsampled parts of the population in area d, respectively. The focus here will be on modeling the individual observations of the variable of interest y_{jd} for unit $j \in s_d$ in area d, $d = 1, \ldots, D$, that is on unit level models. Section 10.2 reviews the use of nonparametric regression tools in SAE starting with the extension of the nested error unit level linear regression model of Battese et al. (1988) to a nonparametric trend and then moving to alternative proposals present in the literature. Section 10.3 illustrates the application of some of the reviewed methods to the estimation of the average household per-capita consumption expenditures at the district level

Analysis of Poverty Data by Small Area Estimation, First Edition. Edited by Monica Pratesi.

Companion Website: www.wiley.com/go/pratesi/poverty

using data from the Living Standards Measurement Study (LSMS) in Albania. Finally, some concluding remarks are provided in Section 10.4.

10.2 Nonparametric Methods in Small Area Estimation

In SAE, the canonical approach for unit level models is to express the relationship between the variable of interest and the auxiliary variables as a linear regression model plus a random effect for the small areas. The inclusion of random effects makes it possible to characterize departures from linearity in the conditional distribution of the response variables given the covariates among the small areas. The resulting nested error regression model, as proposed for SAE by Battese et al. (1988), has the form

$$y_{jd} = \mathbf{x}_{jd}\boldsymbol{\beta} + u_d + e_{jd}, \tag{10.1}$$

where $\mathbf{x}_{jd} = (1, x_{1,jd}, \ldots, x_{p-1,jd})$ is a p-vector of auxiliary variables on unit j of area d, u_d is an area-level random effect, and e_{jd} is a residual error term. The u_d and e_{jd} are mutually independent with zero mean and variance σ_u^2 and σ_e^2, respectively. If we let $\mathbf{u} = (u_1, \ldots, u_D)^T$ and $\mathbf{z}_{jd}$ be a vector with the value 1 in its dth position and zeros everywhere else, then $\mathbf{z}_{jd}\mathbf{u} = u_d$, and model (10.1) can be written in matrix notation as

$$\mathbf{y} = \mathbf{X}\boldsymbol{\beta} + \mathbf{Z}\mathbf{u} + \mathbf{e}, \tag{10.2}$$

where matrices $\mathbf{X}$ and $\mathbf{Z}$ have $\mathbf{x}_{jd}$ and $\mathbf{z}_{jd}$ in row jd, respectively. Then,

$$\mathbf{u} \sim (\mathbf{0}, \boldsymbol{\Sigma}_u), \text{ with } \boldsymbol{\Sigma}_u = \sigma_u^2\mathbf{I}_D, \text{ and } \mathbf{e} \sim (\mathbf{0}, \boldsymbol{\Sigma}_e), \text{ with } \boldsymbol{\Sigma}_e = \sigma_e^2\mathbf{I}_n. \tag{10.3}$$

Under model (10.1), the true small area population means are given by $\bar{y}_{N,d} = \bar{\mathbf{x}}_{N,d}\boldsymbol{\beta} + u_d + \bar{e}_{N,d}$. Nonetheless, since the mean of the errors $\bar{e}_{N,d}$ is close to zero for large N_d, the interest is usually in predicting $\tilde{y}_{N,d} = E(\bar{y}_{N,d}|u_d) = \bar{\mathbf{x}}_{N,d}\boldsymbol{\beta} + u_d$. If variance components σ_u^2 and σ_e^2 are known, standard results from best linear unbiased prediction (BLUP) theory provide that the generalized least squares estimator

$$\widehat{\boldsymbol{\beta}} = (\mathbf{X}^T\mathbf{V}^{-1}\mathbf{X})^{-1}\mathbf{X}^T\mathbf{V}^{-1}\mathbf{y}, \tag{10.4}$$

with $\mathbf{V} = \text{var}(\mathbf{y}) = \mathbf{Z}\boldsymbol{\Sigma}_u\mathbf{Z}^T + \boldsymbol{\Sigma}_e$, and the predictor

$$\hat{\mathbf{u}} = \boldsymbol{\Sigma}_u\mathbf{Z}^T\mathbf{V}^{-1}(\mathbf{y} - \mathbf{X}\widehat{\boldsymbol{\beta}}) \tag{10.5}$$

are optimal among all linear estimators or predictors (see e.g. Rao 2003, section 7.2). Then, as a predictor for $\bar{y}_{N,d}$, the BLUP of $\tilde{y}_{N,d}$ is usually employed and given by $\widehat{\bar{y}}_{N,d} = \bar{\mathbf{x}}_{N,d}\widehat{\boldsymbol{\beta}} + \hat{u}_d$. Since variances are usually unknown, empirical best linear unbiased prediction (EBLUP) predictors of $\widehat{\boldsymbol{\beta}}$, $\hat{\mathbf{u}}$ and $\widehat{\bar{y}}_{N,d}$ are obtained by replacing variances σ_u^2 and σ_e^2 by estimators. Restricted maximum likelihood can be employed to this end.

With respect to measures of precision of such predictors, a substantial amount of literature exists on approximations and estimators for the mean squared error (MSE) of small area estimators under unit level models. In the case of SAE with a linear mean model and independent variance components, Prasad and Rao (1990) derive a second-order approximation for the prediction MSE together with an estimator for the prediction MSE that is correct up to second order. Datta and Lahiri (2000) extend their results for the case of restricted maximum likelihood estimation of the variance components, and Das et al. (2004) further expand it to encompass more general linear mixed models. We refer to (Rao 2003, chapters 6 and 7) for comprehensive introduction and review of this topic.

10.2.1 Nested Error Nonparametric Unit Level Model Using Penalized Splines

Many extensions to the setting above have been considered to handle more complex estimation contexts. In particular, Opsomer et al. (2008) extend model (10.1) to the case in which the small area random effects can be combined with a smooth, nonparametrically specified trend. The basic idea is to avoid assuming a prespecified functional form for the expectation of the response variable, and to estimate it using penalized spline regression (Eilers and Marx 1996). Suppose for simplicity that there is only a univariate auxiliary variable, denoted x. Then, Opsomer et al. (2008) consider the following model

$$y_{jd} = m_0(x_{jd}) + u_d + e_{jd}, \tag{10.6}$$

where the function $m_0(\cdot)$ is unknown, but assumed to be approximated sufficiently well by the spline function

$$m(x;\boldsymbol{\beta},\boldsymbol{\gamma}_1) = \beta_0 + \beta_1 x + \cdots + \beta_q x^q + \sum_{k=1}^{K} \gamma_k (x-\kappa_k)_+^q. \tag{10.7}$$

Here q is degree of the spline, $(t)_+^q = t^q$ if $t > 0$ and 0 otherwise, the $\kappa_k, k = 1, \dots, K$ are a set of fixed constants called knots, $\boldsymbol{\beta} = (\beta_0, \dots, \beta_q)^T$ is the coefficient vector of the parametric portion of the model, and $\boldsymbol{\gamma} = (\gamma_1, \dots, \gamma_K)^T$ is the vector of spline coefficients. If knot locations cover the range of x and K is sufficiently large, the class of functions in (10.7) can approximate most smooth functions with high accuracy, even if the degree of the spline is small. (Ruppert et al. 2003, chapter 5) suggest the use of a knot every four or five observations for the univariate setting, with a maximum number of about 40. If we let $\mathbf{x}_{jd} = (1, x_{jd}, \dots, x_{jd}^q)$ and define $\mathbf{d}_{jd} = ((x_{jd}-\kappa_1)_+^q, \dots, (x_{jd}-\kappa_K)_+^q)$, then the non-parametric nested error model can be written in matrix notation as

$$\mathbf{y} = \mathbf{X}\boldsymbol{\beta} + \mathbf{D}\boldsymbol{\gamma} + \mathbf{Z}\mathbf{u} + \mathbf{e}, \tag{10.8}$$

where the matrix $\mathbf{D}$ has $\mathbf{d}_{jd}$ in row jd. Given that K is usually large, this model is potentially overparametrized if one considers $\boldsymbol{\gamma}$ as model parameters. This issue can be overcome by putting a penalty on their magnitude, and one way in which this can be accomplished is by treating $\boldsymbol{\gamma}$ as an additional random-effect component of the model (Ngo and Wand 2004). In this way, an appropriate amount of smoothing for the spline is also ensured. Therefore, data are assumed to follow model (10.8) where $\boldsymbol{\gamma} \sim (\mathbf{0}, \boldsymbol{\Sigma}_\gamma)$, with $\boldsymbol{\Sigma}_\gamma = \sigma_\gamma^2 \mathbf{I}_K$ and the other random components are as in (10.3). The three random variables are all mutually independent so that, in this case, $\text{var}(\mathbf{y}) = \mathbf{V} = \mathbf{D}\boldsymbol{\Sigma}_\gamma\mathbf{D}^T + \mathbf{Z}\boldsymbol{\Sigma}_u\mathbf{Z}^T + \boldsymbol{\Sigma}_e$. Under this model, we are interested in predicting $\tilde{y}_{N,d} = E(\overline{y}_{N,d}|u_d) = \overline{\mathbf{x}}_{N,d}\boldsymbol{\beta} + \overline{\mathbf{d}}_{N,d}\boldsymbol{\gamma} + u_d$. Similarly to the linear case, the BLUP of $\tilde{y}_{N,d}$ is given by

$$\hat{\tilde{y}}_{N,d} = \overline{\mathbf{x}}_{N,d}\hat{\boldsymbol{\beta}} + \overline{\mathbf{d}}_{N,d}\hat{\boldsymbol{\gamma}} + \hat{u}_d, \tag{10.9}$$

where $\widehat{\boldsymbol{\beta}}$ is as in (10.4), $\widehat{u}_d$ as in (10.5), and $\widehat{\boldsymbol{\gamma}} = \boldsymbol{\Sigma}_\gamma \mathbf{D}^T \mathbf{V}^{-1}(\mathbf{y} - \mathbf{X}\widehat{\boldsymbol{\beta}})$. Their EBLUP versions can again be employed when variance components are unknown, by using for example (restricted) maximum likelihood methods. Giusti et al. (2012) apply this approach to area-level data by extending the Fay and Herriot area-level model using the penalized splines approach.

Opsomer et al. (2008) derive the prediction MSE of the EBLUP of (10.9) correct to second order for the case where the unknown variances are estimated using restricted maximum likelihood. An estimator of the prediction MSE is also provided with bias correct to the same order. To this end, results and the method of proof are extensions of Das et al. (2004) to encompass the case of a spline-based random effect. For testing purposes, a likelihood ratio test (or restricted likelihood ratio test) for testing the presence of small area effects is also proposed based on the approach developed in Claeskens (2004). A nonparametric bootstrap algorithm is also proposed for both estimating the prediction MSE and for testing for small-area effects and non-linearities. The bootstrap procedure is not second order correct but is very easy to use, so that it provides a practical way to perform inference for the nonparametric small area model.

As defined above, the spline function in (10.7) uses the truncated polynomial basis $\{1, x, \ldots, x^q, (x - \kappa_1)_+^q, \ldots, (x - \kappa_K)_+^q\}$ to approximate the function $m_0(\cdot)$, following the description of penalized spline regression in Ruppert et al. (2003). Other bases like B-splines or radial basis functions can be used and are often preferred to truncated polynomials in some estimation contexts (Wood 2006). Nonetheless, for prediction purposes (as is the case in SAE), low-degree truncated polynomials do offer the advantage of requiring less auxiliary information as compared with other, more general bases. In fact, to compute predictor (10.9), the true population means of the powers of x up to q and of the spline basis functions over the small area are assumed known. In the simplest case, that is when $q = 1$, this reduces to knowing, over each small area, the population size N_d and the population mean of x – which would be required for the linear nested error model of Battese et al. (1988) as well – and the population counts and means of x in strata defined by the knots. This requirement is less stringent than the knowledge of individual level auxiliary information for all units in the population (the census). This property can be particularly valuable in the SAE context and is not shared by other basis functions.

As an alternative to the mixed model representation of penalized splines, Jiang et al. (2010) propose using the so-called *fence method* for the selection of the appropriate amount of smoothing for the spline. The basic idea of the fence method is to build a statistical "fence" to isolate a subgroup of what are known as the correct models. Once the fence is constructed, the optimal model is selected from those within the fence according to a criterion which can incorporate quantities of practical interest, such as model simplicity or better prediction. Within this framework, Jiang et al. (2010) propose to choose, other than the amount of penalization, also the number of knots K and the degree of the splines q at the same time.

Model (10.6) is a simple illustration with a univariate covariate specified nonparametrically in the mean function. Semiparametric modeling, that is the combination of nonparametric and parametric terms in the mean function, can be achieved easily by exploiting the mixed-model representation of penalized splines. If, for example, another variable needs to be included nonparametrically as an additive term, resulting in the model $y_{jd} = m_0(x_{1,jd}) + m_0(x_{2,jd}) + u_d + e_{jd}$, say, this can be accomplished by adding another random component to model (10.8). In particular, we now have

$$\mathbf{y} = \mathbf{X}\boldsymbol{\beta} + \mathbf{D}_1\boldsymbol{\gamma}_1 + \mathbf{D}_2\boldsymbol{\gamma}_2 + \mathbf{Z}\mathbf{u} + \mathbf{e}$$

where matrix $\mathbf{X}$ in this case collects the powers of both x_1 and x_2, that is $\mathbf{x}_{jd} = (1, x_{1jd}, \ldots, x_{1jd}^q, x_{2jd}, \ldots, x_{2jd}^q)$, and $\mathbf{D}_1$ and $\mathbf{D}_2$ include the spline basis functions for variable x_1 and x_2, that is it is such that $\mathbf{d}_{2jd} = ((x_{2jd} - \zeta_1)_+^q, \ldots, (x_{1jd} - \zeta_K)_+^q)$, where ζ_k, $k = 1, \ldots, K$ is the set of knots for the second variable. For simplicity, we consider that the number of knots is the same for the two variables. Then, $\boldsymbol{\gamma}_t \sim (\mathbf{0}, \boldsymbol{\Sigma}_{\gamma t})$, with $\boldsymbol{\Sigma}_{\gamma t} = \sigma_{\gamma t}^2 \mathbf{I}_K$ for $t = 1, 2$, and all random components continue to be assumed independent of each other. If other auxiliary variables need to be included in the model as parametric terms, they can be added as columns in the matrix of fixed effects $\mathbf{X}$. Estimation/prediction for small area means is conducted similarly as described above.

Multivariate smoothing can also be included by choosing suitable basis functions, such as radial or tensor product basis functions (see e.g. Wood 2006, chapter 4, for details). This is particularly relevant in the spatial context when auxiliary information takes the form of the geographical coordinates of the location of the unit (Kammann and Wand 2003). Opsomer et al. (2008) use thin plate splines in an environmental application of SAE. Dreassi et al. (2014) incorporate such a bivariate smooth function of the geographical coordinates of units in a two-part random effects SAE model to deal with a zero-inflated target variable that exhibits a point mass at zero and a highly skewed distribution of the nonzero values. Chapter 13 in this book provides a thorough treatment of geoadditive models in SAE and their application to poverty data. Estimation for additive mixed models that use penalized splines can be conducted in a flexible and modular way using the `mgcv` (Wood 2006) or the `gamm4` packages of the `R` software. The analytical and the nonparametric bootstrap based prediction MSE estimators proposed in Opsomer et al. (2008) can be extended easily to the case of more than one spline component.

The approach proposed by Opsomer et al. (2008) is used by Ugarte et al. (2009) to tackle longitudinal data in SAE. In particular, the spline component of the semiparametric mixed effects model is used to estimate a smooth time trend. The main objective of the paper is forecasting future values of the response within the small areas. Penalized splines that use B-spline bases and their representation as mixed effects models are used to obtain the EBLUP for particular observations within the small areas. The prediction MSE of both fitted and forecast values, as well as estimators of those quantities, are also derived following the guidelines provided in Opsomer et al. (2008).

Militino et al. (2012), on the other hand, fit an individual curve in each small area, using penalized splines with B-spline bases and, again, exploit the mixed model representation of the penalized splines for inferential purposes. To account for possible bias due to informativeness of the sampling design, a design-oriented bootstrap correction is proposed. The bootstrap proposed here does not use the model to generate the data, but the sampling design.

10.2.2 Nested Error Nonparametric Unit Level Model Using Kernel Methods

Lombardía and Sperlich (2008) also consider a class of semiparametric mixed effects models for SAE of the type

$$y_{jd} = m_0(t_{jd}) + \mathbf{x}_{jd}\boldsymbol{\beta} + u_d + e_{jd}, \tag{10.10}$$

of which model (10.6) can be seen as a particular case, but that can be handled as described above. We follow here the notation used in Lombardìa and Sperlich (2008) and denote the auxiliary variable that is the argument of the smooth function by t to distinguish it from the other covariates that enter the model parametrically. The main difference with the approach followed by Opsomer et al. (2008) is that the smooth function $m_0(\cdot)$ is now estimated using kernel-based methods. This is accomplished by a two-step procedure in which the canonical fully parametric likelihood approach for random-effect models is combined with a semiparametric regression using a double profile likelihood.

Lombardìa and Sperlich (2008) also introduce a test for the hypothesis of a parametric mixed effects model, against the alternative of a semiparametric model. The critical values are estimated using a parametric bootstrap procedure. Although the kernel function considered in the paper can be taken to be multivariate, the extension to more than one smooth function at a time is not as straightforward. Sperlich and Lombardía (2010) extend this work by considering estimation and testing with local polynomials. In a similar framework, Rueda and Lombardìa (2012) propose semiparametric monotone mixed models, using the assumption that some of the auxiliary variables have a monotone relationship with the response. They also propose two bootstrap procedures to estimate the error of small area estimates.

10.2.3 Generalized Responses

The proposals in Lombardìa and Sperlich (2008) can handle non-continuous response variables, such as binary or count variables. When the response variable is binary, that is y_{jd} takes values 0 or 1, then the small area quantities of interest are usually proportions or counts. Extension of the approach in Opsomer et al. (2008) to generalized response can be achieved using *generalized additive mixed models*, although no explicit expression for a best predictor as in the linear case can be found. An approach such as those proposed in Jiang and Lahiri (2001) and also considered in Molina and Rao (2010) could be extended to encompass nonparametric regression methods in SAE, but further research in this field is required. D'Alò et al. (2012) apply an empirical plug-in predictor using logistic additive mixed model for estimation of the proportion and total of unemployed persons in small areas in Italy.

10.2.4 Robust Approaches

In addition to nonparametric regression as a tool to robustify inference in SAE by relaxing the assumption of linearity in the relationship with the covariates, other approaches have been proposed that do not require explicit distributional assumptions and/or are robust to outliers. Pratesi et al. (2008) extend the M-quantile approach proposed by Chambers and Tzavidis (2006) for SAE by incorporating nonparametric regression. M-quantile regression provides a "quantile-like" generalization of regression based on influence functions. In a linear M-quantile regression model, the τth M-quantile $Q_\tau(y|\mathbf{x}, \psi)$ of the conditional distribution of y given $\mathbf{x}$ is such that $Q_\tau(y|\mathbf{x}, \psi) = \mathbf{x}\boldsymbol{\beta}_\psi(\tau)$, where ψ denotes an influence function and $\boldsymbol{\beta}_\psi(\tau)$ denotes a different set of regression parameters for each τ. M-quantile regression is applied to the estimation of a small area mean as follows: the first step is to estimate the M-quantile coefficients τ_{jd} for each unit in the sample, where the M-quantile coefficient τ_{jd} is the value of τ solving $Q_{\tau_{jd}}(y_{jd}|\mathbf{x}_{jd}, \psi) = y_{jd}$. If a hierarchical structure does explain part of the variability in the population data, units within areas are expected to have similar M-quantile coefficients. Therefore, an estimate of the mean quantile for area d, $\overline{\tau}_d$, is obtained by taking the corresponding average value of the sample M-quantile coefficient of each unit in area d, $\hat{\bar{\tau}}_d = n_d^{-1} \sum_{j \in s_d} \tau_{jd}$. Following this, predictions are obtained as

$$\hat{y}_{jd} = \mathbf{x}_{jd} \hat{\boldsymbol{\beta}}_{\psi}(\hat{\bar{\tau}}_d),$$

where $\hat{\boldsymbol{\beta}}_{\psi}(\hat{\bar{\tau}}_d)$ is the coefficient vector of the M-quantile regression function at $\hat{\bar{\tau}}_d$. See Chapter 9 for all the details on M-quantile-based SAE.

The work by Pratesi et al. (2008) can be seen as the nonparametric regression extension of this approach or, put another way, as the M-quantile version of the proposal in Opsomer et al. (2008). In particular, they apply to SAE the nonparametric M-quantile approach based on penalized splines proposed by Pratesi et al. (2009). For illustration, consider a nonparametric model with a univariate covariate x for the τth M-quantile so that $Q_\tau(y|x, \psi) = m_{\psi,\tau}(x)$, where, as in (10.6) the function $m_{\psi,\tau}(\cdot)$ is unknown, but assumed to be approximated sufficiently well by

$$m_{\psi,\tau}(x; \boldsymbol{\beta}_{\psi}(\tau), \boldsymbol{\gamma}_{\psi}(\tau)) = \beta_{\psi,0}(\tau) + \beta_{\psi,1}(\tau)x + \cdots + \beta_{\psi,q}(\tau)x^q + \sum_{k=1}^{K} \gamma_{\psi,k}(\tau)(x - \kappa_k)_+^q.$$

This is the same spline function of Equation (10.7), the only difference being that we have a different function and, therefore, a different set of coefficients for each τ. Here as well the influence of the knots is limited by putting a constraint on the size of the spline coefficients: in particular, $\sum_{k=1}^{K} \gamma_{\psi,k}^2(\tau)$ is bounded by some constant, while the parametric coefficients $\boldsymbol{\beta}_{\psi}(\tau)$ are left unconstrained. An algorithm based on iteratively reweighted penalized least squares is proposed in Pratesi et al. (2009) to compute the parameter estimates and select an appropriate amount of smoothing. Once parameter estimates are obtained, $\hat{m}_{\psi,q}(x) = m_{\psi,\tau}(x; \hat{\boldsymbol{\beta}}_{\psi}(\tau), \hat{\boldsymbol{\gamma}}_{\psi}(\tau))$ can be computed as an estimate for $Q_\tau(y|x, \psi)$. The approach is indeed very general: other continuous or categorical variables can be easily inserted parametrically in the model, while other continuous variables can be added nonparametrically in an additive model as before. Bivariate smoothing can be handled in a spatial context as illustrated in Pratesi et al. (2009).

As in the linear case, penalized splines M-quantile regression is applied to the estimation of a small area mean by first estimating the M-quantile coefficients τ_{jd} and their area mean, and letting τ_{jd} be the value of τ such that $Q_{\tau_{jd}}(y|x_{ij}, \psi) = y_{jd}$. The unit level coefficients are estimated by defining a fine grid of values on the interval (0, 1) and using the sample data to fit the penalized splines M-quantile regression functions at each value τ on this grid. If a data point lies exactly on the τ th fitted curve, then the coefficient of the corresponding sample unit is equal to τ. Otherwise, to obtain τ_{jd}, a linear interpolation over the grid is used. Finally, the small area estimator of the mean may be taken as the mean of the predictions

$$\hat{y}_{jd} = \mathbf{x}_{jd} \hat{\boldsymbol{\beta}}_{\psi}(\hat{\bar{\tau}}_d) + \mathbf{d}_{jd} \hat{\boldsymbol{\gamma}}_{\psi}(\hat{\bar{\tau}}_d),$$

where $\mathbf{x}_{jd}, \mathbf{d}_{jd}$ contain the parametric and nonparametric portions of the spline basis as in Section 10.2.1 and $\hat{\boldsymbol{\beta}}_{\psi}(\hat{\bar{\tau}}_d)$ and $\hat{\boldsymbol{\gamma}}_{\psi}(\hat{\bar{\tau}}_d)$ are the corresponding coefficient vectors of the fitted splines M-quantile regression function at $\hat{\bar{\tau}}_d$.

A bias adjusted version of this predictor has been proposed in the presence of outliers (see e.g. Salvati et al. 2011) and is given by

$$\hat{\bar{y}}_d = \frac{1}{N_d} \left\{ \sum_{j \in U_d} \hat{y}_{jd} + \frac{N_d}{n_d} \sum_{j \in s_d} (y_{jd} - \hat{y}_{jd}) \right\}. \tag{10.11}$$

Salvati et al. (2011) compare the finite-sample performance of the small area estimator based on penalized splines M-quantile regression with the linear M-quantile small area estimator of Chambers and Tzavidis (2006) and with EBLUP estimators based on Battese et al. (1988) model and on nonparametric regression model (Opsomer et al. 2008).

Rao et al. (2013) note that the EBLUP under the spline model considered in Opsomer et al. (2008) can be sensitive to outliers in the distribution of the area random effects and in the errors. They obtain a robust EBLUP of the small area mean under the spline nested error regression model using results in Fellner (1986) on robust mixed model equations and a two-step iterative method that robustifies estimating equations for variance components.

An alternative approach that tries to robustify small area inference is *model-based direct estimation* (MBDE; Chandra and Chambers 2005, 2009): the MBDE for a small area mean is a weighted average of the sample values from the area, but with weights derived from a linear predictor of the corresponding population mean under a linear model with random area effects. It can be seen, therefore, as a compromise between direct and indirect estimation, like the EBLUP. For the former, the estimated value of the mean for area d is just a weighted average of the sample data from area d. The latter, on the other hand, can only be represented as weighted sums over the entire sample. Unfortunately, when weights are the inverses of sample inclusion probabilities, the direct estimator can be quite inefficient. The MBDE tries to improve upon the efficiency of the design-based direct estimator by using weights that define the EBLUP for the population total under the same linear mixed model with random area effects that underpins the EBLUP for the small area mean. That is, if the weights w_{jd}^{EBLUP} define the EBLUP for the population total of the y_{jd}, then the MBDE of the small area mean of these values is given $\widehat{\overline{y}}_d = \sum_{j\in s_d} w_{jd}^{\text{EBLUP}} y_{jd} / \sum_{j\in s_d} w_{jd}^{\text{EBLUP}}$.

Salvati et al. (2010) extend the MBDE to a nonparametric mixed effects model like the one considered in Opsomer et al. (2008), by taking advantage of the fact that, for given values of the variance components, predictor (10.9) can be written as a linear combination of the y_{jd}'s. In particular, under model (10.8), the vector of sample weights that defines the corresponding EBLUP of the population total of the y_{jd}'s is

$$\mathbf{w}_s^{\text{NPEBLUP}} = \left(w_{jd}^{\text{NPEBLUP}}\right) = \mathbf{1}_n + \widehat{\mathbf{H}}^T\left(\mathbf{X}_\mathbf{N}^\mathbf{T}\mathbf{1}_N - \mathbf{X}^T\mathbf{1}_n\right) + \left(\mathbf{I}_n - \widehat{\mathbf{H}}^T\mathbf{X}^T\right)\widehat{\mathbf{V}}^{-1}\widehat{\mathbf{V}}_{sr}\mathbf{1}_{N-n},$$

where $\widehat{\mathbf{H}}^T = (\mathbf{X}^T\widehat{\mathbf{V}}^{-1}\mathbf{X})^{-1}\mathbf{X}^T\widehat{\mathbf{V}}^{-1}$, $\widehat{\mathbf{V}}$ is $\mathbf{V}$ with estimated variance components, $\widehat{\mathbf{V}}_{sr} = \mathbf{D}\widehat{\boldsymbol{\Sigma}}_\gamma\mathbf{D}_\mathbf{r}^\mathbf{T} + \mathbf{Z}\widehat{\boldsymbol{\Sigma}}_u\mathbf{Z}_\mathbf{r}^\mathbf{T}$, and subscript r denotes the out-of-sample part of the population. The nonparametric model-based direct estimator of the mean of y for small area d is then given by

$$\widehat{\overline{y}}_d = \sum_{j\in s_d} w_{jd}^{\text{NPEBLUP}} y_{jd} / \sum_{j\in s_d} w_{jd}^{\text{NPEBLUP}}.$$

Even if the latter is referred to as a direct estimator, because it is a weighted average of the sample data from the small area of interest, this does not mean that it can be calculated just using these data, since weights w_{jd}^{NPEBLUP} are a function of the data from the entire sample and, therefore, borrow strength from other areas through model (10.8).

10.3 A Comparison for the Estimation of the Household Per-capita Consumption Expenditure in Albania

In this section, we wish to apply some of the methodologies reviewed in the previous sections and compare their behavior in measuring a poverty indicator at small area level. To this end, we employ data from the Living Standards Measurement Study (LSMS) conducted in Albania in 2002. The LSMS was established by the World Bank in 1980 to explore ways of improving the type and quality of household data collected by statistical offices in developing countries. This survey is the basis of poverty assessment in this country as it provides information on living conditions of the people in Albania, including details on income and non-income related dimensions of poverty. In fact, poverty can be defined both in terms of income deprivation and inadequacies in a number of non-income measures of welfare such as education, health and access to basic services and/or infrastructures. In this study we refer to the poverty as income deprivation. In addition, since the economy of Albania is largely rural and income is not accurately measured, income poverty in Albania is estimated on the basis of a consumption-based measure: a household is defined as poor if its per-capita consumption expenditure falls below a minimum level (poverty line) necessary to meet its basic needs.

We have data available on 3591 households selected from both urban and rural areas of the country. The sample is designed to provide reliable direct estimates of Albania as a whole, Tirana and the three main classification of agro-ecological areas – Coast, Central and Mountain. Albania is divided into three different geographical and administrative levels. In particular, there are 12 prefectures, with a prefecture consisting of several districts. There are 36 districts and 374 communes. See Betti et al. (2003) for all details on the survey.

Direct estimates of household per-capita consumption expenditure at district level suffer from lack of precision due to a large variance, particularly for districts with small sample sizes. Tzavidis et al. (2008) analyze this dataset and apply linear M-quantile models to obtain model-based small area estimates at the district level. They use the following set of household level covariates:

- `land`: binary, takes value 1 if the household is owner of agricultural land;
- `tv`: binary, takes value 1 if a TV is in the house;
- `parab`: binary, takes value 1 if a parabolic antenna is in the house;
- `refrig`: binary, takes value 1 if a refrigerator is in the house;
- `air`: binary, takes value 1 if air conditioning is in the house;
- `pc`: binary, takes value 1 if a personal computer is in the house;
- `car`: binary, takes value 1 if the household possesses a car;
- `owner`: binary, takes value 1 if the household is owner of the house;
- `hhsize`: number of persons in the household.

These covariates are also available from the Albanian Population and Housing Census carried out in 2001 and can be employed in small area unit level models. Chapter 13 in this book applies a geoadditive model to this dataset having available a wider set of covariates and the geographical location of each household for sampled households. For the non-sampled units, the geographical location is approximated with the centroid coordinates of the commune where the household is located.

Table 10.1 Parameter estimates for the nested error unit level linear regression model

	Estimate	Std error	t-value	p-value
`Intercept`	13811.102	499.946	27.625	< 0.001
`owner`	−240.580	267.839	−0.898	0.369
`land`	443.525	165.230	2.684	0.007
`tv`	−142.046	357.731	−0.397	0.691
`parab`	−313.322	188.096	−1.666	0.096
`refrig`	261.558	218.417	1.198	0.231
`air`	1275.040	546.982	2.331	0.020
`pc`	1183.033	472.161	2.506	0.012
`car`	339.771	248.743	1.366	0.172
`hhsize`	−1222.769	40.914	−29.887	< 0.001

First, a nested error unit level linear regression model is fit to the data as in (10.1) using district level random effects. The function `lme` of the R package `nlme` or the function `mer` of the R package `lme4` can be used to this end. Table 10.1 shows the parameter estimates obtained for this model. Most covariates are significant. The estimate for the standard deviation σ_d of the random area effect takes value 1385.615, while the one for σ_e takes value 4238.881. The likelihood ratio test statistic of this model compared with a simple linear regression model with no area random effect takes value 299.682, providing evidence of the presence of extra structure in the data not explained by a linear function of the covariates available. More complex models with interactions between pairs of variables were fitted, but such interactions were tested to be non-significant.

Looking at the residual plots for this model in Figure 10.1 suggests the presence of some curvilinear feature in the relationship with variable `hhsize`. The second set of residual plots in Figure 10.1 is obtained from a nested error unit level nonparametric regression model in which a smooth function for the relationship with the variable `hhsize` is estimated. The function `gamm` of the R package `mgcv` is used to this end. Figure 10.2 shows the shape of such estimated functional relationship: there is an inverse association between household size and consumption, as estimated also with the linear model, but such relationship stabilizes after the value 8 for `hhsize`. This behavior is reflected in the better fit of the relatively larger values for `hhsize` detected by a direct comparison of the two sets of residual plots in Figure 10.1. The smooth component uses 4.31 degrees of freedom and is tested to be highly significant [the p-value for the F-test reported in the summary of the `gamm` R function is virtually zero; for details on this test see (Wood 2006, Section 8.4.5)]. Table 10.2 shows the parameter estimates for the parametric components of the additive model; all values are comparable with those from Table 10.1.

Figure 10.3 compares population predictions from the nested error unit level linear regression model and the nonparametric one. As can be seen, the range of predictions is substantially larger for the linear model, which results in a large number of negative predicted values. These occur for relatively larger households, for which predictions made using the nonparametric model are more robust.

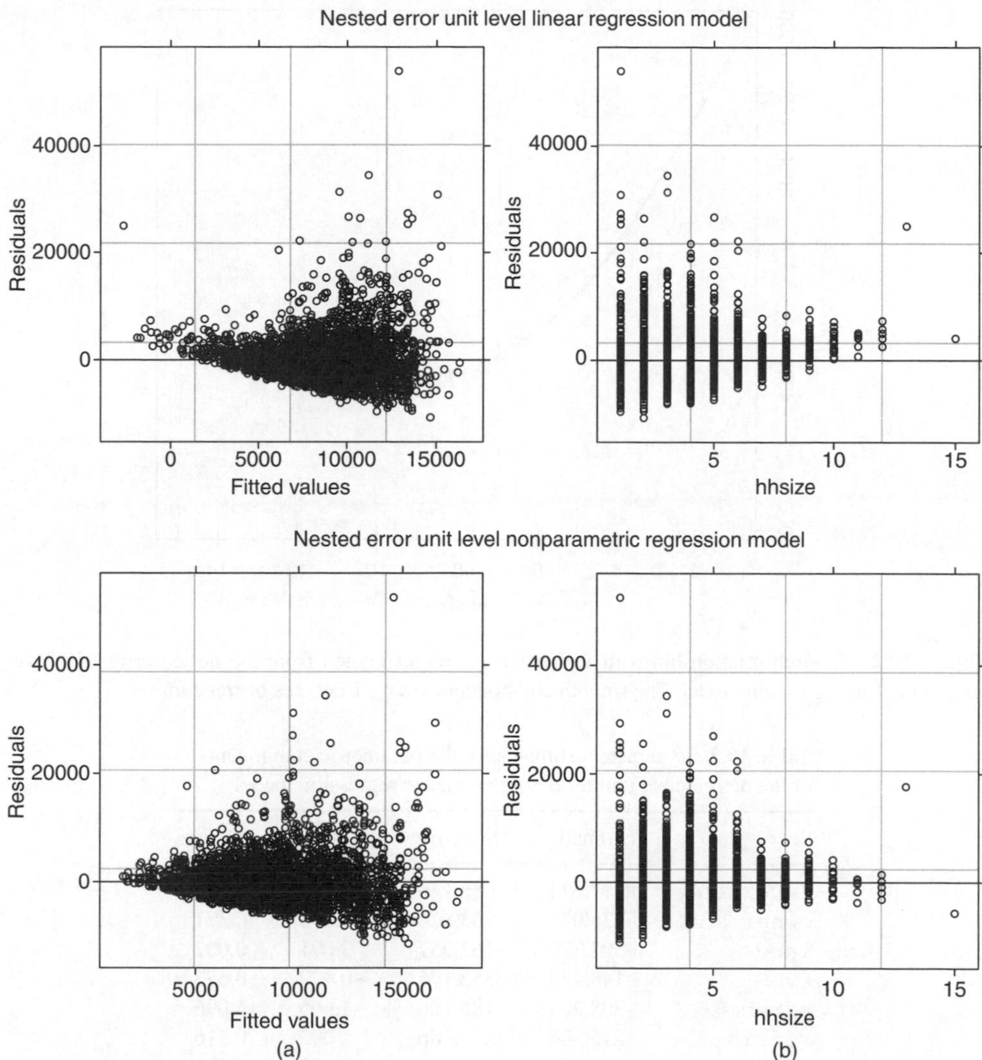

Figure 10.1 Pearson residuals vs. fitted values (a) and vs. `hhsize` values (b) for the nested error unit level linear regression model (top) and for the nonparamateric regression model (bottom). The smooth component for the nonparametric regression model is for variable `hhsize`

Table 10.3 reports the district level estimates of the average household per-capita consumption expenditure, with estimated root mean squared error (RMSE) and coefficient of variation (CV%) using a nested error unit level linear regression model (EBLUP) and a nested error unit level nonparametric regression model (NPEBLUP). The two sets of estimates are very similar and are both highly correlated with the direct estimates (about 93%). Nonetheless, for those districts with a relatively larger proportion of very large `hhsize`, EBLUP is usually smaller than NPEBLUP. From this table, the districts of Bulqizë, Kurbin, and Pukë are the poorest ones, with the first one showing an average household per-capita consumption expenditure

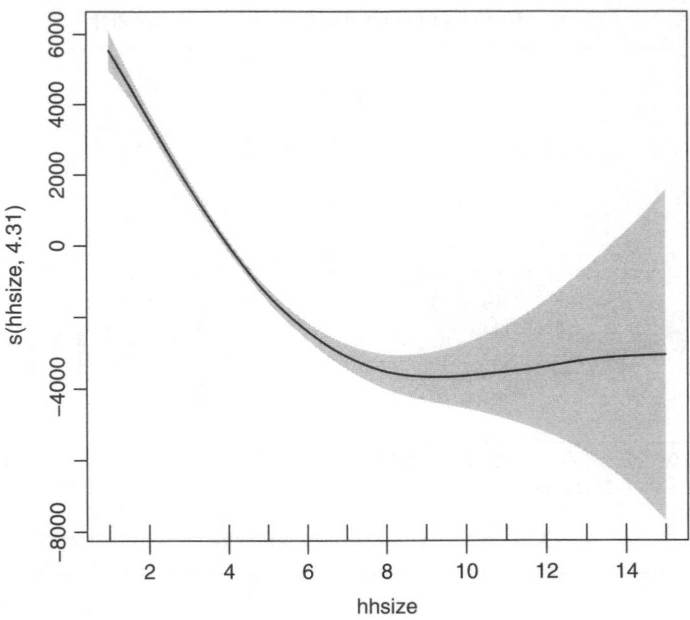

Figure 10.2 Smooth relationship with variable `hhsize` estimated from the nested error unit level nonparametric regression model. The smooth component uses 4.31 degrees of freedom

Table 10.2 Parameter estimates for the parametric components of the nested error unit level nonparametric regression model

	Estimate	Std error	t-value	p-value
`Intercept`	8517.674	456.146	18.673	0.000
`owner`	−236.031	263.826	−0.895	0.371
`land`	503.779	162.957	3.091	0.002
`tv`	−148.478	352.175	−0.422	0.673
`parab`	−308.206	185.126	−1.665	0.096
`refrig`	215.846	215.286	1.003	0.316
`air`	1132.964	538.656	2.103	0.036
`pc`	1172.959	464.814	2.524	0.012
`car`	382.303	244.892	1.561	0.119

that is comparable with the poverty line, set for this dataset to 4891 Leks per month. On the other hand, the districts of Gjirokastër, Vlorë, Sarandë, Lezhë, and Përmet show the highest value of the average per-capita consumption expenditure.

The prediction RMSE for the EBLUP is estimated using formula (7.2.33) of Rao (2003), while that of NPEBLUP is estimated in two alternative ways: the analytical estimator provided in Opsomer et al. (2008, section 3.1) and the estimator introduced in Salvati et al. (2010, section 2.2). The latter is an alternative measure of uncertainty that is obtained by conditioning on the realized values of the area effect and is based on the pseudo-linearization approach to MSE estimation described for example in Chandra and Chambers (2005, 2009). First, when

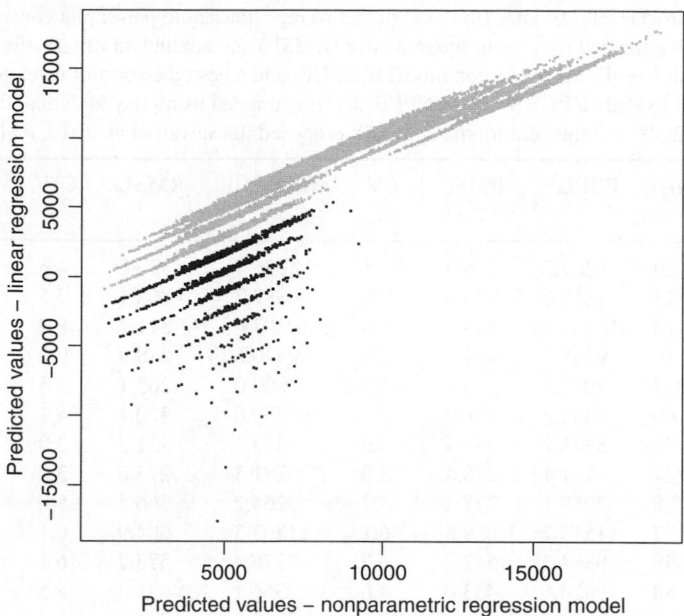

Figure 10.3 Comparison of population predictions using the nested error unit level linear regression model and the nonparametric one. Black dots refer to households with 10 or more units

comparing the standard methods of MSE estimation for small area BLUPs (RMSE for EBLUP and RMSE for NPEBLUP), we can note that estimated RMSEs are always smaller for NPEBLUP. In addition, when comparing the two alternative measures of uncertainty for NPEBLUP, we can note that they are comparable, although the conditional estimator has a more variable pattern. Indeed, the unconditional estimator has often the tendency to over-smoothing (see e.g. simulation results in Salvati et al. 2010). The most evident differences are found in some areas (such as Gjirokastër, Përmet, Devoll) that are relatively wealthier and for which the conditional RMSE estimator is sensibly larger than the unconditional one. We can note that these areas are also found to have high levels of inequality in income distribution (see e.g. results and discussion in Tzavidis et al. 2008) and have therefore an intrinsic more variable distribution of wealth that is better captured with the conditional estimator (see also the simulation results in Chapter 13). These findings are in line with those of Tzavidis et al. (2008) using the same set of covariates, and are also comparable with those of Chapter 13 in this book. R code is provided in the Appendix to perform EBLUP and NPEBLUP estimation and their MSE estimation.

Both Tzavidis et al. (2008) and Chapter 13 note the skew distribution of the household per-capita consumption expenditure, and we also note from Figure 10.1 the presence of a number of outlying observations. The first issue could be tackled fitting a nested error unit level nonparametric regression model for the log transformed variable of interest. However, back transformation of predicted values introduces a non-negligible bias. In addition, assessing the uncertainty of such a predictor is not straightforward. Some ideas are illustrated in Martin and Molina (2014) and Chapter 15 in this book, although it is still not clear on how to extend these to the case of nonparametric regression models. In addition, such a model could still suffer from the presence of outlying observations, so that the implementation of a nonparametric

Table 10.3 District level estimates (in Leks) of the average household per-capita consumption expenditure, with estimated root mean squared error (RMSE) and coefficient of variation (CV%) using a nested error unit level linear regression model (EBLUP) and a nested error unit level nonparametric regression model (NPEBLUP). RMSE for NPEBLUP is estimated using the analytical formula in Opsomer et al. (2008) and the conditional approach proposed in (Salvati et al. 2010, RMSE COND)

	n_d	EBLUP	RMSE	CV %	NPEBLUP	RMSE	CV %	RMSE COND	CV %
Berat	120	8267.2	376.0	4.5	8076.7	370.8	4.6	344.2	4.3
Bulqizë	128	4813.0	376.6	7.8	4887.7	368.1	7.5	307.3	6.3
Delvinë	16	10061.3	849.7	8.4	10207.6	811.7	8.0	779.4	7.6
Devoll	16	9738.7	870.3	8.9	9440.6	848.9	9.0	1226.8	13.0
Dibër	232	7902.6	278.9	3.5	8000.0	265.1	3.3	181.6	2.3
Durrës	160	7832.3	328.0	4.2	7839.0	320.1	4.1	285.6	3.6
Elbasan	152	8378.2	337.4	4.0	8473.7	331.3	3.9	425.7	5.0
Fier	224	9241.1	278.8	3.0	9281.3	274.6	3.0	286.3	3.1
Gramsh	120	7259.1	377.2	5.2	7264.2	366.5	5.0	378.7	5.2
Gjirokastër	32	11513.7	699.8	6.1	11302.3	684.9	6.1	1330.1	11.8
Has	48	9384.2	582.7	6.2	9376.4	573.7	6.1	450.2	4.8
Kavajë	88	9424.6	433.0	4.6	9366.1	425.3	4.5	413.7	4.4
Kolonjë	8	7803.2	1051.6	13.5	7722.8	1025.5	13.3	972.1	12.6
Korë	136	8500.2	354.4	4.2	8356.0	347.9	4.2	301.6	3.6
Krujë	38	7145.0	626.2	8.8	7105.7	614.1	8.6	555.4	7.8
Kuovë	32	7911.0	668.5	8.5	7886.7	654.1	8.3	410.6	5.2
Kukës	184	8033.9	313.6	3.9	8018.2	306.4	3.8	245.8	3.1
Kurbin	64	6184.2	513.2	8.3	6298.9	504.3	8.0	428.6	6.8
Lezhë	64	10432.7	508.0	4.9	10379.4	499.5	4.8	670.9	6.5
Librazhd	200	7359.9	295.3	4.0	7371.2	287.3	3.9	253.7	3.4
Lushnjë	152	8862.4	335.2	3.8	8851.2	326.7	3.7	296.2	3.3
Mal.Madhe	24	8363.3	743.1	8.9	8514.1	724.5	8.5	586.6	6.9
Mallakastër	32	9078.5	668.3	7.4	9105.6	655.4	7.2	724.1	8.0
Mat	32	8168.7	663.9	8.1	8393.8	650.6	7.8	398.3	4.7
Mirditë	16	8118.0	849.2	10.5	8120.9	830.4	10.2	482.4	5.9
Peqin	24	7409.4	740.2	10.0	7465.1	720.4	9.6	940.9	12.6
Përmet	16	10365.0	877.0	8.5	10333.3	863.6	8.4	1263.7	12.2
Pogradec	48	8034.5	563.3	7.0	8049.6	552.1	6.9	453.2	5.6
Pukë	23	6173.3	792.9	12.8	6512.0	777.6	11.9	809.8	12.4
Sarandë	48	11061.4	572.0	5.2	11109.0	552.3	5.0	560.5	5.0
Skrapar	16	7836.5	853.8	10.9	7587.7	837.4	11.0	474.7	6.3
Shkodër	138	8928.6	355.1	4.0	9033.1	345.3	3.8	267.6	3.0
Tepelenë	32	7986.6	664.4	8.3	7726.4	652.8	8.4	499.6	6.5
Tiranë	688	9729.7	164.5	1.7	9742.2	157.5	1.6	197.8	2.0
Tropojë	88	7686.4	437.2	5.7	7491.5	433.5	5.8	363.5	4.9
Vlorë	152	11130.2	339.0	3.0	11122.5	331.6	3.0	431.3	3.9

M-quantile model approach for SAE could be advisable. Table 10.4 is similar in structure to Table 10.3 and shows the district level estimates using a nonparametric M-quantile model with a Chambers–Dunstan type bias correction (NPMQCD) as in Equation (10.11). The prediction MSE for this estimator is estimated analytically using the approach proposed in Chambers and

Table 10.4 District level estimates (in Leks) of the average household per-capita consumption expenditure, with estimated root mean squared error (RMSE) and coefficient of variation (CV%) using a nonparametric M-quantile model with a Chambers–Dunstan type bias correction (NPMQCD). Estimates using the robust approaches proposed by Rao et al. (2013) (NPREBLUP) and by Salvati et al. (2010) (NPMBDE) are displayed for comparison as well

	n_d	NPMQCD	RMSE	CV%	NPREBLUP	NPMBDE
Berat	120	8023.5	407.0	5.1	7777.1	8718.9
Bulqizë	128	4830.1	428.4	8.9	4632.3	4752.3
Delvinë	16	9863.3	1254.7	12.7	9453.3	9563.2
Devoll	16	10518.2	1101.6	10.5	9428.1	11051.4
Dibër	232	8050.9	208.5	2.6	7823.4	8478.6
Durrës	160	7666.7	343.6	4.5	7405.0	7580.1
Elbasan	152	8298.1	481.9	5.8	7694.9	8026.5
Fier	224	9219.2	298.5	3.2	8673.5	9197.1
Gramsh	120	7204.8	431.6	6.0	6826.1	7330.9
Gjirokastër	32	13058.0	1177.7	9.0	11257.8	13375.2
Has	48	9460.8	489.6	5.2	9320.0	9662.1
Kavajë	88	9560.5	446.4	4.7	9155.1	9827.1
Kolonjë	8	6885.3	1079.6	15.7	7164.6	7136.8
Korë	136	8510.9	337.9	4.0	7888.5	9031.1
Krujë	38	6782.3	650.7	9.6	6783.1	6652.8
Kuovë	32	7628.2	615.9	8.1	7733.0	7812.9
Kukës	184	7927.3	239.6	3.0	7613.6	7330.2
Kurbin	64	5895.4	510.2	8.7	6015.5	5759.8
Lezhë	64	10655.5	657.5	6.2	10082.4	11232.0
Librazhd	200	7381.9	279.4	3.8	7066.4	7395.3
Lushnjë	152	8908.5	313.5	3.5	8535.6	9250.2
Malësi e Madhe	24	8117.6	880.6	10.8	7785.7	7807.3
Mallakastër	32	9343.0	817.0	8.7	8445.3	10010.5
Mat	32	8439.3	464.9	5.5	8156.2	7681.9
Mirditë	16	8176.1	757.1	9.3	8054.4	7686.9
Peqin	24	7226.6	1345.9	18.6	6354.1	7350.9
Përmet	16	11155.5	1203.1	10.8	9896.6	11146.5
Pogradec	48	7929.8	577.0	7.3	7550.4	8088.3
Pukë	23	5684.6	858.1	15.1	5979.5	5869.9
Sarandë	48	11350.7	530.2	4.7	10785.0	11190.1
Skrapar	16	7564.8	754.8	10.0	7401.1	7933.8
Shkodër	138	9045.3	322.4	3.6	8613.7	8704.7
Tepelenë	32	7735.7	713.6	9.2	7371.1	8098.2
Tiranë	688	9679.7	207.2	2.1	8841.5	9921.8
Tropojë	88	7279.2	398.1	5.5	7104.6	6846.3
Vlorë	152	11273.5	439.4	3.9	10333.2	11390.6

Tzavidis (2006) and extended to the nonparametric M-quantile model in Salvati et al. [2011, equation (16)]. Under this approach, given the estimates for the area mean quantile for $\hat{\tau}_d$ and for the smoothing parameter, the predictor can be written as a linear combination of the sample values of the variable of interest and these weights are employed to build a prediction MSE estimator using standard methods for robust estimation of the MSE of unbiased weighted linear estimators (Royall and Cumberland 1978). Note that such an approach assumes that $\hat{\tau}_d$ and the smoothing parameter are constant, which leads to a first-order approximation to the actual MSE of (10.11). Resampling techniques as the nonparametric bootstrap proposed in Pratesi et al. (2008) could be used instead, but of course considerably increase the computational time.

The results are comparable with those reported in Tzavidis et al. (2008, Table 5) wherein a similar approach is followed but using a linear M-quantile model instead of a nonparametric one. In other words, in NPMQCD a smooth function is assumed and estimated for the relationship with `hhsize`, while in Tzavidis et al. (2008) a linear function is considered. Relative to the linear case, there is a reduction in estimated RMSE for NPMQCD for essentially all areas, showing the usefulness of introducing a nonparametric trend. Estimates using the robust approaches proposed by Rao et al. (2013) (NPREBLUP) and by Salvati et al. (2010) (NPMBDE) are displayed for comparison as well. R code is provided in the Appendix to implement such robust estimation techniques.

10.4 Concluding Remarks

Estimation of poverty-related variables at fine scales is a challenging statistical problem, because sample sizes are often very small and the relationships between the variables poorly captured by simple parametric models. At the same time, there are clear societal needs for such estimates, given their importance in increasing the understanding of factors affecting poverty and in designing and implementing government policies targeting poverty. In this context, flexible and robust SAE methods have the potential to make a real impact.

The methods described in this chapter are a valuable first step in the development of nonparametric methods with good statistical properties that can be used in estimating small-scale poverty variables. It is also clear that much work remains to be done. For instance, poverty indicators are often non-linear in the variable of interest (like FGT indicators). These can be handled within the M-quantile approach as in Tzavidis et al. (2008), while more needs to be done in extending the empirical best approach of Molina and Rao (2010) to encompass a nonparametric trend. As another example, poverty indicators rely on variables of interest like consumer expenditure or income that are skew and usually violate the normality assumption for the errors. It would also be interesting to develop a bias-adjustment for correcting estimates obtained by fitting the nonparametric model on the (usually log-) transformed variable and then transforming fitted values back as, for example, in Martin and Molina (2014) or to extend methods illustrated in this book in Chapter 15.

Continued collaboration between statisticians and poverty experts will be needed to ensure that the statistical methodologies being developed address the most important issues in the poverty arena, and that these methods make their way into the "toolbox" of researchers and policy-makers.

References

Battese GE, Harter RM, and Fuller WA 1988 An error-components model for prediction of county crop areas using survey and satellite data. *Journal of the American Statistical Association* **83** (401), 28–36.

Betti G, Neri L, and Bellini F 2003 Poverty and inequality mapping in albania: final report. *Washington, DC and Tirana, World Bank and INSTAT*.

Chambers R and TzavidisN 2006 M-quantile models for small area estimation. *Biometrika* **93**, 255–268.

Chandra H and Chambers R 2005 Comparing EBLUP and c-EBLUP for small area estimation. *Statistics in Transition* **7** (3), 637–648.

Chandra H and Chambers R 2009 Multipurpose weighting for small area estimation. *Journal of Official Statistics* **25** (3), 379–395.

Claeskens G 2004 Restricted likelihood ratio lack-of-fit tests using mixed spline models. *Journal of the Royal Statistical Society: Series B (Statistical Methodology)* **66** (4), 909–926.

D'Alò M, Di Consiglio L, Falorsi S, Ranalli MG, and Solari F 2012 Use of spatial information in small area models for unemployment rate estimation at sub-provincial areas in Italy. *Journal of the Indian Society of Agricultural Statistics* **66** (1), 43–54.

Das K, Jiang J, and Rao JNK 2004 Mean squared error of empirical predictor. *The Annals of Statistics* **32** (2), 818–840.

Datta GS and Lahiri P 2000 A unified measure of uncertainty of estimated best linear unbiased predictors in small area estimation problems. *Statistica Sinica* **10** (2), 613–628.

Dreassi E, Petrucci A, and Rocco E 2014 Small area estimation for semicontinuous skewed spatial data: An application to the grape wine production in tuscany. *Biometrical Journal* **56** (1), 141–156.

Eilers PH and Marx BD 1996 Flexible smoothing with B-splines and penalties. *Statistical Science* **11** (2), 89–102.

Fellner WH 1986 Robust estimation of variance components. *Technometrics* **28** (1), 51–60.

Giusti C, Marchetti S, Pratesi M, and Salvati N 2012 Semiparametric Fay-Herriot model using penalized splines. *J. Indian Society of Agricultural Statistics* **66** (1), 1–14.

Jiang J and Lahiri P 2001 Empirical best prediction for small area inference with binary data. *Annals of the Institute of Statistical Mathematics* **53** (2), 217–243.

Jiang J, Nguyen T, and Rao JS 2010 Fence method for nonparametric small area estimation. *Survey Methodology* **36** (1), 3–11.

Kammann E and Wand MP 2003 Geoadditive models. *Journal of the Royal Statistical Society: Series C (Applied Statistics)* **52** (1), 1–18.

Lombardía MJ and Sperlich S 2008 Semiparametric inference in generalized mixed effects models. *Journal of the Royal Statistical Society: Series B (Statistical Methodology)* **70** (5), 913–930.

Martin N and Molina I 2014 Best prediction under a nested error model with log transformation. *ArXiv e-prints* arXiv:1404.5465v1.

Militino AF, Goicoa T, and Ugarte MD 2012 Estimating the percentage of food expenditure in small areas using bias-corrected p-spline based estimators. *Computational Statistics & Data Analysis* **56** (10), 2934–2948.

Molina I and Rao JNK 2010 Small area estimation of poverty indicators. *Canadian Journal of Statistics* **38** **(3)**, 369–385.

Ngo L and Wand MP 2004 Smoothing with mixed model software. *Journal of Statistical Software* **9** (1), 1–54.

Opsomer JD, Claeskens G, Ranalli MG, Kauermann G, and Breidt FJ 2008 Non-parametric small area estimation using penalized spline regression. *Journal of the Royal Statistical Society: Series B (Statistical Methodology)* **70** (1), 265–286.

Pfeffermann D 2013 New important developments in small area estimation. *Statistical Science* **28** (1), 40–68.

Prasad NGN and Rao JNK 1990 The estimation of the mean squared error of small-area estimators. *Journal of the American Statistical Association* **85** (409), 163–171.

Pratesi M, Ranalli MG, and Salvati N 2008 Semiparametric M-quantile regression for estimating the proportion of acidic lakes in 8-digit HUCs of the northeastern US. *Environmetrics* **19** (7), 687–701.

Pratesi M, Ranalli MG, and Salvati N 2009 Nonparametric M-quantile regression using penalised splines. *Journal of Nonparametric Statistics* **21** (3), 287–304.

Rao JNK 2003 *Small Are Estimation*. John Wiley & Sons, Inc.

Rao JNK, Sinha SK, and Dumitrescu L 2013 Robust small area estimation under semi-parametric mixed models. *Canadian Journal of Statistics* **42** (1), 126–141.

Royall RM and Cumberland WG 1978 Variance estimation in finite population sampling. *Journal of the American Statistical Association* **73** (362), 351–358.

Rueda C and Lombardìa MJ 2012 Small area semiparametric additive monotone models. *Statistical Modelling* **12** (6), 527–549.

Ruppert D, Wand MP, and Carroll RJ 2003 *Semiparametric Regression*. Cambridge University Press.

Salvati N, Chandra H, Ranalli MG, and Chambers R 2010 Small area estimation using a nonparametric model-based direct estimator. *Computational Statistics & Data Analysis* **54** (9), 2159–2171.

Salvati N, Ranalli MG, and Pratesi M 2011 Small area estimation of the mean using non-parametric M-quantile regression: a comparison when a linear mixed model does not hold. *Journal of Statistical Computation and Simulation* **81** (8), 945–964.

Sperlich S and Lombardía MJ 2010 Local polynomial inference for small area statistics: estimation, validation and prediction. *Journal of Nonparametric Statistics* **22** (5), 633–648.

Tzavidis N, Salvati N, Pratesi M, and Chambers R 2008 M-quantile models with application to poverty mapping. *Statistical Methods and Applications* **17** (3), 393–411.

Ugarte MD, Goicoa T, Militino AF, and Durbán M 2009 Spline smoothing in small area trend estimation and forecasting. *Computational Statistics & Data Analysis* **53** (10), 3616–3629.

Wood S 2006 *Generalized Additive Models: an Introduction with R*. CRC Press.

Part IV

Spatio-Temporal Modeling of Poverty

11

Area-level Spatio-temporal Small Area Estimation Models

María Dolores Esteban, Domingo Morales and Agustín Pérez
Centro de Investigación Operativa, Universidad Miguel Hernández de Elche, Elche, Spain

11.1 Introduction

Linear models (LMs) look for the relationship between a dependent or response variable y and a vector of auxiliary variables $\boldsymbol{x}$. Three basic hypotheses are assumed for these models: linearity, normality, and independence. The first assumption states that the mean of y is a linear function of the components of $\boldsymbol{x}$. The second assumption specifies a multivariate normal distribution for the vector of observed y-values. The last one is the stochastic independence of the measurements of y. The study of LMs is a classical matter within applied statistics and there are many books dealing at length with them. We name but a few sources: Graybill (1976), Seber (1977), Arnold (1981), Hocking (1985), Searle (1997), and Rencher (2000).

LMs assume that observations are drawn from the same population and are independent. Mixed models have a more complex multilevel or hierarchical structure. Observations in different levels or clusters are assumed to be independent, but observations within the same level or cluster are considered as dependent because they share common properties. For these data, we can speak about two sources of variation: between and within clusters. The possibility of modeling those sources of variation, commonly present in real data, gives a high flexibility, and therefore applicability, to mixed models.

Linear mixed models (LMMs) handle data in which observations are not independent. That is, LMMs correctly model correlated errors, whereas the procedures in the LM family usually do not. Mixed models are generalizations of LMs to better support the analysis of a dependent variable. These models allow to incorporate:

1. *Random effects:* sometimes the number of levels of a categorical explanatory variable is so large (with respect to sample size) that introduction of fixed effects for its levels would

Analysis of Poverty Data by Small Area Estimation, First Edition. Edited by Monica Pratesi.

Companion Website: www.wiley.com/go/pratesi/poverty

lead to poor estimates of the model parameters. If this is the case, the explanatory variable should not be introduced in a LM. Mixed models solve this problem by treating the levels of the categorical variable as random, and then predicting their values.

2. *Hierarchical effects:* response variables are often measured at more than one level; for example, in nested territories in small area estimation problems. This situation can be modeled by mixed models and it is thus an appealing property of them.
3. *Repeated measures:* when several observations are collected on the same individual then the corresponding measurements are likely to be correlated rather than independent. This happens in longitudinal studies, time series data, or matched-pairs designs.
4. *Spatial correlations:* when there is correlation among clusters due to their location; for example, the correlation between nearby domains may give useful information to improve predictions.
5. *Small area estimation:* where the flexibility in effectively combining different sources of information and explaining different sources of errors is of great help. Mixed models typically incorporate area-specific random effects that explain the additional between area variation in the data that is not explained by the fixed part of the model.

Books dealing with LMMs include Searle, Casella and McCullogh (1992), Longford (1995), McCullogh and Searle (2001), Goldstein (2003), Demidenko (2004), and Jiang (2007).

In small area estimation, samples are drawn from a finite population, but estimations are required for subsets (called small areas or domains) where the effective sample sizes are too small to produce reliable direct estimates. An estimator of a small area parameter is called direct if it is calculated just with the sample data coming from the corresponding small area. Thus, the lack of sample data from the target small area affects seriously the accuracy of the direct estimators, and this fact has given rise to the development of new tools for obtaining more precise estimates. See a description of this theory in the monograph of Rao (2003), or in the reviews of Ghosh and Rao (1994), Rao (1999), Pfeffermann (2002), Jiang and Lahiri (2006), and Pfeffermann (2013).

Mixed models increase the effective information used in the estimation process by linking all observations of the sample, and at the same time they can allow for between-area variation. Models of this kind have been used for instance in the United States to estimate per capita income for small areas (Fay and Herriot, 1979), for estimating census undercount (Ericksen and Kadane, 1985; Dick, 1995), and for estimating poor school-age children. It is worth mentioning that "using these estimates, the U.S. Department of Education allocates annually over 7 billion of general funds to counties, and then the states distribute these funds among school districts" (Rao, 2003). The usage of these techniques is not restricted to socioeconomic data; an example in the field of agriculture is the work of Battese *et al.* (1988), who used a LMM to estimate county crop areas.

Indirect estimators, based on implicit linking models, include synthetic and composite estimators. These estimators are generally design-based and their design variances are usually small relative to the design variances of the direct estimators. However, the indirect estimators will be generally design biased and the design bias will not decrease as the overall sample size increases. If the implicit linking model is approximately true, then the design bias will be small, leading to significantly smaller design mean square error (MSE) compared with the MSE of a direct estimator. Reduction in MSE is the main reason for using indirect estimators. Indirect estimators of a target y-based parameter can make use of y-values and x-values from other domains and times and also of the correlation structure between domains and times.

Explicit linking models based on random area-specific effects that account for between area variation, beyond that explained by auxiliary variables included in the model, will be called "small area models". Indirect estimators based on small area models will be called "model-based estimators". We classify small area models into two broad types: (i) aggregate (or area)-level models that relate small area direct estimators to area specific covariates. Such models are necessary if unit (or element) level data are not available; and (ii) unit-level models that relate the unit values of a study variable to unit-specific covariates.

A small area model defines the way that the related data are incorporated in the estimation process. The model-based approach to small area estimation offers several advantages: (1) "optimal" estimators can be derived under the assumed model; (2) area specific measures of variability can be associated with each estimator unlike global measures often used with traditional indirect estimators; (3) models can be validated from the sample data; and (4) a variety of models can be entertained depending on the nature of the response variables and the complexity of data structures (e.g., time series and spatial structures).

This chapter deals only with aggregate-level models and, in particular, with extensions of the basic small area model: the Fay–Herriot model. This model is a special case of a general LMM and is used for continuous responses y.

11.2 Extensions of the Fay–Herriot Model

The Fay–Herriot model is composed of two stages. In the first stage a model, called *sampling* model, is used to represent the sampling error of direct estimators. Let μ_d be the characteristic of interest in the dth area (typically the mean) and y_d be a direct estimator of μ_d. The sampling model indicates that direct estimators $\{y_d\}$ are unbiased and can be expressed as

$$y_d = \mu_d + e_d, \quad d = 1, \dots, D,$$

where D is the total number of areas or domains target of inference. Here, $\{e_d\}$ are sampling errors, which, given μ_d, are independent and normally distributed with known variances, that is $e_d|\mu_d \sim N(0, \sigma_d^2)$, where σ_d^2 is the (assumed known) design-based variance of direct estimator y_d, $d = 1, \dots, D$. In the second stage, the true area characteristics $\{\mu_d\}$ are assumed to vary linearly with a number p of area-level auxiliary variables, that is,

$$\mu_d = \mathbf{x}'_d\boldsymbol{\beta} + u_d, \quad d = 1, \dots, D,$$

where $\mathbf{x}_d$ is a (column) vector containing the aggregated (population) values of p auxiliary variables for area d, $\boldsymbol{\beta}$ is the vector of regression coefficients and $\{u_d\}$ are model errors, typically assumed to be independent and identically distributed (i.i.d.) from $N(0, \sigma_u^2)$ with variance σ_u^2 unknown and independent of $\{e_d\}$. Note that this model, called a *linking* model, links the target quantities μ_d of all the areas through the common regression parameter $\boldsymbol{\beta}$. The Fay–Herriot model can be expressed as a single model in the form

$$y_d = \mathbf{x}'_d\boldsymbol{\beta} + u_d + e_d, \quad d = 1, \dots, D.$$

Many different extensions of this model have been proposed in the literature. For example, a multivariate generalization was studied by González-Manteiga et al. (2008). However, the generalizations to models including spatial and time correlation have played a prominent role.

If there is unexplained spatial correlation in the data, not considering it in the model will lead to erroneous inferences (Cressie, 1993). In small area estimation, when areas are properly delimited regions of the population, closer areas tend to have more similar socioeconomic characteristics. This was taken into account in the basic Fay–Herriot model by Singh et al. (2005), Petrucci and Salvati (2006), and Pratesi and Salvati (2008), who considered an extension of the Fay–Herriot model by assuming that area effects $\{u_d\}$ follow a spatial autoregressive process of order 1 or SAR(1). More concretely, Pratesi and Salvati (2008) assumed that the vector of area effects $(u_1, \ldots, u_D)'$ follows a SAR(1) process with variance parameter σ_1^2, spatial autocorrelation ρ_1 and row-standardized proximity matrix $\boldsymbol{W} = (w_{d,\ell})$, that is,

$$u_d = \rho \sum_{\ell \neq d} w_{d,\ell} u_\ell + \epsilon_d, \quad |\rho| < 1, \quad \epsilon_d \overset{iid}{\sim} N(0, \sigma_u^2), \quad d = 1, \ldots, D. \tag{11.1}$$

The $\boldsymbol{W}$ is obtained from an original proximity matrix $\boldsymbol{W}^0$, whose diagonal elements are equal to zero and the remaining entries are equal to 1 when the two areas corresponding to the row and the column indices are regarded as neighbors and zero otherwise. The row standardization is carried out by dividing each entry of $\boldsymbol{W}^0$ by the sum of the elements in the same row, see Anselin (1988), Cressie (1993) and Banerjee et al. (2004) for more details on the SAR(1) process with the above parametrization. When data from neighbor areas are correlated, considering this kind of spatial correlation in the model leads to more efficient small area estimators, see Molina et al. (2009). Bayesian spatial models have been considered by Moura and Migon (2002) and You and Zhou (2011). Thus, taking into account the spatial correlation among data from different areas allows to borrow even more strength from them.

When available, historical data offer precious information that can be used to improve the estimators at the current instant, that is, it is also possible to borrow strength from time. In this sense, Choudry and Rao (1989) extended the basic Fay–Herriot model including several time instants and considering an autocorrelated structure for sampling errors. More concretely, they considered the model

$$y_{dt} = \boldsymbol{x}_{dt}'\boldsymbol{\beta} + u_d + e_{dt}, \quad d = 1, \ldots, D, \quad t = 1, \ldots, T,$$

where y_{dt} and $\boldsymbol{x}_{dt}$ are, respectively, the response and the vector of auxiliary variables for area d at time instant t, with $\mu_{dt} = \boldsymbol{x}_{dt}'\boldsymbol{\beta} + u_d$ being the target characteristic for the same area and time instant. For each domain d, the errors $\{e_{dt}\}_{t=1}^T$ are assumed to follow an autoregressive process of order 1, AR(1), that is,

$$e_{dt} = \rho e_{d,t-1} + \epsilon_{dt}, \quad |\rho| < 1, \quad \epsilon_{dt} \overset{iid}{\sim} N(0, \sigma_\epsilon^2).$$

This model is not allowing for time variation in the area characteristics $\{\mu_{dt}\}$ that is not explained by auxiliary variables. For this sake, in an application to small area estimation of poverty indicators, Esteban et al. (2011) considered the model

$$y_{dt} = \boldsymbol{x}_{dt}'\boldsymbol{\beta} + u_{dt} + e_{dt}, \quad d = 1, \ldots, D, \quad t = 1, \ldots, T, \tag{11.2}$$

where, for each domain d, the random effects $\{u_{dt}\}_{t=1}^T$ are assumed to follow an AR(1) stochastic process and the random errors $\{e_{dt}\}_{t=1}^T$ are i.i.d. $N(0, \sigma_e^2)$.

Another simple model that borrows information across areas and over time, and which includes unexplained area–time variation, was proposed by Rao and Yu (1994), and is given by

$$y_{dt} = \boldsymbol{x}'_{dt}\boldsymbol{\beta} + u_{1d} + u_{2dt} + e_{dt}, \quad d = 1, \dots, D, \quad t = 1, \dots, T, \tag{11.3}$$

with $\mu_{dt} = \boldsymbol{x}'_{dt}\boldsymbol{\beta} + u_{1d} + u_{2dt}$ being now the target characteristic for area d and time instant t. In this model, the area effects $\{u_{1d}\}$ are constant over time following the usual assumptions in the basic Fay–Herriot model and, for each area d, $\{u_{2dt}\}_{t=1}^{T}$ are time-varying effects that follow an AR(1), but they are independent across areas. Sampling errors $\{e_{dt}\}_{t=1}^{T}$ are also independent across areas and normally distributed with zero mean vector and general covariance matrix Σ_d, assumed to be known. Sampling errors $\{e_{dt}\}$ are also independent of area and area–time random effects, $\{u_d\}$ and $\{u_{dt}\}$. Rao and Yu (1994) estimated the variance parameters by a method of moments and, for known autocorrelation parameter ρ, they gave a second-order approximation to the MSE of the empirical best linear unbiased prediction (EBLUP) obtained from that model. They also proposed several alternative estimators of ρ together with corresponding MSE estimators. In simulation studies, they report significant gains in efficiency of the EBLUP based on the temporal model when the between-time variation relative to sampling variation was small and area variation was large. Their results indicated that the efficiency was growing with the number of available time instants T. Esteban et al. (2012) applied models (11.2) and (11.3) to derive EBLUP estimates of poverty indicators in the Spanish Living Condition Survey.

Other models with temporal correlation have been proposed. Ghosh et al. (1996) proposed a slightly more complicated time correlated area-level model to estimate the median income of four-person families for the 50 American states and the district of Columbia. You and Rao (2000) and Datta et al. (2002) used the Rao-Yu model (11.3), but replacing the AR(1) process by a random walk. Datta et al. (1999) considered a similar model but added extra terms to the linking models to reflect seasonal variation in their application. They applied their model to estimate monthly unemployment rates for nine American states and the district of Columbia. You et al. (2001) applied the Rao–Yu model to estimate monthly unemployment rates for census metropolitan areas in Canada. Finally, Pfeffermann and Burck (1990) and Singh et al. (1991) considered a model with time-varying random slopes obeying an autoregressive process.

In the framework of spatio-temporal models, Marhuenda et al. (2013) generalized the model (11.3) by considering spatial correlation in the domain random effects $\{u_{1d}\}$. They assumed that $(u_{11}, \dots, u_{1D})'$ follows a SAR(1) process with variance parameter σ_1^2, spatial autocorrelation ρ_1 and row-standardized proximity matrix $\boldsymbol{W} = (w_{d,\ell})$, that is,

$$u_{1d} = \rho_1 \sum_{\ell \neq d} w_{d,\ell} u_{1\ell} + \epsilon_{1d}, \quad |\rho_1| < 1, \quad \epsilon_{1d} \overset{iid}{\sim} N(0, \sigma_1^2), \quad d = 1, \dots, D. \tag{11.4}$$

They fit the model by using the Restricted Maximum Likelihood (REML) method and they gave a bootstrap procedure for estimating the MSE of the EBLUP.

In the following sections, we present a new modification of the model (11.3) with moving average MA(1) correlated random effects. We give a REML algorithm to fit the model and two approaches to estimate the MSE of the EBLUP. We show how to apply the MA(1) model to derive EBLUP estimates of poverty proportions. We finally present some simulations to illustrate the behavior of the introduced procedures.

11.3 An Area-level Model with MA(1) Time Correlation

Let us consider the model

$$y = X\beta + Z_1 u_1 + Z_2 u_2 + e, \tag{11.5}$$

where $y = \underset{1\le d\le D}{\text{col}}(y_d), y_d = \underset{1\le t\le m_d}{\text{col}}(y_{dt}), u_1 = \underset{1\le d\le D}{\text{col}}(u_{1,d}), u_2 = \underset{1\le d\le D}{\text{col}}(u_{2,d}), u_{2,d} = \underset{1\le t\le m_d}{\text{col}}(u_{2,dt})$, $e = \underset{1\le d\le D}{\text{col}}(e_d)$, $e_d = \underset{1\le t\le m_d}{\text{col}}(e_{dt})$, $X = \underset{1\le d\le D}{\text{col}}(X_d)$, $X_d = \underset{1\le t\le m_d}{\text{col}}(x_{dt})$, $x_{dt} = \underset{1\le j\le p}{\text{col}'}(x_{dtj})$, $\beta = \underset{1\le j\le p}{\text{col}}(\beta_j)$, $Z_1 = \underset{1\le d\le D}{\text{diag}}(1_{m_d})$, $Z_2 = I_M$, $M = \sum_{d=1}^{D} md$. Let us assume that $u_1 \sim N(0, V_{u_1})$, $u_2 \sim N(0, V_{u_2})$ y $e \sim N(0, V_e)$ are independent with covariance matrices $V_{u_1} = \sigma_1^2 I_D$, $V_{u_2} = \sigma_2^2 \Omega(\theta)$,

$$\Omega(\theta) = \underset{1\le d\le D}{\text{diag}}(\Omega_d(\theta)), V_e = \underset{1\le d\le D}{\text{diag}}(V_{ed}), V_{ed} = \underset{1\le t\le m_d}{\text{diag}}(\sigma_{dt}^2),$$

known sampling error variances σ_{dt}^2's and

$$\Omega_d = \Omega_d(\theta) = \begin{pmatrix} 1+\theta^2 & -\theta & \dots & 0 & 0 \\ -\theta & 1+\theta^2 & \ddots & & 0 \\ \vdots & \ddots & \ddots & \ddots & \vdots \\ 0 & & \ddots & 1+\theta^2 & -\theta \\ 0 & 0 & \dots & -\theta & 1+\theta^2 \end{pmatrix}_{m_d \times m_d}.$$

We can write model (11.5) in the more concise vector form $y = X\beta + Zu + e$, where $Z = (Z_1, Z_2)$ and $u = (u_1', u_2')'$. Alternatively model (11.5) can be written in the equation-by-equation form

$$y_{dt} = x_{dt}\beta + u_{1,d} + u_{2,dt} + e_{dt}, \quad d = 1, \dots, D, \quad t = 1, \dots, m_d \tag{11.6}$$

where y_{dt} is a direct estimator of the target population indicator and x_{dt} contains the aggregated values of p auxiliary variables. The subindex d is for domains and subindex t for time instants.

The best linear unbiased (BLU) estimators and predictors of β and u are

$$\hat{\beta} = (X'V^{-1}X)^{-1}X'V^{-1}y \qquad \text{and} \qquad \hat{u} = V_u Z'V^{-1}(y - X\hat{\beta}),$$

where $V_u = \text{diag}(V_{u_1}, V_{u_2})$ and $V = \text{var}(y) = \underset{1\le d\le D}{\text{diag}}(V_d)$ takes the form

$$V = \sigma_1^2 Z_1 Z_1' + \sigma_2^2 \underset{1\le d\le D}{\text{diag}}(\Omega_d(\theta)) + V_e = \underset{1\le d\le D}{\text{diag}}(\sigma_1^2 1_{m_d} 1_{m_d}' + \sigma_2^2 \Omega_d(\theta) + V_{ed}).$$

The REML log-likelihood, $l_{reml} = l_{reml}(\sigma_1^2, \sigma_2^2, \theta)$, is

$$l_{reml} = -\frac{M-p}{2}\log 2\pi + \frac{1}{2}\log|X'X| - \frac{1}{2}\log|V| - \frac{1}{2}\log|X'V^{-1}X| - \frac{1}{2}\, y'Py,$$

where

$$P = V^{-1} - V^{-1}X(X'V^{-1}X)^{-1}X'V^{-1}, \quad PVP = P, \quad PX = 0.$$

Let $\boldsymbol{\theta} = (\theta_1, \theta_2, \theta_3) = (\sigma_1^2, \sigma_2^2, \theta)$ be the vector of parameters. Let us define $\boldsymbol{V}_1 = \frac{\partial \boldsymbol{V}}{\partial \sigma_1^2} = \underset{1\le d\le D}{\text{diag}}\,(\mathbf{1}_{m_d}\mathbf{1}'_{m_d})$, $\boldsymbol{V}_2 = \frac{\partial \boldsymbol{V}}{\partial \sigma_2^2} = \underset{1\le d\le D}{\text{diag}}\,(\Omega_d(\theta))$ and $\boldsymbol{V}_3 = \frac{\partial \boldsymbol{V}}{\partial \theta} = \sigma_2^2\,\underset{1\le d\le D}{\text{diag}}\,(\dot{\Omega}_d(\theta))$, then

$$\boldsymbol{P}_a = \frac{\partial \boldsymbol{P}}{\partial \theta_a} = -\boldsymbol{P}\frac{\partial \boldsymbol{V}}{\partial \theta_a}\boldsymbol{P} = -\boldsymbol{P}\boldsymbol{V}_a\boldsymbol{P}, \quad a = 1, 2, 3.$$

By taking derivatives of l_{reml} with respect to θ_a, we get the scores

$$S_a = \frac{\partial l_{reml}}{\partial \theta_a} = -\frac{1}{2}\text{tr}(\boldsymbol{P}\boldsymbol{V}_a) + \frac{1}{2}\boldsymbol{y}'\boldsymbol{P}\boldsymbol{V}_a\boldsymbol{P}\boldsymbol{y}, \quad a = 1, 2, 3.$$

By taking again derivatives with respect to θ_a and θ_b, taking expectations and changing the sign, we get the Fisher information matrix components

$$F_{ab} = \frac{1}{2}\text{tr}(\boldsymbol{P}\boldsymbol{V}_a\boldsymbol{P}\boldsymbol{V}_b), \quad a, b = 1, 2, 3.$$

The updating formula of the Fisher-scoring algorithm, for calculating the REML estimators of the variance components, is

$$\boldsymbol{\theta}^{k+1} = \boldsymbol{\theta}^k + \boldsymbol{F}^{-1}(\boldsymbol{\theta}^k)\boldsymbol{S}(\boldsymbol{\theta}^k).$$

Seeds to initialize the algorithm are $\theta_0^{(0)} = \theta_1^{(0)} = S^2/2$, $\theta_3^{(0)} = 0$, where $S^2 = \frac{1}{M-p}(\boldsymbol{y} - \boldsymbol{X}\tilde{\boldsymbol{\beta}})'\boldsymbol{V}_e^{-1}(\boldsymbol{y} - \boldsymbol{X}\tilde{\boldsymbol{\beta}})$ and $\tilde{\boldsymbol{\beta}} = (\boldsymbol{X}'\boldsymbol{V}_e^{-1}\boldsymbol{X})^{-1}\boldsymbol{X}'\boldsymbol{V}_e^{-1}\boldsymbol{y}$. Alternatively, we can use the seeds derived from the model without $\boldsymbol{u}_1$ and with $\theta = 0$. More concretely, $\theta^{(0)} = 0$ and $\sigma_1^{2(0)} = \sigma_2^{2(0)} = \hat{\sigma}_{uH}^2/2$, where $\hat{\sigma}_{uH}^2$ is the Henderson 3 estimator of the resulting Fay–Herriot model. Let $\hat{\boldsymbol{\theta}}$ and $\hat{\boldsymbol{V}} = \boldsymbol{V}(\hat{\boldsymbol{\theta}})$ be the REML estimators of $\boldsymbol{\theta}$ and $\boldsymbol{V}$, respectively. The REML estimator of $\boldsymbol{\beta}$ is

$$\hat{\boldsymbol{\beta}} = (\boldsymbol{X}'\hat{\boldsymbol{V}}^{-1}\boldsymbol{X})^{-1}\boldsymbol{X}'\hat{\boldsymbol{V}}^{-1}\boldsymbol{y}.$$

The asymptotic distributions of $\hat{\boldsymbol{\theta}}$ and $\hat{\boldsymbol{\beta}}$ are

$$\hat{\boldsymbol{\theta}} \sim N_3(\boldsymbol{\theta}, \boldsymbol{F}^{-1}(\boldsymbol{\theta})), \quad \hat{\boldsymbol{\beta}} \sim N_p(\boldsymbol{\beta}, (\boldsymbol{X}'\boldsymbol{V}^{-1}\boldsymbol{X})^{-1}).$$

Asymptotic $(1-\alpha)$-level confidence intervals for θ_a and β_j are

$$\hat{\theta}_a \pm z_{\alpha/2}\, \nu_{aa}^{1/2},\ a = 1, 2, 3, \quad \hat{\beta}_j \pm z_{\alpha/2}\, q_{jj}^{1/2},\ j = 1, \dots, p,$$

where $\boldsymbol{F}^{-1}(\hat{\boldsymbol{\theta}}) = (\nu_{ab})_{a,b=1,2,3}$, $(\boldsymbol{X}'\boldsymbol{V}^{-1}(\hat{\boldsymbol{\theta}})\boldsymbol{X})^{-1} = (q_{ij})_{i,j=1,\dots,p}$, and z_α is the α-quantile of the $N(0, 1)$ distribution. If we observe $\hat{\beta}_j = \beta_0$, then the p-value for testing $H_0:\ \beta_j = 0$ is

$$p = 2P_{H_0}(\hat{\beta}_j > |\beta_0|) = 2P(N(0, 1) > |\beta_0|/\sqrt{q_{jj}}\,).$$

11.4 EBLUP and MSE

This section deals with the problem of predicting $\mu_{dt} = \boldsymbol{x}_{dt}\boldsymbol{\beta} + u_{1,d} + u_{2,dt}$ by using the EBLUP $\hat{\mu}_{dt} = \boldsymbol{x}_{dt}\hat{\boldsymbol{\beta}} + \hat{u}_{1,d} + \hat{u}_{2,dt}$. More specifically, we are interested in predicting $y_{dt} = \boldsymbol{a}'\boldsymbol{y}$, where $\boldsymbol{a} = \underset{1\leq\ell\leq D}{\text{col}}\,(\underset{1\leq k\leq m_\ell}{\text{col}}\,(\delta_{d\ell}\delta_{tk}))$ is a vector with 1 in the position $t + \sum_{\ell=1}^{d-1} m_\ell$ and 0's in the remaining positions. The model-based estimator of $\bar{Y}_{dt}$ is $\hat{\bar{Y}}_{dt}^{eblup} = \hat{\mu}_{dt}$. By adapting to model (11.5) the formula of Prasad and Rao (1990) for moment-based estimator, or the analogous one of Datta and Lahiri (2000) or Das et al. (2004) to maximum likelihood and REML estimators, the MSE of $\hat{\bar{Y}}_{dt}^{eblup}$ is

$$MSE(\hat{\bar{Y}}_{dt}^{eblup}) = g_1(\boldsymbol{\theta}) + g_2(\boldsymbol{\theta}) + g_3(\boldsymbol{\theta}), \tag{11.7}$$

where $\boldsymbol{\theta} = (\sigma_1^2, \sigma_2^2, \theta)$,

$$g_1(\boldsymbol{\theta}) = \boldsymbol{a}'\boldsymbol{Z}\boldsymbol{T}\boldsymbol{Z}'\boldsymbol{a},$$

$$g_2(\boldsymbol{\theta}) = [\boldsymbol{a}'\boldsymbol{X} - \boldsymbol{a}'\boldsymbol{Z}\boldsymbol{T}\boldsymbol{Z}'\boldsymbol{V}_e^{-1}\boldsymbol{X}]\boldsymbol{Q}[\boldsymbol{X}'\boldsymbol{a} - \boldsymbol{X}'\boldsymbol{V}_e^{-1}\boldsymbol{Z}\boldsymbol{T}\boldsymbol{Z}'\boldsymbol{a}],$$

$$g_3(\boldsymbol{\theta}) \approx \text{tr}\left\{\left(\nabla\boldsymbol{b}'\right) V(\nabla\boldsymbol{b}')'E\left[\left(\hat{\theta} - \theta\right)(\hat{\theta} - \theta)'\right]\right\}.$$

and $\boldsymbol{Q} = (\boldsymbol{X}'\boldsymbol{V}^{-1}\boldsymbol{X})^{-1}$, $\boldsymbol{T} = \boldsymbol{V}_u - \boldsymbol{V}_u\boldsymbol{Z}'\boldsymbol{V}^{-1}\boldsymbol{Z}\boldsymbol{V}_u$, $\boldsymbol{b}' = \boldsymbol{a}'\boldsymbol{Z}\boldsymbol{V}_u\boldsymbol{Z}'\boldsymbol{V}^{-1}$. The $MSE(\hat{\bar{Y}}_{dt}^{eblup})$ can be estimated by the plug-in estimator (see the Appendix for details)

$$mse_{dt}^0 = mse(\hat{\bar{Y}}_{dt}^{eblup}) = g_1(\hat{\boldsymbol{\theta}}) + g_2(\hat{\boldsymbol{\theta}}) + 2g_3(\hat{\boldsymbol{\theta}}).$$

The parametric bootstrap approach gives several estimators of $MSEs(\hat{\bar{Y}}_{dt}^{eblup})$. The estimator mse_{dt}^1, defined in step B7 below is the basic parametric bootstrap estimator. By combining the parametric bootstrap with the MSE components appearing in formula (11.7), step B7 introduces estimators mse_{dt}^2 and mse_{dt}^3. The steps of the parametric bootstrap algorithm are:
B1: Calculate the REML estimators $\hat{\sigma}_1$, $\hat{\sigma}_2$, $\hat{\theta}$ and $\hat{\boldsymbol{\beta}}$.
B2: For $d = 1, \ldots, D$, generate $u_{1,d}^*$ i.i.d. $N(0, \hat{\sigma}_1^2)$. Build a vector $\boldsymbol{u}_1^* = (u_{1,1}^*, \ldots, u_{1,D}^*)'$.
B3: For $d = 1, \ldots, D$, generate independent vectors $\boldsymbol{u}_{2,d}^* = (u_{2,d1}^*, \ldots, u_{2,dm_d}^*)'$ with MA(1) distribution of parameters $\hat{\sigma}_2$ and $\hat{\theta}$. Build a vector $\boldsymbol{u}_2^* = \underset{1\leq d\leq D}{\text{col}}(\boldsymbol{u}_{2,d}^*)$.
B4: For $d = 1, \ldots, D$, $t = 1, \ldots, m_d$, generate independent variables $e_{dt}^* \sim N(0, \sigma_{dt}^2)$. Build a vector $\underset{1\leq d\leq D}{\text{col}}\,(\underset{1\leq t\leq m_d}{\text{col}}\,(e_{dt}^*))$.

B5: Build the bootstrap model

$$\boldsymbol{y}^* = \boldsymbol{X}\hat{\boldsymbol{\beta}} + \boldsymbol{Z}_1\boldsymbol{u}_1^* + \boldsymbol{Z}_2\boldsymbol{u}_2^* + \boldsymbol{e}^*. \tag{11.8}$$

B6: Generate B bootstrap vectors $\boldsymbol{y}^{*(b)}$, $b = 1, \ldots, B$, from (11.8). For each vector $\boldsymbol{y}^{*(b)}$, calculate the mean $\boldsymbol{\mu}^{*(b)}$, the EBLUP $\hat{\boldsymbol{\mu}}^{*(b)}$ and the BLUP $\tilde{\boldsymbol{\mu}}^{*(b)}$ with components $\mu_{dt}^{*(b)}$, $\widehat{\mu_{dt}}^{*(b)}$ and $\tilde{\mu}_{dt}^{*(b)}$, respectively.

B7: For $d = 1, \dots, D, t = 1, \dots, m_d$, calculate

$$mse^1_{dt} = \frac{1}{B}\sum_{b=1}^{B}(\widehat{\mu}^{*(b)}_{dt} - \mu^{*(b)}_{dt})^2,$$

$$mse^2_{dt} = \frac{1}{B}\sum_{b=1}^{B}\left\{g^{(*b)}_{1dt}(\widehat{\theta}^{*(b)}) + g^{*(b)}_{2dt}(\widehat{\theta}^{*(b)}) + 2g^{*(b)}_{3dt}(\widehat{\theta}^{*(b)})\right\},$$

$$mse^3_{dt} = g_{1dt}(\widehat{\theta}) + g_{2dt}(\widehat{\theta}) + \frac{1}{B}\sum_{b=1}^{B}(\widehat{\mu}^{*(b)}_d - \tilde{\mu}^{*(b)}_d)^2.$$

11.5 EBLUP of Poverty Proportions

Area-level models are applicable to the estimation of domain parameters such as totals, means, or proportions. This section illustrates how to use model (11.6) for estimating domain poverty proportions. For this sake, let us consider the EU Statistics on Income and Living Conditions (EU-SILC). The EU-SILC provides information regarding the household income received during the year prior to that of the interview. This income includes income from work for others, benefits/losses from self-employed work, social benefits, income from private pension schemes not related to work, capital and property income, transfers between other households, income received by minors and the result of the income tax statement. The income per household consumption unit (or equivalent personal income) is calculated in order to take into account scale economies in households. It is obtained by dividing the total household income by the number of consumption units. These are calculated using the modified OECD scale, which assigns a weight of 1 to the first adult, a weight of 0.5 to remaining adults, and a weight of 0.3 to children under 14 years of age. Once the household equivalent income is calculated, it is assigned to each of its members.

Let E_{dtj} denote the equivalent personal income of individual j of domain d at time period t, $d = 1, \dots, D$, $t = 1, \dots, T$, $j = 1, \dots, N_{dt}$, which is used in calculating the poverty incidence. The poverty threshold, z_t, is calculated each year, using the distribution of the equivalent personal income for the previous year. Following the criteria recommended by Eurostat, this threshold is set at 60% of the median income per household consumption units. The poverty incidence is the proportion of people with equivalent personal income below the poverty threshold. The poverty incidences at the domain levels are

$$p_{dt} = \frac{1}{N_{dt}}\sum_{j=1}^{N_{dt}} y_{dtj}, \quad \text{where} \quad y_{dtj} = I(E_{dtj} < z_t). \tag{11.9}$$

A direct estimator of the total $Y_{dt} = \sum_{j=1}^{N_{dt}} y_{dtj}$ is

$$\hat{Y}^{dir}_{dt} = \sum_{j \in S_{dt}} w_{dtj}\, y_{dtj},$$

where S_{dt} is the domain sample at time period t and the w_{dtj}'s are the sampling weights. A direct estimator of the domain mean is

$$y_{dt} = \hat{Y}_d^{dir}/\hat{N}_{dt}^{dir}, \text{ where } \hat{N}_{dt}^{dir} = \sum_{j \in S_{dt}} w_{dtj}.$$

Let $\sigma^2_{\pi,dt} = \hat{V}_\pi(y_{dt})$ be a unit-level data estimator of the design-based variance of the direct estimator y_{dt}. As domain sample sizes are typically small, the estimators $\sigma^2_{\pi,dt}$ might not be precise enough. The generalized variance function (GVF) method improves the quality of direct estimates of variances by using models. For example, we could fit the model

$$\log \sigma^2_{\pi,dt} = b_0 + b_1 y_{dt} + \epsilon_{dt} \text{ (model A)},$$

where the ϵ_{dt}'s are i.i.d. $N(0, \sigma^2_A)$. The selected σ^2_{dt}'s are the predicted variance values under the model A, which are calculated by using the formula

$$\sigma^2_{dt} = \exp\{\hat{\sigma}^2_A/2\} \cdot \exp\{\hat{b}_0 + \hat{b}_1 y_{dt}\}.$$

The factor $\exp\{\hat{\sigma}^2_A/2\}$ is the usual bias correction term in the log-linear regression analysis. Note that ignoring the correction term leads to underestimation of the true variances when applying the GVF method. For estimating the poverty proportion p_{dt}, we could use the EBLUP based on the model (11.6) with an appropriate set of auxiliary variable $\boldsymbol{x}_{dt}$. In that case, the target variable and the model error variances are the poverty proportion direct estimates y_{dt} and the corresponding GVF estimates σ^2_{dt}, respectively.

11.6 Simulations

11.6.1 Simulation 1

Simulation 1 is designed to empirically investigate the behavior of the REML estimators of the model parameters. For $d = 1, \ldots, D$, $t = 1, \ldots, m_d$, the explanatory and the target variables are

$$x_{dt} = \frac{1}{5}(b_{dt} - a_{dt})U_{dt} + a_{dt}, U_{dt} = \frac{t}{m_d + 1}, a_{dt} = 1, b_{dt} = 1 + \frac{1}{D}(m_d(d-1) + t),$$

$$y_{dt} = \beta_1 + \beta_2 x_{dt} + u_{1,d} + u_{2,dt} + e_{dt}, \beta_1 = 0, \beta_2 = 1,$$

where $u_{1,d} \sim N(0, \sigma^2_1)$ with $\sigma^2_1 = 0.75$, $e_{dt} \sim N(0, \sigma^2_{dt})$ and

$$\sigma^2_{dt} = \frac{1}{5}\frac{(\alpha_1 - \alpha_0)(m_d(d-1) + t - 1)}{M - 1} + \alpha_0, \quad \alpha_0 = 0.8, \alpha_1 = 1.2.$$

For $d = 1, \ldots, D$, the random effects u_{dt} are

$$u_{2,dt} = \epsilon_{dt} - \theta\epsilon_{dt-1}, t = 1, \ldots, m_d,$$

where $\epsilon_{dt} \sim N(0, \sigma_2^2)$, $d = 1, \ldots, D$, $t = 0, 1, \ldots, m_d$, with $\sigma_2^2 = 0.75$ and $\theta = 0.5$. The first simulation experiment follows the steps:

1. Repeat $K = 10^4$ times ($k = 1, \ldots, K$)
   (a) Generate a sample of size $m = \sum_{d=1}^{D} md$ and calculate $\mu_{dt}^{(k)} = \beta_1^{(k)} + \beta_2^{(k)} x_{dt} + u_{1,d}^{(k)} + u_{2,dt}^{(k)}$.
   (b) Calculate $\hat{\tau}^{(k)} \in \{\hat{\beta}_1^{(k)}, \hat{\beta}_2^{(k)}, \hat{\sigma}_1^{2(k)}, \hat{\sigma}_2^{2(k)}, \hat{\theta}^{(k)}\}$ and $\hat{\mu}_{dt}^{(k)}$ using the REML method.
2. For $\hat{\tau} \in \{\beta_1, \beta_2, \sigma_1^2, \sigma_2^2, \theta\}$ and $\hat{\mu}_{dt}$, $d = 1, \ldots, D$, $t = 1, \ldots, m_d$, calculate

$$BIAS(\hat{\tau}) = \frac{1}{K}\sum_{k=1}^{K}(\hat{\tau}^{(k)} - \tau), \quad MSE(\hat{\tau}) = \frac{1}{K}\sum_{k=1}^{K}(\hat{\tau}^{(k)} - \tau)^2.$$

$$BIAS_{dt} = \frac{1}{K}\sum_{k=1}^{K}(\hat{\mu}_{dt}^{(k)} - \mu_{dt}^{(k)}), \quad MSE_{dt} = \frac{1}{K}\sum_{k=1}^{K}(\hat{\mu}_{dt}^{(k)} - \mu_{dt}^{(k)})^2,$$

$$BIAS = \frac{1}{M}\sum_{d=1}^{D}\sum_{t=1}^{m_d} BIAS_{dt}, \quad MSE = \frac{1}{M}\sum_{d=1}^{D}\sum_{t=1}^{m_d} MSE_{dt}.$$

The simulation experiment is carried out for each of the six combinations of sample sizes appearing in Table 11.1.

Table 11.2 and Table 11.3 present the results of the simulation experiment.

Table 11.2 and Table 11.3 show that the bias is always close to zero and that the MSE decreases when the number of domains increases or when the number of time instants increases. This is to say, Table 11.2 and Table 11.3 empirically show that the estimators are consistent.

11.6.2 Simulation 2

The target of simulation 2 is to study the behavior of the MSE estimators. More concretely, we investigate the bias and the MSE of the estimators mse_{dt}^0, mse_{dt}^1, mse_{dt}^2 and mse_{dt}^3 of MSE_{dt}. For this sake, we compare the values of the empirical MSE of $\hat{\mu}_{dt}$, obtained in simulation 1, with the corresponding ones of mse_{dt}^a, $a = 0, 1, 2, 3$. The steps of the simulation experiment are:

1. For $D = 50, 100, 200$, take the values of MSE_{dt} from the output of simulation 1.
2. Repeat $K = 500$ times ($k = 1, \ldots, K$)
   (a) Generate a sample $(y_{dt}^{(k)}, x_{dt})$, $d = 1, \ldots, D$, $t = 1, \ldots, m_d$.
   (b) Calculate the estimators $\hat{\beta}_1^{(k)}$, $\hat{\beta}_2^{(k)}$, $\hat{\sigma}_1^{2(k)}$, $\hat{\sigma}_2^{2(k)}$, $\hat{\theta}^{(k)}$.

Table 11.1 Sample sizes

D	50	100	200	50	100	200
m_d	5	5	5	20	20	20
m	250	500	1000	1000	2000	4000

Table 11.2 *MSE* and *BIAS*

	$m_d = 5$			$m_d = 20$		
D	50	100	200	50	100	200
$MSE(\widehat{\beta}_1)$	0.23895	0.12067	0.06110	0.01996	0.00995	0.00497
$MSE(\widehat{\beta}_2)$	0.14196	0.07167	0.03630	0.00115	0.00058	0.00030
$MSE(\widehat{\sigma}_1^2)$	0.03114	0.01531	0.00770	0.02422	0.01201	0.00597
$MSE(\widehat{\sigma}_2^2)$	0.02056	0.01279	0.00816	0.00256	0.00127	0.00063
$MSE(\widehat{\theta})$	0.03286	0.02274	0.01578	0.00337	0.00162	0.00079
MSE	0.17102	0.17068	0.17047	0.15879	0.15867	0.15860
$BIAS(\widehat{\beta}_1)$	0.00074	0.00071	−0.00100	0.00062	−0.00031	0.00037
$BIAS(\widehat{\beta}_2)$	−0.00039	−0.00002	0.00084	0.00003	0.00012	−0.00001
$BIAS(\widehat{\sigma}_1^2)$	−0.01484	−0.00645	−0.00181	−0.00025	0.00009	−0.00138
$BIAS(\widehat{\sigma}_2^2)$	0.02555	0.00447	−0.00591	−0.00302	−0.00160	−0.00111
$BIAS(\widehat{\theta})$	−0.04283	−0.00699	0.00979	0.00400	0.00218	0.00114
$BIAS$	0.00015	0.00007	0.00006	0.00006	0.00000	0.00002

Table 11.3 E^a and B^a, $a = 0, 1, 2, 3$

	$m_d = 5$			$m_d = 20$		
D	50	100	200	50	100	200
E^0	0.026761	0.009031	0.006375	0.000027	0.000007	0.000007
E^1	0.000199	0.000196	0.000200	0.000168	0.000170	0.000171
E^2	0.125213	0.093907	0.047642	0.000034	0.000006	0.000006
E^3	0.000010	0.000004	0.000002	0.000005	0.000003	0.000006
B^0	0.109646	0.071106	0.051683	0.004809	0.002261	0.000864
B^1	−0.001251	−0.000499	0.000236	−0.000018	−0.000056	−0.000007
B^2	0.290532	0.230065	0.154741	0.005415	0.002308	0.001096
B^3	0.000940	0.000539	0.000213	0.000011	0.000001	−0.000270

(c) For $d = 1, \ldots, D$, $t = 1, \ldots, m_d$, calculate $mse_{dt}^{0(k)} = mse(\widehat{\overline{Y}}_{dt}^{eblup(k)})$.

(d) Repeat $B = 500$ times ($b = 1, \ldots, B$)

   (i) Generate $u_{1,d}^{*(kb)}$, $u_{2,dt}^{*(kb)}$, $e_{dt}^{*(kb)}$, $d = 1, \ldots, D$, $t = 1, \ldots, m_d$, using $\widehat{\beta}_1^{(k)}$, $\widehat{\beta}_2^{(k)}$, $\widehat{\sigma}_2^{2(k)}$, $\widehat{\sigma}_2^{2(k)}$, $\widehat{\theta}^{(k)}$, instead of β_1, β_2, σ_2^2, σ_2^2, θ.

   (ii) Generate a bootstrap sample $\{y_{dt}^{*(kb)}; d = 1, \ldots, D, t = 1, \ldots, m_d\}$, from the model

$$y_{dt}^{*(kb)} = \widehat{\beta}_1^{(k)} + \widehat{\beta}_2^{(k)} x_{dt} + u_{1,d}^{*(kb)} + u_{2,dt}^{*(kb)} + e_{dt}^{*(kb)}.$$

   (iii) Calculate $\mu_{dt}^{*(kb)} = \widehat{\beta}_1^{(k)} + \widehat{\beta}_2^{(k)} x_{dt} + u_{1,d}^{*(kb)} + u_{2,dt}^{*(kb)}$.

(iv) Calculate the bootstrap model estimators, $\widehat{\beta}_1^{*(k)}$, $\widehat{\beta}_2^{*(k)}$, $\widehat{\sigma}_1^{2*(k)}$, $\widehat{\sigma}_2^{2*(k)}$, $\widehat{\theta}^{*(k)}$, and the corresponding BLUP estimators $\widetilde{\beta}_1^{*(k)}$ and $\widetilde{\beta}_2^{*(k)}$.

(v) For $d = 1, \dots, D$, $t = 1, \dots, m_d$, calculate $\widehat{\mu}_{dt}^{*(kb)} = \widehat{\beta}_1^{*(k)} + \widehat{\beta}_2^{*(k)} x_{dt} + \widehat{u}_{1,d}^{*(kb)} + \widehat{u}_{2,dt}^{*(kb)}$.

(e) For $d = 1, \dots, D, t = 1, \dots, m_d$, calculate

$$
\begin{aligned}
mse_{dt}^{1(k)} &= \frac{1}{B}\sum_{b=1}^{B}(\widehat{\mu}_{dt}^{*(kb)} - \mu_{dt}^{*(kb)})^2,\\
mse_{dt}^{2(k)} &= \frac{1}{B}\sum_{b=1}^{B}\left\{g_{1dt}^{*(kb)}(\widehat{\theta}^{*(kb)}) + g_{2dt}^{*(kb)}(\widehat{\theta}^{*(kb)}) + 2g_{3dt}^{*(kb)}(\widehat{\theta}^{*(kb)})\right\},\\
mse_{dt}^{3(k)} &= g_{1dt}^{(k)}(\widehat{\theta}^{(k)}) + g_{2dt}^{(k)}(\widehat{\theta}^{(k)}) + \frac{1}{B}\sum_{b=1}^{B}(\widehat{\mu}_d^{*(kb)} - \widetilde{\mu}_d^{*(kb)})^2.
\end{aligned}
$$

3. Output: for $d = 1, \dots, D, t = 1, \dots, m_d$, calculate

$$
B_{dt}^a = \frac{1}{K}\sum_{k=1}^{K}(mse_{dt}^{a(k)} - MSE_{dt}), \quad E_{dt}^a = \frac{1}{K}\sum_{k=1}^{K}(mse_{dt}^{a(k)} - MSE_{dt})^2,
$$

$$
B^a = \frac{1}{M}\sum_{d=1}^{D}\sum_{t=1}^{m_d} B_{dt}^a, \quad E^a = \frac{1}{M}\sum_{d=1}^{D}\sum_{t=1}^{m_d} E_{dt}^a, \quad a = 0, 1, 2, 3.
$$

Table 11.3 presents the results of the simulation experiment. We observe that B^a and E^a, $a = 0, 1, 2, 3$, tend to zero as D or m_d increase.

Figure 11.1 and Figure 11.2 present the boxplots of the values of $mse_{dt}^{a(k)}$, $a = 0, 1, 2, 3$, for the cases $m_d = 5$, $D = 100$, $t = 5$, $d = 1, 50, 100$ and $m_d = 20$, $D = 100$, $t = 5$, $d = 1, 50, 100$,

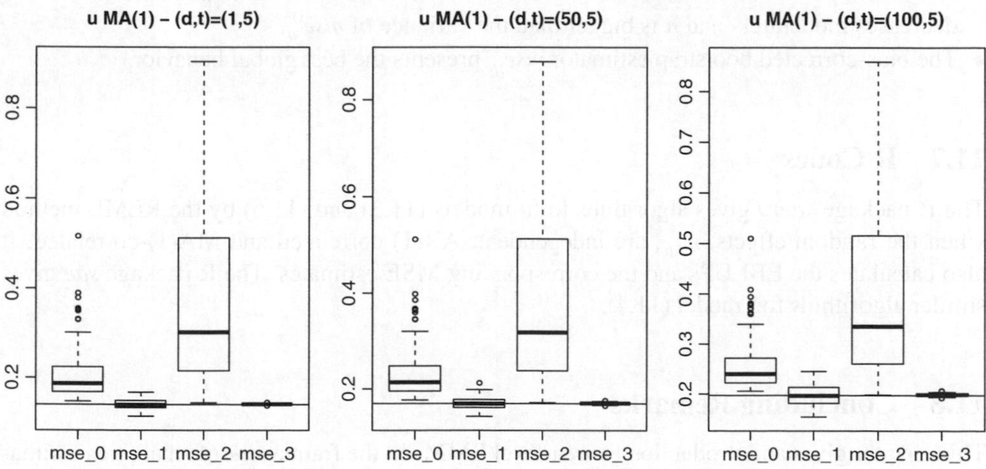

Figure 11.1 Boxplot for $m_d = 5$, $D = 100$, $t = 5$ and $d = 1, 50, 100$

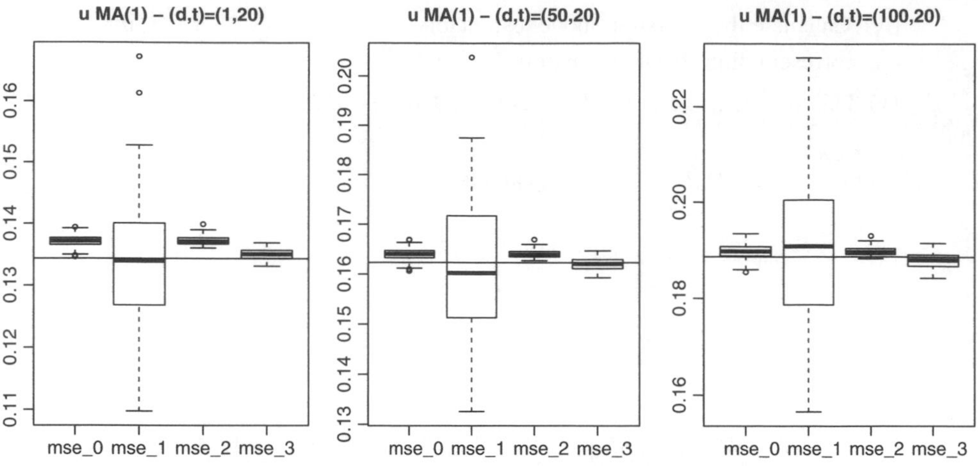

Figure 11.2 Boxplot for $m_d = 20$, $D = 100$, $t = 5$ and $d = 1, 50, 100$

respectively. By observing Figure 11.1 and Figure 11.2, we can draw the following conclusions:

- The analytic estimator mse_{dt}^0 has a high positive bias when the number of time instants is small ($m_d = 5$). This means that function g_3 is not helping too much in approximating the MSE of the EBLUP. However, as the number of instants increases ($m_d = 20$), the bias and the variance of mse_{dt}^0 decreases in a significant way.
- The bootstrap-bagging estimator mse_{dt}^2 increases the effects of the analytic estimator mse_{dt}^0 when $m_d = 5$. However, it has a similar behavior when $m_d = 20$.
- The simple bootstrap estimator mse_{dt}^1 is basically unbiased and has lower variance than mse_{dt}^0 for $m_d = 5$. However, as the number of time instants increases ($m_d = 20$), its variance decreases moderately and it is bigger than the variance of mse_{dt}^0.
- The bias-corrected bootstrap estimator mse_{dt}^3 presents the best global behavior.

11.7 R Codes

The R package *saery* gives algorithms to fit models (11.3) and (11.6) by the REML method when the random effects, u_{2dt}, are independent, AR(1)-correlated and MA(1)-correlated. It also calculates the EBLUPs and the corresponding MSE estimates. The R package *sae* gives similar algorithms for model (11.4).

11.8 Concluding Remarks

This chapter gives an introduction to area-level LMMs in the framework of small area estimation. The seminal Fay–Herriot model is the starting point for giving an overview on area-level spatio-temporal models. In addition, the chapter introduces a new temporal model with domain and time MA(1)-correlated random effects $u_{1,d}$ and $u_{2,dt}$, respectively. The new model takes into account cross sectional and temporal variability of direct estimators of domain parameters. The domain EBLUPs and several MSE estimators are given and investigated by means of Monte Carlo simulations. The chapter also illustrates how to apply the new model to the estimation of domain poverty proportions. The applicability of the introduced area-level model to the estimation of small area parameters is the main message for applied statisticians.

Appendix 11.A: MSE Components

The following section calculates the MSE components $g_1 - g_3$ appearing in (11.7).

11.A.1 Calculation of $g_1(\theta)$

We have $g_1(\theta) = \boldsymbol{a}'\boldsymbol{Z}\boldsymbol{T}\boldsymbol{Z}'\boldsymbol{a}$, where $\boldsymbol{T} = \boldsymbol{V}_u - \boldsymbol{V}_u\boldsymbol{Z}'\boldsymbol{V}^{-1}\boldsymbol{Z}\boldsymbol{V}_u$. This is to say,

$$\boldsymbol{T} = \begin{pmatrix} \sigma_1^2\boldsymbol{I}_D & \boldsymbol{0} \\ \boldsymbol{0} & \sigma_2^2\Omega(\theta) \end{pmatrix} - \begin{pmatrix} \sigma_1^2\boldsymbol{I}_D & \boldsymbol{0} \\ \boldsymbol{0} & \sigma_2^2\Omega(\theta) \end{pmatrix} \begin{pmatrix} \underset{1\le\ell\le D}{\text{diag}}\left(\boldsymbol{1}'_{m_\ell}\right) \\ \boldsymbol{I}_M \end{pmatrix}$$

$$\cdot \underset{1\le\ell\le D}{\text{diag}}(\boldsymbol{V}_\ell^{-1})\left(\underset{1\le\ell\le D}{\text{diag}}\left(\boldsymbol{1}_{m_\ell}\right), \boldsymbol{I}_M\right)\begin{pmatrix} \sigma_1^2\boldsymbol{I}_D & \boldsymbol{0} \\ \boldsymbol{0} & \sigma_2^2\Omega(\theta) \end{pmatrix} = \begin{pmatrix} \boldsymbol{T}_{11} & \boldsymbol{T}_{12} \\ \boldsymbol{T}_{21} & \boldsymbol{T}_{22} \end{pmatrix},$$

and

$$\boldsymbol{T}_{11} = \sigma_1^2\boldsymbol{I}_D - \sigma_1^4 \underset{1\le\ell\le D}{\text{diag}}(\boldsymbol{1}'_{m_\ell}\boldsymbol{V}_\ell^{-1}\boldsymbol{1}_{m_\ell}) = \underset{1\le\ell\le D}{\text{diag}}(T_{11\ell}),$$

$$\boldsymbol{T}_{12} = -\sigma_1^2\sigma_2^2 \underset{1\le\ell\le D}{\text{diag}}(\boldsymbol{1}'_{m_\ell}\boldsymbol{V}_\ell^{-1}\Omega_\ell(\theta)) = \underset{1\le\ell\le D}{\text{diag}}(\boldsymbol{T}_{12\ell}) = \boldsymbol{T}'_{21},$$

$$\boldsymbol{T}_{22} = \sigma_2^2 \underset{1\le\ell\le D}{\text{diag}}(\Omega_\ell(\theta)) - \sigma_2^4 \underset{1\le\ell\le D}{\text{diag}}(\Omega_\ell(\theta)\boldsymbol{V}_\ell^{-1}\Omega_\ell(\theta)) = \underset{1\le\ell\le D}{\text{diag}}(\boldsymbol{T}_{22\ell}).$$

Let us define $\boldsymbol{a}_{dl} = \underset{1\le k\le m_d}{\text{col}}(\delta_{tk})$. Then

$$g_1(\theta) = \boldsymbol{a}'\boldsymbol{Z}\boldsymbol{T}\boldsymbol{Z}'\boldsymbol{a} = \boldsymbol{a}'(\underset{1\le\ell\le D}{\text{diag}}(\boldsymbol{1}_{m_\ell}), \boldsymbol{I}_M)\begin{pmatrix} \boldsymbol{T}_{11} & \boldsymbol{T}_{12} \\ \boldsymbol{T}_{21} & \boldsymbol{T}_{22} \end{pmatrix}\begin{pmatrix} \underset{1\le\ell\le D}{\text{diag}}\left(\boldsymbol{1}'_{m_\ell}\right) \\ \boldsymbol{I}_M \end{pmatrix}\boldsymbol{a}$$

$$= \boldsymbol{a}' \underset{1\le\ell\le D}{\text{diag}}(\boldsymbol{1}_{m_\ell}T_{11\ell}\boldsymbol{1}'_{m_\ell} + \boldsymbol{1}_{m_\ell}\boldsymbol{T}_{12\ell} + \boldsymbol{T}_{21\ell}\boldsymbol{1}'_{m_\ell} + \boldsymbol{T}_{22\ell})\boldsymbol{a}$$

$$= \boldsymbol{a}'_d(\boldsymbol{1}_{m_d}T_{11d}\boldsymbol{1}'_{m_d} + \boldsymbol{1}_{m_d}\boldsymbol{T}_{12d} + \boldsymbol{T}_{21d}\boldsymbol{1}'_{m_d} + \boldsymbol{T}_{22d})\boldsymbol{a}_d.$$

11.A.2 Calculation of $g_2(\theta)$

We have

$$g_2(\theta) = [\boldsymbol{a}'\boldsymbol{X} - \boldsymbol{a}'\boldsymbol{Z}\boldsymbol{T}\boldsymbol{Z}'\boldsymbol{V}_e^{-1}\boldsymbol{X}]\boldsymbol{Q}[\boldsymbol{X}'\boldsymbol{a} - \boldsymbol{X}'\boldsymbol{V}_e^{-1}\boldsymbol{Z}\boldsymbol{T}\boldsymbol{Z}'\boldsymbol{a}] = (\boldsymbol{a}'_1 - \boldsymbol{a}'_2)\boldsymbol{Q}(\boldsymbol{a}_1 - \boldsymbol{a}_2),$$

where $\boldsymbol{a}_1' = \boldsymbol{a}'\boldsymbol{X} = \boldsymbol{a}_d'\boldsymbol{X}_d = \boldsymbol{x}_{dt}$ and

$$\begin{aligned}
\boldsymbol{a}_2' &= \boldsymbol{a}'\boldsymbol{Z}\boldsymbol{T}\boldsymbol{Z}'\boldsymbol{V}_e^{-1}\boldsymbol{X} \\
&= \boldsymbol{a}'(\underset{1\le\ell\le D}{\text{diag}}(\mathbf{1}_{m_\ell}),\boldsymbol{I}_M)\begin{pmatrix}\boldsymbol{T}_{11} & \boldsymbol{T}_{12}\\ \boldsymbol{T}_{21} & \boldsymbol{T}_{22}\end{pmatrix}\begin{pmatrix}\underset{1\le\ell\le D}{\text{diag}}\left(\mathbf{1}_{m_\ell}'\right)\\ \boldsymbol{I}_M\end{pmatrix}\underset{1\le\ell\le D}{\text{diag}}(\boldsymbol{V}_{e\ell}^{-1})\underset{1\le\ell\le D}{\text{col}}(\boldsymbol{X}_\ell) \\
&= \boldsymbol{a}_d'\{\mathbf{1}_{m_d}\boldsymbol{T}_{11d}\mathbf{1}_{m_d}' + \mathbf{1}_{m_d}\boldsymbol{T}_{12d} + \boldsymbol{T}_{21d}\mathbf{1}_{m_d}' + \boldsymbol{T}_{22d}\}\boldsymbol{V}_{ed}^{-1}\boldsymbol{X}_d.
\end{aligned}$$

11.A.3 Calculation of $g_3(\theta)$

We have

$$g_3(\boldsymbol{\theta}) \approx \text{tr}\{(\nabla\boldsymbol{b}')\boldsymbol{V}(\nabla\boldsymbol{b}')'E[(\hat{\boldsymbol{\theta}}-\boldsymbol{\theta})(\hat{\boldsymbol{\theta}}-\boldsymbol{\theta})']\},$$

where

$$\begin{aligned}
\boldsymbol{b}' &= \boldsymbol{a}'\boldsymbol{Z}\boldsymbol{V}_u\boldsymbol{Z}'\boldsymbol{V}^{-1} \\
&= \boldsymbol{a}'(\underset{1\le\ell\le D}{\text{diag}}(\mathbf{1}_{m_\ell}),\boldsymbol{I}_M)\begin{pmatrix}\sigma_1^2\boldsymbol{I}_D & \mathbf{0}\\ \mathbf{0} & \sigma_2^2\Omega(\theta)\end{pmatrix}\begin{pmatrix}\underset{1\le\ell\le D}{\text{diag}}\left(\mathbf{1}_{m_\ell}'\right)\\ \boldsymbol{I}_M\end{pmatrix}\underset{1\le\ell\le D}{\text{diag}}(\boldsymbol{V}_\ell^{-1}) \\
&= \boldsymbol{a}'\underset{1\le\ell\le D}{\text{diag}}([\sigma_1^2\mathbf{1}_{m_\ell}\mathbf{1}_{m_\ell}' + \sigma_2^2\Omega_\ell(\theta)]\boldsymbol{V}_\ell^{-1}) \\
&= \underset{1\le\ell\le D}{\text{col}'}(\delta_{d\ell}\boldsymbol{a}_\ell'[\sigma_1^2\mathbf{1}_{m_\ell}\mathbf{1}_{m_\ell}' + \sigma_2^2\Omega_\ell(\theta)]\boldsymbol{V}_\ell^{-1}).
\end{aligned}$$

It holds that

$$\begin{aligned}
\frac{\partial\boldsymbol{b}'}{\partial\sigma_1^2} &= \underset{1\le\ell\le D}{\text{col}'}(\delta_{d\ell}\boldsymbol{a}_\ell'\mathbf{1}_{m_\ell}\mathbf{1}_{m_\ell}'\boldsymbol{V}_\ell^{-1}) - \underset{1\le\ell\le D}{\text{col}'}(\delta_{d\ell}\boldsymbol{a}_\ell'[\sigma_1^2\mathbf{1}_{m_\ell}\mathbf{1}_{m_\ell}' + \sigma_2^2\Omega_\ell(\theta)]\boldsymbol{V}_\ell^{-1}\boldsymbol{V}_{\ell1}\boldsymbol{V}_\ell^{-1}), \\
\frac{\partial\boldsymbol{b}'}{\partial\sigma_2^2} &= \underset{1\le\ell\le D}{\text{col}'}(\delta_{d\ell}\boldsymbol{a}_\ell'\Omega_\ell(\theta)\boldsymbol{V}_\ell^{-1}) - \underset{1\le\ell\le D}{\text{col}'}(\delta_{d\ell}\boldsymbol{a}_\ell'[\sigma_1^2\mathbf{1}_{m_\ell}\mathbf{1}_{m_\ell}' + \sigma_2^2\Omega_\ell(\theta)]\boldsymbol{V}_\ell^{-1}\boldsymbol{V}_{\ell2}\boldsymbol{V}_\ell^{-1}), \\
\frac{\partial\boldsymbol{b}'}{\partial\theta} &= \sigma_2^2\underset{1\le\ell\le D}{\text{col}'}(\delta_{d\ell}\boldsymbol{a}_\ell'\dot{\Omega}_\ell(\theta)\boldsymbol{V}_\ell^{-1}) - \underset{1\le\ell\le D}{\text{col}'}(\delta_{d\ell}\boldsymbol{a}_\ell'[\sigma_1^2\mathbf{1}_{m_\ell}\mathbf{1}_{m_\ell}' + \sigma_2^2\Omega_\ell(\theta)]\boldsymbol{V}_\ell^{-1}\boldsymbol{V}_{\ell3}\boldsymbol{V}_\ell^{-1}),
\end{aligned}$$

where $\boldsymbol{V}_{\ell1} = \frac{\partial V_\ell}{\partial\sigma_1^2} = \mathbf{1}_{m_\ell}'\mathbf{1}_{m_\ell}$, $\boldsymbol{V}_{\ell2} = \frac{\partial V_\ell}{\partial\sigma_2^2} = \Omega_\ell(\theta)$, $\boldsymbol{V}_{\ell3} = \frac{\partial V_\ell}{\partial\theta} = \sigma_2^2\dot{\Omega}_\ell(\theta)$. We define

$$\begin{aligned}
\frac{\partial\boldsymbol{b}_d'}{\partial\sigma_1^2} &= \boldsymbol{a}_d'\mathbf{1}_{m_d}\mathbf{1}_{m_d}'\boldsymbol{V}_d^{-1} - \boldsymbol{a}_d'[\sigma_1^2\mathbf{1}_{m_d}\mathbf{1}_{m_d}' + \sigma_2^2\Omega_d(\theta)]\boldsymbol{V}_d^{-1}\mathbf{1}_{m_d}\mathbf{1}_{m_d}'\boldsymbol{V}_d^{-1}, \\
\frac{\partial\boldsymbol{b}_d'}{\partial\sigma_2^2} &= \boldsymbol{a}_d'\Omega_d(\theta)\boldsymbol{V}_d^{-1} - \boldsymbol{a}_d'[\sigma_1^2\mathbf{1}_{m_d}\mathbf{1}_{m_d}' + \sigma_2^2\Omega_d(\theta)]\boldsymbol{V}_d^{-1}\Omega_d(\theta)\boldsymbol{V}_d^{-1}, \\
\frac{\partial\boldsymbol{b}_d'}{\partial\theta} &= \sigma_2^2\boldsymbol{a}_d'\dot{\Omega}_d(\theta)\boldsymbol{V}_d^{-1} - \sigma_2^2\boldsymbol{a}_d'[\sigma_1^2\mathbf{1}_{m_d}\mathbf{1}_{m_d}' + \sigma_2^2\Omega_d(\theta)]\boldsymbol{V}_d^{-1}\dot{\Omega}_d(\theta)\boldsymbol{V}_d^{-1}.
\end{aligned}$$

It holds that

$$q_{11} = \frac{\partial \boldsymbol{b}'}{\partial \sigma_1^2} \underset{1\le \ell \le D}{\text{diag}} (\boldsymbol{V}_\ell) \left(\frac{\partial \boldsymbol{b}'}{\partial \sigma_1^2} \right)' = \frac{\partial \boldsymbol{b}_d'}{\partial \sigma_1^2} \boldsymbol{V}_d \left(\frac{\partial \boldsymbol{b}_d'}{\partial \sigma_1^2} \right)'$$
$$= \boldsymbol{a}_d' \mathbf{1}_{m_d} \mathbf{1}_{m_d}' \boldsymbol{V}_d^{-1} \mathbf{1}_{m_d} \mathbf{1}_{m_d}' \boldsymbol{a}_d - 2\boldsymbol{a}_d' \mathbf{1}_{m_d} \mathbf{1}_{m_d}' \boldsymbol{V}_d^{-1} \mathbf{1}_{m_d} \mathbf{1}_{m_d}' \boldsymbol{V}_d^{-1} [\sigma_1^2 \mathbf{1}_{m_d} \mathbf{1}_{m_d}' + \sigma_2^2 \Omega_d(\theta)] \boldsymbol{a}_d$$
$$+ \boldsymbol{a}_d' [\sigma_1^2 \mathbf{1}_{m_d} \mathbf{1}_{m_d}' + \sigma_2^2 \Omega_d(\theta)] \boldsymbol{V}_d^{-1} \mathbf{1}_{m_d} \mathbf{1}_{m_d}' \boldsymbol{V}_d^{-1} \mathbf{1}_{m_d} \mathbf{1}_{m_d}' \boldsymbol{V}_d^{-1} [\sigma_1^2 \mathbf{1}_{m_d} \mathbf{1}_{m_d}' + \sigma_2^2 \Omega_d(\theta)] \boldsymbol{a}_d,$$

$$q_{22} = \frac{\partial \boldsymbol{b}'}{\partial \sigma_2^2} \underset{1\le \ell \le D}{\text{diag}} (\boldsymbol{V}_\ell) \left(\frac{\partial \boldsymbol{b}'}{\partial \sigma_2^2} \right)' = \frac{\partial \boldsymbol{b}_d'}{\partial \sigma_2^2} \boldsymbol{V}_d \left(\frac{\partial \boldsymbol{b}_d'}{\partial \sigma_2^2} \right)'$$
$$= \boldsymbol{a}_d' \Omega_d(\theta) \boldsymbol{V}_d^{-1} \Omega_d(\theta) \boldsymbol{a}_d - 2\boldsymbol{a}_d' \Omega_d(\theta) \boldsymbol{V}_d^{-1} \Omega_d(\theta) \boldsymbol{V}_d^{-1} [\sigma_1^2 \mathbf{1}_{m_d} \mathbf{1}_{m_d}' + \sigma_2^2 \Omega_d(\theta)] \boldsymbol{a}_d$$
$$+ \boldsymbol{a}_d' [\sigma_1^2 \mathbf{1}_{m_d} \mathbf{1}_{m_d}' + \sigma_2^2 \Omega_d(\theta)] \boldsymbol{V}_d^{-1} \Omega_d(\theta) \boldsymbol{V}_d^{-1} \Omega_d(\theta) \boldsymbol{V}_d^{-1} [\sigma_1^2 \mathbf{1}_{m_d} \mathbf{1}_{m_d}' + \sigma_2^2 \Omega_d(\theta)] \boldsymbol{a}_d,$$

$$q_{33} = \frac{\partial \boldsymbol{b}'}{\partial \theta} \underset{1\le \ell \le D}{\text{diag}} (\boldsymbol{V}_\ell) \left(\frac{\partial \boldsymbol{b}'}{\partial \theta} \right)' = \frac{\partial \boldsymbol{b}_d'}{\partial \theta} \boldsymbol{V}_d \left(\frac{\partial \boldsymbol{b}_d'}{\partial \theta} \right)'$$
$$= \sigma_2^4 \boldsymbol{a}_d' \dot{\Omega}_d(\theta) \boldsymbol{V}_d^{-1} \dot{\Omega}_d(\theta) \boldsymbol{a}_d - 2\sigma_2^4 \boldsymbol{a}_d' \dot{\Omega}_d(\theta) \boldsymbol{V}_d^{-1} \dot{\Omega}_d(\theta) \boldsymbol{V}_d^{-1} [\sigma_1^2 \mathbf{1}_{m_d} \mathbf{1}_{m_d}' + \sigma_2^2 \Omega_d(\theta)] \boldsymbol{a}_d$$
$$+ \sigma_2^4 \boldsymbol{a}_d' [\sigma_1^2 \mathbf{1}_{m_d} \mathbf{1}_{m_d}' + \sigma_2^2 \Omega_d(\theta)] \boldsymbol{V}_d^{-1} \dot{\Omega}_d(\theta) \boldsymbol{V}_d^{-1} \dot{\Omega}_d(\theta) \boldsymbol{V}_d^{-1} [\sigma_1^2 \mathbf{1}_{m_d} \mathbf{1}_{m_d}' + \sigma_2^2 \Omega_d(\theta)] \boldsymbol{a}_d,$$

and that

$$q_{12} = \frac{\partial \boldsymbol{b}'}{\partial \sigma_1^2} \underset{1\le \ell \le D}{\text{diag}} (\boldsymbol{V}_\ell) \left(\frac{\partial \boldsymbol{b}'}{\partial \sigma_2^2} \right)' = \frac{\partial \boldsymbol{b}_d'}{\partial \sigma_1^2} \boldsymbol{V}_d \left(\frac{\partial \boldsymbol{b}_d'}{\partial \sigma_2^2} \right)'$$
$$= \boldsymbol{a}_d' \mathbf{1}_{m_d} \mathbf{1}_{m_d}' \boldsymbol{V}_d^{-1} \Omega_d(\theta) \boldsymbol{a}_d - \boldsymbol{a}_d' \mathbf{1}_{m_d} \mathbf{1}_{m_d}' \boldsymbol{V}_d^{-1} \Omega_d(\theta) \boldsymbol{V}_d^{-1} [\sigma_1^2 \mathbf{1}_{m_d} \mathbf{1}_{m_d}' + \sigma_2^2 \Omega_d(\theta)] \boldsymbol{a}_d$$
$$- \boldsymbol{a}_d' [\sigma_1^2 \mathbf{1}_{m_d} \mathbf{1}_{m_d}' + \sigma_2^2 \Omega_d(\theta)] \boldsymbol{V}_d^{-1} \mathbf{1}_{m_d} \mathbf{1}_{m_d}' \boldsymbol{V}_d^{-1} \Omega_d(\theta) \boldsymbol{a}_d$$
$$+ \boldsymbol{a}_d' [\sigma_1^2 \mathbf{1}_{m_d} \mathbf{1}_{m_d}' + \sigma_2^2 \Omega_d(\theta)] \boldsymbol{V}_d^{-1} \mathbf{1}_{m_d} \mathbf{1}_{m_d}' \boldsymbol{V}_d^{-1} \Omega_d(\theta) \boldsymbol{V}_d^{-1} [\sigma_1^2 \mathbf{1}_{m_d} \mathbf{1}_{m_d}' + \sigma_2^2 \Omega_d(\theta)] \boldsymbol{a}_d,$$

$$q_{13} = \frac{\partial \boldsymbol{b}'}{\partial \sigma_1^2} \underset{1\le \ell \le D}{\text{diag}} (\boldsymbol{V}_\ell) \left(\frac{\partial \boldsymbol{b}'}{\partial \theta} \right)' = \frac{\partial \boldsymbol{b}_d'}{\partial \sigma_1^2} \boldsymbol{V}_d \left(\frac{\partial \boldsymbol{b}_d'}{\partial \theta} \right)'$$
$$= \sigma_2^2 \boldsymbol{a}_d' \mathbf{1}_{m_d} \mathbf{1}_{m_d}' \boldsymbol{V}_d^{-1} \dot{\Omega}_d(\theta) \boldsymbol{a}_d - \sigma_2^2 \boldsymbol{a}_d' \mathbf{1}_{m_d} \mathbf{1}_{m_d}' \boldsymbol{V}_d^{-1} \dot{\Omega}_d(\theta) \boldsymbol{V}_d^{-1} [\sigma_1^2 \mathbf{1}_{m_d} \mathbf{1}_{m_d}' + \sigma_2^2 \Omega_d(\theta)] \boldsymbol{a}_d$$
$$- \sigma_2^2 \boldsymbol{a}_d' [\sigma_1^2 \mathbf{1}_{m_d} \mathbf{1}_{m_d}' + \sigma_2^2 \Omega_d(\theta)] \boldsymbol{V}_d^{-1} \mathbf{1}_{m_d} \mathbf{1}_{m_d}' \boldsymbol{V}_d^{-1} \dot{\Omega}_d(\theta) \boldsymbol{a}_d$$
$$+ \sigma_2^2 \boldsymbol{a}_d' [\sigma_1^2 \mathbf{1}_{m_d} \mathbf{1}_{m_d}' + \sigma_2^2 \Omega_d(\theta)] \boldsymbol{V}_d^{-1} \mathbf{1}_{m_d} \mathbf{1}_{m_d}' \boldsymbol{V}_d^{-1} \dot{\Omega}_d(\theta) \boldsymbol{V}_d^{-1} [\sigma_1^2 \mathbf{1}_{m_d} \mathbf{1}_{m_d}' + \sigma_2^2 \Omega_d(\theta)] \boldsymbol{a}_d,$$

$$q_{23} = \frac{\partial \boldsymbol{b}'}{\partial \sigma_2^2} \underset{1\le \ell \le D}{\text{diag}} (\boldsymbol{V}_\ell) \left(\frac{\partial \boldsymbol{b}'}{\partial \theta} \right)' = \frac{\partial \boldsymbol{b}_d'}{\partial \sigma_2^2} \boldsymbol{V}_d \left(\frac{\partial \boldsymbol{b}_d'}{\partial \theta} \right)'$$
$$= \sigma_2^2 \boldsymbol{a}_d' \Omega_d(\theta) \boldsymbol{V}_d^{-1} \dot{\Omega}_d(\theta) \boldsymbol{a}_d - \sigma_2^2 \boldsymbol{a}_d' \Omega_d(\theta) \boldsymbol{V}_d^{-1} \dot{\Omega}_d(\theta) \boldsymbol{V}_d^{-1} [\sigma_1^2 \mathbf{1}_{m_d} \mathbf{1}_{m_d}' + \sigma_2^2 \Omega_d(\theta)] \boldsymbol{a}_d$$
$$- \sigma_2^2 \boldsymbol{a}_d' [\sigma_1^2 \mathbf{1}_{m_d} \mathbf{1}_{m_d}' + \sigma_2^2 \Omega_d(\theta)] \boldsymbol{V}_d^{-1} \Omega_d(\theta) \boldsymbol{V}_d^{-1} \dot{\Omega}_d(\theta) \boldsymbol{a}_d$$
$$+ \sigma_2^2 \boldsymbol{a}_d' [\sigma_1^2 \mathbf{1}_{m_d} \mathbf{1}_{m_d}' + \sigma_2^2 \Omega_d(\theta)] \boldsymbol{V}_d^{-1} \Omega_d(\theta) \boldsymbol{V}_d^{-1} \dot{\Omega}_d(\theta) \boldsymbol{V}_d^{-1} [\sigma_1^2 \mathbf{1}_{m_d} \mathbf{1}_{m_d}' + \sigma_2^2 \Omega_d(\theta)] \boldsymbol{a}_d.$$

Finally,

$$g_3(\boldsymbol{\theta}) = \text{tr}\left\{\begin{pmatrix} q_{11} & q_{12} & q_{13} \\ q_{21} & q_{22} & q_{23} \\ q_{31} & q_{32} & q_{33} \end{pmatrix}\begin{pmatrix} F_{11} & F_{12} & F_{13} \\ F_{21} & F_{22} & F_{23} \\ F_{31} & F_{32} & F_{33} \end{pmatrix}^{-1}\right\}.$$

where F_{ab} is the element (a, b) of the Fisher information matrix in the REML Fisher-scoring algorithm.

Acknowledgements

The research was supported by the Spanish grant MTM2012-37077-C02-01.

References

Anselin, L. (1988). *Spatial Econometrics. Methods and Models*. Kluwer, Boston.

Arnold, S. (1981). *The Theory of Linear Models and Multivariate Analysis*. John Wiley & Sons, Inc., New York.

Battese, G. E., Harter, R. M., Fuller, W. A. (1988). An error component model for prediction of county crop areas using survey and satellite data. *Journal of the American Statistical Association*, **83**, 28–36.

Banerjee, S., Carlin, B., Gelfand, A. (2004). *Hierarchical Modeling and Analysis for Spatial Data*. Chapman and Hall, New York.

Choudry, G.H., Rao, J.N.K. (1989). Small area estimation using models that combine time series and cross sectional data. In: Singh, A.C., Whitridge, P. (Eds), *Proceedings of Statistics Canada Symposium on Analysis of Data in Time*, Ottawa: Statistics Canada, pp. 67–74.

Cressie, N. (1993). *Statistics for Spatial Data*. John Wiley & Sons, Inc., New York.

Das, K., Jiang, J., Rao, J.N.K. (2004). Mean squared error of empirical predictor. *The Annals of Statistics*, **32**, 818–840.

Datta, G.S., Lahiri, P. (2000). A unified measure of uncertainty of estimated best linear unbiased predictors in small area estimation problems. *Statistica Sinica*, **10**, 613–627.

Datta, G.S., Lahiri, P., Maiti, T. (2002). Empirical Bayes estimation of median income of four-person families by state using time series and cross-sectional data. *Journal of Statistical Planning and Inference*, **102**, 83–97.

Datta, G.S., Lahiri, P., Maiti, T., Lu, K.L. (1999). Hierarchical Bayes estimation of unemployment rates for the U.S. states. *Journal of the American Statistical Association*, **94**, 1074–1082.

Demidenko, E. (2004). *Mixed Models, Theory and Applications*. John Wiley & Sons, Inc., Hoboken.

Dick, P. (1995). Modelling net undercoverage in the 1991 Canadian census. *Survey Methodology*, **21**, 45–54.

Ericksen, E.P., Kadane, J.B. (1985). Estimating the population in census year: 1980 and beyond (with discussion). *Journal of the American Statistical Association*, **80**, 98–131.

Esteban, M.D., Morales, D., Pérez, A., Santamaría, L. (2011). Two area-level time models for estimating small area poverty indicators. *Journal of the Indian Society of Agricultural Statistics*, **66**, 75–89.

Esteban, M.D., Morales, D., Pérez, A., Santamaría, L. (2012). Small area estimation of poverty proportions under area-level time models. *Computational Statistics and Data Analysis*, **56**, 2840–2855.

Fay, R.E., Herriot, R.A. (1979). Estimates of income for small places: An application of James-Stein procedures to census data. *Journal of the American Statistical Association*, **74**, 269–277.

Ghosh, M., Nangia, N., Kim, D. (1996). Estimation of median income of four-person families: a Bayesian time series approach. *Journal of the American Statistical Association*, **91**, 1423–1431.

Ghosh, M., Rao, J.N.K. (1994). Small area estimation: An appraisal. *Statistical Science*, **9**, 55–93.

Goldstein, H. (2003). *Multilevel Statistical Models*. Arnold, London.

González-Manteiga, W., Lombardía, M.J., Molina, I., Morales, D., Santamaría, L. (2008). Analytic and bootstrap approximations of prediction errors under a multivariate Fay–Herriot model. *Computational Statistics and Data Analysis*, **52**, 5242–5252.

Graybill, F.A. (1976). *Theory and Application of the Linear Model*. Duxbury, Massachusetts.

Hocking, R.R. (1985). *The Analysis of Linear Models*. Brooks/Cole, Pacific Grove.

Jiang, J (2007). *Linear and Generalized Linear Mixed Models and Their Applications*. Springer, Berlin Heidelberg.

Jiang, J., Lahiri, P. (2006). Mixed model prediction and small area estimation. *Test*, **15**, 1–96.

Longford, N.T. (1995). *Random Coefficient Models*. Clareton Press, Clarendon Press, Oxford, London.

Marhuenda, Y., Molina, I., Morales, D. (2013). Small area estimation with spatio-temporal Fay–Herriot models. *Computational Statistics and Data Analysis*, **58**, 308–325.

McCullogh, C.E., Searle, S.R. (2001). *Generalized, Linear and Mixed Models*. John Wiley & Sons, Inc., New York.

Molina, I., Salvati, N., Pratesi, M. (2009). Bootstrap for estimating the MSE of the Spatial EBLUP. *Computational Statistics*, **24**, 441–458.

Moura, F.A.S., Migon, H.S. (2002). Bayesian Spatial Models for small area estimation of proportions. *Statistical Modelling*, **2**, 183–201.

Petrucci, A., Salvati, N. (2006). Small area estimation for spatial correlation in watershed erosion assessment. *Journal of Agricultural, Biological, and Environmental Statistics*, **11**, 169–182.

Pfeffermann, D. (2002). Small area estimation – new developments and directions. *International Statistical Review*, **70**, 125–143.

Pfeffermann D. (2013). New important developments in small area estimation. *Statistical Science*, **28**, 1–134.

Pfeffermann, D., Burck, L. (1990). Robust small area estimation combining time series and cross-sectional data. *Survey Methodology*, **16**, 217–237.

Prasad, N.G.N., Rao, J.N.K. (1990). The estimation of the mean squared error of small-area estimators. *Journal of the American Statistical Association*, **85**, 163–171.

Pratesi, M., Salvati, N. (2008). Small area estimation: the EBLUP estimator based on spatially correlated random area effects. *Statistical Methods and Applications*, **17**, 113–141.

Rao, J.N.K. (1999). Some recent advances in model-based small area estimation. *Survey Methodology*, **25**, 175–186.

Rao, J.N.K. (2003). *Small Area Estimation*. John Wiley & Sons, Inc., Hoboken.

Rao, J.N.K., Yu, M. (1994). Small area estimation by combining time series and cross sectional data. *Canadian Journal of Statistics*, **22**, 511–528.

Rencher, A.C. (2000). *Linear Models in Statistics*. John Wiley & Sons, Inc., New York.

Searle, S.R. (1997). Built-in restrictions on best linear unbiased predictors (BLUP) of random effects in mixed models. *The American Statistician*, **51**, 19–21.

Searle, S.R., Casella, G., McCullogh, C.E. (1992). *Variance Components*. John Wiley & Sons, Inc., New York.

Seber, G.A.F. (1977). *Linear Regression Analysis*. John Wiley & Sons, Inc., New York.

Singh, A.C., Mantel, H., Thomas, B.W. (1991). Time series generalizations of Fay–Herriot estimator for small areas. In: *Proceedings of Survey Research Methods Section. American Statistical Association*, pp. 455–460.

Singh, B., Shukla, G., Kundu, D. (2005). Spatio-temporal models in small area estimation. *Survey Methodology*, **31**, 183–195.

You, Y., Rao, J.N.K. (2000). Hierarchical Bayes estimation of small area means using multi-level models. *Survey Methodology*, **26**, 173–181.

You, Y., Rao, J.N.K., Gambino, J. (2001). Model-based unemployment rate estimation for the Canadian Labour Force Survey: a hierarchical approach. Technical report, Household Survey Method Division. Statistics Canada.

You, Y., Zhou, Q.M. (2011). Hierarchical Bayes small area estimation under a spatial model with application to health survey data. *Survey Methodology*, **37**, 25–37.

12

Unit Level Spatio-temporal Models

Maria Chiara Pagliarella[1] and Renato Salvatore[2]

[1] *Department of Economics and Statistics, University of Siena, Siena, Italy*
[2] *Department of Economics and Jurisprudence, University of Cassino and Southern Lazio, Cassino (FR), Italy*

Small area statistics are important tools for planning policies in specific regional and administrative areas, as well as for satisfying the general request for information on social and economics conditions in local areas (Elbers *et al.*, 2003; European Commission, 2011). In the last few decades, we have observed an increasing demand of information about poverty and living conditions at small area level (Berger and Skinner, 2003). This is because the national official surveys do not provide accurate estimates of totals and means in small geographical areas (Di Consiglio *et al.*, 2003). This is because traditional direct estimators have large standard errors, even when the sampling design itself is planned starting from an expected high level of precision of the estimates for larger areas (Elbers *et al.*, 2003; Chandra *et al.*, 2007; Lehtonen and Veijanen, 2012). The small area statistics can be defined as a collection of statistical methods that "borrow strength" from related or similar small areas, whose information is collected by the sample itself (i.e., it is the basis of the so-called indirect estimation) (Rao, 2003). To do this, statistics models connect variables of interest (in small areas) with vectors of supplementary data, such as demographic, behavioral, economic notices, coming from administratvive, census and specific sample surveys records (Tarozzi and Deaton, 2009). Small area efficient statistics provide, in addition to this, excellent statistics for local estimation of population, farms, and other characteristics of interest in post-censual years (Prasad and Rao, 1999; Sinha and Rao, 2009). The main reason that the statistical bureau offices are exploiting the field of estimation theory is that indirect estimation models can improve the efficiency of direct survey estimates in the small domains (Pfefferman, 2002; Rao, 2003; Molina and Rao, 2010; European Commission, 2011).

Analysis of Poverty Data by Small Area Estimation, First Edition. Edited by Monica Pratesi.

Companion Website: www.wiley.com/go/pratesi/poverty

The most commonly used tecniques for small area estimation are the empirical Bayes procedure, the hierarchical Bayes, and the empirical best linear unbiased prediction (EBLUP) approach.

Inside the general statistical models, in the context of the small area estimation methods, unit level models belong to the classic model framework of the general linear mixed model (Battese *et al.*, 1988; Jiang, 2007). This category of linear models was widely used in recent years in order to realize effective small area statistics (Chambers *et al.*, 2009). One of the targets of the small area estimation is to derive empirical best linear unbiased predictors of totals and means of survey variables, through two categories of models: area level (Fay–Herriot) (Fay and Herriot, 1979), and unit level models, as well as their generalizations. The main feature of the unit level models is that they work with some data that are available at the level of the survey units. When this type of records is not available, small area estimation is also reliable, with the adoption of the Fay–Herriot model. Extensions of both types of these classic small area models are discussed in a wide and analytic literature (see e.g., Rao 2003). In particular, two different types of extensions of small area estimation models will be considered jointly here. The first considers the spatial autocorrelation amongst the small areas. With the second, we can share common regression and autocorrelation parameters in the model, from time series data. Spatio-temporal models (at unit level), exploit these two kinds of abilities in a common (general) model, in which spatial and temporal autocorrelations allow the maximum gain of efficiency of the survey estimates to be achieved.

One of the relevant aspects that arises in model-based small area estimation, is the question of handling the neighborhood similarities in the regression model (Cressie and Wikle, 2011). Due to these common characteristics, the neighboring areas have spatial dependence (Pratesi and Salvati, 2008). Ignoring this spatial correlation in the data can lead to erroneous inferences (Marhuenda *et al.*, 2013). Models that account for this type of dependence can have better performances, compared with the efficiency of the direct estimators (Coelho and Pereira, 2011). Simultaneous dependence and conditional dependence among the random-area effects have been studied, following the structure of simultaneous autoregressive (SAR) and conditional autoregressive (CAR) models (Banerjee *et al.*, 2004). The major part of the literature on spatial small area models focuses on the area level Fay–Herriot model, with the spatial structure defined by the simultaneous dependence. There are two fundamental different ways of modeling the lattice information from the areas, in both SAR and CAR models. In the first, we have a continuous indexing set, in which some geostatistical data are treated (Wall, 2004). A spatial weight matrix containing row standardized distances between the centroids of the areas, or based on the length of the common boundary of them, is incorporated in the model. Alternatively, a row standardized proximity matrix is considered, based on a boolen neighborhood index that specifies whether the areas are neighbors or not (Cressie, 1993; Cressie and Wikle, 2011). The model parameter of spatial dependence (correlation) is then estimated on the basis of one of these types of matrices (Wall, 2004). Spatial small area estimation models are defined by a three-stage (area level) or a two-stage (unit level) linking model (Petrucci and Salvati, 2004), including in one of these stages the spatial dependence regression model on the vector of the random-area effects.

The so-called spatio-temporal models extend spatial models to time series data (Singh *et al.*, 2005). These models strengthen survey direct estimators by considering space–time correlated random-area effects (Pfefferman *et al.*, 1988). There are many case studies that have demonstrated that EBLUP estimators receive a significant gain in efficiency when the estimated between-time variation is relatively small (Pratesi and Salvati, 2004). This property is particularly relevant. For example, in area level models, when we have the situation in which the estimated area sampling variances are very heterogeneous with respect to the random-area time variation (Rao and Yu, 1994). Spatial models can include temporal correlation in different ways. The most common models include time-varying effects in within-area by time observations (areas or units). The classic Fay–Herriot model can be extended after including a space–time effect to the standard random-area effect in the regression equation (Marhuenda *et al.*, 2013). In most studies, the time-varying effects usually follow an AR(1) process. Otherwise, random walk processes, seasonal effects, and more complicated temporal models with random slopes have been studied. In order to consider time correlation in small area estimation, a further approach is based on the state space models. These models produce estimates at area and time updating the estimates over time by the Kalman filter equations (Gu *et al.*, 2014). At the time instant t, the EBLUP estimators of the state vector (in these types of models denominated the model "transition equation"), that define the so-called mesurement equation (i.e., the vector of the model fixed and random effects) are obtained on the basis of the data observed up to time $(t-1)$.

A common feature af all the methods and models in small area estimation is that the empirical predictor can improve direct survey estimates (Prasad and Rao, 1990; Das *et al.*, 2004; Longford, 2007). This is because the mean squared error (MSE) of the empirical predictors have reduced standard errors. In general, almost every research work on small area statistics is devoted to the prediction of the MSE of the model-based estimators. Spatio-temporal models have slightly more complicated mathematical expressions of the linear approximations of the MSE of the EBLUP. Parametric bootstrap estimators of the MSE are often studied, as well as a variety of simulation experiments, in order to analyze the gain in efficiency of even more complex models (Pfefferman and Tiller, 2001; Hall and Maiti, 2006; Gonzlez-Manteiga *et al.* 2008; Molina *et al.*, 2009).

Official statistical institutions need to produce comprehensive data about the geographical distribution of poverty over national and interregional territories. This is because the policy analysis needs to understand the various determinants of poverty phenomena in some areas, as well as mapping the evolution and changes over time in order to reduce the poverty by effective economic and social policy strategies. Several studies is recent years have shown that there are recursive poverty determinants related to some fundamental social area characteristics. The number of people engaged in the agriculture of small land cultivation, the number of young and of women that have access to productive resources and finance, the role of education, as well as the household size, are the most important. Generally, geographical and demographic factors influence both chronic and transient poverty. From a methodological point of view, the analysis of the data through statistical models can improve the comprehension of the relations between the poverty determinants and the role of geographical information, as well as the correct evaluation and the tracking of their changes over time. A more meaningful statistical analysis is characterized by analyzing the data taking in consideration spatio-temporal interactions. These interactions can be evaluated to be part of the statistical model in different ways: the extension of time series analysis to space, or the study of the implementation of the space–time interaction in the model. From an economic standpoint the objectives of the spatio-temporal statistical models can be the prediction of the poverty indicators, as well as the dimension reduction. The role of the small area estimation models inside the spatio-temporal data analysis can be twofold. The first, and the more explicit, is to deliver small area statistics on the main poverty indicators and factors, such as poverty proportion, poverty gap, poverty incidence, as well as the household wealth and other poverty determinants. Small area models can exploit spatio-temporal data interactions, improving the traditional effectiveness of the basic area level and unit level framework. These models are able to collect a powerful system of statistical tools that can be a further help in making political and economic decision strategies. Nevertheless, the ability of the small area models in describing economic and social changes among the areas can be a useful provision in poverty studies. In addition, the careful examination of the spatio-temporal information implemented in the small area models can lead to a more plentiful analysis of the poverty measures among the areas of study.

This chapter is dedicated to the unit level spatio-temporal models, in the context of the EBLUP estimation.

12.1 Unit Level Models

The small area model-based estimates that we call empirical best linear unbiased predictors (EBLUP), a rise from considering the general linear mixed model form, as follows:

$$\mathbf{y} = \mathbf{X}\boldsymbol{\beta} + \mathbf{Z}\mathbf{v} + \mathbf{e} \tag{12.1}$$

where $\mathbf{y} = col\left(\mathbf{y}_i\right)$, $i = 1,\ldots n$, $\mathbf{y}_i = col(y_{ij})$, $j = 1,\ldots n_i$, is a vector of the response (survey) variable, and $\mathbf{X} = col(\mathbf{X}_i), \mathbf{X}_i = col(\mathbf{x}'_{ij})$ is a matrix of known model covariates. In the case of linear mixed models with block-diagonal covariance structure, $\mathbf{Z} = diag(\mathbf{Z}_1,\ldots,\mathbf{Z}_n)$ is the design matrix, $\mathbf{v}_1,\ldots,\mathbf{v}_n \overset{\text{iid}}{\sim} N(\mathbf{0},\mathbf{G})$ are the components of the vector of random (area) effects, $e_i \overset{\text{ind}}{\sim} N(\mathbf{0},\mathbf{R})$ are the model errors. Further, the general expression of the model covariance is $Var(\mathbf{y}) = \mathbf{V} = \mathbf{V}(\theta) = \mathbf{Z}\mathbf{D}\mathbf{Z}' + \mathbf{R}$, with $\mathbf{D} = \mathbf{I}_m \otimes \mathbf{G}$, and $\mathbf{R} = \oplus_{i=1}^m \mathbf{R}_i$, $\mathbf{V} = \oplus_{i=1}^m \mathbf{V}_i, \boldsymbol{\theta} = (\theta_1,\ldots\theta q)$. Here $\boldsymbol{\theta}$ is the vector of covariance parameters of the model, generally with components in either $\mathbf{G}$ and $\mathbf{R}$. The symbols $\otimes$ and $\oplus$ denote, respectively, the Kronecker and the direct product. In linear models, the components of the vector $\boldsymbol{\theta}$ are estimated by method of moments, minimum norm quadratic estimation method, variance least squares, and, in the case of normal random effects and errors, by the maximum likelihood and restricted (residual) maximum likelihood estimation method. In the classic literature, the last is treated as the reference method (Jiang, 2007).

A general form for the EBLUP estimator t_i (at area i) is:

$$t_i\left(\hat{\theta}_R\right) = \mathbf{p}'_i\hat{\boldsymbol{\beta}} + \mathbf{m}'_i\hat{\mathbf{v}}, \tag{12.2}$$

being $\mathbf{p}_i, \mathbf{m}_i$ some vectors or matrices (e.g., in the case of multivariate EBLUP) that define t_i. Here the subscript R indicates that the estimate $\hat{\theta}_R$ is obtained through the restricted maximum likelihood estimation method (REML). Starting from $\hat{\mathbf{V}} = \mathbf{V}(\hat{\theta}_R)$, we have the following estimate of the vector of the fixed and the random effects of the model:

$$\hat{\boldsymbol{\beta}} = \left(\mathbf{X}'\hat{\mathbf{V}}^{-1}\mathbf{X}\right)^{-1}\mathbf{X}'\hat{\mathbf{V}}^{-1}\mathbf{y}, \quad \hat{\mathbf{v}} = \hat{\mathbf{G}}\mathbf{Z}'\hat{\mathbf{V}}^{-1}\left(\mathbf{y} - \mathbf{X}\hat{\boldsymbol{\beta}}\right).$$

Here $\hat{\boldsymbol{\beta}}$ is the $p \times 1$ vector of the best linear unbiased estimator (Blue) of $\boldsymbol{\beta}$.

The MSE of the EBLUP estimator $t(\hat{\theta}_R)$ in the so-called "naive approach" form, that is ignoring the variability of the estimates of the covariance parameters, is as follows:

$$mse\left[t_i\left(\hat{\theta}_R\right)\right] = g_{1i}\left(\hat{\theta}_R\right) + g_{2i}\left(\hat{\theta}_R\right), \tag{12.3}$$

with

$$g_{1i}\left(\hat{\theta}_R\right) = \mathbf{m}_i\left(\mathbf{G} - \mathbf{G}\mathbf{Z}_i'\mathbf{V}_i^{-1}\mathbf{Z}_i\mathbf{G}\right)\mathbf{m}_i',$$

$$g_{2i}\left(\hat{\theta}_R\right) = \left(\mathbf{p}_i - \mathbf{m}_i\mathbf{G}_i\mathbf{Z}_i'\mathbf{V}_i^{-1}\mathbf{X}_i\right)\mathbf{M}\left(\mathbf{p}_i - \mathbf{m}_i\mathbf{G}_i\mathbf{Z}_i'\mathbf{V}_i^{-1}\mathbf{X}_i\right)' = \mathbf{d}_i\mathbf{M}\mathbf{d}_i',$$

being $\mathbf{M} = \left(\sum \mathbf{X}_i'\mathbf{V}_i^{-1}\mathbf{X}_i\right)^{-1}$. When $\mathbf{p}_i$ and $\mathbf{m}_i$, in (12.2), are some known matrices, the off-diagonal elements of $mse[t_i(\hat{\theta}_R)]$ are the estimates of the mean product error between the components of the vector of the empirical predictor. The "true" value of the expected MSE of the empirical predictor is obtained by an approximated linear expression (by a Taylor expansion), considering an additional third component in (12.3): $2 \times g_{3i}(\hat{\theta}_R)$, with:

$$g_{3i}\left(\hat{\theta}_R\right) \approx tr\left[\left(\partial \mathbf{d}_i/\partial\theta\right)\mathbf{V}_i\left(\partial \mathbf{d}_i/\partial\theta\right)'\bar{\mathbf{V}}\left(\hat{\theta}_R\right)\right]. \tag{12.4}$$

Here $\bar{\mathbf{V}}(\hat{\theta}_R)$ is the asymptotic covariance matrix of the vector $\hat{\theta}_R$. Detailed formulae of (12.1), (12.2), and (12.3) for the state space unit level spatio-temporal model are given in Appendices 12.A and 12.B.

Unit level models have the special characteristic that the model in (12.1) can be adapted following the basic line of the nested error regression model. The observations $(j, j = 1, \ldots, n_i)$ are the units in their own area i. The covariance parameters vector is bivariate, and has two components, namely the variance of random-area effect and the within-area variance. Instead of the area level model, in the unit level model the sampling variances of each area do not take place in the matrix **R**. The drawback of the unit level models can be the lack of the design consistency of the EBLUP estimates. If we make the design weights on the model at unit level, we can achieve design consistency (the so-called bechmarking property) of the EBLUP estimates, that is, the pseudo-EBLUP estimation (Rao, 2003).

Let us now take M as the number of areas $(m = 1, \ldots, M)$ for which we estimate fixed and random effects of the model in (12.1). Each area has $n_i$observations (the units). Starting from the expression of the general linear mixed model, we have the following substitutions in the general setting of (12.1):

$$\mathbf{y} = \underset{m}{\text{col}}\left(\underset{i\in m}{\text{col}}\left(y_{mi}\right)\right), \quad \mathbf{X} = \underset{m}{\text{col}}\left(\underset{i\in m}{\text{col}}\left(\mathbf{x}_{mi}'\right)\right),$$

$$m = 1, \ldots, M, i = 1, \ldots, n_m, \quad \sum_m n_m = n,$$

$$\mathbf{Z} = \mathbf{I}_M \otimes \mathbf{1}_{n_m}, \quad \mathbf{v} = \underset{m}{\text{col}}\left(\mathbf{v}_m\right),$$

$$\mathbf{G} = \sigma_v^2, \quad \mathbf{R} = \sigma_e^2 \times \mathbf{I}_M \oplus_{m=1}^{M} \mathbf{I}_{n_m} = \underset{m,i}{diag}\left(\sigma_e^2\right) \tag{12.5}$$

with $\mathbf{v}_m \overset{\text{iid}}{\sim} N(0, \sigma_v^2)$, $\mathbf{e}_{mi} \overset{\text{iid}}{\sim} N(0, \sigma_e^2)$, and $\boldsymbol{\theta} = (\sigma_v^2, \sigma_e^2)$.

With known population means for the vector $\bar{\mathbf{X}}_m$, we can find as the EBLUP estimator of the mean of $\mathbf{y}$ at each small area m the following predicted value:

$$\hat{\bar{y}}_m^{EBLUP} = \gamma_m\left[\bar{y}_m + \left(\bar{\mathbf{X}}_m - \bar{\mathbf{x}}_m\right)'\hat{\beta}\right] + \left(1 - \gamma_m\right)\bar{\mathbf{X}}_m'\hat{\beta}, \tag{12.6}$$

being γ_m the so-called "shrinkage" factor $\sigma_v^2/(\sigma_v^2 + \sigma_e^2/n_i)$. Here $\bar{y}_m$ and $\bar{\mathbf{x}}_m$ are, respectively, the sample mean of the survey variable and the vector of the sample means of the model covariates.

12.2 Spatio-temporal Time-varying Effects Models

Spatial models consider a contiguity matrix to describe the neighborhood structure between small areas. In the field of unit level spatial models, the standard procedure is to incorporate spatial dependence in the random error structure through a SAR process. This resulting standard model is a two-stage random effects linear model. As the first stage we consider as model (12.1), the following:

$$\mathbf{y} = \underset{m}{\text{col}}\left(\underset{i\in m}{\text{col}}\left(y_{mi}\right)\right), \quad \mathbf{X} = \underset{m}{\text{col}}\left(\underset{i\in m}{\text{col}}\left(\mathbf{x}'_{mi}\right)\right),$$
$$m = 1,\ldots,M, i = 1,\ldots,n_m, \quad \sum_m n_m = n,$$
$$\mathbf{v} = \underset{m}{\text{col}}\left(\underset{i\in m}{\text{col}}\left(v_m\right)\right). \tag{12.7}$$

Because of the spatial SAR dependence, if we consider a standardardized contiguity known matrix $\mathbf{W}$, we have

$$\mathbf{v} = \rho_s \mathbf{W}\mathbf{v} + \mathbf{u}, \quad \mathbf{u} \overset{\text{iid}}{\sim} N_m\left(\mathbf{0}, \sigma_u^2 \mathbf{I}\right),$$

being $\mathbf{G} = \mathbf{G}_s = \sigma_u^2 \times [(\mathbf{I} - \rho_s \mathbf{W})'(\mathbf{I} - \rho_s \mathbf{W})]^{-1} = \sigma_u^2 \times (\mathbf{I} - \rho \mathbf{W})^{-1}[(\mathbf{I} - \rho \mathbf{W})']^{-1}$ the covariance matrix of the effects $\mathbf{v} \sim N(\mathbf{0}, \mathbf{G}_s)$. As the second stage, we can rewrite the model (12.7):

$$\mathbf{y} = \mathbf{X}\boldsymbol{\beta} + \mathbf{Z}\left(\mathbf{I} - \rho_s \mathbf{W}\right)^{-1}\mathbf{u} + \mathbf{e}. \tag{12.8}$$

Further we have:

$$\mathbf{R} = \sigma_e^2 \times \mathbf{I}_M \oplus_{m=1}^{M} \mathbf{I}_{n_m} = \underset{m,i}{diag}\left(\sigma_e^2\right),$$
$$\boldsymbol{\theta} = \left(\sigma_u^2, \rho_s, \sigma_e^2\right) \quad \left|\rho_s\right| < 1.$$

Now the vector of covariance parameters has three components: besides the traditional random-area and error variances, we have now the spatial autocorrelation parameter ρ_s. It is a measure of the overall spatial autocorrelation, and reflects the appropriateness of the matrix W to modeling the random area variation by its entries, that is, the contiguities or standardized distances between the areas considered. For $\rho_s = 0$, the model (12.8) reduces to (12.5).

A time response survey variable can be modeled as an extension of the area level and unit level models. The simpler way of modeling time series data in these type of models is as follows: (i) to consider only time correlated sampling errors in area level models, or (ii) to specify time autocorrelated within-area errors in unit level models. A more realistic assumption is to include random-area time variation in the model, and assume time-independent sampling errors (in area level models) or independent normal distributed residual errors (in unit level models). Time-varying random-area effects usually follow an AR(1) autoregressive process. Besides this, we can also include another random-area effect, independent of the time considered. An example of the latter is the Rao–Yu area level model (Rao and Yu, 1994). More complex models for time series data can be studied in the context of the unit level models. An example is to assume different independent within-area by time residual errors, at each time instant.

When we decide to investigate the impact of the level of correlation between some neighboring areas by time series data, we need to work with a spatio-temporal model. Similar to the well-known Rao–Yu model, as a model designed for the survey units, we can define the random vector of the model as a stacked column vector $\mathbf{v} = (\mathbf{v}_1', \mathbf{v}_2')' \sim N(\mathbf{0}, \mathbf{G})$, in which $\mathbf{v}_1 = \mathrm{col}(\mathbf{v}_{1m})$ is the vector of the random-area effect. Again in this model, as in the model (10), we incorporate SAR spatial dependence through the random effect linear dependence by defining the weight spatial matrix $\mathbf{W}$: $\mathbf{v}_1 = \rho_s \mathbf{W}\mathbf{v}_1 + \mathbf{u}_1$, with $\mathbf{u}_1 \overset{iid}{\sim} N_m(\mathbf{0}, \sigma_s^2\mathbf{I})$. Because we have to consider repeated measurement of the survey units over time, now the vectors of the area–time random effects within the areas follow an AR(1) process, with the autocorrelation parameter ρ_t. While they are independent and identically distributed across the areas, we have for the generic small area m: $v_{2mt} = \rho_t v_{2m(t-1)} + u_{2mt}$, with $u_{2mt} \overset{iid}{\sim} N(0, \sigma_t^2)$.

Stacking with the same order the repeated measurement on units, we have now the following specification of the general model in (12.1):

$$\mathbf{y} = \underset{t}{\mathrm{col}}\left(\underset{m}{\mathrm{col}}\left(\underset{i\in m}{\mathrm{col}}\left(y_{tmi}\right)\right)\right), \quad \mathbf{X} = \underset{t}{\mathrm{col}}\left(\underset{m}{\mathrm{col}}\left(\underset{i\in m}{\mathrm{col}}\left(\mathbf{x}_{tmi}'\right)\right)\right), \tag{12.9}$$

with $t = 1,\ldots,T, m = 1,\ldots,M, i = 1,\ldots,n_m$, and $\sum_m n_m = n$.

The design matrix of the random effects has the following structure:

$$\begin{aligned}\mathbf{Z} &= (\mathbf{Z}_1, \mathbf{Z}_2)\\ &= \left[\mathbf{1}_T \otimes \left(\mathbf{I}_M \otimes \mathbf{1}_{n_m}\right), \left(\mathbf{I}_M \otimes \mathbf{1}_{n_m}\right) \otimes \mathbf{I}_T\right],\end{aligned}$$

while the covariance matrix of the vector of the model random effects has two diagonal blocks, $\mathbf{G} = diag(\mathbf{G}_s, \mathbf{G}_t)$, with $\mathbf{v} \sim N_{M\times T}[\mathbf{0}, diag(\mathbf{G}_s, \mathbf{G}_t)]$:

$$\begin{aligned}\mathbf{G}_s &= \sigma_s^2 \times \left[(\mathbf{I} - \rho_s\mathbf{W})'(\mathbf{I} - \rho_s\mathbf{W})\right]^{-1},\\ \mathbf{G}_t &= \mathbf{I}_M \otimes \mathbf{\Omega},\\ \mathbf{\Omega} &= \frac{\sigma_t^2}{1-\rho_t^2}\begin{pmatrix} 1 & \rho_t & \cdots & \rho_t^{T-1}\\ \rho_t & 1 & & \rho_t^{T-2}\\ \vdots & & \ddots & \vdots\\ \rho_t^{T-1} & \rho_t^{T-2} & \cdots & 1\end{pmatrix}_{T\times T}.\end{aligned} \tag{12.10}$$

Again ρ_s in (12.10) is the spatial autocorrelation parameter, to be estimated as a component of the vector of the model covariance parameters $\boldsymbol{\theta} = (\sigma_s^2, \rho_s, \sigma_t^2, \rho_t, \sigma_e^2)$, with $|\rho_s, \rho_t| < 1$. The matrix $\mathbf{R}$ of the model (12.1) is now:

$$\begin{aligned}\mathbf{R} &= \sigma_e^2 \times \mathbf{1}_T \otimes \left(\mathbf{I}_M \oplus_{m=1}^{M} \mathbf{I}_{n_m}\right)\\ &= \mathbf{1}_T \otimes \left(\underset{m,i}{\mathrm{diag}}\left(\sigma_e^2\right)\right), \quad \text{with} \quad e_{mi} \overset{iid}{\sim} N\left(0, \sigma_e^2\right).\end{aligned} \tag{12.11}$$

When exhaustive information by geostatistical data on survey units is available, that is when we have also information from repeated measurements, a more complex time-varying effects spatio-temporal model can be applied for the case studies. There are situations in which we need to estimate different time-dependent residual errors. This can drive to fit a more analytic model, if the data have the more realistic different behavior for the several time instants. Instead of (12.11), we can fit the model (12.9) through the parameters vector $\theta = (\sigma_s^2, \rho_s, \sigma_t^2, \rho_t, \sigma_{e1}^2, \ldots, \sigma_{eT}^2)$, $|\rho_s, \rho_t| < 1$, as follows:

$$\mathbf{R}_t = diag\left(\sigma_{e1}^2, \ldots, \sigma_{eT}^2\right),$$

$$\mathbf{R} = (\mathbf{I}_M \oplus_{m=1}^M \mathbf{I}_{n_m}) \otimes \mathbf{R}_t, \quad e_{tmi} \overset{iid}{\sim} N\left(0, \sigma_{et}^2\right)$$

12.3 State Space Models with Spatial Structure

State space models are one of the most important tools for time series analysis. When we have a multiple set of surveys that are repeated in two or more time instants, the state space models analysis allow us to take advantage from the interpretation of the updating information process, collected by the repeated measurements on units in the small areas. An advantage of this type of model, with respect to the time-varying effects model, is that we can obtain more comprehensive information from the different states of the EBLUP estimates of the model in (12.1). The EBLUP estimates, that in the state space model are the components of the *state vector*, must undergo an estimation process, that is the Kalman filter recursive algorithm, that provides more complete information about the updating-by-time model estimates. Following the general model (12.1), the parameters of the state space model can be estimated, using the REML estimation method, through a recursive Kalman filter algorithm (Pfefferman and Tiller, 2001; Singh *et al.*, 2005).

Let us take as the unit level model the following expression for the model (12.1), at the time instant t:

$$\mathbf{y}_t = \mathbf{X}_t\boldsymbol{\beta}_t + \mathbf{Z}\mathbf{v}_t + \mathbf{e}_t. \tag{12.12}$$

Now consider

$$\mathbf{y}_t = \mathbf{U}_t\boldsymbol{\alpha}_t + \mathbf{e}_t, \tag{12.13}$$

with

$$\mathbf{U}_t = \left(\mathbf{X}_t, \mathbf{Z}\right), \quad E\left(\mathbf{e}_t\right) = \mathbf{0},$$

$$E\left(\mathbf{e}_t\mathbf{e}_t'\right) = \boldsymbol{\Sigma}_t, \quad \mathbf{e}_t \overset{ind}{\sim} N_m\left(\mathbf{0}, \mathbf{R}_t\right)$$

$$\mathbf{R}_t = diag\left(\sigma_e^2\right).$$

If we consider the simultaneous spatial dependence among small areas, a spatial weight matrix is incorporated directly in the the design matrix $\mathbf{Z}$: $\mathbf{Z} = (\mathbf{I} - \rho_s\mathbf{W})^{-1}$. The equation in (12.13) is called the *measurement equation*, and describes the relationship between the observed variables $(\mathbf{y}_t, \mathbf{U}_t)$ and latent state vector $\boldsymbol{\alpha}_t = (\boldsymbol{\beta}_t', \mathbf{v}_t')'$. The following relation is called the *transition equation*, and explains the temporal relationship between latent state variables at adjacent measurement time occasions:

$$\boldsymbol{\alpha}_t = \mathbf{T}\boldsymbol{\alpha}_{t-1} + \zeta_t,$$

$$\zeta_t \sim N_{p+m}\left(\mathbf{0}, \mathbf{Q}\right), \tag{12.14}$$

with $\mathbf{Q} = diag(\mathbf{0}_p, \sigma_v^2 \mathbf{I}_m)$, and $\mathbf{T} = diag(\mathbf{I}_p, \rho_t \mathbf{I}_m)$. Further, we have:

$$\mathbf{v}_t = \rho_t \mathbf{v}_{t-1} + \boldsymbol{\eta}_t, \quad \boldsymbol{\eta}_t \overset{ind}{\sim} N_m\left(\mathbf{0}, \sigma_v^2 \mathbf{I}_m\right).$$

The time-dependent random errors $\mathbf{e}_t$ and $\boldsymbol{\eta}_t$ are uncorrelated contemporaneously and over time. Now the covariance parameter vector, to be estimated, is $\theta = (\rho_s, \sigma_v^2, \rho_t, \sigma_e^2)'$.

Throughout the time series data, the Kalman recursion implements the equations sequentially, with the Kalman filter equations

$$\begin{aligned}\hat{\boldsymbol{\alpha}}_t &= \hat{\boldsymbol{\alpha}}_{t|t-1} + \boldsymbol{\Sigma}_{t|t-1} \mathbf{U}_t' \mathbf{V}_t^{-1} \left(\mathbf{y}_t - \mathbf{U}_t \hat{\boldsymbol{\alpha}}_{t|t-1}\right) \\ \boldsymbol{\Sigma}_t &= \boldsymbol{\Sigma}_{t|t-1} - \boldsymbol{\Sigma}_{t|t-1} \mathbf{U}_t' \mathbf{V}_t^{-1} \mathbf{U}_t \boldsymbol{\Sigma}_{t|t-1}.\end{aligned} \tag{12.15}$$

Here $\boldsymbol{\Sigma}_t$ is the covariance matrix associated with the latent state vector $\boldsymbol{\alpha}_t$. If we consider $\boldsymbol{\Sigma}_{t-1} = E(\hat{\boldsymbol{\alpha}}_{t|t-1} - \boldsymbol{\alpha}_{t|t-1})(\hat{\boldsymbol{\alpha}}_{t|t-1} - \boldsymbol{\alpha}_{t|t-1})'$ the covariance matrix of the prediction errors at time $(t-1)$, the following

$$\boldsymbol{\Sigma}_{t|t-1} = \mathbf{T} \boldsymbol{\Sigma}_{t-1} \mathbf{T}' + \mathbf{Q} \tag{12.16}$$

is the covariance matrix of the prediction errors $\hat{\boldsymbol{\alpha}}_{t|t-1} - \boldsymbol{\alpha}_t$. Note that we have $\hat{\boldsymbol{\alpha}}_{t|t-1} = \mathbf{T}\hat{\boldsymbol{\alpha}}_{t-1}$. In the Kalman filter equation (12.15), $\mathbf{V}_t = \mathbf{R}_t + \mathbf{U}_t \boldsymbol{\Sigma}_{t|t-1} \mathbf{U}_t'$ is the updated covariance matrix of the unit level model (12.12) at the time instant t. It is derived from the estimated covariance matrix of the prediction errors $\hat{\boldsymbol{\alpha}}_{t|t-1} - \boldsymbol{\alpha}_t$.

At first observation, we have to initalize the Kalman filter recursion of the state vector (12.15), and then initialize the update of the transition equation (12.14). To do this, we must estimate the covariance matrix of the prediction errors (12.16) by considering the first time instant, based on the REML estimate of the vector of covariance parameters $\hat{\theta}$. In practice, we need to know the initial covariance matrix of the fixed and random effects of the model (12.13), $\boldsymbol{\Sigma}_{(t-1)=1}$, at the time $t = 1$. We have the following partition for this covariance matrix:

$$\boldsymbol{\Sigma}_1 = \begin{bmatrix} \boldsymbol{\Sigma}_{11} & \boldsymbol{\Sigma}_{12} \\ \boldsymbol{\Sigma}_{21} & \boldsymbol{\Sigma}_{22} \end{bmatrix} \tag{12.17}$$

with

$$\underset{(p\times p)}{\boldsymbol{\Sigma}_{11}} = \left(\mathbf{X}_1 \mathbf{V}_1^{-1} \mathbf{X}_1'\right)^{-1}, \quad \underset{(p\times m)}{\boldsymbol{\Sigma}_{12}} = \boldsymbol{\Sigma}_{21}' = -\sigma_v^2 \left(\mathbf{X}_1 \mathbf{V}_1^{-1} \mathbf{X}_1'\right)^{-1} \mathbf{X}_1' \mathbf{V}_1^{-1} \mathbf{Z},$$

$$\underset{(m\times m)}{\boldsymbol{\Sigma}_{22}} = \sigma_v^2 \mathbf{I}_m - \sigma_v^4 \mathbf{Z}' \mathbf{V}_1^{-1} \mathbf{Z} + \sigma_v^4 \mathbf{Z}' \mathbf{V}_1^{-1} \mathbf{X}_1 \left(\mathbf{X}_1 \mathbf{V}_1^{-1} \mathbf{X}_1'\right)^{-1} \mathbf{X}_1' \mathbf{V}_1^{-1} \mathbf{Z}.$$

The spatio-temporal unit level model estimates of fixed and random effects, at the time instant $t = 1$, are:

$$\hat{\boldsymbol{\beta}}_1 = \left(\mathbf{X}_1 \hat{\mathbf{V}}_1^{-1} \mathbf{X}_1'\right)^{-1} \mathbf{X}_1' \hat{\mathbf{V}}_1^{-1} \mathbf{y}_1, \quad \hat{\mathbf{v}}_1 = \sigma_v^2 \mathbf{Z}' \hat{\mathbf{V}}_1^{-1} \left(\mathbf{y}_1 - \mathbf{X}_1 \hat{\boldsymbol{\beta}}_1\right),$$

with

$$\hat{\mathbf{V}}_1 = \hat{\mathbf{R}}_1 + \hat{\sigma}_v^2 \mathbf{A}^{-1}, \quad \mathbf{A} = \left(\mathbf{I} - \hat{\rho}_s \mathbf{W}\right)' \left(\mathbf{I} - \hat{\rho}_s \mathbf{W}\right).$$

Finally, the vector of the EBLUP estimates is, at the instant t:

$$\hat{\mathbf{y}}_t^{EBLUP} = \mathbf{U}_t \hat{\boldsymbol{\alpha}}_t = \mathbf{y}_t - \mathbf{R}_t \hat{\mathbf{V}}_t^{-1} [\mathbf{y}_t - \mathbf{U}_t \hat{\boldsymbol{\alpha}}_{t|t-1}]. \tag{12.18}$$

The parameter vector $\boldsymbol{\theta}$ is estimated by maximizing the loglikelihood function for the state space model. This is done by summing the log-density function all over the time instants. In the state space literature it is called the prediction error decomposition (PED) function. In the case of the REML, we have to find the maximizer $\hat{\boldsymbol{\theta}}$ of the following (restricted) PED function, as the state space edition of the restricted maximum loglikelihood function (see Appendix 12.A):

$$l_R = const - \frac{1}{2}\log\left|\mathbf{X}_1\mathbf{V}_1^{-1}\mathbf{X}_1'\right| - \frac{1}{2}\sum_t \log\left|\mathbf{V}_t\right| - \frac{1}{2}(\mathbf{y}_1 - \mathbf{X}_1\hat{\boldsymbol{\beta}}_1)'\mathbf{V}_1^{-1}(\mathbf{y}_1 - \mathbf{X}_1\hat{\boldsymbol{\beta}}_1)$$

$$-\frac{1}{2}\sum_{t\geq 2}(\mathbf{y}_t - \mathbf{U}_t\hat{\boldsymbol{\alpha}}_{t|t-1})'\mathbf{V}_t^{-1}(\mathbf{y}_t - \mathbf{U}_t\ \hat{\boldsymbol{\alpha}}_{t|t-1}).$$

As mentioned above, the estimation of the MSE of the EBLUP plays a central role in model-based small area estimation. The general formulas (12.3) and (12.4) have special expressions for the spatio-temporal unit level models, and in particular in the case of the state space model with spatial structure (see Appendix 12.B).

12.4 The Italian EU-SILC Data: an Application with the Spatio-temporal Unit Level Models

We now discuss the application of the models described in the previous paragraphs.

The focus of the study is on household wealth, as an indicator of poverty. We take into account the data of some poverty indicators in the South of Italy. This territory, as the official statistics data describe, is the part of the country that has registered the major incidence of the whole set of the poverty indicators over years. To evaluate the distribution of the average household income, as the response variable in the model, among the small areas (the administrative Provinces of the South of Italy), we take into account as relevant poverty determinants some model covariates, such as the level of the education and the age of the householder. We compare the potential of the most used unit level small area model frameworks, starting from the nested error regression model up to the state space model with spatial structure. The evaluation of the impact of the space–time dependence on the average household income among and inside the Provinces is a main economic policy question, that the models listed above can highlight. In particular, the state space model has special characteristics so that we can get different estimates of fixed effects over time, together with a whole and comprehensive evaluation of the state-of-the-art of the model at each time instant.

The dataset used is selected from the Italian Survey on Income and Living Conditions (SILC), which is the Italian version of the EU Statistics on Income and Living Conditions (EU-SILC). EU-SILC survey data come from an example of survey samples that have been harmonized for EU countries by EUROSTAT. The Italian survey consists of a selection of about 26 000 households for a total of nearly 70 000 individuals, over 800 municipalities of varying demographic sizes. The SILC survey started in 2004 with an annual periodicity, producing longitudinal and cross-sectional data. Although EUROSTAT standards require estimates at national level, SILC is designed to ensure reliable estimates even at regional level (NUTS 2).

Here we work with a repeated-by-time collection of 724 householders in the years 2005–2008 ($T = 4$ time instants) from IT-SILC data, leaving out all the units with partially missing information. We consider in the application $M = 22$ small domains, the administrative

Table 12.1 Unit level models

1	Unit level nested error regression model
2	Spatial unit level model
3	Spatio-temporal unit level model (A)
4	Spatio-temporal unit level model (B)
5	Spatial state space unit level model

Provinces (NUTS 3) of the Regions of the South Italy (Abruzzo, Molise, Puglia, Campania, Basilicata, Calabria). The response variable is the average household income. The covariates are the education level [0 pre-primary education; 1 primary education; 2 lower secondary education; 3 (upper) secondary education; 4 post-secondary non tertiary education; 5 first stage of tertiary education not leading directly to an advanced research qualification; 6 second stage of tertiary education leading to an advanced research qualification], and the age of the householder (>16). The estimation method is the REML, for all the models involved. The models that we compare are shown in Table 12.1.

The first is the basic unit level model (model 1). The spatial weight matrix for the other models is obtained with a row standardization of the distance matrix between the representative towns in the small area. This is, traditionally, the city with the same name of the Province territory. A step by step adaptive fitting is performed by the models (2, 3, and 4), considering first a simple spatial unit level model (model 2). Then spatio-temporal time varying effects models are fitted to the data (models 3 and 4). Finally a "spatial" state space approach is studied (model 5). Notice that the more complex the model, the more time expensive is the estimation process (Selukar, 2011; Pebesma, 2012).

Table 12.2 contains information on the parameter estimates of all the models. Table 12.3 and Table 12.4 show the EBLUP estimates from the models in Table 12.1. Futher more the root MSE for the estimates are also shown. In particular, for models 3, 4, and 5, we show the time-averaged values of the root MSE. Table 12.3 and Table 12.4 also illustrate the different values of the resulting estimates of the coefficient of variation of the EBLUP.

Table 12.2 Estimates of the parameters for the unit level models of Table12.1

Parameter	Model 1	Model 2	Model 3	Model 4	Model 5
σ_v^2	22956	—	—	—	9034
σ_e^2	14889	11843	9366	—	7971
ρ_s	—	0.38	0.49	0.39	0.37
σ_u^2	—	17730	—	—	—
σ_s^2	—	—	1564	1623	—
σ_t^2	—	—	803	418	—
ρ_t	—	—	0.52	0.43	0.82
$\sigma_{e_1}^2$	—	—	—	10047	—
$\sigma_{e_2}^2$	—	—	—	7402	—
$\sigma_{e_3}^2$	—	—	—	9267	—
$\sigma_{e_4}^2$	—	—	—	5584	—

Table 12.3 EBLUP, root MSE of the EBLUP estimates, and their coefficient of variation for the first 11 Provinces of South Italy

Models	AV	BA	BN	BR	CB	CE	CZ	CH	CS	FG	IS
					Average small area estimate (EBLUP)						
Model 1	24937	28113	19223	17675	25876	34112	21997	25443	26218	23853	31001
Model 2	24785	27819	19545	17994	24532	33901	22439	25387	26574	23986	30956
Model 3	25932	27493	20056	18102	25383	34247	22193	25720	26438	23604	30796
Model 4	24877	27981	19392	17768	25777	33991	22004	25366	26119	23750	31036
Model 5	24776	28009	19340	17669	25808	34200	22034	25384	26361	23773	30939
					Average root MSE of the EBLUP estimates						
Model 1	2045.5	2190.5	1038.2	1032.8	2000.4	1997.3	1860.3	1735.9	1993.4	2011.1	1774.3
Model 2	2013.2	2201.6	1027.0	1018.5	1903.6	1879.4	1849.6	1699.4	1832.0	1942.1	1712.9
Model 3	1911.4	1989.6	900.2	948.0	1701.8	1619.3	1577.0	1480.9	1784.6	1674.0	1438.5
Model 4	1897.4	1991.3	909.7	954.8	1703.6	1618.1	1569.4	1474.7	1774.7	1660.5	1439.2
Model 5	1872.0	1899.6	911.5	944.6	1667.3	1601.8	1566.9	1444.0	1739.6	1599.9	1409.5
					Average coefficient of variation (%)						
Model 1	8.20	7.79	5.40	5.84	7.73	5.86	8.46	6.82	7.60	8.43	5.72
Model 2	8.12	7.91	5.25	5.66	7.76	5.54	8.24	6.69	6.89	8.10	5.53
Model 3	7.37	7.24	4.49	5.24	6.70	4.73	7.11	5.76	6.75	7.09	4.67
Model 4	7.63	7.12	4.69	5.37	6.61	4.76	7.13	5.81	6.79	6.99	4.64
Model 5	7.56	6.78	4.71	5.35	6.46	4.68	7.11	5.69	6.60	6.73	4.56

In general, the "spatial" state space model offers a wide list of precious information about the model and the estimation process. This amount of information is not comparable with the traditional time-varying effects models.

Tables of results from the models involved highlight some of their features and related issues. The first one is that to incorporate in the model spatial or a spatio-temporal autocorrelation can lead to an appreciable reduction of the MSE of the estimators. In the spatio-temporal time-varying effects models 3 and 4, the particular parametrization of model 4 has the advantage of the best possible model fitting. Further, we have more precise EBLUP estimates (Table 12.3 and Table 12.4).

Model 5 seems to offer a further improvement of the result estimates with respect to the other previous models. This is because the space–time model, in its spatial version, has the following primary features: first, the spatial state space model, as the models 2, 3 and 4, considers the spatial dependence among the small areas (the Provinces). However, the state space model seems to exploit at best this spatial autocorrelation. The update of the model via the Kalman filter recursions seems to take great advantage from the pegging work of the SAR spatial dependence. Further, the spatial state space model has the special property that at each time instant we have different estimates of the vector of the fixed effects [see (12.12)].

From the point of view of the analysis of the average household income, this can be a useful tool. In fact, the estimation of the vector of both fixed and random effects at each time instant deals with the estimation of the fixed and random effects, given the previous observations. In

Table 12.4 EBLUP, root MSE of the EBLUP estimates, and their coefficient of variation for the second 11 Provinces of South Italy

Models	AQ	LE	MT	NA	PE	PZ	RC	SA	TA	TE	VV
					Average small area estimate (EBLUP)						
Model 1	30278	24293	36864	27998	36783	26667	37565	20441	25971	31520	24447
Model 2	30572	24279	34549	27773	35601	28132	37289	20357	25782	31692	24161
Model 3	29964	24867	35521	27485	35899	28004	37712	20898	25442	32311	23984
Model 4	30256	24309	36776	27891	36357	26609	37495	20570	25953	31482	24276
Model 5	30148	24401	36620	27912	36437	27833	37444	20504	25842	31387	24379
					Average root MSE of the EBLUP estimates						
Model 1	2209.4	2203.6	1892.3	2210.9	1276.1	1765.8	2874.9	1646.6	1375.7	2450.0	1734.2
Model 2	2176.0	1984.7	1885.3	2167.0	1279.4	1738.7	2699.3	1665.8	1333.6	2224.8	1689.5
Model 3	2029.3	1791.6	1483.7	1792.6	988.3	1563.1	2410.9	1355.2	1068.3	2014.0	1383.1
Model 4	2027.3	1790.1	1472.2	1790.8	989.0	1554.7	2376.1	1345.0	1071.3	2009.9	1380.7
Model 5	2022.8	1773.7	1455.4	1781.3	991.8	1546.4	2298.0	1327.7	1070.5	1992.6	1356.3
					Average coefficient of variation (%)						
Model 1	7.30	9.07	5.13	7.90	3.47	6.62	7.65	8.06	5.30	7.77	7.09
Model 2	7.12	8.17	5.46	7.80	3.59	6.18	7.24	8.18	5.17	7.02	6.99
Model 3	6.77	7.20	4.18	6.52	2.75	5.58	6.39	6.48	4.20	6.23	5.77
Model 4	6.70	7.36	4.00	6.42	2.72	5.84	6.34	6.54	4.13	6.38	5.69
Model 5	6.71	7.27	3.97	6.38	2.72	5.56	6.14	6.48	4.14	6.35	5.56

particular, the random effects are updated with the latest estimates through the time autocorrelation parameter. This means that we can evaluate the impact on the different small areas average household income (the EBLUP in the model) directly on the random-area effect by time, simply through the value of the time autocorrelation estimate in the model.

12.5 Concluding Remarks

There is a growing demand for information about the geographic distribution of poverty, as well as for information on the dynamics of poverty over space. Small area estimation models can help policy strategies, in terms of clarifying what site-specific issues can be recognized by using spatio-temporal interactions that are provided by available data. Differences of poverty measures among small areas can lead to interventions in order to ease more effectively the overall impact of poverty. As explained throughout this chapter, the simultaneous treatment of the spatial and temporal dependence in the same statistical model can lead to a general improvement of the quality of the small area statistics. Even the implementation of the spatio-temporal information requires a well-structured model framework, the information that we can evaluate downline of the model estimation is more plentiful with respect to the basic small area models. We focused our analysis on unit level spatio-temporal models. In particular, the chapter is dedicated to the discussion of the time-varying effects model in small area estimation, as an example of simultaneous implementation of the space–time information in the linear mixed effects model framework. Further, we analyzed the extension of the state space model treated as a small area model considering spatial dependence. This is an example of extension of the time series analysis model to the spatial data. The last model provided the best survey parameter estimation performance (the average income) among the models studied. When the estimated time dependence (the time autocorrelation parameter in the model) is relatively high, the state space model seems to exploit best the borrowing of information by the neighbouring areas. The application study pointed out that the state space model with spatial structure seems to borrow best the whole information for the same area at some other time, and borrows at the same time from some other areas. However, some challenges about the spatio-temporal small area models are in evidence. The first is the handling of the MSE estimates of the EBLUP, in particular for the state space model. There is a need for easier approximate formulae. The second is the implementation in the state space small area model of non-stationary time series data. Further, but very important in the practical evaluation of poverty measures, specific spatio-temporal model diagnostics can be developed. Leverages, influential measures of the space and time estimated parameters on the EBLUP of the state space model, are relevant issues to analyze.

Appendix 12.A: Restricted Maximum Likelihood Estimation

In the REML estimation method, for the general linear mixed model we have the following loglikelihood function:

$$l_R(\boldsymbol{\beta}, \boldsymbol{\theta}) = const - \frac{1}{2}\log\left(\left|\mathbf{X}'\mathbf{V}^{-1}\mathbf{X}\right|\right) - \frac{1}{2}\log\left(|\mathbf{V}|\right) - \frac{1}{2}(\mathbf{y} - \mathbf{X}\boldsymbol{\beta})'\mathbf{V}^{-1}(\mathbf{y} - \mathbf{X}\boldsymbol{\beta}).$$

Consider now the orthogonal projection matrix $\mathbf{P} = \mathbf{V}^{-1} - \mathbf{V}^{-1}\mathbf{X}(\mathbf{X}'\mathbf{V}^{-1}\mathbf{X})^{-1}\mathbf{X}'\mathbf{V}^{-1} = \mathbf{V}^{-1} - \mathbf{V}^{-1}\mathbf{Q}$, with $\mathbf{Q}$ the affine projection matrix using the distance induced by the matrix $\mathbf{V}^{-1}$. If $rank(\mathbf{X}) = p$, and for a $N \times (N - p)$ matrix $\mathbf{C}$ of full rank, with $\mathbf{C}'\mathbf{X} = \mathbf{0}$ and $\mathbf{C}'\mathbf{y} = \mathbf{z}$, we have the following orthogonal transformation of the likelihood for the REML estimation:

$$l'_R(\boldsymbol{\theta}) = const - \frac{1}{2}\log\left(|\mathbf{C}'\mathbf{V}\mathbf{C}|\right) - \frac{1}{2}\mathbf{z}'\left(\mathbf{C}'\mathbf{V}\mathbf{C}\right)^{-1}\mathbf{z}. \tag{12.A.1}$$

We have to maximize (12.A.1) as the affine orthogonal projection in the residual loglikelihood by knowing the covariance matrix of the General Least Squares estimates of the fixed effects parameters. In fact, for the matrix $\mathbf{C}$, we have:

$$\mathbf{C}\left(\mathbf{C}'\mathbf{V}\mathbf{C}\right)^{-1}\mathbf{C}' = \mathbf{V}^{-1} - \mathbf{V}^{-1}\mathbf{X}\left(\mathbf{X}'\mathbf{V}^{-1}\mathbf{X}\right)^{-1}\mathbf{X}'\mathbf{V}^{-1} = \mathbf{P}$$

because of the equivalence of the following projection matrices for $\mathbf{C}$ and $\mathbf{X}$:

$$\mathbf{P}_{\mathbf{V}^{1/2}\mathbf{C}} = \mathbf{I} - \mathbf{P}_{\mathbf{V}^{-1/2}\mathbf{X}}.$$

The score equations for the likelihood in (12.A.1) are, in terms of $\mathbf{z}$ and $\mathbf{y}$:

$$\begin{aligned} S_R(\theta) &= \frac{\partial l'_R}{\partial \boldsymbol{\theta}} \\ &= -\frac{1}{2}\left[\frac{\partial vec\left(\mathbf{C}'\mathbf{V}\mathbf{C}\right)}{\partial \boldsymbol{\theta}}\right]' vec\left[\left(\mathbf{C}'\mathbf{V}\mathbf{C}\right)^{-1}\right] \end{aligned}$$

$$+\frac{1}{2}\left(\mathbf{z}'\otimes\mathbf{z}'\right)\left[\left(\mathbf{C}'\mathbf{V}\mathbf{C}\right)^{-1}\otimes\left(\mathbf{C}'\mathbf{V}\mathbf{C}\right)^{-1}\right]\frac{\partial vec\left(\mathbf{C}'\mathbf{V}\mathbf{C}\right)}{\partial\theta}$$

$$=\frac{1}{2}\left(\frac{\partial vec\mathbf{V}}{\partial\theta}\right)'\left(\mathbf{Py}\otimes\mathbf{Py}-vec\mathbf{P}\right)$$

$$=\frac{1}{2}\left(\frac{\partial vec\mathbf{V}}{\partial\theta}\right)'\left[\left(\mathbf{P}\otimes\mathbf{P}\right)vec\left(\mathbf{yy}'\right)-vec\mathbf{P}\right]=\mathbf{0}.$$

The components of the vector in the last term of $S_R(\theta)$ are:

$$\mathbf{y}'\mathbf{P}\frac{\partial\mathbf{V}}{\partial\theta_s}\mathbf{Py}-tr\left(\mathbf{P}\frac{\partial\mathbf{V}}{\partial\theta_s}\right),\quad s=1,\ldots,q.$$

Appendix 12.B: Mean Squared Error Estimation of the Unit Level State Space Model

Under certain basic features of the data and of the model (the so-called regularity conditions), the MSE of the EBLUP can be estimated as shown in the following.

First, the following must be satisfied: (a) the elements of $\mathbf{X}_t(t=1,\ldots,T)$ are uniformly bounded; (b) m and T are finite; (c) $\mathbf{U}_t\mathbf{\Sigma}_{t|t-1}\mathbf{U}_t'\mathbf{V}_t^{-1}\mathbf{U}_t=o(1)$, $\partial(\mathbf{U}_t\mathbf{\Sigma}_{t|t-1}\mathbf{U}_t'\mathbf{V}_t^{-1}\mathbf{U}_t)/\partial\theta_i=o(1)$, and further $\partial^2(\mathbf{U}_t\mathbf{\Sigma}_{t|t-1}\mathbf{U}_t'\mathbf{V}_t^{-1})/\partial\theta_i\partial\theta_j=o(1)$, for $i,j=1,2,3,4$, that is the dimension of the vector $\boldsymbol{\theta}$; (d) $(\hat{\boldsymbol{\theta}}-\boldsymbol{\theta})=o(m^{-1/2})$. Given the EBLUP expression (12.18), and assuming that we know $\boldsymbol{\theta}$, that is the vector of covariance population parameters, the joint expression of the MSE of the Blup estimator is:

$$mse\left(\hat{\mathbf{y}}_t^{Blup}\right)=mse\left[\hat{\mathbf{y}}_t(\theta)\right]=g_{(1,2),t}(\theta)=\mathbf{U}_t(\theta)\mathbf{\Sigma}_t(\theta)\mathbf{U}_t'(\theta).$$

It is easy now to consider the so-called "naive" MSE estimator as:

$$mse\left(\hat{\mathbf{y}}_t^{EBLUP}\right)=mse\left[\hat{\mathbf{y}}_t\left(\hat{\theta}\right)\right]=g_{(1,2),t}(\theta)=\mathbf{U}_t\left(\hat{\theta}\right)\mathbf{\Sigma}_t\left(\hat{\theta}\right)\mathbf{U}_t'\left(\hat{\theta}\right).\qquad(12.B.1)$$

This estimator is biased, due to ignoring the variability of the estimates of the vector of covariance parameters $\hat{\boldsymbol{\theta}}$. The expression in (12.B.1) underestimates the true value of $mse[\hat{\mathbf{y}}_t(\hat{\theta})]$. This true value can be approximated through the linearization derived by the Taylor expansion of the expected value of $[\hat{\mathbf{y}}_t(\hat{\theta})-\mathbf{y}_t][\hat{\mathbf{y}}_t(\hat{\theta})-\mathbf{y}_t]'$. In practice, it is possible to derive the second-order approximation of the MSE of the EBLUP $\hat{\mathbf{y}}_t^{EBLUP}$, as follows:

$$mse\left[\hat{\mathbf{y}}_t\left(\hat{\theta}\right)\right]=E\left[\hat{\mathbf{y}}_t\left(\hat{\theta}\right)-\mathbf{y}_t\right]\left[\hat{\mathbf{y}}_t\left(\hat{\theta}\right)-\mathbf{y}_t\right]'=g_{(1,2),t}(\theta)+g_{(3),t}(\theta)+o\left(m^{-1}\right).$$

Consider the following convention on matrix derivatives: $\mathbf{A}^\theta=\frac{\partial\mathbf{A}}{\partial\theta}$. For example we can have: $\mathbf{A}_i^\theta=\frac{\partial\mathbf{A}_i}{\partial\theta},\mathbf{A}_i^{\theta\theta'}=\frac{\partial^2\mathbf{A}_i}{\partial\theta\partial\theta'},\mathbf{A}_i^{\theta=0}=\left(\frac{\partial\mathbf{A}_i}{\partial\theta}\right)_{\theta=0}$. Then we have as (12.4):

$$g_{(3),t}(\theta)=\mathbf{L}_t'(\theta)\mathbf{I}_\theta^{-1}(\theta)\mathbf{K}(\theta)\mathbf{V}_t\mathbf{I}_\theta^{-1}(\theta)\mathbf{L}_t(\theta),$$

with

$$\mathbf{K}(\theta)=\frac{1}{2}\sum_t trace\left(\mathbf{V}_t^{-1}\mathbf{V}_i^\theta\mathbf{V}_t^{-1}\mathbf{V}_j^\theta\right),i,j=1,2,3,4$$

$$\mathbf{L}_t(\theta)=\mathrm{col}\left(L_{ti}(\theta)\right)=\mathrm{col}\left[\partial\left(\mathbf{U}_t\mathbf{\Sigma}_{t|t-1}\mathbf{U}_t'\mathbf{V}_t^{-1}\right)/\partial\boldsymbol{\theta}_i\right],$$

and $\mathbf{I}_\theta(\theta) = -E(l_R^{\theta\theta'})$ the negative expected hessian, the Fisher information matrix of the model. Now, a more complete expression for (12.B.1) can be (Singh *et al.* 2005):

$$mse[\hat{\mathbf{y}}_t(\hat{\theta}] = g_{(1,2),t}(\theta) + g_{(3),t}(\theta) + g^*_{(3),t}(\theta) - g_{(4),t}(\theta) - g_{(5),t}(\theta) + o\left(m^{-1}\right),$$

with

$$g^*_{(3),t}(\theta) = \mathbf{L}'_t(\theta)\left[\mathbf{I}_\theta^{-1}(\theta) \otimes \mathbf{V}_t\right]\mathbf{L}_t(\theta),$$

$$g_{(4),t}(\theta) = \left\{\frac{1}{2}\mathbf{I}_\theta^{-1}(\theta)\,\mathrm{col}\left[trace\left(\mathbf{I}_\beta(\theta) \times \mathbf{I}_\beta^{\theta_i}(\theta)\right)\right]\right\} \otimes \mathbf{I}_m \times g^\theta_{(1,2),t}(\theta),$$

$$g_{(5),t}(\theta) = \frac{1}{2}trace\left[\mathbf{I}_4 \otimes \mathbf{R}_t\mathbf{V}_t^{-1}, \mathbf{V}_t^{\theta\theta'}\left(\mathbf{I}_\theta^{-1}(\theta) \otimes \mathbf{V}_t^{-1}\mathbf{R}_t\right)\right]'.$$

References

Banerjee S, Carlin B, and Gelfand A 2004 *Hierarchical Modeling and Analysis for Spatial Data.* Chapman and Hall.

Battese G, Harter R, and Fuller W 1988 An error-components model for prediction of county crop areas using survey and satellite data. *Journal of the American Statistical Association*, **83**, p. 28.

Berger YG and Skinner CJ 2003 Variance estimation for a low-income proportion. *Applied Statistics*, **52**, p. 457.

Chambers R, Tzavidis N, and Salvati N 2009 *Borrowing strength over space in small area estimation: Comparing parametric, semiparametric and non-parametric random effects and M-quantile small area models.* Centre for Statistical and Survey Methodology, University of Wollongong, Working Paper 12.

Chandra H, Salvati N, and Chambers R 2007 Small area estimation for spatially correlated populations - a comparison of direct and indirect model-based methods. *Statistics in Transition*, **8** (2), p. 331.

Coelho PS and Pereira LN 2011 A Spatial Unit Level model for Small Area Estimation. *Revstat - Statistical Journal*, **9** (2), p. 155.

Cressie N 1993 *Statistics for Spatial Data.* John Wiley & Sons, Inc.

Cressie N and Wikle C 2011 *Statistics for Spatio-Temporal Data.* John Wiley & Sons, Inc.

Das K, Jiang J, and Rao JNK 2004 Mean squared error of empirical predictor. *Annals of Statistics*, **32**, p. 818.

Di Consiglio L, Falorsi S, Paladini P, Righi P, Scavalli E, and Solari F 2003 Stimatori per Piccole Aree per le Stime di Povertà Regionali. *Rivista di Statistica Ufficiale*, **2**, The Italian National Statistical Institute.

Elbers C, Lanjouw JO, and Lanjouw P 2003 Micro-level estimation of poverty and inequality. *Econometrica*, **71**, p. 355.

European Commission 2011 *Small Area Methods for Poverty and Living condition Estimates (SAMPLE)*, EU-FP7-SSH-2007-1- Grant Agreement 217565, http://www.sample-project.eu/.

Fay RE and Herriot RA 1979 Estimates of income for small places: an application of James-Stein procedures to census data. *Journal of the American Statistical Association*, **74**, p. 269.

González-Manteiga W, Lombardía MJ, Molina I, Morales D, and Santamaría L 2008 Analytic and bootstrap approximations of prediction errors under a multivariate Fay–Herriot model. *Computational Statistics and Data Analysis,* **52** (12), p. 5242.

Gu F, Preacher KJ, Wu W, and Yung YF 2014 A computationally efficient state space approach to estimating multilevel regression models and multilevel confirmatory factor models. *Multivariate Behavioral Research*, **49** (2), p. 119.

Hall P and Maiti T 2006 On parametric bootstrap methods for small area prediction. *Journal of the Royal Statistical Society, Series B: Statistical Methodology*, **68**, p. 221.

Jiang J 2007 *Linear and Generalized Linear Mixed Models and Their Applications.* Springer-Verlag.

Lehtonen R and Veijanen A 2012 Small area poverty estimation by model calibration. *Journal of the Indian Society of Agricultural Statistics*, **66**, p. 125.

Longford NT 2007 On standard errors of model-based small area estimators. *Survey Methodology*, **33**, p. 69.

Marhuenda Y, Molina I, and Morales D 2013 Small area estimation with spatio-temporal Fay–Herriot models. *Computational Statistics and Data Analysis*, **58**, p. 308.

Molina I and Rao JNK 2010 Small area estimation of poverty indicators. *Canadian Journal of Statistics*, **38** (3), p. 369.

Molina I, Salvati N, and Pratesi M 2009 Bootstrap for estimating the Mse of the spatial EBLUP. *Computational Statistics*, **24**, p. 441.

Pebesma E 2012 Spacetime: Spatio-temporal data in R. *Journal of Statistical Software*, **51** (7).

Petrucci A and Salvati N 2004 *Small Area Estimation considering spatially correlated errors: the unit level random effects model*. Dipartimento di Statistica "G. Parenti", Florence, Working Paper 10.

Pfefferman D 2002 Small area estimation - new developments and directions. *International Statistical Review,* **70**(1), p. 125.

Pfefferman D, Feder M, and Signorelli D 1988 Estimation of autocorrelations of survey errors with application to trend estimation in small areas. *Journal of Business & Economic Statistics*, **16**, p. 339.

Pfefferman D and Tiller RB 2001 *Bootstrap approximation to prediction MSE for state-space models with estimated parameters*. Technical Report, Hebrew University, Jerusalem.

Prasad N and Rao JNK 1990 The estimation of the mean-squared error of small area estimators. *Journal of the American Statistical Association*, **85**, p. 163.

Prasad N and Rao JNK 1999 On robust small area estimation using a simple random effects model. *Survey Methodology*, **25** (1), p. 67.

Pratesi M and Salvati N 2004 *Small area estimation: the EBLUP estimator with autoregressive random area effects*. Report no. 261, Department of Statistics and Mathematics, University of Pisa.

Pratesi M and Salvati N 2008 Small area estimation: the EBLUP estimator based on spatially correlated random area effects. *Statistical Methods and Applications*, **17**, p. 113.

Rao JNK 2003 *Small Area Estimation*. John Wiley & Sons, Inc.

Rao JNK and Yu M 1994 Small area estimation by combining time series and cross-sectional data. *Canadian Journal of Statistics,* **22**, p. 511.

Selukar R 2011 State space modeling using SAS. *Journal of Statistical Software*, **41** (12), p. 1.

Singh BB, Shukla GK, and Kundu D 2005 Spatio-temporal models in small area estimation. *Survey Methodology*, **31** (2), p. 183.

Sinha SK and Rao JNK 2009 Robust small area estimation. *The Canadian Journal of Statistics*, **37** (3), p. 381.

Tarozzi A and Deaton A 2009 Using census and survey data to estimate poverty and inequality for small areas. *The Review of Economic and Statistics*, **91** (4), p. 773.

Wall M 2004 A close look at the spatial structure implied by the CAR and SAR models. *Journal of Statistical Planning and Inference*, **121**, p. 311.

13

Spatial Information and Geoadditive Small Area Models

Chiara Bocci[1] and Alessandra Petrucci[2]

[1] *IRPET – Regional Institute for Economic Planning of Tuscany, Florence, Italy*

[2] *Department of Statistics, Informatics, Applications, University of Florence, Florence, Italy*

13.1 Introduction

In social sciences, maps are useful tools to describe the spatial pattern of poverty in a country, especially when they represent the phenomenon in small geographic units, such as municipalities or districts. Information on the spatial distribution of poverty is of interest to policy makers and researchers for a number of reasons: first, it can be used to quantify suspected regional disparities in living standards and identify the areas falling behind in the process of economic development; secondly, the information facilitates the formulation of efficient policies and programs such as education, health, credit and food aid whose purpose, at least partly, is to alleviate poverty; and thirdly, the information may shed light on geographic factors, such as topography or market access, that are associated with poverty.

Extracting interesting and useful patterns from spatial datasets is more difficult than extracting the corresponding patterns from traditional numeric and categorical data due to the complexity of spatial data types, spatial relationships, and spatial autocorrelation. The complexity of spatial data and the intrinsic spatial relationships limit the usefulness of conventional techniques for extracting spatial patterns. Therefore, the area definition and the assignment of the data to appropriate areas can pose problems in the estimation process.

It is worth stressing the usefulness of the geographical location for the analysis of non stationary spatial phenomena. The "global" dependence models, such as the classical regression model, assume the independence of the data from the spatial location, generate spatially autocorrelated residuals and often lead to wrong conclusions. Adequate use of the geographic

Analysis of Poverty Data by Small Area Estimation, First Edition. Edited by Monica Pratesi.

Companion Website: www.wiley.com/go/pratesi/poverty

information in more complex models which take into account the spatial variability can help to understand the underlying phenomenon.

In literature, semi-parametric models have been proposed to simultaneously incorporate the spatial distribution of the study variable and the other covariate effects. The advantage of the semi-parametric approach is that it imposes parametric structure where it may be reasonable, while leaving the structure of the model unrestricted for another set of variables. Thus, the semi-parametric approach is a particularly easy and flexible approach for modeling broad spatial trends while also permitting the effects of other explanatory variables to vary by location. Geoadditive models (Kammann and Wand 2003), in particular, merge an additive model (Hastie and Tibshirani 1990), that accounts for the relationship between the variables, and a kriging model (Cressie 1993), that accounts for the spatial distribution, under the linear mixed model framework.

In addition, as pointed out by Neri et al. (2005), in order to produce poverty maps large datasets are required which include reasonable measures of income or consumption expenditure and which are representative and of sufficient size at low levels of aggregation to yield statistically reliable estimates. Household budget surveys or living standard surveys covering income and consumption usually used to calculate distributional measures are rarely of such a sufficient size; whereas census or other large sample surveys large enough to allow disaggregation have little or no information regarding monetary variables. Then, the required small area estimates are usually based on a combination of sample surveys and administrative data. Moreover, economic variables such as income and consumption are usually skewed, thus the relationship between the response variable and the auxiliary variables may not be linear in the original scale, but can be linear in a transformed scale, for example the logarithm scale. In such case, small area estimation (SAE) methods based on log-transformed models are required (Chandra and Chambers 2011).

Combining the need of SAE methods for skewed variables with the flexibility of a semi-parametric model, this study discusses a recent approach to identify and include the spatial pattern in SAE. Thus, the estimated spatial patterns reflect the propensity of the considered characteristic in a region, after controlling for other unit-level effects. In particular, the work focuses on socio-economic data collected by the World Bank program on Living Standard Measurement Study (LSMS) (Grosh and Glewwe 2000). The program is designed to assist policy makers in their efforts to identify how policies could be designed and improved to positively affect outcomes in health, education, economic activities, housing and utilities, and so on.

We combine the model parameters estimated using the dataset of the 2002 Living Standard Measurement Study with the 2001 Population and Housing Census covariate information and we apply a geoadditive SAE model in order to estimate the district level mean of the household per-capita consumption expenditure for the Republic of Albania.

The chapter is structured as follows. In the next section the generic geoadditive model is presented. Section 13.3 describes how to combine a geoadditive model with a SAE model for skewed data and in Section 13.4 we discuss two possible estimators of the small area mean. Section 13.5 describes the datasets used in the analysis and presents the empirical results. The final section summarizes the main findings and discusses possible future works.

13.2 Geoadditive Models

Geoadditive models, introduced by Kammann and Wand (2003), analyze the spatial distribution of a study variable while accounting for possible linear or non-linear covariate effects. Under the additivity assumption they can handle such covariate effects by merging an additive model (Hastie and Tibshirani 1990) – that accounts for the relationship between the variables – and a kriging model (Cressie 1993) – that accounts for the spatial correlation – and by expressing both as a linear mixed model. The linear mixed model representation is a useful instrument because it allows estimation using mixed model methodology and software.

Let x_i and t_i, $1 \leq i \leq n$, be linear and non-linear predictors of y_i at spatial location $\mathbf{s}_i$, $\mathbf{s} \in \mathbb{R}^2$. A geoadditive model for such data can be formulated as

$$y_i = \alpha + \beta_x x_i + g(t_i) + h(\mathbf{s}_i) + \epsilon_i, \quad \epsilon_i \sim N(0, \sigma_\epsilon^2), \tag{13.1}$$

where g is an unspecified univariate smooth function and h is an unspecified bivariate smooth function, that can be represented through penalized splines:

$$g(t_i) = \beta_{0t} + \beta_t t_i + \sum_{k=1}^{K_t} \gamma_k^t z_k^t(t_i)$$

$$h(\mathbf{s}_i) = \beta_{0\mathrm{s}} + \mathbf{s}_i \boldsymbol{\beta}_{\mathrm{s}} + \sum_{k=1}^{K_{\mathrm{s}}} \gamma_k^{\mathrm{s}} z_k^{\mathrm{s}}(\mathbf{s}_i),$$

where $\gamma_k^t \overset{\text{i.i.d.}}{\sim} N(0, \sigma_t^2)$, $\gamma_k^{\mathrm{s}} \overset{\text{i.i.d.}}{\sim} N(0, \sigma_{\mathrm{s}}^2)$, $(z_1^t, \ldots, z_{K_t}^t)$ is a set of K_t univariate spline basis functions and $(z_1^{\mathrm{s}}, \ldots, z_{K_{\mathrm{s}}}^{\mathrm{s}})$ is a set of K_{s} bivariate spline basis functions.

Exploiting the connection between penalized splines and linear mixed models illustrated in Wand (2003), model (13.1) can be written as

$$\mathbf{y} = \mathbf{X}\boldsymbol{\beta} + \mathbf{Z}\boldsymbol{\gamma} + \boldsymbol{\varepsilon}, \tag{13.2}$$

with

$$\mathrm{E}\begin{bmatrix}\boldsymbol{\gamma}\\ \boldsymbol{\varepsilon}\end{bmatrix} = \begin{bmatrix}\mathbf{0}\\ \mathbf{0}\end{bmatrix}, \quad \mathrm{Cov}\begin{bmatrix}\boldsymbol{\gamma}\\ \boldsymbol{\varepsilon}\end{bmatrix} = \begin{bmatrix}\sigma_t^2 \mathbf{I}_{K_t} & \mathbf{0} & \mathbf{0}\\ \mathbf{0} & \sigma_{\mathrm{s}}^2 \mathbf{I}_{K_{\mathrm{s}}} & \mathbf{0}\\ \mathbf{0} & \mathbf{0} & \sigma_\epsilon^2 \mathbf{I}_n\end{bmatrix}$$

where

$$\mathbf{X} = [1 \; x_i \; t_i \; \mathbf{s}_i]_{1 \leq i \leq n},$$

$$\boldsymbol{\beta} = [\beta_0 \; \beta_x \; \beta_t \; \boldsymbol{\beta}_{\mathrm{s}}^T]^T,$$

$$\boldsymbol{\gamma} = [\gamma_1^t \ldots \gamma_{K_t}^t \; \gamma_1^{\mathrm{s}} \ldots \gamma_{K_{\mathrm{s}}}^{\mathrm{s}}]^T,$$

$\beta_0 = \alpha + \beta_{0t} + \beta_{0\mathrm{s}}$ and $\mathbf{Z}$ is obtained by concatenating the matrices containing the spline basis functions to handle g and h, respectively

$$\mathbf{Z} = [\mathbf{Z}_t | \mathbf{Z}_{\mathrm{s}}].$$

Choosing a low-rank thin plate spline formulation for $(z_1^t, \dots, z_{K_x}^t)$ and $(z_1^{\mathbf{s}}, \dots, z_{K_{\mathbf{s}}}^{\mathbf{s}})$, we have

$$\mathbf{Z}_t = [z_k^t(t_i)]_{1\le i\le n, 1\le k\le K_t} = [C(t_i - \kappa_k)]_{1\le i\le n, 1\le k\le K_t} \cdot [C(\kappa_h - \kappa_k)]^{-1/2}_{1\le h,k\le K_t},$$
$$\mathbf{Z}_{\mathbf{s}} = [z_k^{\mathbf{s}}(\mathbf{s}_i)]_{1\le i\le n, 1\le k\le K_{\mathbf{s}}} = [C(\mathbf{s}_i - \boldsymbol{\kappa}_k)]_{1\le i\le n, 1\le k\le K_{\mathbf{s}}} \cdot [C(\boldsymbol{\kappa}_h - \boldsymbol{\kappa}_k)]^{-1/2}_{1\le h,k\le K_{\mathbf{s}}},$$

where $C(\mathbf{v}) = \|\mathbf{v}\|^2 \log \|\mathbf{v}\|$, $(\kappa_1, \dots, \kappa_{K_t})$ are univariate spline knot locations for x and $(\boldsymbol{\kappa}_1, \dots, \boldsymbol{\kappa}_{K_{\mathbf{s}}})$ are bivariate spline knot locations for $\mathbf{s}$.

The amount of smoothing for both the additive component and the geostatistical component of the model can be quantified through the variance component ratios $\sigma_\epsilon^2/\sigma_t^2$ and $\sigma_\epsilon^2/\sigma_{\mathbf{s}}^2$.

The addition of other explicative variables is straightforward: smoothing components are added in the random effects term $\mathbf{Z}\boldsymbol{\gamma}$, while linear components can be incorporated as fixed effects in the $\mathbf{X}\boldsymbol{\beta}$ term. Moreover, the mixed model structure provides a unified and modular framework that allows the model to be easily extended to include various kinds of generalization and evolution (Ruppert et al. 2009). In particular, under this framework the geoadditive model and the classic SAE model can be easily combined (Opsomer et al. 2008).

13.3 Geoadditive Small Area Model for Skewed Data

Suppose that there are T small areas for which we want to estimate a quantity of interest Y that assumes strictly positive values with a skewed distribution, a common situation when analyzing economic variables such as income and consumption, for a population of size N. Let y_{it} denote the value of the response variable for the ith unit, $i = 1, \dots, n$, in small area t, $t = 1, \dots, T$ and let $\mathbf{x}_{it}$ be a vector of p covariates associated with the same unit, that are related to the log-transformed value of Y, then we can build a SAE mode for $log(y_{it})$ (Chandra and Chambers 2011)

$$y_{it}^* = log(y_{it}) = \mathbf{x}_{it}\boldsymbol{\beta} + u_t + \epsilon_{it}, \quad \epsilon_{it} \sim N(0, \sigma_\epsilon^2), \quad u_t \sim N(0, \sigma_u^2), \tag{13.3}$$

where $\boldsymbol{\beta}$ is a vector of p unknown coefficients, u_t is the random area effect associated with small area t and ϵ_{it} is the individual level random error. The two error terms are assumed to be mutually independent, both across individuals as well as across areas.

If we define the matrix $\mathbf{D} = [d_{it}]$ with

$$d_{it} = \begin{cases} 1 & \text{if observation } i \text{ is in small area } t, \\ 0 & \text{otherwise} \end{cases} \tag{13.4}$$

and $\mathbf{y}^* = [y_{it}^*]$, $\mathbf{X} = [\mathbf{x}_{it}^T]$, $\mathbf{u} = [u_t]$ and $\varepsilon = [\epsilon_{it}]$, then the matrix notation of (13.3) is

$$\mathbf{y}^* = \mathbf{X}\boldsymbol{\beta} + \mathbf{D}\mathbf{u} + \varepsilon, \tag{13.5}$$

with

$$\mathrm{E}\begin{bmatrix}\mathbf{u}\\ \varepsilon\end{bmatrix} = \begin{bmatrix}0\\0\end{bmatrix}, \quad \mathrm{Cov}\begin{bmatrix}\mathbf{u}\\ \varepsilon\end{bmatrix} = \begin{bmatrix}\sigma_u^2\mathbf{I}_T & 0\\ 0 & \sigma_\epsilon^2\mathbf{I}_n\end{bmatrix}.$$

The covariance matrix of $\mathbf{y}^*$ is $\text{Var}(\mathbf{y}^*) \equiv \mathbf{V} = \sigma_u^2 \mathbf{D}\mathbf{D}^T + \sigma_\epsilon^2 \mathbf{I}_n$ and the best linear unbiased predictions (BLUPs) of the model coefficients are

$$\begin{aligned}\beta &= (\mathbf{X}^T\mathbf{V}^{-1}\mathbf{X})^{-1}\mathbf{X}^T\mathbf{V}^{-1}\mathbf{y}^*,\\ \mathbf{u} &= \sigma_u^2\mathbf{D}^T\mathbf{V}^{-1}\left(\mathbf{y}^* - \mathbf{X}\boldsymbol{\beta}\right).\end{aligned}$$

If the variance components σ_u^2 and σ_ϵ^2 are unknown, they are estimated by restricted maximum likelihood (REML) or maximum likelihood (ML) methods and the model coefficients are obtained with the empirical best linear unbiased predictions (EBLUPs).

The formulation (13.5) is a linear mixed model, analogous to the geoadditive model (13.2), thus it is straightforward to compose the geoadditive SAE model, which is a particular specification of the non-parametric SAE model introduced by Opsomer et al. (2008). Consider again the response $y_{it}^* = log(y_{it})$ and the vector of p linear covariates $\mathbf{x}_{it}$, and suppose that both are measured at a spatial location $\mathbf{s}_{it}$, $\mathbf{s} \in \mathbb{R}^2$. The geoadditive SAE model for such data is a linear mixed model with two random effects components:

$$\mathbf{y}^* = \mathbf{X}\beta + \mathbf{Z}\boldsymbol{\gamma} + \mathbf{D}\mathbf{u} + \varepsilon, \tag{13.6}$$

with

$$\text{E}\begin{bmatrix}\boldsymbol{\gamma}\\ \mathbf{u}\\ \varepsilon\end{bmatrix} = \begin{bmatrix}0\\0\\0\end{bmatrix}, \quad \text{Cov}\begin{bmatrix}\boldsymbol{\gamma}\\ \mathbf{u}\\ \varepsilon\end{bmatrix} = \begin{bmatrix}\sigma_\gamma^2\mathbf{I}_K & 0 & 0\\ 0 & \sigma_u^2\mathbf{I}_T & 0\\ 0 & 0 & \sigma_\epsilon^2\mathbf{I}_n\end{bmatrix}.$$

Now $\mathbf{X} = [\mathbf{x}_{it}^T, \mathbf{s}_{it}^T]_{1\le i\le n}$ has $p+2$ columns, $\boldsymbol{\beta}$ is a vector of $p+2$ unknown coefficients, $\mathbf{u}$ are the random small area effects, $\boldsymbol{\gamma}$ are the thin plate spline coefficients (seen as random effects) and ε are the individual level random errors. Matrix $\mathbf{D}$ is still defined by (13.4) and $\mathbf{Z}$ is the matrix of the thin plate spline basis functions

$$\mathbf{Z} = [C(\mathbf{s}_i - \boldsymbol{\kappa}_k)]_{1\le i\le n, 1\le k\le K}[C(\boldsymbol{\kappa}_h - \boldsymbol{\kappa}_k)]_{1\le h,k\le K}^{-1/2},$$

with K knots $\boldsymbol{\kappa}_k$ and $C(\mathbf{v}) = \|\mathbf{v}\|^2 \log \|\mathbf{v}\|$.

Again, the unknown variance components are estimated via REML or ML estimators and are indicated with $\hat{\sigma}_\gamma^2$, $\hat{\sigma}_u^2$ and $\hat{\sigma}_\epsilon^2$. The estimated covariance matrix of $\mathbf{y}^*$ is

$$\hat{\mathbf{V}} = \hat{\sigma}_\gamma^2\mathbf{Z}\mathbf{Z}^T + \hat{\sigma}_u^2\mathbf{D}\mathbf{D}^T + \hat{\sigma}_\epsilon^2\mathbf{I}_n \tag{13.7}$$

and the EBLUP estimators of the model coefficients are

$$\hat{\beta} = (\mathbf{X}^T\hat{\mathbf{V}}^{-1}\mathbf{X})^{-1}\mathbf{X}^T\hat{\mathbf{V}}^{-1}\mathbf{y}^*, \tag{13.8}$$

$$\hat{\boldsymbol{\gamma}} = \hat{\sigma}_\gamma^2\mathbf{Z}^T\hat{\mathbf{V}}^{-1}\left(\mathbf{y}^* - \mathbf{X}\hat{\beta}\right), \tag{13.9}$$

$$\hat{\mathbf{u}} = \hat{\sigma}_u^2\mathbf{D}^T\hat{\mathbf{V}}^{-1}\left(\mathbf{y}^* - \mathbf{X}\hat{\beta}\right). \tag{13.10}$$

Based on model (13.6) and estimators (13.7)–(13.10), Chandra and Chambers (2011) derived the second-order bias corrected estimator of $\text{E}(y_{it}|\mathbf{x}_{it}, \mathbf{z}_{it}, \mathbf{d}_{it})$ and, consequently, the predicted values of y_{it}

$$\hat{y}_{it} = \hat{\lambda}_{it}^{-1}\exp\left(\mathbf{x}_{it}^T\hat{\beta} + \frac{\hat{v}_{iit}}{2}\right) \tag{13.11}$$

where $\hat{v}_{iit}$ is the ith diagonal element of $\hat{\mathbf{V}}$, and $\hat{\lambda}_{it}$ is the bias adjustment factor for the log-back transformation: $\hat{\lambda}_{it} = 1 + 0.5[\hat{a}_{it} + 0.25\hat{V}(\hat{v}_{iit})]$ where $\hat{a}_{it} = \mathbf{x}_{it}^T\hat{V}(\hat{\boldsymbol{\beta}})\mathbf{x}_{it}$, $\hat{V}(\hat{\boldsymbol{\beta}})$ is the usual estimator of $\text{Var}(\hat{\boldsymbol{\beta}})$ and $\hat{V}(\hat{v}_{iit})$ is the estimated asymptotic variance of $\hat{v}_{iit}$.

13.4 Small Area Mean Estimators

For a given small area t, we are interested in predicting the area mean $\overline{Y}_t = N_t^{-1}\sum_{i=1}^{N_t} y_{it}$. Notice that the mean is computed over all the individuals and it may be decomposed as:

$$\overline{Y}_t = N_t^{-1}\left(\sum_{i\in S_t} y_{it} + \sum_{i\in R_t} y_{it}\right) \tag{13.12}$$

where N_t is the population size of area t, S_t is the area specific sample and R_t is its complement to the area population.

A first approach is to substitute the predicted values (13.11) inside (13.12), obtaining the empirical predictor

$$\hat{\overline{y}}_t^{EP} = N_t^{-1}\left(\sum_{i\in S_t} y_{it} + \sum_{i\in R_t} \hat{y}_{it}\right). \tag{13.13}$$

Estimation of the mean squared error (MSE) of (13.13) is not straightforward since this predictor is a non-linear function of Y values, thus to obtain a measure of the variability of (13.13) we need to rely on bootstrap estimation.

Alternatively, Chandra and Chambers (2011) extended the model-based direct estimator (MBDE) described in Chandra and Chambers (2009) to the situation where the SAE model holds on the log scale, using weights derived via model calibration (Wu and Sitter 2001). They developed the area mean estimator and the corresponding MSE estimator as in the following.

The generic MBDE is defined as a weighted average of the sample values Y in area t,

$$\hat{\overline{y}}_t^{MBDE} = \sum_{i\in S_t} w_{it} y_{it}.$$

where the specific weights w_{it} formulation depends on the underlying SAE model.

In particular, following Chandra and Chambers (2011) and under the log-linear geoadditive SAE model (13.6), we define $\mathbf{J}_U = [\mathbf{1}_U\ \hat{\mathbf{y}}_U]$, where $\mathbf{1}_U$ denotes the unit vector of size N and $\hat{\mathbf{y}}_U$ is the vector of the predicted values (13.11), and $\hat{\mathbf{O}}_U = [\hat{\omega}_{jk}]_{1\le j,k\le N}$ where

$$\hat{\omega}_{jk} = \widehat{\text{cov}}\left(y_j, y_k | \hat{y}_j, \hat{y}_k\right) = e^{\left(\mathbf{x}_j^T\hat{\beta}+\mathbf{x}_k^T\hat{\beta}\right)+(\hat{v}_{jj}+\hat{v}_{kk})/2}\left(e^{\hat{v}_{jk}} - 1\right) \tag{13.14}$$

and $\hat{v}_{jk}$ is the element j,k of matrix $\hat{\mathbf{V}}$. If we partition $\mathbf{J}_U$ and $\hat{\mathbf{O}}_U$ according to sample (s) and non-sample (r) units as

$$\mathbf{J}_U = \begin{bmatrix}\hat{\mathbf{J}}_s\\ \hat{\mathbf{J}}_r\end{bmatrix}, \quad \hat{\mathbf{O}}_U = \begin{bmatrix}\hat{\mathbf{O}}_{ss} & \hat{\mathbf{O}}_{sr}\\ \hat{\mathbf{O}}_{rs} & \hat{\mathbf{O}}_{rr}\end{bmatrix}.$$

the empirical model-based model-calibrated (MBMC) weights can be written as:

$$\mathbf{w}^{mbmc} = [w_{it}^{mbmc}] = \mathbf{1}_s + \hat{\mathbf{H}}_s^T\left(\mathbf{J}_U^T\mathbf{1}_U - \mathbf{J}_s^T\mathbf{1}_s\right) + \left(\mathbf{I}_s - \hat{\mathbf{H}}_s^T\mathbf{J}_s^T\right)\hat{\mathbf{O}}_{ss}^{-1}\hat{\mathbf{O}}_{sr}\mathbf{1}_r \tag{13.15}$$

with $\hat{\mathbf{H}}_s = (\mathbf{J}_s^T\hat{\mathbf{O}}_{ss}^{-1}\mathbf{J})^{-1}\mathbf{J}_s^T\hat{\mathbf{O}}_{ss}^{-1}$.

Using the weights defined in (13.15), there can be two type of MBDE model-calibrated estimators, the Hajek type and the Horvitz–Thompson type. Chandra and Chambers (2011) considered both types and then suggested the use of the Horvitz–Thompson type, that is

$$\hat{\bar{y}}_t^{HT-MBDE} = N_t^{-1} \sum_{i \in S_t} w_{it}^{mbmc} y_{it}. \tag{13.16}$$

Finally, for the estimation of the MSE of (13.16) we followed Chambers et al. (2011) and Chandra and Chambers (2011) and adapted standard methods for estimating the MSE of a weighted domain mean estimate (Royall and Cumberland 1978). We refer to Chambers et al. (2011) for the complete formulation of the MSE estimator.

13.5 Estimation of the Household Per-capita Consumption Expenditure in Albania

The spatial distribution of poverty and inequality at disaggregated geographical level are useful tools in measuring poverty. There are many different definitions and concepts of well-being. This study focuses on the aspect of inequality in the distribution of income, consumption or other attributes across the population. Since the economy of Albania is largely rural and income is not accurately measured, income poverty in Albania is estimated on the basis of a consumption-based measure (World Bank and INSTAT 2003). Deriving poverty estimates also depends on how one sets the poverty line, that is a threshold below which a given household or individual will be classified as poor. The choices in setting the poverty line can impact upon the estimates and hence have an effect on policymaking. For this reason, the distribution of the per-capita consumption expenditure at small area (district) level is important.

13.5.1 Data

The Republic of Albania is divided into three geographical levels: there are 12 prefectures, 36 districts, and 374 communes. In the 2000s the two main sources of statistical information available in Albania were the 2001 Population and Housing Census (PHC) and the 2002 Living Standard Measurement Study (LSMS), both conducted in Albania by the INSTAT (Albanian Institute of Statistics).

The 2002 LSMS provides individual level and household level socio-economic data from 3599 households drawn from urban and rural areas in Albania. The sample was designed to be representative of Albania as a whole, Tirana, other urban/rural locations, and the three main agro-ecological areas (Coastal, Central, and Mountain).

Four survey instruments were used to collect information for the 2002 Albania LSMS: a household questionnaire, a diary for recording household food consumption, a community questionnaire, and a price questionnaire. The household questionnaire included all the core LSMS modules as defined in Grosh and Glewwe (2000), plus additional modules on migration, fertility, subjective poverty, agriculture, and non-farm enterprises. Geographical referencing data on the longitude and latitude of each household were also recorded using portable GPS devices (World Bank and INSTAT 2003).

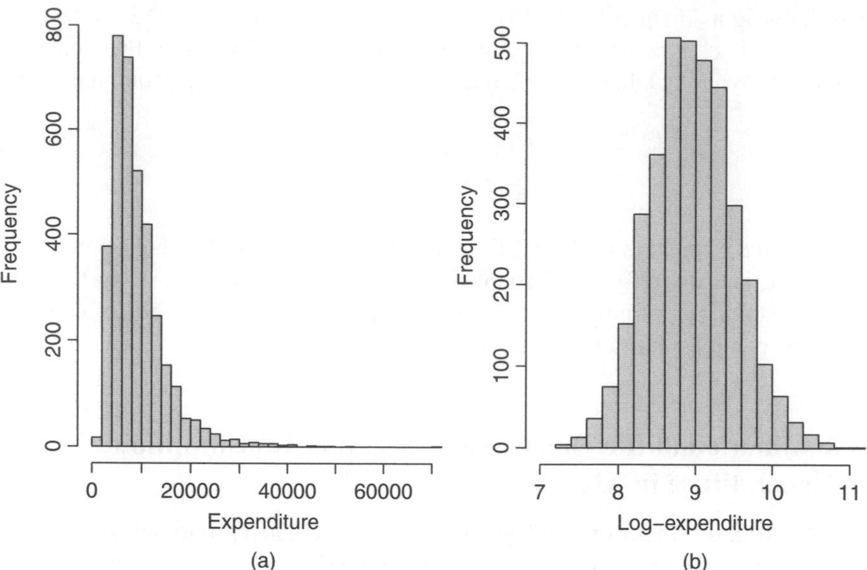

Figure 13.1 Distribution of the household per-capita consumption expenditure (in Lek), both in original scale (a) and in logarithmic scale (b)

We model the household per-capita consumption expenditure (in Lek) in each district using a geoadditive SAE log-transformed model as presented in the previous sections (Figure 13.1). The covariates selected to fit the model are chosen following prior studies on poverty assessment in Albania (Betti et al. 2003, Neri et al. 2005) and through the application of a stepwise procedure. We selected the following household level covariates:

- *size of the household* (in terms of the number of components);
- *information on the components of the household*: age of the householder, marital status of the householder (married, not married), age of the spouse of the householder, number of children 0–5 years, age of the first child, number of components without work, highest level of education in the household (high, medium, low);
- *information on the house*: building with 2–15 units (yes, no), built with brick or stone (yes, no), built before 1960 (yes, no), number of rooms per person, house surface ($<40\,\text{m}^2$, $40\text{–}69\,\text{m}^2$, $>69\,\text{m}^2$), toilet inside (yes, no);
- *presence of facilities in the dwelling*: TV (yes, no), parabolic (satellite) dish (yes, no), refrigerator (yes, no), washing machine (yes, no), air conditioning (yes, no), computer (yes, no), car (yes, no);
- *ownership of agricultural land* (yes, no).

All these variables are available both in LSMS and PHC surveys (see Neri et al. (2005) for comparability between the two sources); in addition, the geographical location of each household is available for the LSMS data. For the non sampled data, the geographical location is approximated with the centroid coordinates of the commune where the household is located.

13.5.2 Results

Estimates of the average per-capita consumption expenditure in each of the 36 district areas are derived using the geoadditive SAE log-transformed model presented in (13.6) and the MBDE model-calibrated estimator (13.16).

After the preliminary analysis of various combinations of parametric and non-parametric specifications for the selected covariates, the chosen model is composed of a bivariate thin plate spline on the universal transverse Mercator (UTM) coordinates, a linear term for all the other variables and a random intercept component for the area effect. The spline knots, shown in Figure 13.2b, are selected setting $K = 100$ and using the *clara* space filling algorithm of Kaufman and Rousseeuw (1990) that is available in the `R` package `cluster`. The model is then fitted by REML using the `lme` function in the `R` package `nlme`.

The estimated parameters are presented in Table 13.1, along with their confidence interval at 95% and the p-values. With the exclusion of the intercept and of the coordinate coefficients (which are required by the model structure), almost all the parameters are highly significant.

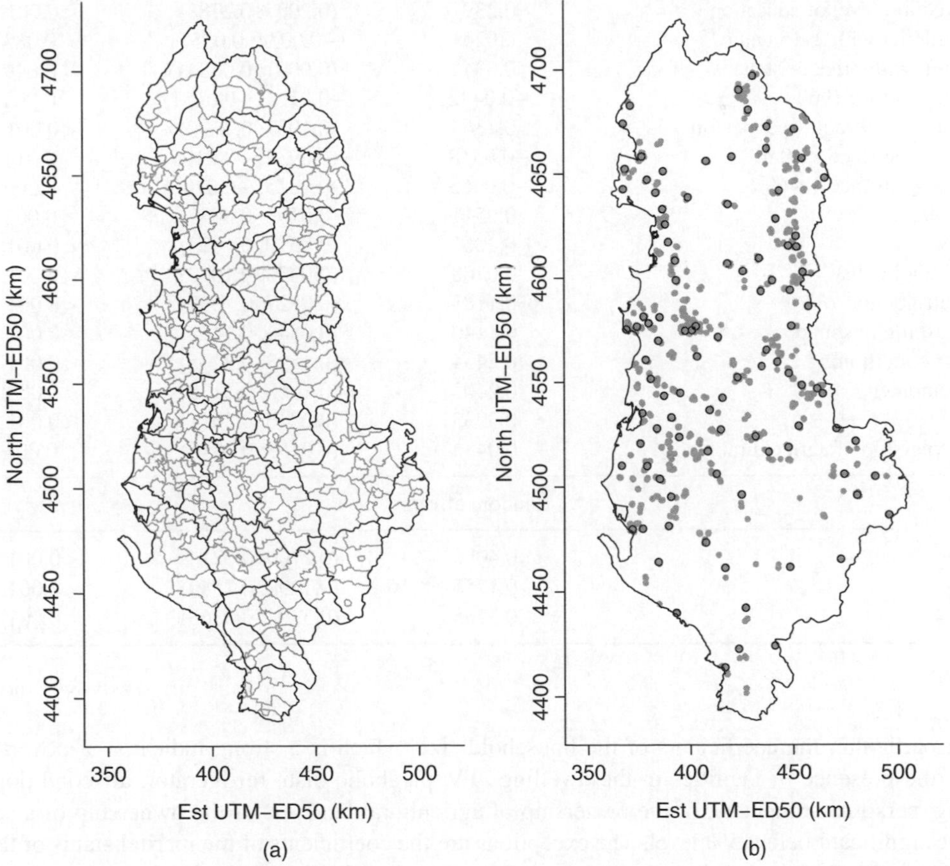

Figure 13.2 (a) Communes (gray line) and districts (black line) in Albania. (b) Geographical location of the LSMS sample units; black circles indicate the selected spline knots

Table 13.1 Estimated parameters of the geoadditive SAE log-transformed model for the household per-capita consumption expenditure at district level.

Parameter	Estimate	Confidence interval	p-value
	Fixed effects		
Intercept	7.1143	(−34.3241; 48.5527)	0.736
X coordinate	−0.0594	(−0.7807; 0.6618)	0.872
Y coordinate	0.0393	(−0.8700; 0.9487)	0.932
Household size	−0.0775	(−0.0913; −0.0638)	<0.001
Age of the householder	0.0029	(0.0014; 0.0044)	<0.001
Marital status of the householder	0.0745	(0.0004; 0.1485)	0.049
Age of the spouse or husband	−0.0021	(−0.0035; −0.0008)	0.001
Number of children 0–5 years	−0.0202	(−0.0382; −0.0023)	0.027
Age of the first child	−0.0023	(−0.0037; −0.0009)	0.001
Number of components without work	−0.0661	(−0.0784; −0.0537)	<0.001
High level of education	0.0913	(0.0648; 0.1178)	<0.001
Medium level of education	0.2397	(0.2007; 0.2788)	<0.001
Building with 2–15 units	0.0261	(−0.0034; 0.0557)	0.083
Built with brick or stone	0.0342	(0.0001; 0.0684)	0.049
Built before 1960	−0.0442	(−0.0734; −0.0151)	0.003
Number of rooms per person	0.1364	(0.1037; 0.1690)	<0.001
House surface <40m^2	−0.0518	(−0.0932; −0.0105)	0.014
House surface 40–69^2	−0.0365	(−0.0625; −0.0105)	0.006
Toilet inside	0.0511	(0.0190; 0.0833)	0.002
TV	0.1066	(0.0510; 0.1623)	<0.001
Parabolic dish	0.0768	(0.0473; 0.1062)	<0.001
Refrigerator	0.1183	(0.0827; 0.1539)	<0.001
Washing machine	0.1140	(0.0843; 0.1438)	<0.001
Air conditioning	0.2434	(0.1593; 0.3275)	<0.001
Computer	0.2403	(0.1668; 0.3138)	<0.001
Car	0.3233	(0.2846; 0.3621)	<0.001
Ownership of agricultural land	0.0484	(0.0153; 0.0815)	0.004
	Random effects		
σ_γ	0.4096	(0.2700; 0.6214)	<0.001
σ_u	0.1756	(0.1290; 0.2389)	<0.001
σ_e	0.3285	(0.3208; 0.3363)	<0.001

In particular, the coefficients of the household size, which is a strong indicator of poverty, of the presence of facilities in the dwelling (TV, parabolic dish, refrigerator, air conditioning, personal computer), of the ownership of agricultural land and of the ownership of a car are significant below 5% level. The exceptions are the coefficients of the marital status of the householder, of the number of children 0–5 years and of the 'built with brick or stone' variable that are significant at 5% level, and the coefficient of 'building with 2–15 units' that is significant at 10% level.

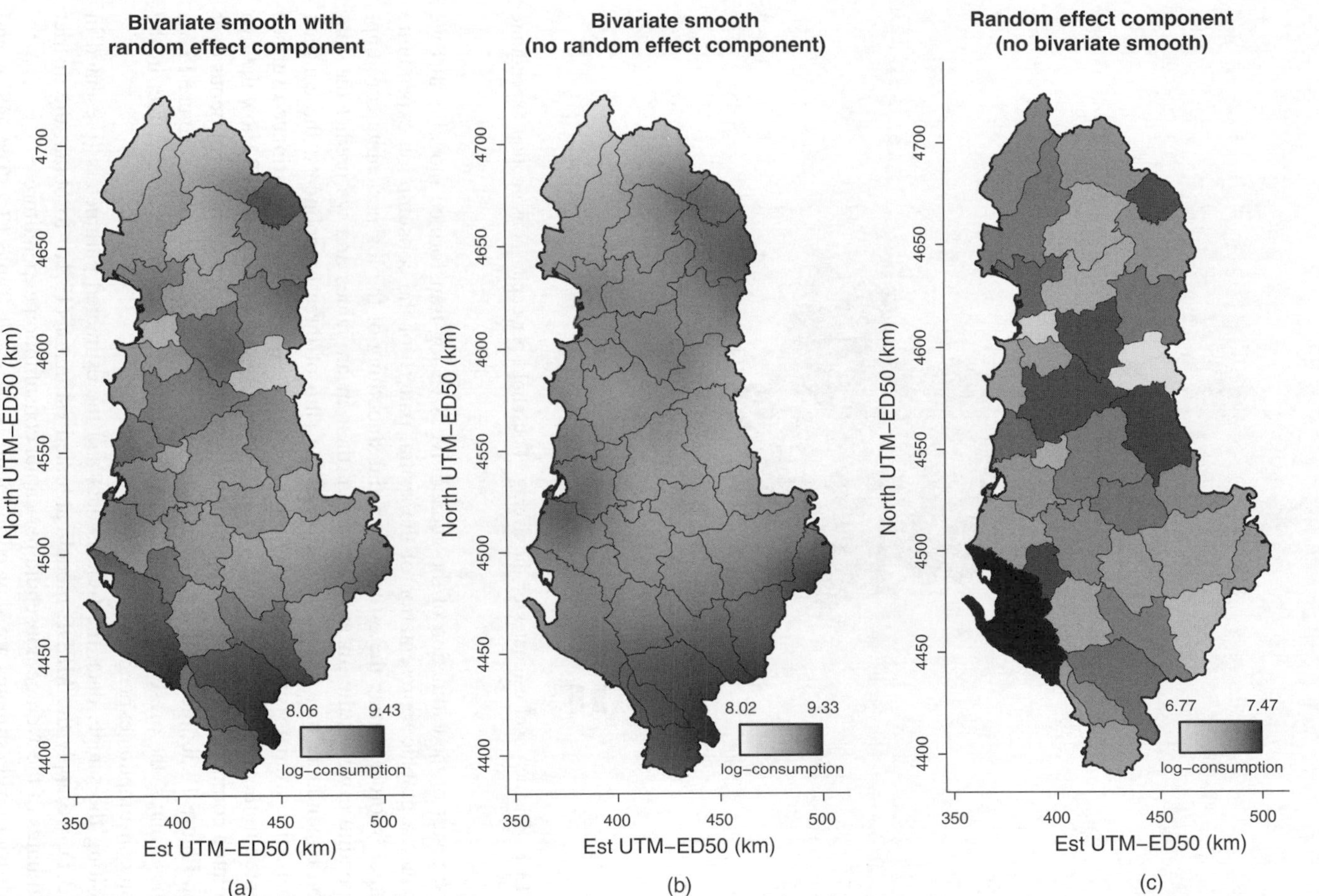

Figure 13.3 Spatial smoothing and district random effects of the household log per-capita consumption expenditure. Map (a) shows the smoothing obtained with the geoadditive SAE model as the sum of two components: the bivariate smoothing, map (b), and the small area random effects, map (c). Linear covariates are set at their mean values

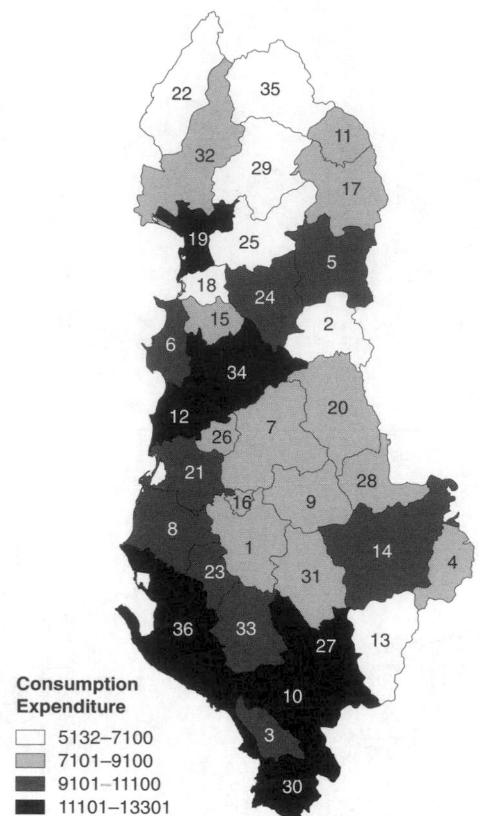

Figure 13.4 District level estimates of the average household per-capita consumption expenditure (Lek)

A first result of the estimation of the geoadditive SAE log-transformed model is that we can produce a graphic representation of the spatial pattern of the consumption expenditure in Albania. In order to map the estimated spatial smoothing of the log per-capita consumption expenditure we set the linear covariates at their mean values and we predict the consumption expenditure in a specific location as the sum of three components: the constant linear part, the bivariate spline smoother, and the small area effect. The estimated surface is presented in the map shown in Figure 13.3a. The other maps in Figure 13.3 show the two space-related components separately: the bivariate spline smoother, which is continuous over space, in Figure 13.3b; the small area effect, which is constant in each area, in Figure 13.3c. From these maps, the presence of both a spatial dynamic and a district level effect in the Albanian consumption expenditure is evident.

Exploiting the sample data, the Census data and the estimated parameters (presented in Table 13.1), we apply the MBDE model-calibrated estimator (13.16) to obtain the district level estimates of the average household per-capita consumption expenditure.

The district level estimates are showed in Figure 13.4 and in Table 13.2. The MSEs, and consequently the coefficients of variations CVs, presented in Table 13.2 are calculated using the MSE estimator derived in Chambers et al. (2011).

Table 13.2 District level estimates of the average household per-capita consumption expenditure (in Lek), with estimated root mean squared error (RMSE) and coefficient of variation (CV).

			MBMC estimator		
Code	District name	n_t	Mean	RMSE	CV%
1	Berat	120	7689.50	1245.46	16.20
2	Bulqize	128	5715.08	3163.96	55.36
3	Delvine	16	9522.54	1526.79	16.03
4	Devoll	16	8985.51	481.15	5.35
5	Diber	232	9105.86	335.42	3.68
6	Durres	160	9610.78	543.62	5.66
7	Elbasan	152	8700.42	485.69	5.58
8	Fier	224	9997.52	895.89	8.96
9	Gramsh	120	8085.67	1187.56	14.69
10	Gjirokast	32	13300.59	4153.01	31.22
11	Has	48	8832.13	550.58	6.23
12	Kavaje	88	11100.94	2000.32	18.02
13	Kolonje	8	5484.72	3838.64	69.99
14	Korce	136	9600.71	519.58	5.41
15	Kruje	39	8644.48	536.48	6.21
16	Kucove	32	7958.47	1258.37	15.81
17	Kukes	184	7940.48	1196.45	15.07
18	Kurbin	64	6979.61	2157.36	30.91
19	Lezhe	64	11598.52	2495.46	21.52
20	Librazhd	200	8288.37	950.52	11.47
21	Lushnje	152	9903.80	989.66	9.99
22	Malesi e Madhe	24	6662.54	2917.82	43.79
23	Mallakaster	32	10599.91	1764.43	16.65
24	Mat	32	9753.41	877.08	8.99
25	Mirdite	16	6309.29	2673.89	42.38
26	Peqin	24	7154.06	1870.39	26.14
27	Permet	16	12684.53	3711.93	29.26
28	Pogradec	48	8628.75	464.87	5.39
29	Puke	24	6296.39	2858.42	45.40
30	Sarande	48	12015.02	4548.29	37.86
31	Skrapar	16	7752.54	1294.09	16.69
32	Shkoder	140	8799.77	292.86	3.33
33	Tepelene	32	10781.27	1874.17	17.38
34	Tirane	684	11495.64	2549.94	22.18
35	Tropoje	88	5131.95	3940.45	76.78
36	Vlore	152	12430.18	3490.06	28.08

The map presents a clear geographical pattern, with the higher values of the average household per-capita consumption expenditure in the south and south-west of the country (coastal area) and the lower value in the mountainous area (north and north-east of the country). The examination of Table 13.2 and of the corresponding map reveals the districts with lower values of household per-capita consumption expenditure: Bulqize, Puke, and Kurbin show an average per-capita consumption expenditure below 7000 Lekë per month. On the other hand, the districts of Gjirokaster and Sarande have a very high average per-capita consumption expenditure and the same is also true for the coastal district of Vlore and the southern district of Permet. Finally, the capital district (Tirane) shows a slightly lower value. All the wealthiest districts of Albania have quite high values for the coefficients of variation. This can indicate that those districts have a higher level of inequality, with a high variability of the household income distribution in the district population.

These results have been previously observed in the analysis of Tzavidis et al. (2008), which combined traditional indicators of poverty and SAE. Moreover, our estimates are consistent with the view of the World Bank and with the results obtained from other applications on the same datasets (Neri et al. 2005).

13.6 Concluding Remarks and Open Questions

The interest in spatial data analysis is increased in every area of statistical research. Particular interest is given to the possible ways in which spatially referenced data can support local policy makers, especially in areas of social and economical interventions. Geographical information is frequently available in many areas of observational sciences, and the use of specific techniques of spatial data analysis can improve our understanding of the studied phenomena.

The empirical evidence suggests that, despite being overlooked in the previous studies, the spatial location is an important component to understand the distribution of the consumption expenditure. In particular, the results of our analysis show that the consumption expenditure presents both spatial dynamics and area specific effects. Thus, the region morphology can explain, to some degree, the spatial patterns of the household per-capita consumption expenditure that remain after controlling for all the descriptive household level covariates effect. The map of the estimated district means presents an evident geographical pattern, with the higher values in the south and south-west of the country and the lower value in the mountainous area (north and north-east), confirming the results of previous applications on the same datasets presented in the literature.

Differently from other methods of analysis that exploit some spatial information, the geoadditive SAE model produces not only the map of estimated mean values, but also a spatial interpolation of all the observation. Thus, with this model we can produce an estimated value in any point of the country.

When we produce estimates of a parameter of interest over some pre-specified area, we should always consider the modifiable area unit problem (MAUP). With the geoadditive model we obtain a continuous surface estimation over the entire area, without defining the area a priori, so the MAUP can not occur. In our application the geoadditive model is associated with a SAE model, so in this case we need to define the areas before estimating the model, however the possible MAUP – if it occurs – will be only related to the definition of the small area and not to the spatial interpolation of the studied phenomenon.

Concluding, we recall that the condition under which the geostatistics methodologies can be applied is the knowledge of the location of all population units at the point level. As we find out in our study, this requirement is not so easily accomplished, especially if working with socio-economic data. Usually it is much easier to know the areas to which the population units belong (i.e., census districts, blocks, municipalities, enumeration areas, etc.) and the classic approach is to refer the data with respect to these areas centroids. An aspect to be explored is the use of more precise spatial location data: an imputation approach which considers a more realistic hypothesis on spatial distribution. Further investigations will be done in this direction.

Acknowledgement

The authors would like to acknowledge Dr Gero Carletto, Senior Economist in the Development Research Group of the World Bank and member of the Living Standards Measurement Study team, for providing the datasets of the 2002 Living Standard Measurement Study and the 2001 Population and Housing Census of Albania.

References

Betti G, Ballini F, and Neri L 2003 *Poverty and Inequality Mapping in Albania. Final Report to the World Bank.*

Chambers R, Chandra H, and Tzavidis N 2011 On bias-robust mean squared error estimation for pseudo-linear small area estimators. *Survey Methodology* **37**, 153–170.

Chandra H and Chambers R 2009 Multipurpose weighting for small area estimation. *Journal of Official Statistics* **25**, 379–395.

Chandra H and Chambers R 2011 Small area estimation under transformation to linearity. *Survey Methodology* **37**, 39–51.

Cressie N 1993 *Statistics for Spatial Data (revised edition).* Wiley, New York.

Grosh ME and Glewwe P 2000 A Guide to Living Standards Measurement Study Surveys. *Household Accounting: Experience and Concepts in Compilation.* Statistics Division, Department of Economic and Social Affairs, The United Nations.

Hastie TJ and Tibshirani R 1990 *Generalized Additive Models.* Chapman & Hall, London.

Kammann EE and Wand MP 2003 Geoadditive models. *Applied Statistics* **52**, 1–18.

Kaufman L and Rousseeuw PJ 1990 *Finding Groups in Data: An Introduction to Cluster Analysis.* John Wiley & Sons, Inc., New York.

Neri L, Ballini F, and Betti G 2005 Poverty and inequality mapping in transition countries. *Statistics in Transition* **7**, 135–157.

Opsomer JD, Claeskens G, Ranalli MG, Kauermann G, and Breidt FJ 2008 Non-parametric small area estimation using penalized spline regression. *Journal of the Royal Statistical Society, Series B* **70**, 265–286.

Royall RM and Cumberland WG 1978 Mean squared error estimation in finite population sampling. *Journal of the American Statistical Association* **73**, 351–358.

Ruppert D, Wand MP, and Carroll RJ 2009 Semiparametric regression during 2003–2007. *Electronic Journal of Statistics* **3**, 1193–1256.

Tzavidis N, Salvati N, Pratesi M, and Chambers R 2008 M-quantile models with application to poverty mapping. *Statistical Methods and Applications* **17**, 393–411.

Wand MP 2003 Smoothing and mixed models. *Computational Statistics* **18**, 223–249.

World Bank and INSTAT 2003 *Albania Living Standard Measurement Survey 2002. Basic Information Document.*

Wu C and Sitter RR 2001 A model calibration approach to using complete auxiliary information from survey data. *Journal of the American Statistical Association* **96**, 185–193.

Part V

Small Area Estimation of the Distribution Function of Income and Inequalities

Part V

14

Model-based Direct Estimation of a Small Area Distribution Function

Hukum Chandra[1], Nicola Salvati[2] and Ray Chambers[3]
[1]*Indian Agricultural Statistics Research Institute, New Delhi, India*
[2]*Department of Economics and Management, University of Pisa, Pisa, Italy*
[3]*Centre for Statistical and Survey Methodology, University of Wollongong, Wollongong, Australia*

14.1 Introduction

Most small area estimation (SAE) methods focus on estimation of linear parameters, for example small area means or totals, with comparatively little attention paid to estimation of small area distributions. However, estimation of the small area distribution can be useful if there are extreme values in the small area sample data, or if their distribution is highly skewed. In such cases, users of survey data are often interested in the finite population distribution of the survey variable and in the measures (e.g. medians, quartiles, percentiles) that characterize the shape of this distribution at small area level. Assuming a linear mixed model, Tzavidis et al. (2010) defined two estimators for a small area distribution based on the approaches of Chambers and Dunstan (1986) and Rao *et al.* (1990). These two estimators use predicted values for non-sampled units in the small area of interest to obtain an empirical predictor for the small area distribution. As result, they work well if the assumed model is correctly specified. An alternative approach is to fit a generalized linear mixed model to the indicator for the survey variable of interest being less than or equal to a pre-specified value t and to then predict the small area mean for this indicator, which is just the value of the small area distribution at t, by the small area average of these predicted values; see Saei and Chambers (2003), Mantiega *et al.* (2007) and Chandra *et al.* (2012). Molina and Rao (2010) extend

Analysis of Poverty Data by Small Area Estimation, First Edition. Edited by Monica Pratesi.

Companion Website: www.wiley.com/go/pratesi/poverty

this idea to so-called 'empirical best' predictors of small area poverty indicators. However, all these methods for estimation of small area distribution are model dependent and can lead to biased and inefficient estimates for the small area distribution function when model assumptions fail. Further, mean squared error (MSE) estimation for these estimators remains an outstanding problem. See Salvati *et al.* (2012) for a brief review of these methods of SAE for a distribution function as well as their comparative empirical performances. The model-based direct estimator (MBDE) for the small area distribution function overcomes some of these limitations and can be useful in studies on poverty and living conditions. In particular, the MBDE can be much better than the other estimators in realistic poverty applications where fitted models are approximations at best because this estimator is a compromise between the traditional direct estimator and the model-based SAE methods that 'borrow strength' via statistical models. MBDE is a weighted sum of the sample data from the small area of interest, with weights that are derived from a spline-based calibrated estimator of the population distribution function (Harms and Duchesne, 2006) under a regression model with random area effects. A bias-robust estimator of the MSE of the MBDE is defined based on the pseudo-linearization approach of Chambers *et al.* (2011). Salvati *et al.* (2012) proposed this MBDE for the small area distribution function and its MSE estimation. This chapter summarizes the work of Salvati *et al.* (2012) on MBDE for the small area distribution function along with its MSE estimation followed by an application to real data. In particular, we present the use of MBDE for the estimation of gender-wise distribution of equivalized income and the Head Count Ratio (HCR) in the Toscana, Lombardia and Campania provinces of Italy. The next section first introduces the concept of calibrated sample weights for a finite population distribution function and then defines the MBDE estimator for this function and its MSE estimator. The application based on the MBDE estimator of the small area distribution function is given in Section 14.3.

14.2 Estimation of the Small Area Distribution Function

Let us consider a finite population U of size N and let y be the variable of interest. A common target of inference is then the proportion of values y_j that are bounded by a given constant (e.g. the proportion of households whose monthly per capita expenditure is below the poverty line). More generally, the target of inference is the value of the finite population distribution function for a variable y at a specified value t. This is, $F_N(t) = N^{-1} \sum_{j=1}^{N} I(y_j \leq t)$, that is the proportion of the population whose values for y are less than or equal to t, where $I(y_j \leq t)$ is the indicator function that takes the value 1 if $y_j \leq t$ and 0 otherwise and t is a specified constant. Clearly, once an estimator of the finite population distribution function is obtained, its inverse can be evaluated to obtain the associated estimator of the finite population quantile function. Note that we wish to estimate the small area distribution function, so we assume that a finite population U containing N units can be partitioned into D non-overlapping domains, referred to from now on as small areas, or simply areas or domains, indexed by $i = 1, \ldots, D$, with area i containing N_i units, so $N = \sum_{i=1}^{D} N_i$. Let y_{ij} denote the value of the variable of interest y for unit j ($j = 1, \ldots, N_i$) in area i ($i = 1, \ldots, D$). The area-specific distribution function of y for area i is

$$F_i(t) = N_i^{-1} \sum_{j=1}^{N_i} I(y_{ij} \leq t); i = 1, \ldots, D. \tag{14.1}$$

Let s denote a sample of n units drawn from population U by some specified sampling design, and assume that values of the variable of interest y are available for each of these n sample units. The non-sample component of U, containing $N - n$ units, is denoted by r. In what follows, a subscript of i denotes quantities specific to area i ($i = 1, \ldots, D$). For example, s_i and r_i denote the n_i sample and $N_i - n_i$ non-sample units, respectively, for area i. With this notation, the conventional estimators of the area i distribution function, $F_i(t)$, are the Horvitz–Thompson (HT) estimator

$$\hat{F}_i^{HT}(t) = N_i^{-1} \sum_{j \in s_i} \pi_j^{-1} I(y_j \leq t) \tag{14.2}$$

and the Hajek estimator

$$\hat{F}_i^{Hajek}(t) = \sum_{j \in s_i} \pi_j^{-1} I(y_j \leq t) / \sum_{j \in s_i} \pi_j^{-1}. \tag{14.3}$$

Here π_j denotes the sample inclusion probability of unit j. Both estimators (14.2) and (14.3) are area-specific design-based direct estimators and do not depend on an assumed model for their validity (Cochran, 1977). Unfortunately, empirical evidence presented in Rueda *et al.* (2007) shows that these estimators can be substantially biased, while the fact that they only use information from the area i sample makes them too unstable for SAE. As discussed earlier in Section 14.1, several indirect estimators of (14.1) have been suggested in the literature, see for example, Tzavidis *et al.* (2010), Saei and Chambers (2003), Mantiega *et al.* (2007), Chandra *et al.* (2012), Molina and Rao (2010), Salvati *et al.* (2012) and references therein. However, these indirect methods of SAE of the distribution function are model dependent and can lead to biased and inefficient estimates for the small area distribution function when model assumptions fail. In addition, MSE estimation for these estimators remains an outstanding problem. In contrast, a direct estimate for a small area is simple to interpret, since the estimated value of the variable of interest for the area is just a weighted average of the sample data from the same area. This is not true of an indirect estimator which is a weighted sum over the entire sample. Unfortunately, when weights are the inverses of sample inclusion probabilities, conventional direct estimators like (14.2) and (14.3) can be quite inefficient. The MBDE of a small area mean improves upon the efficiency of these conventional direct estimators by using the weights that define the empirical best linear unbiased prediction (EBLUP) for the population total under a model with random area effects. See Chandra and Chambers (2009, 2011) and Salvati *et al.* (2010). In order to define the MBDE for a small area distribution function, sample weights that are calibrated to the known finite population distribution of the auxiliary variables in $\mathbf{x}$ are used and are based on a model with random area effects. We now briefly describe the MBDE method for estimating the small area distribution function. For detailed discussion see Salvati *et al.* (2012). For simplicity, we restrict our discussion below to a single scalar covariate x, noting that the extension to multiple scalar covariates is straightforward. Note that the application reported in the next section uses multiple scalar covariates.

Harms and Duchesne (2006) defined the calibrated estimator of a finite population distribution function $F_N(t)$ as a weighted empirical distribution function of form

$$\hat{F}_N^{HD}(t) = N^{-1} \sum_{j \in s} w_j I(y_j \leq t) \tag{14.4}$$

where the sample weights w_j in (14.4) are calibrated to the known finite population distribution of x. In particular, let $0 < \alpha_1 < \alpha_2 < \cdots < \alpha_K < 1$ denote an ordered set of constants. Then the weights used in (14.4) sum to N and, for $k = 1, \ldots, K$, also satisfy

$$\sum_{j \in s} w_j I\left\{x_j \leq Q_x\left(\alpha_k\right)\right\} = N\alpha_k \tag{14.5}$$

where $Q_x(\alpha_k)$ is the known α_k-quantile of the finite population distribution of x. That is, the weights used in (14.4) are calibrated to both the population size N and to the population totals of the auxiliary variables defined by the indicators $I\left\{x_j \leq Q_x\left(\alpha_k\right)\right\}$. Standard results from calibration theory (Deville and Särndal, 1992; Chambers, 1996) can be used to show that if these calibrated weights w_j are then chosen to minimize their chi-square distance from the weights used in the HT estimator (14.2), as is commonly done, then (14.4) is a regression estimator of $F_N(t)$ under the linear model

$$I(y_j \leq t) = \beta_{0t} + \sum_{k=1}^{K} \beta_{kt} I\left\{x_j \leq Q_x\left(\alpha_k\right)\right\} + \varepsilon_{jt} \tag{14.6}$$

where the ε_{jt} are uncorrelated errors with zero expectation and variance $\sigma^2_{\varepsilon t}$ (Chambers, 2005). However, (14.6) is also easily seen to be a p-spline model with knots at the α_k th quantiles of the finite population distribution of x. That is, $\widehat{F}_N^{HD}(t)$ is actually a p-spline estimator of $F_N(t)$.

We now define $g_{jk} = I\{x_j \leq Q_x(\alpha_k)\}$ and let $\mathbf{g}_{Uk} = (g_{jk}; j = 1, \ldots, N)$ be the corresponding population N-vector, so $\mathbf{G}_U = [\mathbf{1}_N, \mathbf{g}_{U1}, \ldots, \mathbf{g}_{UK}]$ denotes the population level matrix of values of these variables, where $\mathbf{1}_N$ denotes a N-vector of ones. Also, define $a_{jt} = I(y_j \leq t)$ and put $\mathbf{a}_{Ut}$ equal to the N-vector of population values of the a_{jt}. The population level version of model (14.6) is then

$$\mathbf{a}_{Ut} = \mathbf{G}_U \boldsymbol{\beta}_t + \boldsymbol{\varepsilon}_{Ut}. \tag{14.7}$$

Given the appropriate sample and non-sample components of $\mathbf{a}_{Ut}$, $\mathbf{G}_U$ and the covariance matrix $\mathbf{V}_{Ut} = \sigma^2_{\varepsilon t}\mathbf{I}_U$ of $\boldsymbol{\varepsilon}_{Ut}$, the vector of sample weights w_{jt}^{DF} that define the EBLUP of the population total of the a_{jt} under (14.7) is then

$$\mathbf{w}_{st}^{DF} = (w_{jt}^{DF}; j \in s) = \mathbf{1}_n + \widehat{\mathbf{H}}_{st}^T(\mathbf{G}_U^T \mathbf{1}_N - \mathbf{G}_s^T \mathbf{1}_n) + (\mathbf{I}_n - \widehat{\mathbf{H}}_{st}^T \mathbf{g}_s^T)\widehat{\mathbf{V}}_{sst}^{-1}\widehat{\mathbf{V}}_{srt}\mathbf{1}_{N-n} \tag{14.8}$$

where $\widehat{\mathbf{H}}_{st} = (\mathbf{G}_s^T \widehat{\mathbf{V}}_{sst}^{-1}\mathbf{G}_s)^{-1}\mathbf{G}_s^T \widehat{\mathbf{V}}_{sst}^{-1}$. Under (14.7), $\widehat{\mathbf{V}}_{sst} = \widehat{\sigma}^2_{\varepsilon t}\mathbf{I}_n$ and $\widehat{\mathbf{V}}_{srt} = \mathbf{0}$, so these weights simplify to

$$\mathbf{w}_s^{DF} = (w_j^{DF}; j \in s) = \mathbf{1}_n + \mathbf{G}_s(\mathbf{G}_s^T\mathbf{G}_s)^{-1}(\mathbf{G}_U^T\mathbf{1}_N - \mathbf{G}_s^T\mathbf{1}_n) = \mathbf{1}_n + \mathbf{G}_s(\mathbf{G}_s^T\mathbf{G}_s)^{-1}\mathbf{G}_{N-n}^T\mathbf{1}_{N-n}.$$

The model (14.7) is easily adapted to SAE by including random area effects. That is, expression (14.7) is replaced by a random effects model of form

$$\mathbf{a}_{Ut} = \mathbf{G}_U\boldsymbol{\beta}_t + \mathbf{Z}_U\mathbf{u}_t + \boldsymbol{\varepsilon}_{Ut} \tag{14.9}$$

where $\mathbf{Z}_U$ was defined following (14.4) and $\mathbf{u}_t \sim N(\mathbf{0}, \boldsymbol{\Omega}_t)$ is a D-vector of random area effects. As usual, we assume that $\mathbf{u}_t$ and $\boldsymbol{\varepsilon}_{Ut}$ are independently distributed, so that $Var(\mathbf{a}_{Ut}) = \mathbf{V}_{Ut} = \mathbf{Z}_U\boldsymbol{\Omega}_t\mathbf{Z}_U^T + \sigma^2_{\varepsilon t}\mathbf{I}_N$. The sample weights w_{jt}^{DF} that define the EBLUP of the population total of the d_{jt} under (14.9) are then still given by (14.8), but now with $\widehat{\mathbf{V}}_{sst} = \mathbf{Z}_s\widehat{\boldsymbol{\Omega}}_t\mathbf{Z}_s^T + \widehat{\sigma}^2_{\varepsilon t}\mathbf{I}_n$

and $\widehat{\mathbf{V}}_{srt} = \mathbf{Z}_s \widehat{\mathbf{\Omega}}_t \mathbf{Z}_r^T$, where $\widehat{\mathbf{\Omega}}_t$ and $\widehat{\sigma}^2_{\varepsilon t}$ are the estimated values of the variance components of (14.9).

In practice, one first needs to decide on the calibration constraints (14.5) before (14.9) can be fitted and (14.8) calculated. This in turn requires that one has chosen the values $0 < \alpha_1 < \alpha_2 < \cdots < \alpha_K < 1$. The ordered half-sample cross-validation procedure described in Chambers (2005) is adapted for this purpose. To illustrate, fix $K = 1$ and then search for the value α_t^{opt} that maximizes the concordance between the sample values of a_{jt} and the sample values of $g_j = I\{x_j \leq Q_x(\alpha)\}$. The steps in this procedure are as follows:

1. Order the sample x-values: $x_{(1)}, x_{(2)}, x_{(3)}, \ldots\ldots, x_{(n-1)}, x_{(n)}$.
2. Create two sets $E = \{x_{(1)}, x_{(3)}, \ldots..\}$ and $V = \{x_{(2)}, x_{(4)}, \ldots..\}$;
3. For given α and t, fit the model (14.9) and then compute the weights (14.8), treating E as the 'sample' and V as the 'nonsample'. Denote the corresponding value of (14.4) based on these weights by $\widehat{F}_N^{HD(n)}(t, \alpha)$.
4. The optimal value α_t^{opt} then satisfies

$$\left\{ \widehat{F}_N^{HD(n)}(t, \alpha_t^{opt}) - n^{-1} \sum_{j \in s} I(y_j \leq t) \right\}^2 = \min_{0<\alpha<1} \left\{ \widehat{F}_N^{HD(n)}(t, \alpha) - n^{-1} \sum_{j \in s} I(y_j \leq t) \right\}^2.$$

We note that although this procedure only identifies a single 'most concordant' calibration constraint to use in (14.5), there is nothing to stop it being extended to identification of multiple calibration constraints. However, some care must then be taken to ensure that the resulting values of $Q_x(\alpha)$ are separated sufficiently in the interval spanned by the sample values of the auxiliary x. Failure to do this could result in the sample design matrix defined by (14.9) not being of full rank.

Finally, given the weights (14.8), we write down the MBDE for the area i distribution function $F_i(t)$ as

$$\widehat{F}_i^{MBDE}(t) = \sum_{j \in s_i} w_{jt}^{DF} I(y_j \leq t) \Big/ \sum_{j \in s_i} w_{jt}^{DF}; i = 1, \ldots, D. \tag{14.10}$$

We refer to (14.10) as a direct estimator because it is a weighted average of the sample data from the area of interest. However, this does not mean that it can be calculated from these data alone. The weights (14.8) are a function of the data from the entire sample. That is, they 'borrow strength' from other areas via the model (14.9).

It should also be pointed out that since the weights (14.8) depend on t, there is no guarantee that (14.10) defines a monotone function of t, that is one where $t_1 < t_2$ implies $\widehat{F}_i^{MBDE}(t_1) \leq \widehat{F}_i^{MBDE}(t_2)$. This issue will usually not be relevant when one wishes to estimate the distribution of interest at points that are well separated, but can be a problem when the aim is to invert (14.10) as a function of t in order to estimate quantiles. In such a situation we recommend that (14.10) be first transformed to be monotone in t, for example using the approach described in He (1997).

MSE estimation of the MBDE estimator (14.10) can be carried out via the bias-robust MSE estimator for the MBDE described in Chandra and Chambers (2009, 2011). Also see Chambers *et al.* (2011) and Salvati *et al.* (2012). This estimator of the MSE of the MBDE (14.10) is given by

$$\widehat{M}\left\{ \widehat{F}_i^{MBDE}(t) \right\} = \widehat{V}_{it} + \widehat{B}_{it}^2,\ i = 1, \ldots, D \tag{14.11}$$

where $\widehat{V}_{it}$ is a heteroscedasticity-robust estimator of the conditional prediction variance of $\widehat{F}_i^{MBDE}(t)$, $\widehat{B}_{it}$ is an estimator of the corresponding conditional prediction bias, and the conditioning is with respect to the value of the area effect. We use

$$\widehat{V}_{it} = N_i^{-2} \sum_{j \in s_i} \left\{ \left(N_i w_{jt}^{DF(i)} - 1 \right)^2 + \left(N_i - n_i \right) n^{-1} \right\} \left(d_{jt} - \widehat{\mu}_{jt} \right)^2. \tag{14.12}$$

where $w_{jt}^{DF(i)} = w_{jt}^{DF} \Big/ \sum_{k \in s_i} w_{kt}^{DF}$ and $\widehat{\mu}_{jt}$ is an unbiased linear estimator of the conditional expected value $\mu_{jt} = E(d_{jt} | g_j, \mathbf{u}_t)$. In particular, $\widehat{\mu}_{jt}$ be computed as the 'unshrunken' version of the EBLUP for μ_{jt}, that is

$$\widehat{\mu}_{jt} = \widehat{\beta}_{0t} + g_j \widehat{\beta}_{1t} + \mathbf{z}_j^T \left(\mathbf{Z}_s^T \mathbf{Z}_s \right)^{-1} \mathbf{Z}_s^T \left(\mathbf{I}_s - \widehat{\mathbf{H}}_{st}^T \mathbf{g}_s^T \right)^T \mathbf{1}_n.$$

For the conditional bias of the MBDE, we use a simple 'plug-in' estimator of the form

$$\widehat{B}_{it} = \sum_{j \in s_i} w_{jt}^{DF(i)} \widehat{\mu}_{jt} - N_i^{-1} \sum_{j \in U_i} \widehat{\mu}_{jt}. \tag{14.13}$$

See Salvati *et al.* (2012) for a more detailed discussion.

14.3 Model-based Direct Estimator for the Estimation of the Distribution Function of Equivalized Income in the Toscana, Lombardia and Campania Provinces of Italy

In this section the MBDE theory presented in the preceding sections is illustrated by using it to estimate the gender-specific distribution of equivalized income in the Toscana, Lombardia and Campania provinces of Italy. In particular, the MBDE estimator (14.10) for the small area distribution is used, with sample weights (14.8) based on the model (14.9), with the aim of estimating the province by gender distributions of equivalized income. Salvati *et al.* (2014) estimate mean equivalized income, HCR and poverty gap for each local labour system of Toscana, Lombardia and Campania by using area level small area models. Fabrizi *et al.* (2014) illustrate the use of a M-quantile model to produce reliable maps of economic well-being as measured by the average equivalized income in each local labour system of Toscana. The application presented in this chapter is the first study on the estimation of the small distribution function in three Italian regions that allows for comparisons by gender-provinces.

In Italy, the EU Statistics on Income and Living Conditions (EU-SILC) is conducted yearly by ISTAT and provides accurate country-level information on variables such as household disposable income and household composition. EU-SILC is a rotating panel survey with a 75% sample overlap in successive years. Each year the incoming sample is drawn according to a stratified two-stage sample design where municipalities are the primary sampling units (PSUs) while households are the secondary sampling units (SSUs). The PSUs are divided into strata according to their population size; the SSUs are selected by means of systematic sampling in each PSU. All members of each sampled household are interviewed through an individual questionnaire and one individual in each household (usually the head of the household) is interviewed through a household questionnaire. We note that some provinces (generally the

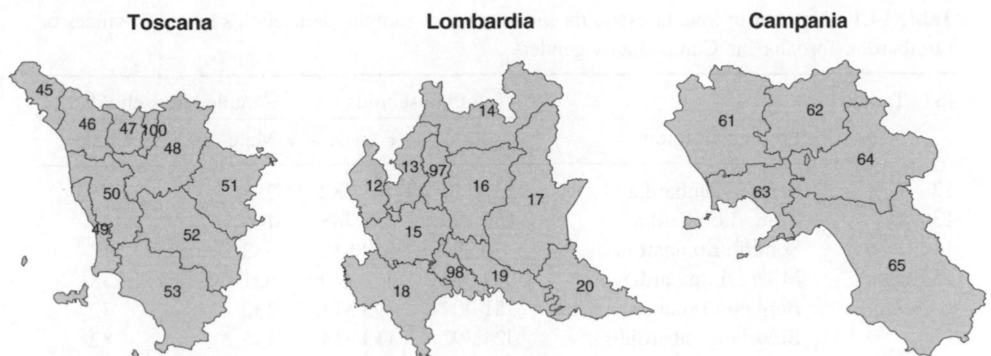

Figure 14.1 Map of Toscana, Lombardia and Campania regions in Italy

smaller ones) may have very few sampled municipalities, and many municipalities have no sample at all. Due to these small sample sizes, direct survey estimates can have very large standard errors at provincial level. Further, they may not even be computable at municipality level. Therefore SAE techniques are required in order to produce the estimates at this level.

In this application of SAE, we wish to estimate of the distribution of the equivalized income and the associated HCR by gender in the 10 Provinces of the Toscana region, the 11 Provinces of the Lombardia region and the 5 Provinces of the Campania region. It is noteworthy that gender by Province is our small area of interest and we therefore have a total of 52 small areas. Figure 14.1 shows the Provinces of the Toscana, Lombardia and Campania regions. In our data, gender is the sex of the head of the household. Further, the choice of these three regions (i.e. Toscana, Lombardia and Campania) out of the 20 regions that make up Italy is motivated by the geographical differences that characterize Italy. In particular, we use these data to investigate the so-called 'north–south' divide in the country, since the three regions can be considered as representative of Northern, Central and Southern/Insular Italy. Moreover, we focus on the differences by gender in these three regions. Table 14.1 reports the total number of households and the number of sampled households for each province by gender of the head of the household.

In general, two types of variables are required for application of the SAE methods described in the previous section. The dependent (or target) variable for which small area estimates are required is drawn from the 2006 wave of EU-SILC and the auxiliary (covariate) variables known for the entire population (i.e. known for both sample and out of sample households) are drawn from the 2001 Population Census of Italy. Here, we used data from the 2006 wave of EU-SILC so the income reference year is 2005 and the household equivalized income is computed by using the modified OECD scale (Hagenaars *et al.*, 1994). This scale has been officially adopted by Eurostat for the definition of equivalized income used in the computation of poverty indicators such as the at-risk-of-poverty rate (EUROSTAT, 2012). It is calculated for each household as the household total disposable net income divided by the equivalized household size, which is defined by giving a weight of 1.0 to the adult head of household, 0.5 to other household members aged 14 or more and 0.3 to each child aged less than 14 in the household. With this definition, the same household equivalized disposable income can then be assigned to all members of the same household. The auxiliary (covariate) information is

Table 14.1 Number of total households and number of sampled households in the provinces of Lombardia, Toscana and Campania by gender

ISTAT code	Province/Region	Total households		Sampled households	
		Male	Female	Male	Female
12	Varese/Lombardia	237 117	83 782	255	81
13	Como/Lombardia	154 799	55 788	106	29
14	Sondrio/Lombardia	50 299	19 518	32	17
15	Milano/Lombardia	1 099 962	455 440	621	238
16	Bergamo/Lombardia	281 907	93 871	232	72
17	Brescia/Lombardia	324 392	113 314	186	82
18	Pavia/Lombardia	150 533	61 253	19	10
19	Cremona/Lombardia	98 468	36 853	61	17
20	Mantova/Lombardia	108 808	37 441	166	61
97	Lecco/Lombardia	90 555	30 766	87	42
98	Lodi/Lombardia	58 282	19 696	52	15
45	Massa-Carrara/Toscana	56 202	24 608	86	33
46	Lucca/Toscana	104 495	41 622	76	37
47	Pistoia/Toscana	76 782	27 684	89	44
48	Firenze/Toscana	265 771	110 484	313	150
49	Livorno/Toscana	96 083	37 646	73	38
50	Pisa/Toscana	112 586	37 673	106	40
51	Arezzo/Toscana	93 291	30 589	118	40
52	Siena/Toscana	75 700	25 699	87	36
53	Grosseto/Toscana	63 189	24 531	39	24
100	Prato/Toscana	64 487	19 130	83	48
61	Caserta/Campania	219 634	60 050	91	19
62	Benevento/Campania	77 968	24 473	50	19
63	Napoli/Campania	741 812	227 498	598	246
64	Avellino/Campania	116 500	35 840	73	23
65	Salerno/Campania	277 071	82 009	134	50

drawn from the 2001 Population Census of Italy. The Population Census of Italy has a very comprehensive questionnaire collecting information on each household and on each individual living in Italian territory. To generate the estimates on poverty and living conditions in the three regions of interest we selected census variables that are also available in the EU-SILC survey. These variables include information referring to the head of the household, namely age and years in education and information referring to the household, that is number of household members. These variables have previously been used as covariates in small area models used for estimating income distributions and poverty indicators (Fabrizi *et al.*, 2014). It is noteworthy that use of covariates from the 2001 Population Census of Italy to model income data from the 2006 wave of EU-SILC may raise issues of comparability. In particular, the use of lagged census information may lead to bias in the small area estimates. However, variables whose totals are known from the census have been shown to be powerful predictors of household income according to tests conducted within the framework of the SAMPLE program (Small

Area Methods for Poverty and Living Conditions Estimates. http://www.sample-project.eu/). Finally, it is important to point out that EU-SILC and Census data are confidential. These data were provided by the Italian National Institute of Statistics (ISTAT) to the researchers of the SAMPLE project, funded by the European Commission under the 7th Framework Programme (FP7), with strong confidentiality restrictions, and the analysis reported here was carried out in such a way that these restrictions were met.

The results generated from the application of the MBDE method of SAE are mapped in Figure 14.2, Figure 14.3 and Figure 14.4 for the three regions, namely Toscana, Lombardia and Campania, respectively. For each region, gender-specific maps show the Provincial distributions of the 0.25, 0.50 and 0.75 quantiles. These maps can be interpreted as follows. If we consider the 0.25 quantile map for female in the Toscana region (Figure 14.2), the last income class (dark grey) varies from 0.34 to 0.62, which means that the percentage of households with female as head with an equivalized income less than the 0.25 population income quantile (equal to 10 257 euros) is between 34% and 62%. At each quantile therefore the lighter the shade of a Province, the richer it is. Note that we used the same cut-points in all three maps, allowing them to be directly compared. The estimates of the distribution of equivalized income for each province show that there are intra-regional differences as well as gender differences. In general female households are poorer than male households in all three regions. In Lombardia (Figure 14.3), all the provinces show a high level of equivalized household income except the province of Pavia that has lower income. For female households the poorest provinces are Cremona and Varese. Such intra-regional variability is also present in Toscana (Figure 14.2). The provinces of Siena and Firenze appear to be as wealthy as the wealthier provinces of Lombardia (in particular, the province of Siena is comparable with the province of Milano) whereas the provinces of Lucca and Massa-Cararra have lower levels of income. These results indicate that Toscana and Lombardia have similar levels of equivalized household income although one may say that Lombardia is somewhat wealthier. Looking now at the results of the southern region of Campania (Figure 14.4), it is clear that provinces in this region have smaller levels of equivalized household income than provinces in Lombardia and Toscana. The intra-regional differences and the gender differences in Campania are not so pronounced.

Table 14.2 reports the HCR estimates by province with their respective standard errors. The HCR is computed as the proportion of households with the equivalized income less than the 60% of the median of the equivalized income of all the households in the sample. The provinces with lowest HCR are Lodi for male households and Pavia for female households (both provinces in Lombardia), whereas Benevento (in Campania) shows the highest values of HCR for both male and female households. The estimated coefficients of variation (CVs) for these estimates vary between 8% (Pavia, female) and 250% (Pavia, male). In this context, we note that Statistics Canada provides quality guidelines for publishing tables: estimates with a CV less than 16.6% are considered reliable for general use; estimates with CVs between 16.6% and 33.3% should be accompanied by a warning to users; and estimates with coefficients of variation larger than 33.3% are deemed to be unreliable. In this application there are 3 Provinces with an estimated CV of less than 16%, 22 Provinces with an estimated CV of less than 33.3%, with the remaining (about half) with CVs greater than 33.3%. The direct estimates of the HCR produce a large CV: it is greater than 33.3% for most of the small area of interest. For this reason the MBDE estimates seem much more accurate with respect to the direct estimates.

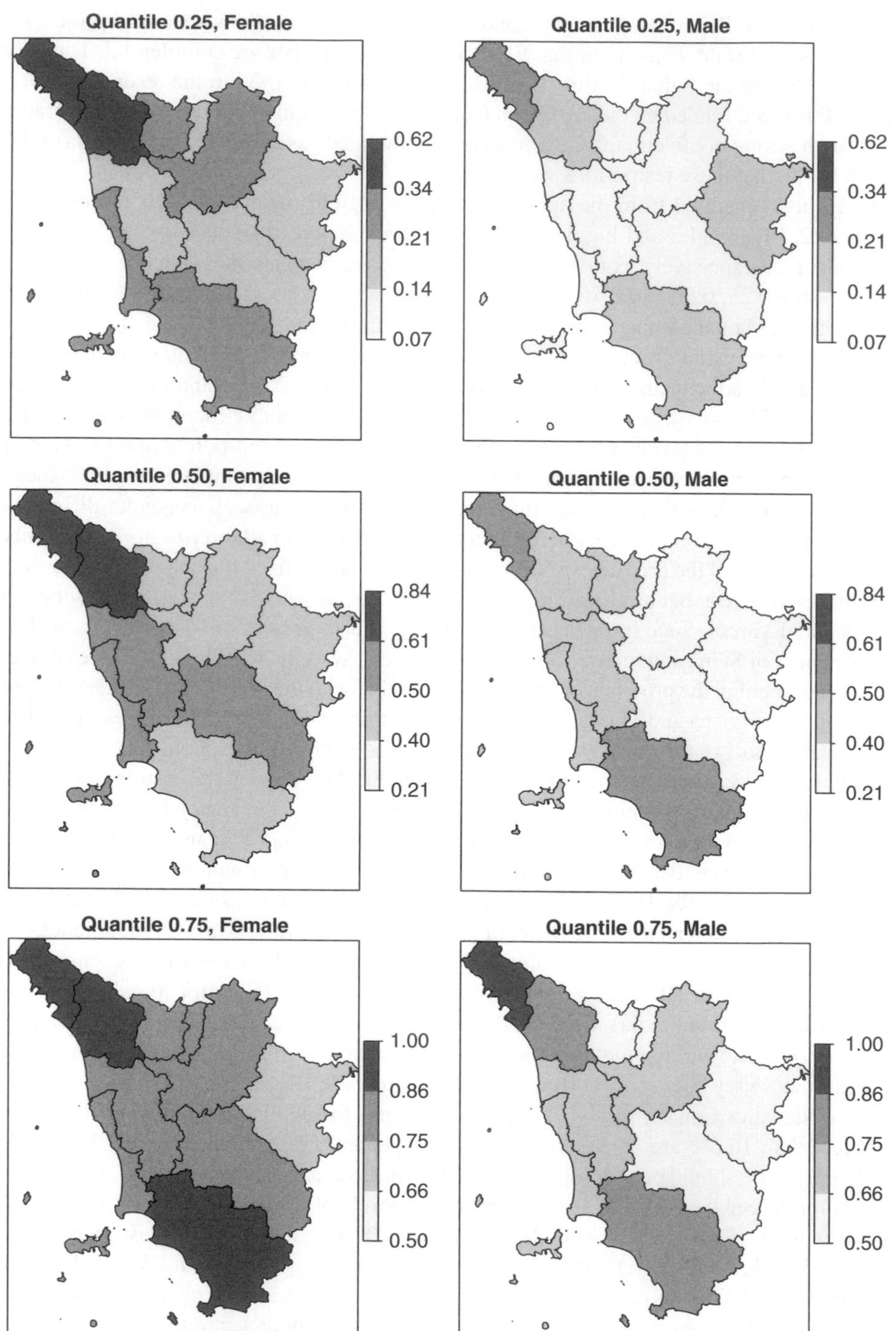

Figure 14.2 Distribution function of the equivalized income by province and gender based on EU-SILC and Census data for income year 2005 in the Toscana region of Italy

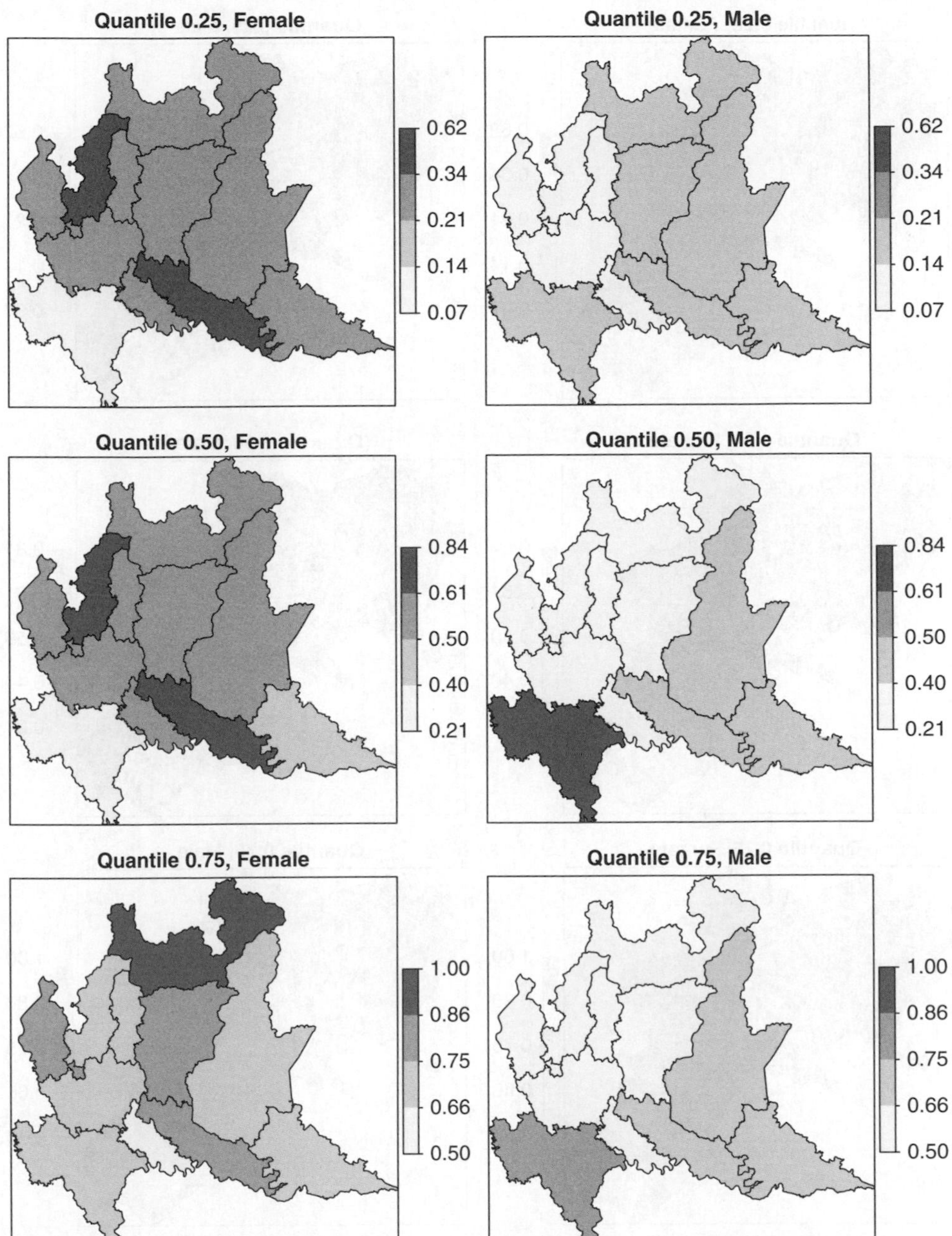

Figure 14.3 Distribution function of the equivalized income by province and gender based on EU-SILC and Census data for income year 2005 in the Lombardia region of Italy

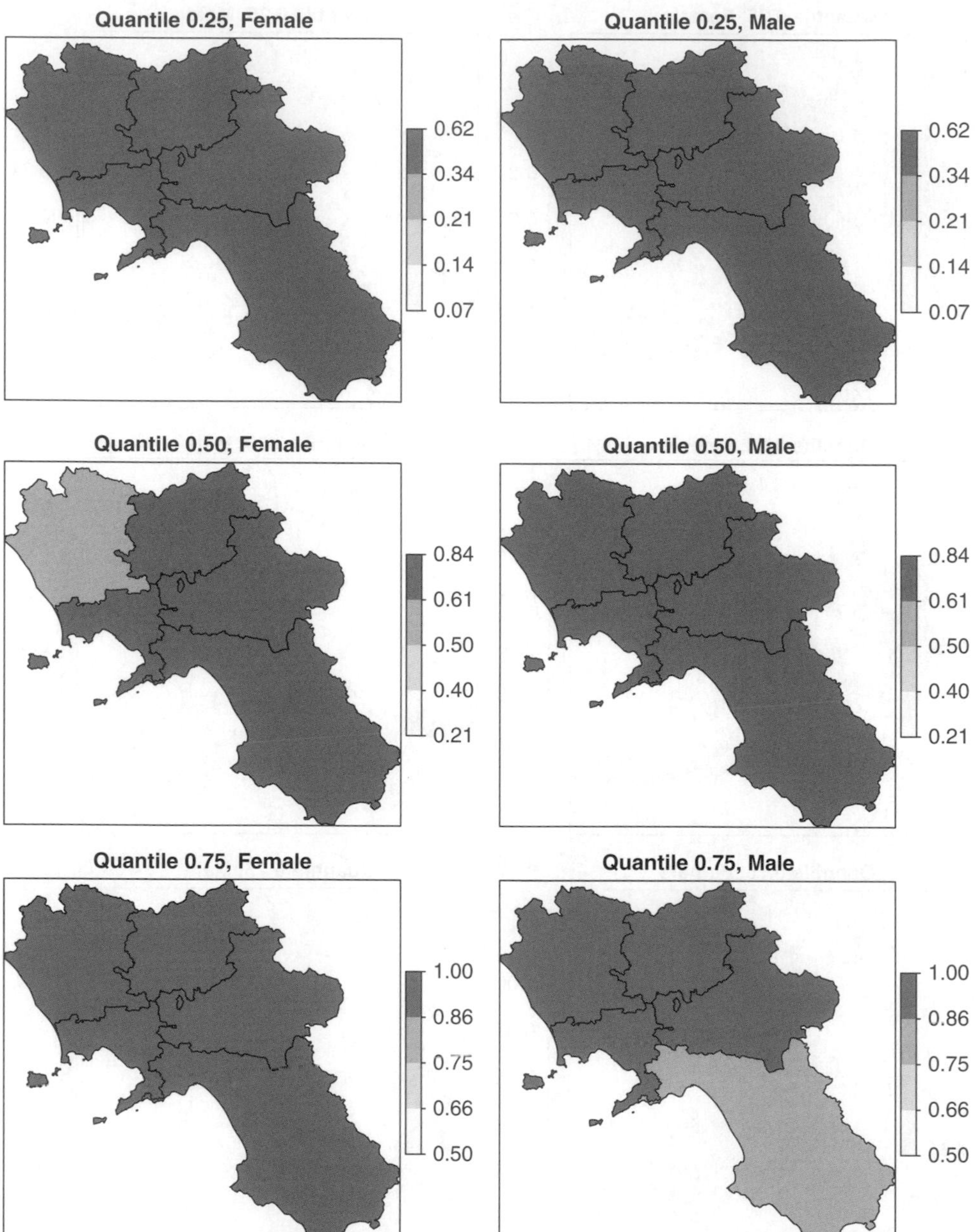

Figure 14.4 Distribution function of the equivalized income by province and gender based on EU-SILC and Census data for income year 2005 in the Camapania region of Italy

Table 14.2 Head count ratio (HCR) (and standard errors) in the Provinces of Lombardia, Toscana and Campania by gender

ISTAT code	Province/Region	HCR Male	HCR Female
12	Varese/Lombardia	0.078 (0.040)	0.201 (0.053)
13	Como/Lombardia	0.097 (0.046)	0.299 (0.057)
14	Sondrio/Lombardia	0.076 (0.091)	0.155 (0.114)
15	Milano/Lombardia	0.083 (0.036)	0.209 (0.017)
16	Bergamo/Lombardia	0.100 (0.035)	0.175 (0.035)
17	Brescia/Lombardia	0.133 (0.035)	0.237 (0.034)
18	Pavia/Lombardia	0.110 (0.074)	0.090 (0.155)
19	Cremona/Lombardia	0.072 (0.053)	0.245 (0.076)
20	Mantova/Lombardia	0.087 (0.039)	0.196 (0.043)
97	Lecco/Lombardia	0.057 (0.043)	0.239 (0.051)
98	Lodi/Lombardia	0.015 (0.038)	0.255 (0.021)
45	Massa-Carrara/Toscana	0.152 (0.086)	0.237 (0.086)
46	Lucca/Toscana	0.125 (0.070)	0.383 (0.097)
47	Pistoia/Toscana	0.059 (0.077)	0.282 (0.070)
48	Firenze/Toscana	0.065 (0.071)	0.154 (0.031)
49	Livorno/Toscana	0.040 (0.073)	0.204 (0.051)
50	Pisa/Toscana	0.054 (0.069)	0.124 (0.067)
51	Arezzo/Toscana	0.092 (0.071)	0.129 (0.034)
52	Siena/Toscana	0.034 (0.069)	0.094 (0.058)
53	Grosseto/Toscana	0.147 (0.074)	0.192 (0.034)
100	Prato/Toscana	0.086 (0.069)	0.118 (0.059)
61	Caserta/Campania	0.363 (0.123)	0.341 (0.132)
62	Benevento/Campania	0.426 (0.114)	0.547 (0.151)
63	Napoli/Campania	0.356 (0.102)	0.424 (0.109)
64	Avellino/Campania	0.364 (0.121)	0.539 (0.144)
65	Salerno/Campania	0.317 (0.105)	0.370 (0.120)

14.4 Final Remarks

This chapter presents an MBDE estimator for the value of the area-specific finite-population distribution of a response variable. MBDE is based on sample weights that are calibrated to the finite-population distribution of an auxiliary variable *x*, and also allow for random area effects. This estimator has the potential to lead to significantly better small area estimates in financial, economic, environmental and public health applications because it is a compromise between the traditional direct estimator and the model-based SAE methods that 'borrow strength' via statistical models. Salvati *et al.* (2012) compared the MBDE with a number of competing estimators and the results indicated that the MBDE can sometimes be much better than these alternatives, particularly in realistic applications where fitted models are approximations at best.

In the chapter we demonstrate how MBDE can be employed successfully for estimating the gender-specific distribution of equivalized income and the HCR in the Toscana, Lombardia and Campania provinces of Italy. From the results we can note the so-called 'north–south' divide in the country: the richest provinces are in the north of Italy, whereas the poorest provinces are in the south of Italy.

Finally, we note that one aspect of the methods described in this chapter that we have not considered is their computational efficiency. Further research is therefore necessary in order to understand this aspect.

References

Chambers, R. (1996) Robust case-weighting for multipurpose establishment surveys. *Journal of Official Statistics*, 12, 3–32.

Chambers, R. (2005) Imputation vs. estimation of finite population distributions. Southampton Statistical Sciences Research Institute, Methodology Working Papers, M05/06.

Chambers, R. and Dunstan, R. (1986) Estimating distribution functions from survey data. *Biometrika,* 73, 597–604.

Chambers, R., Chandra, H. and Tzavidis, N. (2011) On bias-robust mean squared error estimation for pseudo-linear small area estimators. *Survey Methodology*, 37, 153–170.

Chandra, H. and Chambers, R. (2009) Multipurpose weighting for small area estimation. *Journal of Official Statistics*, 25, 379–395.

Chandra, H. and Chambers, R. (2011) Small area estimation under transformation to linearity. *Survey Methodology*, 37, 39–51.

Chandra, H., Chambers, R. and Salvati, N. (2012) Small area estimation of proportions in business surveys. *Journal of Statistical Computation and Simulation*, 82**,** 783–795.

Cochran, W.G. (1977) *Sampling Techniques*, 3rd edition. John Wiley & Sons, Inc., New York.

Deville, J.C. and Särndal, C.E. (1992). Calibration estimators in survey sampling. *Journal of the American Statistical Association*, 87, 376–382.

EUROSTAT (2012). Glossary: Equivalised disposable income. Statistics Explained (2012/9/4). http://epp.eurostat.ec.europa.eu/statistics_explained/index.php/Main_Page [accessed-2-Sep-2015].

Fabrizi, E., Giusti, C., Salvati, N. and Tzavidis, N. (2014) Mapping average equivalized income using robust small area methods. *Papers in Regional Science*, 93, 685–702.

Hagenaars, A., de Vos, K. and Zaidi, M.A. (1994) Poverty Statistics in the Late 1980s: Research Based on Micro-data. Office for Official Publications of the European Communities, Luxembourg.

Harms, T. and Duchesne, P. (2006) On calibration estimation for quantiles. *Survey Methodology*, 32, 37–52.

He, X. (1997) Quantile curves without crossing. *American Statistician*, 51, 186–192.

Manteiga, G. W., Lombardìa, M.J., Molina, I., Morales, D. and Santamarìa, L. (2007) Estimation of the mean squared error of predictors of small area linear parameters under a logistic mixed model. *Computational Statistics and Data Analysis*, 51, 2720–2733.

Molina, I. and Rao, J.N.K. (2010) Small area estimation of poverty indicators. *Canadian Journal of Statistics,* 38, 369–385.

Rao, J.N.K., Kovar, J.G. and Mantel, H.J. (1990) On estimating distribution functions and quantiles from survey data using auxiliary information. *Biometrika*, 77, 365–375.

Rueda, M., Martínez, S., Martínez, H. and Arcos, A. (2007) Estimation of the distribution function with calibration methods. *Journal of Statistical Planning and Inference*, 137, 435–448.

Saei, A. and Chambers, R. (2003) Small area estimation under linear and generalized linear mixed models with time and area effects. Southampton Statistical Sciences Research Institute, Methodology Working Papers, M03/15.

Salvati, N., Chandra, H. and Chambers, R. (2012) Model-based direct estimation of small-area distributions. *Australian and New Zealand Journal of Statistics*, 54, 103–123.

Salvati, N., Chandra, H., Ranalli, M.G. and Chambers, R. (2010) Small area estimation using a nonparametric model-based direct estimator. *Computational Statistics and Data Analysis*, 54, 2159–2171.

Salvati, N., Giusti, C. and Pratesi, M. (2014) The use of spatial information for the estimation of poverty indicators at small area level. In *Poverty and Social Exclusion, New Methods and analysis* edited by G. Betti and A. Lemmi, Chapter 14, pp. 261–282. Routledge, London, New York.

Tzavidis, N., Marchetti, S. and Chambers, R. (2010) Robust prediction of small area means and quantiles. *Australian and New Zealand Journal of Statistics*, 52, 167–186.

15

Small Area Estimation for Lognormal Data

Emily Berg[1], Hukum Chandra[2] and Ray Chambers[3]
[1]*Department of Statistics, Iowa State University, Ames, USA*
[2]*Indian Agricultural Statistics Research Institute, New Delhi, India*
[3]*Centre for Statistical and Survey Methodology, University of Wollongong, Wollongong, Australia*

15.1 Introduction

Survey data on income and expenditures underlie many analyses of economic poverty. For example, the World Bank defines a household to be in poverty if the household expenditure is below a pre-specified threshhold (Haslett and Jones, 2010). Haslett and Jones (2010) and Haslett, Isidro, and Jones (2010) remark that analysts often apply a log transformation to income or expenditure data because the distribution of the data on the original scale is highly right skewed. Ferraz and Moura (2012) apply skew-normal models to area-level income data, and Slud and Maiti (2006) apply estimators based on lognormal models to area-level data from the US Census Bureau Small Area Income and Poverty Estimation project.

Closely related to indicators of economic poverty, measures of health and nutrition frequently have skewed distributions. Nandram and Choi (2010) construct small area estimates for Body Mass Index (BMI) under an assumption of a lognormal distribution. Griswold *et al.* (2004) review alternative models, including the lognormal model, for analyzing skewed data related to health care costs and expenditures.

Accounting for skewness or outliers can lead to improved estimators of population means and totals. Fuller (1991) and Rivest (1994) study Winsorized means for skewed populations. Chambers (1986) develops procedures for outliers. Karlberg (2000) constructs an estimator of an optimal predictor of the population total under a lognormal distribution.

Analysis of Poverty Data by Small Area Estimation, First Edition. Edited by Monica Pratesi.

Companion Website: www.wiley.com/go/pratesi/poverty

Extreme sample values can substantially distort small domain estimates. The assumptions of the linear mixed effects model with constant variance are often violated when the sample data contain such values, causing linear model predictors to be biased or inefficient. In such a situation, defining a model and estimator that adequately reflect the skewed distribution of the response given the covariates can improve model-based small area estimates.

In this chapter, we focus on small area inference methods for a unit level 'income-type' response, that is one that is strictly positive and right-skewed. In particular, we compare three predictors based on the assumption that these data follow a lognormal distribution. The three predictors are analogous to a synthetic estimator, a model-based direct estimator, and a best unbiased predictor for the case where the data follow a linear unit-level mixed model on the log scale.

The methods developed in this chapter are motivated, in general, by the type of data structures in Haslett and Jones (2010) and Haslett, Isidro and Jones (2010). Haslett and Jones (2010) compare alternative small area predictors for data from the Philippine Family Income and Expenditure Survey (FIES) using data from the Census of Population and Housing (CPH) as auxiliary information. Haslett and Jones (2010) analyze household expenditure data from the Nepal Living Standards Survey. In both applications, a log transformation is considered as a mechanism to improve the symmetry of the distribution of the response.

The numerical investigations in this chapter are designed to reflect the properties of data from specific agricultural surveys, largely, because we have experience with such data. Though not explicitly related to poverty, productivity of cropland and food supply are implicitly linked to factors underlying hunger and food security. Furthermore, the simulations in this chapter demonstrate that the proposed methods can improve efficiency of small area estimates for a wide variety of sample sizes and parameter configurations, indicating the potential utility of the proposed methods beyond the agricultural context, such as in studies of poverty.

The remainder of this chapter is organized as follows. In Section 15.2, we review literature on small area estimation for skewed data. In Section 15.3, we define three small area predictors based on a lognormal model for the individual units in the population. The three predictors are nonlinear analogs to the model-based direct estimator, the regression synthetic estimator, and the empirical best linear unbiased prediction (EBLUP) under the linear model. The model-based direct estimator for the lognormal model is developed in Chandra and Chambers (2011). The regression synthetic estimator is an extension of the Karlberg (2000) procedure for estimating the population total. An empirical best (EB) predictor for a unit-level lognormal mixed model is a version of the EBLUP appropriate for the lognormal model (15.4). Section 15.4 presents simulation studies addressing several themes. One is a comparison of the three predictors and mean squared error (MSE) estimators described in section 15.3. The second is an evaluation of the robustness of the EB predictor. The third is a comparison of the EB predictor for the lognormal model to a predictor based on a gamma distribution. Section 15.5 contains a discussion.

15.2 Literature on Small Area Estimation for Skewed Data

Generalized linear mixed models (GLMMs) have been used for small area estimation for skewed data. Jiang and Lahiri (2006) review applications of the GLMM in small area estimation. Ghosh *et al.* (1998) consider hierarchical Bayes (HB) inference for the GLMM and apply the methodology to count data. Jiang (2003) uses a generalized method of moments procedure to estimate the parameters of a GLMM and obtains predictors using a Monte Carlo approximation for the conditional expectation. Torabi and Shokoohi (2012) use data cloning to conduct likelihood inference for spatio-temporal data based on a GLMM, where the random effects are correlated between time points and spatial locations.

Dreassi *et al.* (2014) extend a GLMM for a gamma distribution to a situation where a large number of observations is zero. They use a logistic model for the probability of a zero observation and specify a gamma distribution for the non-zero part of the population. Using Baysian methods, they obtain estimates and credible intervals for average grape production in small areas in Italy. We discuss the Dreassi et al. (2014) model in more detail in Section 15.4. Other models for data with large numbers of zero observations are discussed in Chandra and Sud (2012) and Pfeffermann *et al.* (2008).

Several studies apply GLMMs to estimate small area means of count data. Trevisani and Torelli (2007) explore alternative models for counts and use HB for inference. Ferrante and Trivisano (2010) apply multivariate Poisson lognormal models to estimate the number of individuals recruited by firms in domains defined by cross-classifying geographic areas with three categorical characteristics of the firms. Tzavidis *et al.* (2013) use M-quantile regression based on a Poisson distribution to obtain robust inferences for small areas.

Ghosh and Maiti (2004) consider an alternative approach based on natural exponential quadratic variance function models. Their method uses a generalized linear model for the distribution of a unit-level response conditional on an area mean and a conjugate prior for the distribution of the area mean. The Ghosh and Maiti (2004) procedure assumes that the covariates are available at the area level.

Wang and Fuller (2003) modify an EBLUP to account for a bias due to asymmetric error distributions. The procedure applies a benchmarking adjustment to a predictor based on an ordinary least squares estimator of the regression coefficients. The distributions of the estimators of the area-level variances are approximated by a multiple of a chi-squared distribution, where the degrees of freedom are estimated from the sample fourth moments.

Ferraz and Moura (2012), Fabrizi and Trivisano (2010), and Bell and Huang (2006) use HB methods to specify explicit distributions that account for skewness and outliers in the context of an area-level model. Fabrizi and Trivisano (2010) extend the Fay and Herriot (1979) procedure to a model in which the area random effects have a skewed exponential power distribution. Ferraz and Moura (2012) specify a skew-normal distribution for the sampling errors. Bell and Huang (2006) evaluate the impacts of an assumption of a t distribution for the area random effects or sampling errors.

A log transformation may appropriately convert the data to approximate normality. Nandram and Choi (2010) specify a unit-level lognormal model for BMI data and use HB methods for inference. They select a log transformation from a class of Box–Cox transformations using a minimum posterior predictive loss approach. Their model accounts for non-ignorable nonresponse and an informative sample design.

Slud and Maiti (2006) consider area-level models for log-transformed data. Using the moments of the lognormal distribution, they obtain a closed-form expression for an approximately unbiased MSE estimator. They apply the method to obtain predictors of the proportion of children in poverty using data from the US Census Bureau Small Area Income and Poverty Estimation project. The lognormal model of Slud and Maiti (2006) may be viewed as an area-level counterpart to the unit-level lognormal model defined in Section 15.3, and the predictor that Slud and Maiti (2006) suggest is closely related to the empirical Bayes predictor described in section 15.3.4.

15.3 Small Area Predictors for a Unit-Level Lognormal Model

15.3.1 The Linear Unit-Level Mixed Model

In preparation for our study of prediction for the lognormal model, we describe three procedures for the linear model. See Berg and Chandra (2012) for a similar discussion. In particular, we focus on the linear unit-level mixed model of Battese, Harter, and Fuller (1988),

$$y_{jd} = \mathbf{x}_{jd}\boldsymbol{\beta} + u_d + e_{jd}, \tag{15.1}$$

where y_{jd} is the response for unit j in area d, $\mathbf{x}_{jd}$ is the corresponding vector of covariates, and $(u_d, e_{jd}) \sim \mathrm{N}(\mathbf{0}, \mathrm{diag}(\sigma_v^2, \sigma_\epsilon^2))$ are the model random effects. The parameters σ_v^2 and σ_ϵ^2 are typically referred to as the variance components of this mixed model. The quantity of interest is the population mean of y_{jd} in area d, defined

$$\bar{y}_{N_d} = \bar{\mathbf{x}}'_{N_d}\boldsymbol{\beta} + u_d + \bar{e}_{N_d},$$

where $\bar{\mathbf{x}}_{N_d}$ is the mean of $\mathbf{x}_{jd}$ for the population of units in area d, and N_d is the number of units in area d.

Three types of predictors have been suggested for this setting. A regression synthetic estimator for the linear mixed model (15.1) is of the form,

$$\widehat{y}_d^{syn} = N_d^{-1}\left\{\sum_{j=1}^{n_d} y_{jd} + \sum_{j=n_d+1}^{N_d} \mathbf{x}'_{jd}\widehat{\boldsymbol{\beta}}_{wls}\right\}, \tag{15.2}$$

where $\widehat{\boldsymbol{\beta}}_{wls}$ is a weighted least squares estimator of $\boldsymbol{\beta}$ under (15.1), $j = 1, \ldots, n_d$ is the index for the sampled units, and $j = n_d + 1, \ldots, N_d$ is the index for the nonsampled units for area d. In the synthetic estimator (15.2), a nonsampled unit y_{jd} in $\bar{y}_{N_d}$ is replaced with an estimator of $E[y_{jd}]$. Rao (2003, chapter 4) discusses several other synthetic estimators.

Chandra and Chambers (2009) propose a class of model-based direct estimators (MBDEs). They begin with the linear mixed model (15.1) and derive the optimal linear predictor of the population total $\sum_{d=1}^{D} N_d\bar{y}_{N_d}$. The MBDE is a weighted sum of the sample units in area d,

$$\widehat{y}_d^{MBDE} = N_d^{-1}\sum_{j=1}^{n_d} w_{jd}y_{jd}. \tag{15.3}$$

The weight w_{jd} in (15.3) is such that $\sum_{d=1}^{D} N_d\widehat{y}_d^{MBDE}$ is the EBLUP of the population total under (15.1). Chandra and Chambers (2009) and Chambers, Chandra, and Tzavidis (2011) discuss MSE estimation for the MBDE (15.3).

The EBLUP of the area d mean is

$$\widehat{y}_d^{EBLUP} = N_d^{-1}\left\{\sum_{j=1}^{n_d} y_{jd} + \sum_{j=n_d+1}^{N_d}\left[\mathbf{x}'_{jd}\widehat{\boldsymbol{\beta}}_{wls} + \widehat{\gamma}_d\left(\bar{y}_{s_d} - \bar{\mathbf{x}}'_{s_d}\widehat{\boldsymbol{\beta}}_{wls}\right)\right]\right\}, \tag{15.4}$$

where $\widehat{\gamma}_d = (\widehat{\sigma}_v^2 + n_d^{-1}\widehat{\sigma}_\epsilon^2)^{-1}\widehat{\sigma}_v^2$, $(\bar{y}_{s_d}, \bar{\mathbf{x}}'_{s_d}) = n_d^{-1}\sum_{j=1}^{n_d}(y_{jd}, \mathbf{x}'_{jd})$, and $(\widehat{\boldsymbol{\beta}}'_{wls}, \widehat{\sigma}_v^2, \widehat{\sigma}_\epsilon^2)$ is the vector of estimators of $(\boldsymbol{\beta}', \sigma_v^2, \sigma_\epsilon^2)'$. The EBLUP (15.4) is equal to the sum of the synthetic estimator (15.2) and the area-specific term $(1 - N_d^{-1}n_d)\widehat{\gamma}_d(\bar{y}_{s_d} - \bar{\mathbf{x}}'_{s_d}\widehat{\boldsymbol{\beta}}_{wls})$. The $\widehat{\gamma}_d$ depends on the relative magnitudes of estimates of σ_v^2 and σ_ϵ^2. The EBLUP approaches the synthetic estimator as the ratio of the estimated between-areas variance, $\widehat{\sigma}_v^2$, to the estimated within-areas variance, $\widehat{\sigma}_\epsilon^2$, decreases or as the sample size decreases. If $\widehat{\gamma}_d = 1$, the EBLUP is equal to the survey regression estimator, $\bar{y}_{s_d} + (\bar{\mathbf{x}}_{N_d} - \bar{\mathbf{x}}_{s_d})'\widehat{\boldsymbol{\beta}}_{wls}$. Prasad and Rao (1990) and Rao (2003, p. 139) define estimators of the MSE of the EBLUP.

When the assumptions of the linear model (15.1) are not satisfied, linear predictors can have large MSEs. We consider an alternative framework that may be more appropriate for skewed response variables, such as employment and income, which are important for analyses of poverty. We assume that the units in the population have lognormal distributions and write the loglinear mixed model for the variable of interest, y_{jd}, as

$$\log(y_{jd}) := l_{jd} = \beta_0 + z'_{jd}\boldsymbol{\beta}_1 + u_d + e_{jd}, \tag{15.5}$$

where $(u_d, e_{jd}) \sim \mathrm{N}(\mathbf{0}, \mathrm{diag}(\sigma_u^2, \sigma_e^2))$, and z_{jd} is a vector of covariates. Assume that z_{jd} is available for the population of N_d values in area d, and let $(\mathbf{y}, \mathbf{z}) = \{y_{jd}; d = 1, \ldots, D, j \in s_d\} \cup \{z_{jd} : d = 1, \ldots, D, j \in U_d\}$ be the available data. Let $\widehat{\boldsymbol{\theta}} = (\widehat{\beta}_0, \widehat{\boldsymbol{\beta}}'_1, \widehat{\sigma}_u^2, \widehat{\sigma}_e^2)'$ be an estimator of the parameter vector $\boldsymbol{\theta} = (\beta_0, \boldsymbol{\beta}'_1, \sigma_u^2, \sigma_e^2)'$. Methods for estimating $\boldsymbol{\theta}$ include maximum likelihood (ML), restricted maximum likelihood (REML), and the method of moments (MOM). We define three predictors of the population mean,

$$\bar{y}_{N_d} = N_d^{-1}\sum_{j=1}^{N_d} y_{jd}. \tag{15.6}$$

15.3.2 A Synthetic Estimator

A synthetic estimator for the area mean under the lognormal model is

$$\widehat{\bar{y}}_{N_d}^{karlberg} = f_d\bar{y}_{n_d} + (1 - f_d)\left(\frac{1}{N_d - n_d}\right)\left(\sum_{j\in\bar{s}_d}\widehat{y}_{jd}^{kb}\right), \tag{15.7}$$

where

$$\widehat{y}_{jd}^{kb} = (\widehat{c}_{jd}^{karlberg})^{-1}E[y_{jd} \mid \widehat{\boldsymbol{\theta}}, z_{jd}], \tag{15.8}$$

$$E[y_{jd} \mid \widehat{\boldsymbol{\theta}}, z_{jd}] = \exp\{\widehat{\beta}_0 + z'_{jd}\widehat{\boldsymbol{\beta}}_1 + 0.5(\widehat{\sigma}_u^2 + \widehat{\sigma}_e^2)\}, \tag{15.9}$$

and

$$\widehat{c}_{jd}^{karlberg} = \exp\left[0.5\left\{\left(1, z'_{jd}\right)\widehat{V}\{(\widehat{\beta}_0, \widehat{\boldsymbol{\beta}}'_1)'\}(1, z'_{jd})' + 0.25\widehat{V}\{(\widehat{\sigma}_u^2 + \widehat{\sigma}_e^2)\}\right\}\right]. \tag{15.10}$$

The $\widehat{V}\{(\widehat{\beta}_0, \widehat{\boldsymbol{\beta}}'_1)'\}$ and $\widehat{V}\{(\widehat{\sigma}_u^2 + \widehat{\sigma}_e^2)\}$ in (15.10) are estimates of the variances of $(\widehat{\beta}_0, \widehat{\boldsymbol{\beta}}'_1)'$ and $\widehat{\sigma}_u^2 + \widehat{\sigma}_e^2$, respectively. Estimated variances for ML and REML estimators are obtained from the inverse information matrix. (See Rao, 2003, p. 139.) The form for the conditional mean of y_{jd} given $\hat{\theta}$ and z_{jd} is based on the moment generating function of a normal distribution, and $\hat{c}_{jd}^{karlberg}$ is a correction for back transformation bias.

The MSE of the Karlberg-type predictor is

$$E[(\hat{\bar{y}}_{N_d}^{karlberg} - \bar{y}_{N_d})^2] = E\left\{\frac{(1-f_d)^2}{(N_d-n_d)^2}\left(\sum_{j\in\bar{s}_d}\hat{y}_{jd}^{kb} - y_{jd}^{kb} + y_{jd}^{kb} - y_{jd}\right)^2\right\}$$

$$= E\left[(1-f_d)^2(N_d-n_d)^{-2}\left\{\sum_{j\in\bar{s}_d} d_{jd}^2 + \sum_{j\neq k\in\bar{s}_d} d_{jd}d_{kd}\right\}\right], \quad (15.11)$$

where $d_{jd} = \hat{y}_{jd}^{kb} - y_{jd}^{kb} + y_{jd}^{kb} - y_{jd}$, $\hat{y}_{jd}^{kb}$ is defined in (15.8), $y_{jd}^{kb} = E[y_{jd} \mid \theta, z_{jd}]$, and $E[y_{jd} \mid \theta, z_{jd}]$ is given by (15.9) with θ substituted for $\hat{\theta}$. Note that

$$E[d_{jd}^2] = V\{y_{jd}\} + E[(\hat{y}_{jd}^{kb} - y_{jd}^{kb})^2] + 2E[(\hat{y}_{jd}^{kb} - y_{jd}^{kb})(y_{jd}^{kb} - y_{jd})], \quad (15.12)$$

and

$$E[d_{jd}d_{kd}] = Cov\{y_{jd}, y_{kd}\} + E[(\hat{y}_{jd}^{kb} - y_{jd}^{kb})(\hat{y}_{kd}^{kb} - y_{kd}^{kb})] +$$
$$+E[(\hat{y}_{jd}^{kb} - y_{jd}^{kb})(y_{kd}^{kb} - y_{kd})] + E[(\hat{y}_{kd}^{kb} - y_{kd}^{kb})(y_{jd}^{kb} - y_{jd})], \quad (15.13)$$

where

$$V\{y_{jd}\} = \exp(2\beta_0 + 2z'_{jd}\beta_1 + \sigma_u^2 + \sigma_e^2)[\exp(\sigma_u^2 + \sigma_e^2) - 1] \quad (15.14)$$

and

$$C\{y_{jd}, y_{kd}\} = \exp(2\beta_0 + (z'_{jd} + z'_{kd})\beta_1 + \sigma_u^2 + \sigma_e^2)[\exp(\sigma_u^2) - 1]. \quad (15.15)$$

By Taylor's theorem, approximations for $E[(\hat{y}_{jd}^{kb} - y_{jd}^{kb})^2]$ and $E[(\hat{y}_{jd}^{kb} - y_{jd}^{kb})(\hat{y}_{kd}^{kb} - y_{kd}^{kb})]$ are

$$E[(\hat{y}_{jd}^{kb} - y_{jd}^{kb})^2] \approx \mathbf{g}'_{jd}(\theta)V\{\hat{\theta}\}\mathbf{g}_{jd}(\theta) \quad (15.16)$$

and

$$E[(\hat{y}_{jd}^{kb} - y_{jd}^{kb})(\hat{y}_{jd}^{kb} - y_{jd}^{kb})] \approx \mathbf{g}'_{jd}(\theta)V\{\hat{\theta}\}\mathbf{g}_{kd}(\theta), \quad (15.17)$$

where $\mathbf{g}_{jd}(\theta) = y_{jd}^{kb}(1, z'_{jd}, 0.5, 0.5)'$, and $V\{\hat{\theta}\}$ is the variance of $\hat{\theta}$. The effects on the MSE of the last term in (15.12) and the last two terms in the second line of (15.13) are of order smaller than D^{-1}, where D is the number of areas. An estimator of the MSE of the Karlberg-type predictor for area d is therefore

$$\widehat{MSE}_d^{kb} = (1-f_d)^2(N_d-n_d)^{-2}\left\{\sum_{j\in\bar{s}_d}\hat{d}_{2jd} + \sum_{j\neq k\in\bar{s}_d}\hat{d}_{2jkd}\right\}, \quad (15.18)$$

where

$$\widehat{d}_{2jd} = \exp(2\widehat{\beta}_0 + 2z'_{jd}\widehat{\beta}_1 + \widehat{\sigma}_u^2 + \widehat{\sigma}_e^2)[\exp(\widehat{\sigma}_u^2 + \widehat{\sigma}_e^2) - 1]$$
$$+(\widehat{y}_{jd}^{kb})^2(1, z'_{jd}, 0.5, 0.5)\widehat{V}\{\widehat{\theta}\}(1, z'_{jd}, 0.5, 0.5)' \quad (15.19)$$

and

$$\widehat{d}_{2jkd} = \exp(2\widehat{\beta}_0 + (z_{jd} + z_{kd})'\widehat{\beta}_1 + \widehat{\sigma}_u^2 + \widehat{\sigma}_e^2)[\exp(\widehat{\sigma}_u^2) - 1]$$
$$+(\widehat{y}_{jd}^{kb})(1, z'_{jd}, 0.5, 0.5)\widehat{V}\{\widehat{\theta}\}(1, z'_{kd}, 0.5, 0.5)'(\widehat{y}_{kd}^{kb}). \quad (15.20)$$

15.3.3 A Model-Based Direct Estimator

Chandra and Chambers (2011) derive a nonlinear version of the predictor (15.3) based on the first two moments of y_{jd}. The Chandra and Chambers (2011) predictor is

$$\widehat{\overline{y}}_{N_d}^{TrMBDE} = N_d^{-1} \sum_{j \in s_d} \hat{w}_{jd}^{Tr} y_{jd}, \quad (15.21)$$

which has the same form as (15.3), where the weight $\hat{w}_{jd}^{Tr}$ is based on the moments of the lognormal model (15.5) and is defined below. Chandra and Chambers (2011) discuss MSE estimation for the transformed MBDE (TrMBDE) predictor. Also see Salvati *et al.* (2012) and Chambers, Chandra and Tzavidis (2011).

To derive the weight $\hat{w}_{jd}^{Tr}$, Chandra and Chambers (2011) use a linear 'fitted value' model approximation,

$$E[\mathbf{y}_N] \approx \widehat{\boldsymbol{M}}_U \boldsymbol{\gamma} \quad (15.22)$$

and

$$C\{\mathbf{y}_N, \mathbf{y}'_N\} \approx \begin{pmatrix} \widehat{\boldsymbol{V}}_{ss} & \widehat{\boldsymbol{V}}_{ss} \\ \widehat{\boldsymbol{V}}_{sr} & \widehat{\boldsymbol{V}}_{rr} \end{pmatrix}, \quad (15.23)$$

where $\boldsymbol{\gamma}$ is a vector of unknown parameters and

$$\widehat{\boldsymbol{M}}_U = (\widehat{\boldsymbol{M}}'_s, \widehat{\boldsymbol{M}}'_r)' = ((\mathbf{1}'_s, \mathbf{1}'_r)', (\widehat{\mathbf{y}}_s^{karlberg})', (\widehat{\mathbf{y}}_r^{karlberg})').$$

Here $\mathbf{y}_N = (\mathbf{y}'_s, \mathbf{y}'_r)'$, $\mathbf{y}_s$ and $\mathbf{y}_r$ are the vectors of sampled and nonsampled units, respectively, and $\widehat{\mathbf{y}}_s^{karlberg}$ and $\widehat{\mathbf{y}}_r^{karlberg}$ are the vectors containing $\widehat{y}_{jd}^{kb}$ for the sampled and nonsampled units, respectively. The elements of the covariance matrices $\widehat{\boldsymbol{V}}_{ss}$ and $\widehat{\boldsymbol{V}}_{sr}$ are estimates of the variances and covariances defined in (15.14) and (15.15). If one treats the $\hat{y}_{jd}$, $\widehat{\sigma}_u^2$, and $\widehat{\sigma}_e^2$ as fixed, then (15.22) defines a linear model for the mean of y_{jd}, and under this 'fitted value' model the best linear unbiased prediction (BLUP) of the population mean $N^{-1}\sum_{d=1}^{D}\sum_{j=1}^{N_d} y_{jd}$ is $N^{-1}(\widehat{\boldsymbol{w}}^{Tr})'\mathbf{y}_s$, where

$$\widehat{\boldsymbol{w}}^{Tr} = \mathbf{1}_s + \widehat{\boldsymbol{H}}'_s(\widehat{\boldsymbol{M}}'_U\mathbf{1}_U - \widehat{\boldsymbol{M}}'_s\mathbf{1}_s) + (\mathbf{I}_s - \widehat{\boldsymbol{H}}'_s\widehat{\boldsymbol{M}}'_s)\widehat{\boldsymbol{V}}_{ss}^{-1}\widehat{\boldsymbol{V}}_{sr}\mathbf{1}_r$$

and

$$\widehat{\boldsymbol{H}}_s = (\widehat{\boldsymbol{M}}_s'\widehat{\boldsymbol{V}}_{ss}^{-1}\widehat{\boldsymbol{M}}_s)^{-1}\widehat{\boldsymbol{M}}_s'\widehat{\boldsymbol{V}}_{ss}^{-1}.$$

The $\hat{w}_{jd}^{Tr}$ of (15.21) is the element of $\widehat{\boldsymbol{w}}^{Tr}$ associated with unit (j, d). The 'Tr' in the label TrMBDE distinguishes the estimator (15.21) from the MBDE that assumes a linear model for the population.

15.3.4 An Empirical Bayes Predictor

In general, a minimum mean squared error (MMSE) predictor of $\bar{y}_{N_d}$ is $E[\bar{y}_{N_d} \mid (\boldsymbol{y}, \boldsymbol{z})]$. (See, for example, Rao 2003, chapter 9.) An expression for this MMSE predictor under the lognormal model (15.5) is

$$\tilde{\bar{y}}_{N_d}^{MMSE}(\boldsymbol{\theta}) = \frac{1}{N_d}\left\{\sum_{j\in s_d} y_{ij} + \sum_{j\in \bar{s}_d} y_{jd}^{MMSE}(\boldsymbol{\theta})\right\}, \tag{15.24}$$

where

$$y_{jd}^{MMSE}(\boldsymbol{\theta}) = \exp\{\beta_0 + \boldsymbol{z}_{jd}'\boldsymbol{\beta}_1 + \gamma_d(\bar{l}_{s_d} - \beta_0 - \bar{\boldsymbol{z}}_{s_d}'\boldsymbol{\beta}_1) + 0.5\sigma_e^2(\gamma_d n_d^{-1} + 1)\}, \tag{15.25}$$

$(\bar{l}_{s_d}, \bar{\boldsymbol{z}}_{s_d}') = n_d^{-1}\sum_{j\in s_d}(l_{jd}, \boldsymbol{z}_{jd}')$, and $\gamma_d = \sigma_u^2(\sigma_u^2 + n_d^{-1}\sigma_e^2)^{-1}$. An expression for the MSE of (15.24) is given in Appendix 15.A. See Berg and Chandra (2015) for a derivation of (15.24) and of the MSE. This MMSE predictor is a function of the true $\boldsymbol{\theta}$. We replace the true $\boldsymbol{\theta}$ in (15.24) with the REML estimator to obtain the EB predictor,

$$\hat{\bar{y}}_{N_d}^{EB} = \tilde{\bar{y}}_{N_d}^{MMSE}(\widehat{\boldsymbol{\theta}}) = \frac{1}{N_d}\left\{\sum_{j\in s_d} y_{jd} + \sum_{j\in \bar{s}_d} \hat{y}_{jd}^{EB}\right\}, \tag{15.26}$$

where $\hat{y}_{jd}^{EB} = \hat{y}_{jd}^{MMSE}(\widehat{\boldsymbol{\theta}})$.

Berg and Chandra (2015) use a linear approximation to obtain an estimator of the MSE of the EB predictor (15.26). The MSE estimator defined in Berg and Chandra (2015) is used for the simulation studies in Section 15.4. An alternative MSE estimator based on the parametric bootstrap is proposed in Appendix 15.A.

15.3.4.1 Estimation of the Cumulative Distribution Function

In some applications, an estimate of the cumulative distribution function (CDF) for an area is of interest. One important example in allocating aid in developing countries is the poverty incidence, defined as the proportion of households with income below a pre-specified threshhold (Haslett and Jones, 2010). A closed form expression for the MMSE predictor of this CDF can be obtained under the assumptions of the lognormal distribution.

The CDF defined by the values y_{jd} associated with area d is

$$F_{N_d}(t) = N_d^{-1}\sum_{j=1}^{N_d} I[y_{jd} \le t], \tag{15.27}$$

where $I[\cdot]$ is an indicator function. The MMSE predictor of $F_{N_d}(t)$ is

$$E[F_{N_d}(t) \mid (y,z)] = N_d^{-1}\left\{\sum_{j\in s_d} I\left[y_{jd} \le t\right] + \sum_{j\notin s_d} E[I[y_{jd} \le t] \mid (y,z)]\right\}. \tag{15.28}$$

For $t > 0$, the conditional expectation for a nonsampled unit is

$$\begin{aligned} E[I[y_{jd} \le t] \mid (y,z)] &= P(y_{jd} \le t \mid (y,z)) \\ &= P(\log(y_{jd}) \le \log(t) \mid (y,z)) = \Phi\left(\frac{\log(t) - \tilde{\mu}_{jd}}{\tilde{\sigma}_{jd}}\right), \end{aligned} \tag{15.29}$$

where $\Phi(\cdot)$ is the CDF of a standard normal distribution,

$$\tilde{\mu}_{jd} = \beta_0 + z'_{jd}\beta_1 + \gamma_d(\bar{\ell}_{s_d} - \beta_0 - \bar{z}'_{s_d}\beta_1), \tag{15.30}$$

and

$$\tilde{\sigma}^2_{jd} = \gamma_d\sigma_e^2 n_d^{-1} + \sigma_e^2. \tag{15.31}$$

We therefore define an EB predictor of $F_{N_d}(t)$ by

$$\widehat{E}[F_{N_d}(t) \mid (y,z)] = N_d^{-1}\left\{\sum_{j\in s_d} I\left[y_{jd} \le t\right] + \sum_{j\notin s_d} \Phi\left(\frac{\log(t) - \widehat{\mu}_{jd}}{\widehat{\sigma}_{jd}}\right)\right\}, \tag{15.32}$$

where $\widehat{\mu}_{jd}$ and $\widehat{\sigma}_{jd}$ are obtained by replacing the unknown parameters in (15.30) and (15.31) with estimators. Note that (15.32) is a parametric version of the semi-parametric CDF predictor of Chambers and Dunstan (1986). Prediction of finite population CDFs is discussed in more detail in chapter 16 of Chambers and Clark (2012).

15.4 Simulations

We address three main questions through simulation. First, we compare the properties of the predictors and MSE estimators discussed in Section 15.3. Secondly, we summarize simulations examining the bias and robustness of the EB predictor (15.26). Thirdly, we compare this EB predictor to a predictor based on a gamma distribution.

15.4.1 Comparison of Synthetic, TrMBDE, and EB Predictors

First, we consider a model-based simulation study with data generated from the lognormal model (15.5) with two parameter configurations. For both parameter sets,

$$(\mu_z, \beta_0, \beta_1, \sigma_z) = (3.253, -1.62, 0.9, 1.58) \text{ where } z_{jd}\tilde{}N(\mu_z, \sigma_z).$$

For parameter set 1, $(\sigma_u^2, \sigma_e^2) = (0.30, 0.60)$, and for parameter set 2, $(\sigma_u^2, \sigma_e^2) = (0.12, 0.78)$. These parameters give distributions that approximate the mean and variance of the number of chickens per segment in a 1960s United States Department of Agriculture (USDA) area

Table 15.1 Population and sample sizes for each of 10 areas

Area	1	2	3	4	5	6	7	8	9	10
N_d	81	81	161	161	323	323	645	645	1290	1290
n_d	5	5	10	10	20	20	40	40	80	80

survey discussed in Fuller (1991). The configuration of areas and sample sizes in Table 15.1 is based loosely on data for the Boone/Raccoon River Watershed from the (USDA) Conservation Effects Assessment Project (Goebel, 2009). We compare the Monte Carlo MSEs of the EB predictor (15.26) to the synthetic estimator (15.7) and the MBDE (15.21) and evaluate the performance of their associated MSE estimators.

The relative Monte Carlo MSE of predictor $\widehat{\bar{y}}_{N_d}$ for area d is defined as

$$\text{RelMSE}_d = \frac{MSE_{MC}(\widehat{\bar{y}}_{N_d})}{MSE_{MC}(\widehat{\bar{y}}^{EB}_{N_d})}. \tag{15.33}$$

Table 15.2 contains averages of RelMSE_d across areas with the same sample size, with $\widehat{\bar{y}}_{N_d}$ taken to be the Karlberg predictor (15.7) and the TrMBDE predictor (15.21). The relative MSE of the TrMBDE predictor decreases as the sample size increases, and for each sample size, the relative MSE of the TrMBDE predictor is smaller for $\sigma_u^2\sigma_e^{-2} = 0.5$ (parameter set 1) than for $\sigma_u^2\sigma_e^{-2} = 0.15$ (parameter set 2). The behaviour of the relative MSEs of the Karlberg predictor go in the opposite direction to those of the TrMBDE predictor. The relative MSE for the Karlberg predictor increases as the sample size increases for each parameter set, and for a given sample size, the relative MSE of the Karlberg predictor for parameter set 2 is smaller than the corresponding relative MSE for parameter set 1.

As discussed in Berg and Chandra (2015), the patterns in the relative MSEs can be understood through an analogy with the linear unit-level model (15.1). The ratio of the MSE of a mixed-model predictor to the MSE of the sample mean for a linear, unit-level model is approximately $\sigma_v^2(\sigma_v^2 + \sigma_\epsilon^2 n_d^{-1})^{-1}$. Because the TrMBDE predictor has the form of a direct estimator, an analogy with the linear model suggests that the ratio of the MSE of the TrMBDE predictor to the MSE of the EB predictor would decrease as the sample size increases and the ratio $\sigma_v^2\sigma_\epsilon^{-2}$

Table 15.2 Average ratios of Monte Carlo MSEs of Karlberg and TrMBDE predictors to Monte Carlo MSE of the EB predictor for two parameter configurations and five sample sizes

n_d	Parameter set 1 ($\sigma_u^2/\sigma_e^2 = 0.5$)		Parameter set 2 ($\sigma_u^2/\sigma_e^2 = 0.15$)	
	Karlberg	TrMBDE	Karlberg	TrMBDE
5	1.97	1.72	1.15	2.10
10	3.04	1.42	1.42	1.89
20	4.97	1.39	1.99	1.51
40	8.93	1.35	3.03	1.44
80	13.47	1.21	4.84	1.31

Table 15.3 Average empirical coverage of normal theory 95% prediction intervals for two parameter configurations and five sample sizes

n_d	Parameter set 1 ($\sigma_u^2/\sigma_e^2 = 0.5$)			Parameter set 2 ($\sigma_u^2/\sigma_e^2 = 0.15$)		
	Karlberg	TrMBDE	EB	Karlberg	TrMBDE	EB
5	0.95	0.91	0.95	0.94	0.90	0.94
10	0.96	0.92	0.95	0.95	0.90	0.95
20	0.96	0.93	0.95	0.95	0.90	0.94
40	0.96	0.93	0.95	0.96	0.91	0.95
80	0.97	0.92	0.95	0.96	0.90	0.95

increases. Conversely, the ratio of the MSE of a linear model BLUP to the MSE of a synthetic estimator is approximately $\sigma_\epsilon^2 n_d^{-1}(\sigma_v^2 + \sigma_\epsilon^2 n_d^{-1})^{-1}$. Because the Karlberg predictor is a type of synthetic estimator, patterns in the relative MSEs for the Karlberg predictor are consistent with expectations based on the analogy to prediction under a linear mixed model.

We construct normal theory 95% prediction intervals for the three predictors (Table 15.3). Although the distributions of the prediction errors are not normal, normal theory confidence intervals are of interest because of their simplicity for practical situations. The empirical coverages of the Karlberg-type predictor and the EB predictor are close to the nominal 95% level. The empirical coverages of the TrMBDE predictor are between 90% and 93%. Because sample sizes are small and the MSE estimators are based on approximations, bias in the confidence interval coverages is expected.

We evaluate the design properties of the model-based estimates through a design-based simulation, using the same finite population described in Chandra and Chambers (2011). The synthetic finite population of 81982 farms is constructed by sampling with replacement from the sample of 1652 farms that participated in the Australian Agricultural Grazing Industries Survey (AAGIS), with probability proportional to the sample weights of the original sample. The variable of interest (y) is annual farm costs, and the auxiliary variable (z) is the log of the farm area. For the Monte Carlo study, 1000 stratified random samples of size 1652 are selected, where the strata are 29 agricultural regions and the strata sample sizes range from 6 to 117. These agricultural regions are taken to be the small areas of interest in this study and mean predictors are constructed for each of them. These predictors, generically denoted as $\hat{\bar{y}}_{N_d}$, are evaluated on the basis of their Monte Carlo relative design bias (RDB) and relative design root mean squared error (RDRMSE) defined by

$$\text{RDB}_d = \frac{E_{MC}[\hat{\bar{y}}_{N_d} - \bar{y}_{N_d}]}{\bar{y}_{N_d}} \quad \text{and} \quad \text{RDRMSE}_d = \frac{\sqrt{E_{MC}[(\hat{\bar{y}}_{N_d} - \bar{y}_{N_d})^2]}}{\bar{y}_{N_d}}, \tag{15.34}$$

respectively, where $\bar{y}_{N_d}$ is the finite population mean for area d. Table 15.4 shows the average values of RDB and RDRMSE, where the average is across the 29 agricultural regions.

Of the three predictors (TrMBDE, Karlberg and EB), the Karlberg predictor has the largest average RDB because the Karlberg predictor does not account for the area effects, and the variance of the area random effects (σ_u^2) is significant for this population. The biases for the

Table 15.4 Average RDB and average RDRMSE values for predictors based on the AAGIS data

Measure	TrMBDE	Karlberg	EB
RDB	2.81	5.08	−1.14
RDRMSE	24.31	22.51	13.10

A bias correction based on a Taylor expansion was applied to the EB predictor for the design-based simulations. The bias correction is described in Berg and Chandra (2015) and discussed briefly in Section 15.4.2. The result of the design-based simulation for the EB predictor without bias correction is similar.

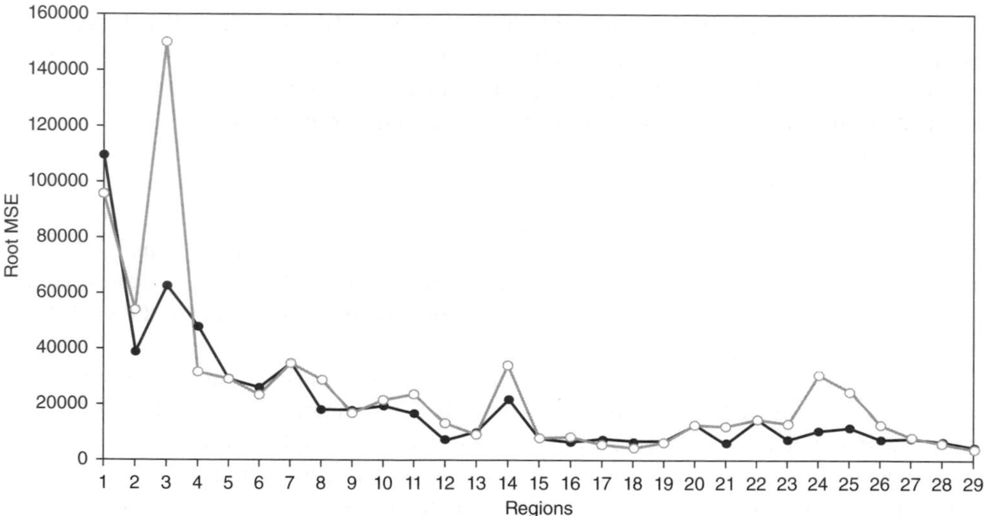

Figure 15.1 Black line: square root of the Monte Carlo mean of the MSE estimator for the EB predictor. Grey line: square root of the Monte Carlo MSE of the EB predictor

Karlberg and EB predictors are negligible relative to the MSE for this finite population. The EB predictor has a significantly smaller average RDRMSE than the other predictors.

To examine the properties of the EB predictor for the synthetic population more closely, Figure 15.1 displays the Monte Carlo design RMSE of the EB predictor and the square root of the Monte Carlo mean of the MSE estimator for the EB predictor for each area. The relatively large value of the Monte Carlo design RMSE for area 3 is the consequence of an extreme value in the finite population. For most of the areas, the average of the Monte Carlo mean of the MSE estimator is somewhat smaller than the corresponding MSE of the EB predictor. The relatively large Monte Carlo bias of the MSE estimator for areas 3, 14, and 24 is due to outliers in the population for these areas. These outliers have a larger impact on the empirical MSE than on the Monte Carlo mean of the estimated MSE. This result suggests a need for effective methods for outliers, discussed in Section 15.5. Overall, however, it appears that for this population the MSE estimator suggested by Berg and Chandra (2015) for the EB predictor (15.26) is a reasonable approximation for its design MSE.

15.4.2 Bias and Robustness of the EB Predictor

Because the EB predictor is a nonlinear function of estimators of parameters, it is biased. Berg and Chandra (2015) examine the nature of the bias through simulation and propose two bias corrected estimators. One of the bias corrected estimators is based on a linear approximation and the other bias correction uses raking. In the simulations, the Monte Carlo bias of the EB predictor is less than 3% of the Monte Carlo MSE. The bias corrections reduce the bias of the EB predictor, with negligible increase in MSE.

Berg and Chandra (2015) also examine the robustness of the EB predictor to two modest departures from the assumptions of the lognormal model (15.5). For the first, a squared term is included in the expectation function so that the simulated model has the form

$$\log(y_{jd}) = \beta_0 + \beta_1 z_{jd} + \beta_2 z_{jd}^2 + u_d + e_{jd}. \tag{15.35}$$

For the second, the data generating model is

$$y_{ij} = (\beta_0 + \beta_1 z_{jd} + u_d + e_{jd})^2, \tag{15.36}$$

such that a square root transformation is the correct transformation. For both simulation models, the estimation model is the lognormal model (15.5) with a univariate z_{jd}. The Monte Carlo MSE of the EB predictor is smaller than the Monte Carlo MSE of the Karlberg and TrMBDE predictors for the specific forms of model misspecification considered. For the simulation where the square root transformation is the correct transformation, the empirical coverages of normal theory 95% prediction intervals for the EB predictor are larger than the nominal level. For the simulation where the squared term is included in the expectation function, the empirical coverages of normal theory 95% prediction intervals are between 94% and 96%.

To evaluate whether the EB predictor is resistant to outliers, a simulation study was conducted in which the e_{jd} are generated from a mixture of two normal distributions. More specifically, $e_{jd} \sim p\text{N}(0, \sigma_{e,1}^2) + (1-p)\text{N}(0, \sigma_{e,2}^2)$, where $\sigma_{e,2}^2 = 10\sigma_{e,1}^2$ and $p = 0.1$. The EB predictor is more efficient than the synthetic or TrMBDE predictors for this simulation study, but the empirical coverage of normal theory 95% prediction intervals for the EB predictor is substantially lower than the nominal level. The Monte Carlo mean of the estimates of σ_e^2 based on the mixture model is approximately $p\sigma_{e,1}^2 + (1-p)\sigma_{e,2}^2$. The simulation suggests that the lognormal predictor is not resistant to outliers on the log scale.

15.4.3 Comparison of Lognormal and Gamma Distributions

We compare predictors based on a gamma model to predictors based on a lognormal model through simulation. For the gamma model, we assume

$$f(y_{jd} \mid \boldsymbol{\delta}) = \frac{(\mu_{jd}\nu^{-1})^{\nu} y_{jd}^{\nu-1} \exp(-\nu y_{jd}/\mu_{jd})}{\Gamma(\nu)},$$
$$\log(\mu_{jd}) = \beta_0 + \beta_1 z_{jd} + u_d, \tag{15.37}$$

where $z_{jd} = \log(x_{jd})$, $u_d \sim N(0, \sigma_u^2)$, and β_0, β_1, σ_u^2, and ν are fixed parameters to estimate. The vector $\boldsymbol{\delta}$ in (15.37) is the vector of fixed parameters and random effects $u_1, \ldots, u_D$. For the gamma model (15.37),

$$E_g[y_{jd} \mid \boldsymbol{\delta}] = \mu_{jd}, \tag{15.38}$$

and

$$V_g\{y_{jd} \mid \boldsymbol{\delta}\} = \frac{1}{\nu}\mu_{jd}^2. \tag{15.39}$$

To apply the gamma model (15.37) to obtain small area predictors, we use the Bayesian method described in Dreassi *et al.* (2014). For priors, we assume $(\beta_0, \beta_1) \sim N(\mathbf{0}, 10^6\mathbf{I}_2)$, where $\mathbf{I}_2$ is a two-dimensional identity matrix. We assume σ_u and $\nu^{-0.5}$ have independent half-Cauchy distributions with mean 0 and scale parameter 25. Dreassi *et al.* (2014) provide further discussion of this choice of the prior for the variance and scale parameters in the gamma model. We implement Gibbs sampling using the software JAGS. The approximation for the posterior is based on samples from three chains, each of length 1000, after a burn-in period of 1000.

To compare the two predictors, we simulate data from both the lognormal and gamma distributions. For each simulation model, we calculate the empirical Bayes predictor for the lognormal model and the Bayes predictor for the gamma model. For both simulation models, $\log(x_{jd}) \sim N(3, 1.53)$, $(\beta_0, \beta_1) = (-1.62, 0.9)$, and $\sigma_u^2 = 0.51$. For the gamma model, $\nu = 1.51$, and for the lognormal model $\sigma_e^2 = 0.6$. The area population and sample sizes are as in Table 15.1.

To give a sense of the nature of the generated data, Figure 15.2 contains plots of two generated populations. The top two plots are of $\log(y)$ vs. $\log(x)$, and the bottom two plots are of y vs. x. The plots in the left panel are based on data generated from the lognormal model, and the plots on the right panel are based on data from the gamma model. Conditional on x_{jd} and u_d, both models generate data with constant coefficients of variation. The coefficient of variation for the gamma model is $\nu^{-0.5} = 0.57$, and the coefficient of variation for the lognormal model is $\exp(\sigma_e^2) - 1 = 0.91$.

Table 15.5 and Table 15.6 contain the Monte Carlo MSEs and empirical coverages of 95% prediction intervals for both predictors based on data generated from the lognormal and gamma models, respectively. The results are averaged across areas for each sample size. As expected, the lognormal predictor is more efficient than the gamma predictor when the true underlying model is lognormal. Likewise, the gamma predictor is more efficient than the lognormal predictor when the true underlying model is gamma. The lognormal predictor gives empirical coverages close to the nominal level for data generated from the lognormal and is conservative for data generated from a gamma distribution. The empirical coverages for the gamma predictor are approximately 90% when the data are generated from the lognormal model.

To further test the lognormal EB predictor, we generated data from a gamma distribution with $\nu = 0.75$ and examined the properties of the lognormal and EB predictors. The gamma distribution with $\nu = 0.75$ generates some values that are close to zero and are transformed to relatively extreme negative values on the log scale. Because of the impact of outliers on estimates of variance components, the EB procedure for the lognormal model can lead to extreme predictions for such populations. As a consequence, the Monte Carlo MSE of the lognormal EB predictor is approximately 4–40 times the Monte Carlo MSE of the gamma predictor for

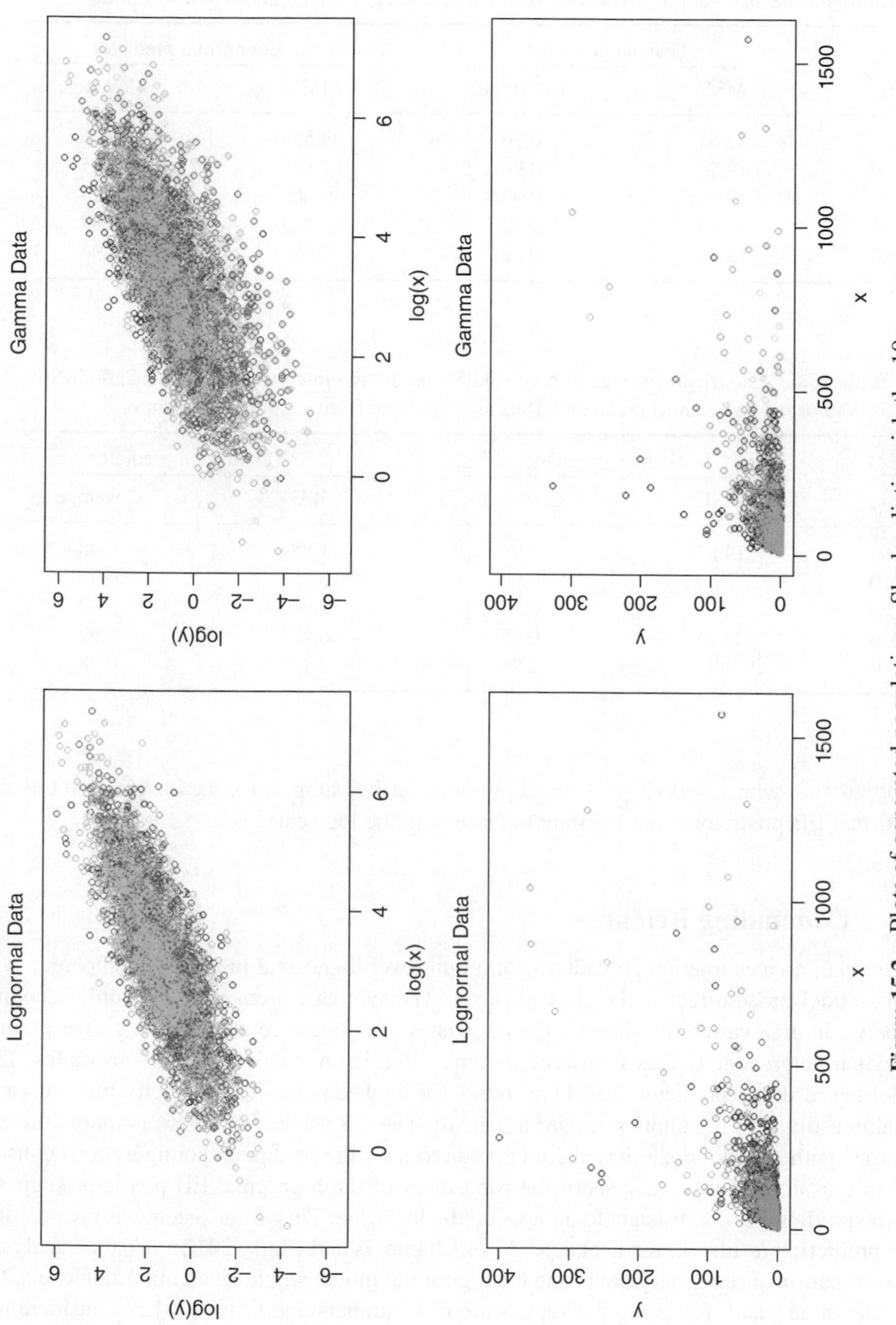

Figure 15.2 Plots of generated populations. Shades distinguish the 10 areas.

Table 15.5 Empirical coverage of nominal 95% prediction intervals and Monte Carlo MSE for lognormal and gamma predictors. Data are generated from a lognormal distribution

	Gamma predictor		Lognormal predictor	
n_d	MSE	Coverage	MSE	Coverage
5	32.31	0.90	28.55	0.96
10	19.76	0.90	17.83	0.95
20	10.41	0.90	9.00	0.95
40	5.41	0.90	4.08	0.97
80	3.37	0.89	2.89	0.96

Table 15.6 Empirical coverage of nominal 95% prediction intervals and Monte Carlo MSE for lognormal and gamma predictors. Data are generated from a gamma distribution

	Gamma predictor		Lognormal predictor	
n_d	MSE	Coverage	MSE	Coverage
5	14.81	0.94	21.88	0.98
10	9.17	0.94	13.48	0.97
20	4.80	0.94	7.58	0.97
40	3.36	0.95	6.47	0.97
80	1.29	0.96	3.13	0.98

the population with $\nu = 0.75$. This result provides further support for the observation that the lognormal EB predictor is not resistant to outliers in the log scale.

15.5 Concluding Remarks

Three small area estimation procedures for a unit-level lognormal model are compared: synthetic, model-based direct, and empirical Bayes. The synthetic predictor (15.7) only accounts for between-area variability through the covariates and therefore can lead to biased predictors when heterogeneity exists between the areas that is unexplained by the covariates. The model-based direct predictor (15.21) accounts for between-area heterogeneity but can yield unstable estimators if sample sizes are too small. The EB predictor (15.26) is more efficient than the synthetic and model-based direct predictors for the parameter configurations considered in our simulations. Analysis of the robustness of the lognormal EB predictor suggests that the predictor is not resistant to outliers on the log scale. This is consistent with population level prediction results set out in chapter 17 of Chambers and Clark (2012). In a data analysis, an examination of residuals from fitting the lognormal model would be an important check for extreme values, and, following the suggestion of Chambers and Clark (2012), consideration should be given to replacing the non-robust REML parameter estimates used in the EB predictor with outlier-robust alternatives. The MBDE (15.21) might be preferred for situations in which predictors are required to have the form of a weighted sum of sampled units. However in this case as well use of outlier robust parameter estimation is recommended.

Berg and Chandra (2015) provide an analysis of soil erosion data for which the lognormal is a reasonable approximation for the distribution of the response given covariates. An application of the proposed methodology data on income or expenditures, with the objective of gaining insight into areas with poverty, may be useful future work. A comparison to existing methods, such as the procedures discussed in Haslett and Jones (2010) or Molina and Rao (2010), would be interesting.

For real applications, the data may exhibit departures from the lognormal assumption. The lognormal predictor may be inefficient for data with outliers on the log scale because the procedures used to estimate the model parameters are not resistant to outliers. The log transformation may also convert values near zero to extreme negative values, leading to residuals with a left skew distribution.

Applying the proposed methodology to income and expenditure data, with the objective of obtaining estimates related to economic poverty, may require extensions of the lognormal predictor to accommodate data with outliers or skewed distributions on the log scale. A scale mixture of lognormal distributions may be appropriate for a situation in which unit-level residuals based on a lognormal model are symmetric but have extreme values relative to a lognormal distribution. As discussed above, use of a robust estimating equation may also be useful for such situations. A log skew-normal distribution may appropriately describe data for which the log transformation leads to extreme negative values and residuals with a left skew distribution. Further investigation of the gamma distribution for modelling such data is also an area of current work. Extensions to multivariate data are of interest.

Appendix 15.A: Mean Squared Error Estimation for the Empirical Best Predictor

The MSE of the MMSE predictor for the lognormal model can be written

$$M_{1d}(\boldsymbol{\theta}) = T_{1d}(\boldsymbol{\theta}) - T_{2d}(\boldsymbol{\theta}) + T_{3d}(\boldsymbol{\theta}) - T_{4d}(\boldsymbol{\theta}), \tag{15A.1}$$

where

$$T_{1d}(\boldsymbol{\theta}) = \left\{ \sum_{j\in\bar{s}_d} \exp\left(\beta_0 + \mathbf{z}'_{jd}\boldsymbol{\beta}_1\right) \right\}^2 \exp(2\sigma_u^2 + \sigma_e^2),$$

$$T_{2d}(\boldsymbol{\theta}) = \left\{ \sum_{j\in\bar{s}_d} \exp\left(\beta_0 + \mathbf{z}'_{jd}\boldsymbol{\beta}_1\right) \right\}^2 \exp(\gamma_d\sigma_u^2 + \sigma_u^2 + \sigma_e^2),$$

$$T_{3d}(\boldsymbol{\theta}) = \left\{ \sum_{j\in\bar{s}_d} \exp\left(2\beta_0 + 2\mathbf{z}'_{jd}\boldsymbol{\beta}_1\right) \right\} \exp(2\sigma_u^2 + 2\sigma_e^2)$$

and

$$T_{4d}(\boldsymbol{\theta}) = \left\{ \sum_{j\in\bar{s}_d} \exp\left(2\beta_0 + 2\mathbf{z}'_{jd}\boldsymbol{\beta}_1\right) \right\} \exp(2\sigma_u^2 + \sigma_e^2).$$

As discussed in Berg and Chandra (2015), the expression (15A.1) is also useful for obtaining an approximately unbiased estimator of the MSE of the EB predictor based on a Taylor approximation.

The naive estimator $M_{1d}(\widehat{\boldsymbol{\theta}})$ is biased due to the nonlinear transformation. The estimator $M_{1d}(\widehat{\boldsymbol{\theta}})$ also does not account for the effect of estimating the parameters on the MSE of the predictor. We define a parametric bootstrap estimator that accounts for the bias of the estimator of $M_{1d}(\widehat{\boldsymbol{\theta}})$ and the variability due to estimation of the parameters. The bootstrap procedure is similar to the bootstrap procedure in Zhang and Chambers (2004).

For $b = 1, \ldots, B$, complete the following steps:

1. For $d = 1, \ldots, D$, and $j = 1, \ldots, N_d$, generate $y_{jd}^{*(b)}$ from the lognormal model with parameter $\widehat{\boldsymbol{\theta}}$. Select a sample using the sampling procedure for the original sample, and let $\boldsymbol{y}_s^{(b)}$ and $\mathbf{x}_s^{(b)}$ denote the set of response variables and covariates based on bootstrap sample b.
2. Obtain a bootstrap estimate of $\boldsymbol{\theta}$ based on $\boldsymbol{y}_s^{(b)}$ and $\mathbf{x}_s^{(b)}$, and denote the bootstrap estimate of $\boldsymbol{\theta}$ by $\widehat{\boldsymbol{\theta}}^{(b)}$. Let $\widehat{\bar{y}}_{N_d}^{(b)}(\widehat{\boldsymbol{\theta}}^{(b)})$ denote the EB predictor based on bootstrap sample b. Let $\widehat{M}_{1d}^{(b)}(\widehat{\boldsymbol{\theta}}^{(b)})$ be an estimate of the leading term in the MSE based on $\boldsymbol{y}_s^{(b)}$, $\mathbf{x}_s^{(b)}$, and $\widehat{\boldsymbol{\theta}}^{(b)}$.
3. Let $\widehat{\bar{y}}_{N_d}^{(b)}(\widehat{\boldsymbol{\theta}})$ denote the EB predictor based on bootstrap sample b and the estimate of $\boldsymbol{\theta}$ based on the original sample. Let $\widehat{M}_{1d}^{(b)}(\widehat{\boldsymbol{\theta}})$ be an estimate of the leading term in the MSE based on $\boldsymbol{y}_s^{(b)}$, $\mathbf{x}_s^{(b)}$, and the original $\widehat{\boldsymbol{\theta}}$.

A bootstrap estimator of the MSE is

$$\widehat{MSE}_d^{bs} = M_{1d}(\widehat{\boldsymbol{\theta}}) \frac{\sum_{b=1}^{B} M_{1d}^{(b)}(\widehat{\boldsymbol{\theta}})}{\sum_{b=1}^{B} M_{1d}^{(b)}(\widehat{\boldsymbol{\theta}}^{(b)})} + B^{-1} \sum_{b=1}^{B} (\widehat{\bar{y}}_{N_d}^{(b)}(\widehat{\boldsymbol{\theta}}^{(b)}) - \widehat{\bar{y}}_{N_d}^{(b)}(\widehat{\boldsymbol{\theta}}))^2. \tag{15A.2}$$

The ratio applied to the first term in (15A.2) is an estimate of the bias of the estimator of the leading term. The second term in (15A.2) is an estimate of the variance due to the use of the estimator $\widehat{\boldsymbol{\theta}}$ instead of the unknown $\boldsymbol{\theta}$ to construct the EB predictor. The double bootstrap is not required because the closed-form expression (15.A.1) is available for the leading term in the MSE.

References

Battese, G.E., Harter, R.M. and Fuller, W.A. 1988 An error-components model for prediction of county crop areas using survey and satellite data. *Journal of the American Statistical Association*, **83**, 28–36.

Bell, W.R. and Huang, E.T. 2006 Using the *t*-distribution to deal with outliers in small area estimation. In *Proceedings of Statistics Canada Symposium on Methodological Issues in Measuring Population Health*. Statistics Canada, Ottawa, Canada.

Berg, E. and Chandra, H. 2012 Small area prediction for a unit level lognormal model. Federal Committee on Statistical Methodology Research Conference.

Berg, E. and Chandra, H. 2015 Small area estimation for a unit-level lognormal model. *Computational Statistics and Data Analysis*, **78**, 159–175.

Chambers, R.L. 1986 Outlier robust finite population estimation. *Journal of the American Statistical Association*, **81**, 1063–1069.

Chambers, R.L., Chandra, H. and Tzavidis, N. 2011 On bias-robust mean squared error estimation for pseudo-linear small area estimators. *Survey Methodology*, **37**, 153–170.

Chambers, R.L. and Clark, R.G. 2012 *An Introduction to Model-Based Survey Sampling with Applications*. Oxford University Press.

Chambers R.L. and Dunstan, R. 1986 *Estimating distribution functions from survey data*. *Biometrika*, **73**, 597–604.

Chandra, H. and Chambers, R. 2009 Multipurpose weighting for small area estimation. *Journal of Official Statistics*, **25**, 379–395.

Chandra, H. and Chambers, R. 2011 Small area estimation under transformation to linearity. *Survey Methodology*, **37**, 39–51.

Chandra, H. and Sud, U.C. 2012 Small area estimation for zero-inflated data. *Communications in Statistics—Simulation and Computation*, **41**, 632–643.

Dreassi, E., Petrucci, A. and Emillia, R. 2014 Small area estimation for semicontinuous skewed spatial data: an application to the grape wine production in Tuscany. *Biometrical Journal*, **56**, 141–156.

Fabrizi, E. and Trivisano, C. 2010 Robust linear mixed models for small area estimation. *Journal of Statistical Planning and Inference*, **140**, 433–443.

Fay, III R.E. and Herriot, R.A. 1979 Estimates of income for small places: an application of James-Stein procedures to census data. *Journal of the American Statistical Association*, **74**, 269–277.

Ferrante, M.R. and Trivisano, C. 2010 Small area estimation of the number of firms' recruits by using multivariate models for count data. *Survey Methodology*, **36**, 171–180.

Ferraz, V.R.S. and Moura, F.A.S. 2012 Small area estimation using skew normal models. *Computational Statistics and Data Analysis*, **56**, 2864–2874.

Fuller, W.A. 1991 Simple estimators for the mean of skewed populations. *Statistica Sinica*, **1**, 137–158.

Ghosh and Maiti, 2004 Small area estimation based on natural exponential family quadratic variance function models and survey weights, *Biometrika*, vol **91**, 1, pp 95–112.

Ghosh, M., Natarajan, K., Stroud, T.W.F. and Carlin, B.P. 1998 Generalized linear models for small-area estimation. *Journal of the American Statistical Association*, **93**, 273–282.

Goebel, J. 2009 Statistical Methodology for the NRI-CEAP Cropland Survey. Natural Resources Conservation Service. Internal report.

Griswold, M., Giovanni, P., Potsky, A. and Lipscomb, J. 2004 Analyzing health care costs: a comparison of statistical methods motivated by Medicare colorectal cancer charges. *Biostatistics*, **1**, 1–23.

Haslett, S.J., Isidro, M.C. and Jones, G. 2010 Comparison of survey regression techniques in the context of small area estimation of poverty. *Survey Methodology*, **36**, 156–170.

Haslett, S.J. and Jones, G. 2010 Small-area estimation of poverty: The Aid Industry Standard and its alternatives. *Australian and New Zealand Journal of Statistics*, **52**, 341–362.

Jiang, J. 2003 Empirical method of moments and its applications. *Journal of Statistical Planning and Inference*, **115**, 69–84.

Karlberg, G. (2000). Population total prediction under a lognormal superpopulation model. *Metron-International Journal of Statistics*, **58**, 53–80.

Jiang, J. and Lahiri, P. 2006 Mixed model prediction and small area estimation. *Test*, **15**, 1–96.

Molina, I. and Rao, J.N.K. 2010 Small area estimation of poverty indicators. *Canadian Journal of Statistics*, **38**, 369–385.

Nandram, B. and Choi, J.W. 2010 A Bayesian analysis of body mass index data from small domains under nonignorable nonresponse and selection. *Journal of the American Statistical Association*, **105**, 120–135.

Pfeffermann, D., Terryn, B. and Moura, F.A. 2008 Small area estimation under a two-part random effects model with application to estimation of literacy in developing countries. *Survey Methodology*, **34**, 235–249.

Prasad, N.G.N. and Rao, J.N.K. 1990 The estimation of mean squared errors of small area estimators. *Journal of the American Statistical Association*, **85**, 163–171.

Rao, J.N.K. 2003 *Small Area Estimation*. John Wiley & Sons, Inc.

Rivest, L.P. 1994 Statistical properties of Winsorized mean for skewed distributions. *Biometrika*, **81**, 373–383.

Salvati, N., Tzavidis, N., Pratesi, M. and Chambers, R. 2012 Small area estimation via M-quantile geographically weighted regression. *Test*, **21**, 1–28.

Slud, E.V. and Maiti, T. 2006 Mean-squared error estimation in transformed Fay-Herriot models. *Journal of the Royal Statistical Society, Series B*, **68**, 239–257.

Torabi, M. and Shokoohi, F. 2012 Likelihood inference in small area estimation by combining time-series and cross-sectional data. *Journal of Multivariate Analysis*, **111**, 213–221.

Trevisani, M. and Torelli, N. 2007 Hierarchical Bayesian Models for Small Area Estimation with count data. Dipartimento di Scinze Economiche e Statiche, Universita degli Studi di Trieste, Working Paper.

Tzavidis, N., Giovanna, R.M., Salvati, N., Dreassi, E. and Chambers, R. 2013 Poisson M-quantile Regression for Small Area Estimation. Centre for Statistical and Survey Methodology, University of Wollongong, Working Paper 14-13, 28.

Wang, J. and Fuller, W.A. 2003 The mean squared error of small area predictors constructed with estimated area variances. *Journal of the American Statistical Association*, **98**, 716–723.

Zhang, L. and Chambers, R. 2004. Small area estimates for cross-classifications. *Journal of the Royal Statistical Society, Series B*, **66**, 479–496.

16

Bayesian Beta Regression Models for the Estimation of Poverty and Inequality Parameters in Small Areas

Enrico Fabrizi[1], Maria Rosaria Ferrante[2] and Carlo Trivisano[2]
[1]*DISES, Università Cattolica del S. Cuore, Piacenza, Italy*
[2]*Dipartimento di Scienze Statistiche "Paolo Fortunati", Università di Bologna, Bologna, Italy*

16.1 Introduction

Many parameters that describe poverty, social exclusion and inequality can take values in the (0, 1) interval. This class includes headcount ratios such as the at-risk-of-poverty or material deprivation rates, the class of poverty measures introduced by Foster *et al.* (1984), the Gini inequality index, just to cite a few. Many of these parameters are routinely calculated in EU countries using the EU Statistics on Income and Living Conditions (EU-SILC) data.

Let us assume that we need estimates of this type of parameters in small subpopulations for which only small or no samples are available. In this chapter, we discuss area level models (Rao 2003, chapter 5) for the estimation of these parameters in small domains. The idea of area models is that of complementing often imprecise direct estimators with auxiliary information available at the area level obtained from external sources.

Given the nature of the target parameters we consider Beta regression models, as the Beta distribution is very flexible over the (0, 1) range and it allows for asymmetric sampling distributions, that are likely to occur when estimating the parameters in question using small samples. Beta regression models have received considerable attention in recent years, both in

Analysis of Poverty Data by Small Area Estimation, First Edition. Edited by Monica Pratesi.

Companion Website: www.wiley.com/go/pratesi/poverty

the frequentist and bayesian literature. Some basic references are: Kieschnick and McCullough (2003), Ferrari and Cribari-Neto (2004), Ferrari and Pinheiro (2012) (frequentist); Branscum *et al.* (2007), Da Silva *et al.* (2011), Bayes *et al.* (2012), and Liu *et al.* (2014) (bayesian).

In this chapter we adopt a bayesian approach with approximate inference for relevant posterior distributions relying on Markov Chain Monte Carlo (MCMC) algorithms. We neither provide an exhaustive review of bayesian inference on Beta regressions, nor do we aim to tackle all the problems that may arise when dealing with small area estimation of parameters in the range (0, 1). We focus on a few problems that we think are particularly relevant for small area applied researchers and apply the proposed solutions to real survey data.

The data set we deal with in this chapter is a subset of the Italian 2010 sample of the EU-SILC survey (European Parliament and Council 2003), as only the Emilia-Romagna and Tuscany administrative regions are considered. The target small areas are given by the health districts. Their population ranges between 35 400 and 377 000 in the considered region and whose administrations play an important role in the implementation of many social and health expenditure programs related to the contrast of poverty and social exclusion.

The problems we consider are: (i) the estimation of the at-risk-of-poverty rate; (ii) the joint estimation of two rates based on increasing thresholds (namely the material deprivation and severe material deprivation rates); (iii) the joint estimation of two correlated parameters (for illustrative purposes we consider at-risk-of-poverty rate and the Gini inequality index based on the distribution of the equivalized disposable income). See Fusco *et al.* (2010) for detailed definitions of equivalized disposable income and related concepts.

When estimating the at-risk-of-poverty rate we face the problem of areas with no poors in the sample that leads us to consider zero-mixture Beta regressions, a class of models that will be considered also in the estimation of other parameters; to reach goals (ii) and (iii) we introduce multivariate extensions of the Beta regression model: in the first case we discuss a multivariate logistic-normal model for the expected values of the Beta distributions already introduced in Fabrizi *et al.* (2011), while in the second setting, we exploit the correlation between direct estimators using the theory of copula functions (see Nelsen 1999).

The rest of the chapter is organized as follows: in the next section we briefly introduce the data and describe direct estimation. The three inference problems just described are dealt with in Sections 16.3, 16.4, and 16.5 respectively. In Section 16.6 we discuss the implementation of the methods in `R` and provide a guide to the codes that may be found in the Appendix. Section 16.7 offers some concluding remarks.

16.2 Direct Estimation

We analyze data from the 2010 wave of the EU-SILC targeting the population of the two contiguous Emilia-Romagna and Tuscany administrative regions, whose overall population is about 8 million according to the 2011 Italian Census. EU-SILC is a rotating panel survey with a 75% overlapping of samples in successive years. The fresh part of the sample is drawn according to a stratified two-stage sample design where municipalities (LAU 2 level, partitions of the regions) are the primary sampling units (PSUs), while households are the secondary sampling units (SSUs). The PSUs are divided into strata according to their population size; the SSUs are selected by means of systematic sampling in each PSU.

Areas for estimation are given by 64 health districts in the two regions. The EU-SILC sample for our study regions includes 2692 respondent households where 6316 individuals live. In terms of households the area specific sample sizes m_d $(d = 1, \dots, D = 64)$ range from 2 to 122 (the median is 38), while in terms of individuals the sample sizes n_d go from 4 up to 253 (median 84).

The parameters that we target in the following sections are the at-risk-of-poverty rate (π_d), the material and severe material deprivation rates (δ_d, Δ_d), and the Gini concentration index (γ_d). π_d and γ_d are based on disposable equivalized income. A person is said to be poor if his/her equivalized income y_{dj} $(d = 1, \dots, D, j = 1, \dots, n_d)$ is below a poverty threshold pt defined as the 60% of the median equivalized income at the national level. The direct estimator of π_d that we consider is given by

$$p_d = \frac{\sum_{j=1}^{n_d} w_{dj} \mathbf{1}(y_{dj} < pt)}{\sum_{j=1}^{n_d} w_{dj}} \tag{16.1}$$

where w_{dj} is the individual weight (the one officially published by Eurostat with the EU-SILC data). Material deprivation is defined using a battery of nine household level questions with yes/no answer, each focused on measuring the ability/inability to afford items considered by most people to be desirable or even necessary to lead an adequate life. If we let sc_{dj} be the number of declared "inabilities" (a deprivation "score" ranging from 0 to 9) a person is said to be materially deprived if $sc_{dj} \geq 3$. The proportion of materially deprived in the population is estimated by

$$md_d = \frac{\sum_{j=1}^{n_d} w_{dj} \mathbf{1}(sc_{dj} \geq 3)}{\sum_{j=1}^{n_d} w_{dj}} \tag{16.2}$$

Deprivation is said to be severe if $sc_{dj} \geq 4$. Consistently, the proportion of severely deprived people in the population is estimated by:

$$MD_d = \frac{\sum_{j=1}^{n_d} w_{dj} \mathbf{1}(sc_{dj} \geq 4)}{\sum_{j=1}^{n_d} w_{dj}} \tag{16.3}$$

As far as the Gini index γ_d is concerned, the most popular direct estimator (see for instance Alfons and Templ 2013) is given by

$$g_d = \frac{2 \sum_{j=1}^{n_d} \left(w_{dj} y_{dj} \sum_{h=1}^{j} w_{dh} \right) - \sum_{j=1}^{n_d} y_{dj} w_{dj}^2}{\left(\sum_{j=1}^{n_d} w_{dj} \right) \left(\sum_{j=1}^{n_d} y_{dj} w_{dj} \right)} - 1 \tag{16.4}$$

where the values $y_{dj}, j = 1, \dots, n_d$ are assumed to be sorted in ascending order.

Although characterized by large variance, estimators (16.1)–(16.3) can be assumed to be approximately unbiased. This assumption is less tenable for g_d which is known to be negatively biased in small samples (Deltas 2003). Davidson (2009) notes, although in a simple random sampling context where the weights plays no role, that the main term in the bias of g_d can be removed by a $n_d(n_d - 1)^{-1}$ multiplication. This is in line with Jasso (1979) that suggests to use $n_d(n_d - 1)$ as the denominator of the Gini mean difference instead of the more common option n_d^2. Langel and Tillé (2013) note that an extension of the Davidson's bias correction to weighted estimation is not trivial. Here we provide an heuristic solution to this problem noting first (see Langel and Tillé 2013) that

$$g_d = \frac{1}{2\hat{\bar{Y}}_d} \frac{\sum_{j=1}^{n_d} \sum_{k=1}^{n_d} w_{dj} w_{dk} |y_{dj} - y_{dk}|}{\hat{N}_d^2} \tag{16.5}$$

where $\hat{N}_d = \sum_{j=1}^{n_d} w_{dj}$, $\hat{\bar{Y}}_d = \hat{N}_d^{-1} \sum_{j=1}^{n_d} w_{dj} y_{dj}$. In line with Jasso's suggestion and the fact that the household is the survey's sampling unit, we replace $\hat{N}_d^2$ in the denominator of (16.5) with $\sum_{h=1}^{m_d} w_{dh} \sum_{k \neq h}^{m_d} w_{dk}$, where $w_{dh} = \sum_{j=1}^{n_h} w_{dj}$ and n_h is the number of individuals in the household. We thus obtain

$$\tilde{g}_d = \frac{1}{2\hat{\bar{Y}}_d} \frac{\sum_{j=1}^{n_d} \sum_{k=1}^{n_d} w_{dj} w_{dk} |y_{dj} - y_{dk}|}{\hat{N}_d^2 - \sum_{h=1}^{m_d} w_{dh}^2}. \tag{16.6}$$

The estimator $\tilde{g}_d$ is not exactly unbiased, but it may be expected to have a lower bias with unpredictable sign (Deltas 2003). We assessed that this is the case using also a simulation exercise based on the the synthetic population *SimPopulation* of Alfons and Kraft (2012). Results of this simulation exercise are not reported here for brevity but are available upon request.

To estimate the variances of p_d, md_d, MD_d, $\tilde{g}_d$ and the covariances that will be needed in the models of Sections 16.4 and 16.5, we use a bootstrap algorithm that is described in Fabrizi *et al.* (2011). The bootstrap based "raw" variances will only be used in the estimation of design effects and other parameters of variance smoothing models that will be described later.

Auxiliary information is based on publicly available archives at the municipal level; we aggregate information to obtain district level variables. These covariates include average taxable income claimed by private residents, percentage of residents filling tax forms, dependency ratio and other demographic indexes, percentage of resident foreigners (immigrants), and population density.

16.3 Small Area Estimation of the At-risk-of-poverty Rate

In this section we present an area level model for the at-risk-of-poverty rate. The problem of estimating small area proportions using hierarchical Beta regression models is considered also in Liu *et al.* (2014). We consider the case of direct estimates $p_d = 0$ (that is not covered by Liu *et al.* 2014), but not that of $p_d = 1$ that is much less likely to occurr in practice. An extension of the proposed methodology to this case is quite easy.

16.3.1 The Model

A simple model for $p_d \in (0, 1)$ consists in assuming:

$$p_d | \pi_d, \hat{\phi}_d \sim Beta(\pi_d(\hat{\phi}_d - 1), (1 - \pi_d)(\hat{\phi}_d - 1)) \tag{16.7}$$

that implies $E(p_d | \pi_d, \hat{\phi}_d) = \pi_d$, $V(p_d | \pi_d, \hat{\phi}_d) = \hat{\phi}_d^{-1} \{\pi_d(1 - \pi_d)\}$. Since under simple random sampling (*SRS*) of households we would have had $V_{srs}(p_d) = n_d^{-1} \{\pi_d(1 - \pi_d)\}$, $\hat{\phi}_d$ may be interpreted as a SRS-equivalent sample size for the actual design of the survey.

Unfortunately, (16.7) is not suitable to accomodate $p_d = 0$, a case that occurs in our data as we have 5 districts without poor households in the sample and that, in general, is quite likely for estimation problems of this type. We note that although $p_d = 0$, it is reasonable to assume that $\pi_d > 0$, that is there are no areas without poor households at the population level. Given a sample of m_d households with probability of being poor π_d and whose poverty status we assume independent, we have that $P(p_d = 0|\pi_d > 0) = (1 - \pi_d)^{m_d}$. We then specify the mixture density

$$p(p_d|\pi_d^*, \hat{\phi}_d) = (1 - \pi_d^*)^{m_d}\mathbf{1}(p_d = 0) + \{1 - (1 - \pi_d^*)^{m_d}\}dBeta(\pi_d^*(\hat{\phi}_d - 1), \\ (1 - \pi_d^*)(\hat{\phi}_d - 1))\mathbf{1}(p_d > 0) \tag{16.8}$$

where $dBeta(a, b)$ is the density function of a $Beta(a, b)$ random variable. Note that (16.8) is different from the zero-inflated model proposed by Ospina and Ferrari (2012) and considered also in Wieczorek and Hawala (2011) for small area estimation, as $P(p_d = 0)$ is a function of π_d^* and it is not modeled separately. Our parameter of interest can be defined as $\pi_d = \pi_d^*\{1 - (1 - \pi_d^*)^{m_d}\}$, i.e. $E(p_d|\pi_d, \hat{\phi}_d) =$ under model 16.8.

From (16.8) it follows (Ospina and Ferrari 2012) that

$$\begin{aligned} V(p_d|\pi_d^*, \hat{\phi}_d) &= E_{\mathbf{1}(p_d>0)}\{V(p_d|p_d > 0)\} + V_{\mathbf{1}(p_d>0)}\{E(p_d|p_d > 0)\} \\ &= \frac{\pi_d^*(1 - \pi_d^*)}{\hat{\phi}_d}\{1 - a(\pi_d^*)\} + a(\pi_d^*)\{1 - a(\pi_d^*)\}\pi_d^{2*} \end{aligned} \tag{16.9}$$

where $a(\pi_d^*) = (1 - \pi_d^*)^{m_d}$. The "equivalent sample size" $\hat{\phi}_d$ enters only in the first addend of (16.9), that is in $V(p_d|p_d > 0)$. To calculate it we use a bootstrap estimator $\hat{V}_{boot}(p_d|p_d > 0)$ of $V(p_d|p_d > 0)$ based on the algorithm mentioned in Section 16.2; areas with $p_d = 0$ are not considered and for the rest, all bootstrap samples s_d^* for which $p_d^* = 0$ are discarded. Then we specify the following variance smoothing model:

$$\frac{p_d(1 - p_d)}{\hat{V}_{boot}(p_d|p_d > 0)} = \beta_f n_d + e_d$$

where $e_d \sim D(0, \tau^2)$. β_f that can be interpreted as the reciprocal of the design effect, is estimated using least squares, then we define $\hat{\phi}_d = n_d\hat{\beta}_f$. To obtain more stable estimates of β_f we ran the bootstrap algorithm for Italian provinces. Every province includes two or more health districts; overall there are 108 provinces in Italy.

For the structural part we specify:

$$logit(\pi_d^*) = \mathbf{x}_d^T\boldsymbol{\beta}_{\pi 1} + v_d \tag{16.10}$$

where $\mathbf{x}_d$ is a vector of auxiliary information known for area d, $\boldsymbol{\beta}_{\pi 1}$ is a vector of regression parameters for which we assume independent zero-meaned normal prior distributions with large variances. v_d is an area specific effect for which we assume $v_d \overset{ind}{\sim} N(0, \sigma_v^2)$; for the prior we assume $\sigma_v \sim$ half $- t(\nu = 2, A = 1)$ according to the recommendation of Gelman (2006). The choice $\nu = 2$ allows for a diffuse prior, close to the more popular half-Cauchy ($\nu = 1$); we prefer $\nu = 2$ for consistency with the multivariate priors that are discussed in the following sections. We chose $A = 1$ after careful consideration of the scale of the parameters' distribution and some sensitivity analysis.

16.3.2 Data Analysis

The data that we used to simulate from the relevant posterior distributions can be found in the R script `Code for hr and gini models.r`, while description of details on the MCMC algorithms will be provided in Section 16.6. Here we briefly review the results of the estimation exercise. We check the adequacy of the specified model using the posterior predictive approach, that is the idea that new observations generated from the posterior distribution should be "similar" to the observed data. Following You and Rao (2002) we use the following discrepancy measure considered also in Fabrizi *et al.* (2011):

$$\text{dis}_d = P(p_d < p_d^\star) \tag{16.11}$$

where $p_d^\star$ is generated from the posterior predictive. The discrepancy dis_d should be far from 0 and 1, otherwise we would have evidence of systematic under- or over-estimation. In our case the value of the discrepancy average over the areas is 0.47, with 0.28 and 0.64 corresponding to its first and third quantile over the set of m areas. In no district do we have clear evidence of misfit from the model (min $\text{dis}_d = 0.08$, max $\text{dis}_d = 0.95$). As we consider the mixture density (16.8) we can use dis_d to check the fit also for areas with $p_d = 0$: in all these cases the associated dis_d are over 0.1, thus indicating an adequate fit.

In Figure 16.1a Bayes estimators obtained under quadratic loss, that is $E(\pi_d|data)$, are plotted against direct estimates p_d. As expected there is a positive correlation and a shrinkage toward the mean effect due to the corrections that the model operates on the more extremes direct estimates (typically based on very small samples). When $p_d = 0$ we may observe how the model uses the auxiliary information to discriminate between areas with different characteristics.

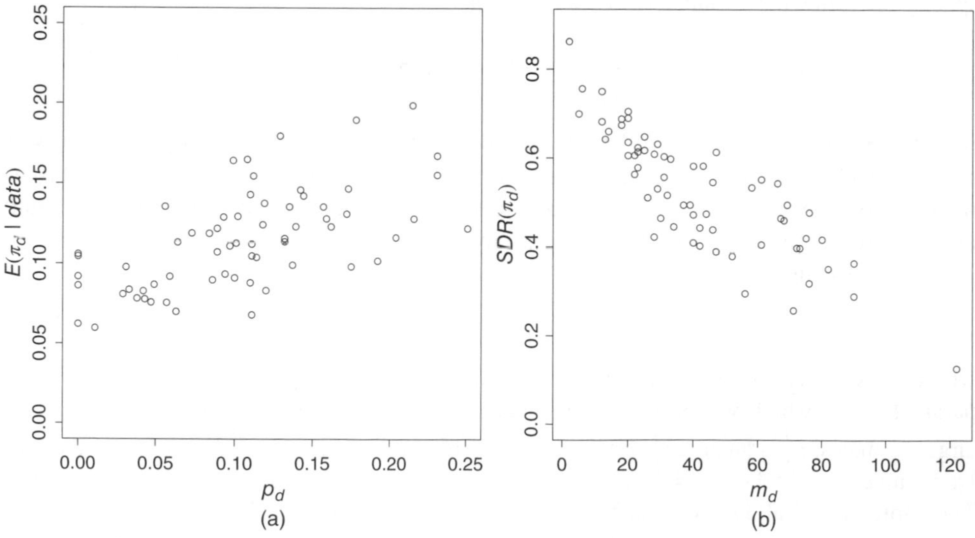

Figure 16.1 Direct versus model-based estimates of π_d (a) and sample size versus reduction in estimator standard deviation (b)

The efficiency improvements the use of the small area model allows for are measured by the reduction in the estimators' standard deviation from direct to model based:

$$SDR(\pi_d) = 1 - \sqrt{\frac{V(\pi_d|data)}{\hat{\phi}_d^{-1} E(\pi_d(1-\pi_d)|data)}} \tag{16.12}$$

Under the square root we have the posterior variance in the numerator; the variance function in the denominator summarizes the information on π_d we have from the design. The use of (16.9) that accomodates the equality to 0 of direct estimates would make numerators and denominators incomparable, unduly inflating the variance associated with the direct estimator. Efficiency improvements are summarized in Figure 16.1b where $SDR(\pi_d)$ is plotted versus the sample size in terms of households (m_d). The gain in efficiency is large when m_d is small and moderate when m_d is large, in line with expectation and the design consistency of estimators based on area level models. On average we have a standard deviation reduction of 0.53.

16.4 Small Area Estimation of the Material Deprivation Rates

Material deprivation and severe material deprivation rates are based on increasing thresholds. This implies that in the population $\Delta_d < \delta_d$ and at the sample level not only $MD_d \leq md_d$ but also that md_d and MD_d are strongly correlated. In this section we present a multivariate model for these two rates that follows closely Fabrizi *et al.* (2011) and that guarantees that the Bayes estimators are monotonically increasing, that is $E(\Delta_d|data) < E(\delta_d|data)$. We start noting that, if we define $r_{d1} = MD_d$, $r_{d2} = md_d - MD_d$, their correlation will be much weaker than that between MD_d and md_d and may be, for practical purposes, overlooked. We propose a model assuming that, within each area, the differences between successive rates are exchangeable.

For illustrative purposes we consider a case with only two rates. Fabrizi *et al.* (2011) consider a case with three; the generalization to any given number K of rates based on increasing thresholds is straightforward.

16.4.1 The Model

Since we can have $r_{d1} = 0$ and/or $r_{d2} = 0$ for some d, we specify, in line with (16.8) and differently from Fabrizi *et al.* (2011) the following densities:

$$\begin{aligned} p(r_{dk}|\rho^*_{dk}, \hat{\psi}_{dk}) = (1-\rho^*_{dk})^{m_d}\mathbf{1}(r_{dk}=0) + \{1-(1-\rho^*_{dk})^{m_d}\} dBeta(\rho^*_{dk}(\hat{\psi}_{dk}-1), \\ (1-\rho^*_{dk})(\hat{\psi}_{dk}-1))\mathbf{1}(r_{dk}>0) \end{aligned} \tag{16.13}$$

with $k = 1, 2$. With arguments similar to those of the previous section, we can write equations parallel to (16.9) and interpret $\hat{\psi}_{dk}$ as equivalent sample sizes that are estimated using a variance smoothing model based on bootstrap estimation conditional on $r_{dk} > 0$. In particular, we consider the estimating equation

$$\frac{r_{dk}(1-r_{dk})}{V_{boot}(r_{dk}|r_{dk}>0)} = \beta_{gk} n_d + e_d$$

where $e_d \sim D(0, \tau^2)$, β_{gk} are estimated using least squares and $\hat{\psi}_{dk} = \hat{\beta}_{gk} n_d$.

We then assume that $\rho^*_{dk} = \exp(\zeta_{dk})/[1 + \sum_{k=1}^{2} \exp(\zeta_{dk})]$ and, once defined $\boldsymbol{\zeta}_d = (\zeta_{d1}, \zeta_{d2})^T$, that $\boldsymbol{\zeta}_d \sim MVN(\boldsymbol{\mu}_d, \Sigma_a)$ that implies that ρ^*_{dk} are assumed to be distributed according to the logistic-normal distribution discussed in Aitchinson and Shen (1980). The vector parameter $\boldsymbol{\mu}_d = (\mu_{d1}, \mu_{d2})^T$ is related to auxiliary information through the equation

$$\mu_{dk} = \mathbf{x}_d^T \boldsymbol{\beta}_{\mu k} \tag{16.14}$$

where $\mathbf{x}_d$ is a vector of auxiliary information for area d, $\boldsymbol{\beta}_{\mu k}$ are vectors of regression parameters for which we assume independent zero-meaned normal prior distributions with large variances. Σ_a is a 2×2 positive definite matrix. For Σ_a we specify a prior within the family proposed by Huang and Wand (2013) with the purpose of keeping the analytical and computational tractability of the inverse Wishart but improving the noninformativity properties:

$$\Sigma_a | a_1, a_2 \sim \text{Inv-Wishart}\,(\nu + 1, 2\nu\text{diag}\,(a_1^{-1}, a_2^{-1}))$$

$$a_i \sim \text{Inv-Gamma}\left(\frac{1}{2}, \frac{1}{A_i}\right), i = 1, 2. \tag{16.15}$$

This prior marginally induces $\sigma_i \sim \text{half} - t(\nu, A_i)$. For the reasons discussed in Section 16.3 we set $A_1 = A_2 = 1$. $\nu = 2$ is chosen as it induces a marginal uniform prior for the correlation between the two random effects. We note that inference on the parameters of interest δ_d and Δ_d is easy to obtain (especially using MCMC methods) as they are simple functions of the $\rho_{dk} = \rho^*_{dk}\{1 - (1 - \rho^*_{dk})^{m_d}\}, k = 1, 2$: $\Delta_d = \rho_{d1}$ and $\delta_d = \rho_{d1} + \rho_{d2}$.

16.4.2 Data Analysis

The R code needed for the analysis of the models for the material deprivation rates can be found in the file `code for matdep models.r`. Here we present some simple analyses of the results. bayesian p-values based on (16.11) are on average very close to 0.5 for both rates and within the range $(0.1, 0.9)$ for all but a few cases that are associated with very small samples. The model fit can therefore be judged adequate. In Figure 16.2 Bayes estimators under quadratic loss, that is $E(\delta_d|data)$, $E(\Delta_d|data)$, are plotted against direct estimates (md_d, MD_d). The two sets of estimates are of course positively correlated and more strongly so for Figure 16.2a (MD_d, Δ_d); the outlier to the extreme right of Figure 16.2b is associated with an area where $m_d = 2$ and we obtain a rather extreme direct estimate (≥ 0.4) for which the model operates a large correction. On the left margin of both Figure 16.2a and b we can see how the model uses auxiliary information to associate different Bayes estimates to areas for which either $md_d = 0$ or $MD_d = 0$.

In Figure 16.3a we report the reduction in the standard errors from direct to Bayes estimators of the material deprivation rate, defined as

$$SDR(\Delta_d) = 1 - \sqrt{\frac{V(\Delta_d|data)}{\hat{\psi}_{d1}^{-1} E(\Delta_d(1 - \Delta_d)|data)}} \tag{16.16}$$

against the sample sizes in terms of households m_d. Something similar is done in Figure 16.3b (note that $\delta_d = \rho_{d1} + \rho_{d2}$) once we define:

$$SDR(\delta_d) = 1 - \sqrt{\frac{V(\delta_d|data)}{\hat{\psi}_{d1}^{-1} E(\rho_{d1}(1 - \rho_{d1})|data) + \hat{\psi}_{d2}^{-1} E(\rho_{d2}(1 - \rho_{d2})|data)}} \tag{16.17}$$

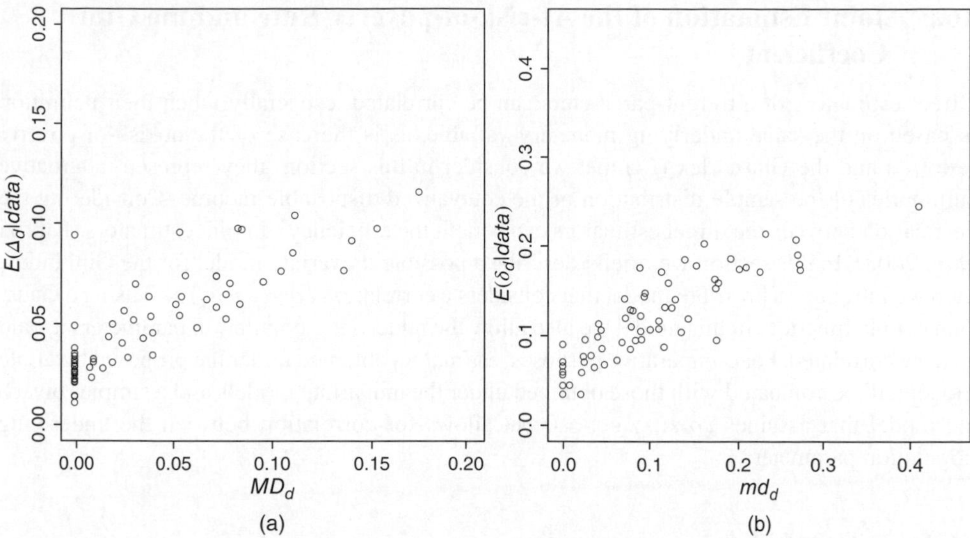

Figure 16.2 Direct versus model-based estimates

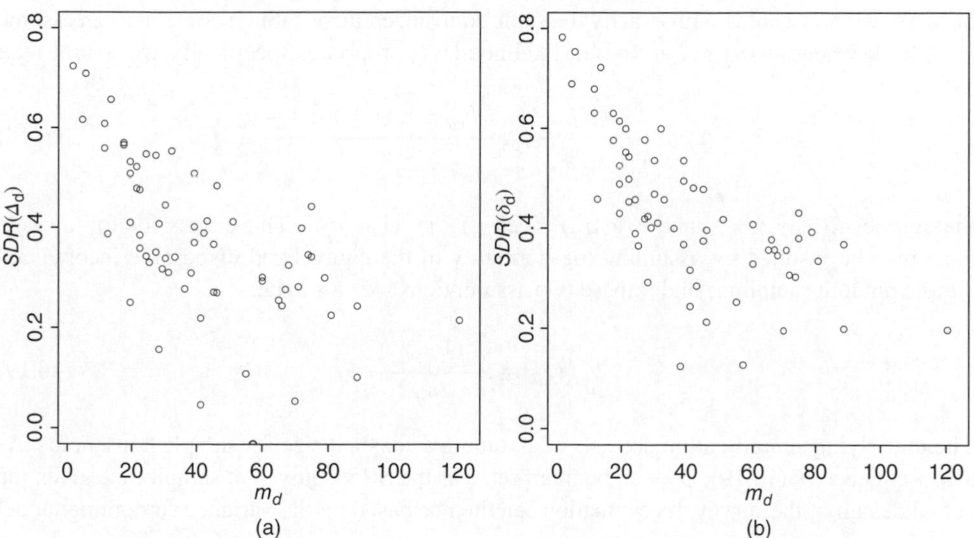

Figure 16.3 Sample size versus reduction in estimator standard deviation

The variance function in the denominators of (16.16) and (16.17) describes the amount of information on δ_d and Δ_d we have from the design and is derived using arguments similar to those leading to (16.12). From Figure 16.3a and b we note that the reduction of standard errors is very high for small samples and decreases as long as m_d increases, in line with the expected consistency of direct estimators. On average we have a 41.6% reduction for δ_d and a 35.4% reduction for Δ_d.

16.5 Joint Estimation of the At-risk-of-poverty Rate and the Gini Coefficient

Direct estimators of different parameters can be correlated, especially when their definition is based on the same underlying monetary variable, as is the case of the at-risk-of-poverty rate (p_d) and the Gini index ($\tilde{g}_d$) that we consider in this section: they represent alternative summaries of the sample distribution of the equivalized disposable income. Considering the correlation between the direct estimators can benefit the efficiency of point estimators (Fabrizi *et al.* 2008). In this section we briefly describe a possible univariate model for the Gini index, then we introduce a bivariate model that considers a correlation $corr(p_d, \tilde{g}_d) > 0$ using a Gaussian copula function. In this model we also allow the underlying population parameters π_d and γ_d to be correlated. For comparative purposes, estimators obtained under the proposed bivariate model will be compared with those obtained under the univariate models and a simpler bivariate model that assumes $corr(p_d, \tilde{g}_d) = 0$ but allows for correlation between the underlying population parameters.

16.5.1 *The Models*

We have already discussed a possibile univariate model for π_d in Section 16.3. A model for the Gini index can be specified along similar lines, with the advantage that, since it is unlikely to observe two households with exactly the same equivalized disposable income, we can assume $\tilde{g}_d \in (0, 1)$ whenever $m_d > 2$ and a Beta likelihood is appropriate. Specifically, we assume that:

$$\tilde{g}_d \sim Beta\left(\frac{2\hat{\phi}_{gd}}{1+\gamma_d} - \gamma_d, \frac{2\hat{\phi}_{gd} - \gamma_d(1+\gamma_d)}{1+\gamma_d}\frac{1-\gamma_d}{\gamma_d}\right) \tag{16.18}$$

that implies $E(\tilde{g}_d|\gamma_d) = \gamma_d$ and $V(\tilde{g}_d|\gamma_d) = (2\hat{\phi}_{gd})^{-1}\{\gamma_d^2(1-\gamma_d^2)\}$. This expression for the variance may be justified by assuming log-normality of the equivalized disposable income and simple random sampling; under these two assumptions we can write:

$$V_{srs}(\tilde{g}_d) \cong \frac{\gamma_d^2(1-\gamma_d^2)}{2n_d}. \tag{16.19}$$

The underlying simplification consists of assuming a constant size for sample households. As a consequence of (16.19), $\hat{\phi}_{dg}$ can be interpreted as the SRS-equivalent sample size given the actual design of the survey. Its estimation can then be based on the variance smoothing model

$$\frac{\tilde{g}_d^2(1-\tilde{g}_d^2)}{2V_{boot}(\tilde{g}_d)} = \beta_{fg}n_d + e_d \tag{16.20}$$

where $e_d \sim D(0, \tau^2)$ and β_{fg} can be interpreted as half the reciprocal of the design effect; we can estimate it using least squares and define $\hat{\phi}_{gd} = n_d\hat{\beta}_{fg}$.

We can justify (16.19) noting first that, under log-normality, that is assuming $\log(y_{jd}) \sim N(\mu_d, \sigma_d^2)$, $\gamma_d = \mathrm{erf}(\sigma_d/2)$, a quantity that under simple random sampling may be consistently estimated by $\hat{\gamma}_d = \mathrm{erf}(s_d/2)$ with $s_d = \sqrt{(n_d-1)^{-1}\sum_{j=1}^{n_d}(y_{dj}-\bar{y}_d)^2}$. From normal distribution theory results we have $V(s_d) \cong \sigma_d^2(1 - c_{ud}^2)$, $c_{ud} = \sqrt{\frac{2}{n_d-1}} \frac{\Gamma(n_d/2)}{\Gamma((n_d-1)/2)}$. As $1 - c_{ud}^2 \cong (2n_d)^{-1}$, using a Delta-method argument we can write

$$V\left\{\operatorname{erf}\left(\frac{s_d}{2}\right)\right\} \cong \frac{\sigma_d^2}{2\pi n_d} \exp\left\{-\frac{\sigma_d^2}{2}\right\} \cong \frac{\sigma_d^2}{2\pi n_d}\left(1 - \frac{\sigma_d^2}{2}\right)$$

where the last approximated equality is justified by a first-order McLaurin expansion. As a consequence

$$V\left\{\operatorname{erf}\left(\frac{s_d}{2}\right)\right\} \cong \frac{\sigma_d^2(1 - \sigma_d^2/2)}{2\pi n_d}.$$

From the expansion of the erf function we can obtain $\sigma_d \cong \gamma_d\sqrt{\pi}$; this leads to $V(\hat{\gamma}_d) \cong (2n_d)^{-1}\gamma_d^2(1 - \pi\gamma_d^2/2)$. This latter expression can be negative for large γ_d so we replace it with

$$V(\hat{\gamma}_d) \cong \frac{\gamma_d^2(1 - \gamma_d^2)}{2n_d}.$$

We note that the accuracy of this approximation for the variance of the estimated Gini index under simple random sampling (and log-normality) is not important in its own right, as we use it mainly to motivate the structure of the variance in (16.18) and to motivate the variance smoothing function. For the data we considered in the estimation of the parameters in (16.20), that is the provincial level estimates (see Section 16.3.1) the fit is very good with a $R^2 = 0.89$.

To specify a joint distribution for $(p_d, \tilde{g}_d)$ is not viable since, marginally, we use a Beta for $\tilde{g}_d$ and a complicated mixture distribution for p_d. Moreover, even if we had two Betas, it is notorious that this distribution does not lend itself to natural multivariate generalizations characterized by general correlation structure. This motivates our recourse to a copula function. Specifically, we consider the family of inversion copulae and among those the Gaussian copula (Clemen and Reilly 1999), which is parametrized by the correlation matrix $\mathbf{R}$ of a multivariate normal distribution. Given a vector of random variables $(X_1, \ldots, X_k)$ the density function of a Gaussian copula can be written in explicit form as

$$f(x_1, \ldots, x_k) = \frac{g(x_1) \times \ldots \times g(x_k)}{|\mathbf{R}|^{\frac{1}{2}}} \exp\left\{-\frac{1}{2}\mathbf{z}_k^T(\mathbf{R}^{-1} - \mathbf{I}_k)\mathbf{z}_k\right\} \tag{16.21}$$

where $\mathbf{z}_k^T = (\Phi^{-1}\{G_1(x_1)\}, \ldots, \Phi^{-1}\{G_d(x_k)\}$, $G_i(x_i)$ is the cumulative distribution function of the ith random variable in the vector and $g_i(x_i)$ the associated density function. We apply this general function to $X_1 = p_d$, $X_2 = \tilde{g}_d$, $k = 2$.

To assess the correlation matrix $\mathbf{R}$ we first estimate the Spearman correlation rank correlations $\hat{\rho}_{boot}$ using the bootstrap samples obtained by the algorithm described in Section 16.2; moreover we assume that the correlation between the two estimators is constant across areas: a single $\hat{\rho}_{boot}$ is then computed averaging the correlations computed over the largest areas (sample size greater or equal to the median). We exploit the invariance property of ranks under monotone transformations to obtain Spearman rank correlation on the copula density scale. Then we obtain the product-moment correlation using $r = 2\sin(\pi\hat{\rho}_{boot})$. This method does not guarantee a positive definite $\mathbf{R}$ in all cases, but as it does it often and especially in the bivariate case that we consider here, we will not discuss the problem in more detail (see for an alternative solution Elfadaly and Garthwaite 2013 for an alternative solution).

The marginal likelihoods (16.8) and (16.18) we endowed with a correlation structure using the copula (16.21) are, respectively, indexed on the parameters π_d and γ_d that are interpretable as the at-risk-of-poverty rate and Gini index at the domain population level. They both have the (0, 1) interval as support, so we can model them jointly along the lines of Section 16.4. Let $\text{logit}(\pi_d) = \xi_{1d}$, $\text{logit}(\gamma_d) = \xi_{2d}$; for $\boldsymbol{\xi}_d = (\xi_{1d}, \xi_{2d})^T$ we assume $\boldsymbol{\xi}_d \sim MVN(\boldsymbol{\nu}_d, \Sigma_b)$. The vector $\boldsymbol{\nu}_d = (\nu_{1d}, \nu_{2d})$ is connected to auxiliary information through the equations:

$$\nu_{1d} = \mathbf{x}_i^t \boldsymbol{\beta}_{\nu 1}$$
$$\nu_{2d} = \mathbf{x}_i^t \boldsymbol{\beta}_{\nu 2}$$

As far as the prior distributions are concerned, we assume independent diffuse normal priors for the components of $\boldsymbol{\beta}_{\nu 1}$ and $\boldsymbol{\beta}_{\nu 2}$. Σ_b is a 2×2 positive definite matrix; in line with Section 16.4.1 we choose the prior (16.15) for Σ_b. Using similar arguments, we adopt also the same values for the hyperparameters.

16.5.2 Data Analysis

In line with Sections 16.3.2 and 16.4.2, the purpose of this section is to illustrate the results of applying the models just illustrated to the data we considered throughout this chapter. In the first place we compare univariate and bivariate models using the deviance information criterion (DIC). Results of model comparison can be found in Table 16.1.

The model allowing for correlation at both the direct estimators level (using the gaussian copula) and the population parameters (through the multivariate normal prior) and that we label *mm* ranks first, with minimum DIC and minimum measure of model complexity *pD*, followed by the one that allows for correlation only at the population parameter level (label *um*) and the univariate models (label *uu*). We check the adequacy of the selected model as best in terms of DIC (i.e. *mm*) by calculating the area-specific bayesian *p*-values based on (16.11): the average value is 0.48 for the at-risk-of-poverty rate and 0.47 for the Gini index. Only a few values are outside the range (0.1,0.9): they are associated with very small areas for which the direct estimates are extreme and therefore largerly corrected by the model.

In Figure 16.4 we summarize results for the bivariate model (at both levels) concerning the estimation of the Gini index. Figure 16.4a shows a clear shrinkage effect; the range of the posterior means $E(\gamma_d|data)$ is more in line with what we expect from the knowledge of the

Table 16.1 Model comparison using DIC and the measure of model complexity *pD*

Model	DIC	*pD*
uu	−353.3	43.91
um	−361.2	51.03
mm	−370.9	41.66

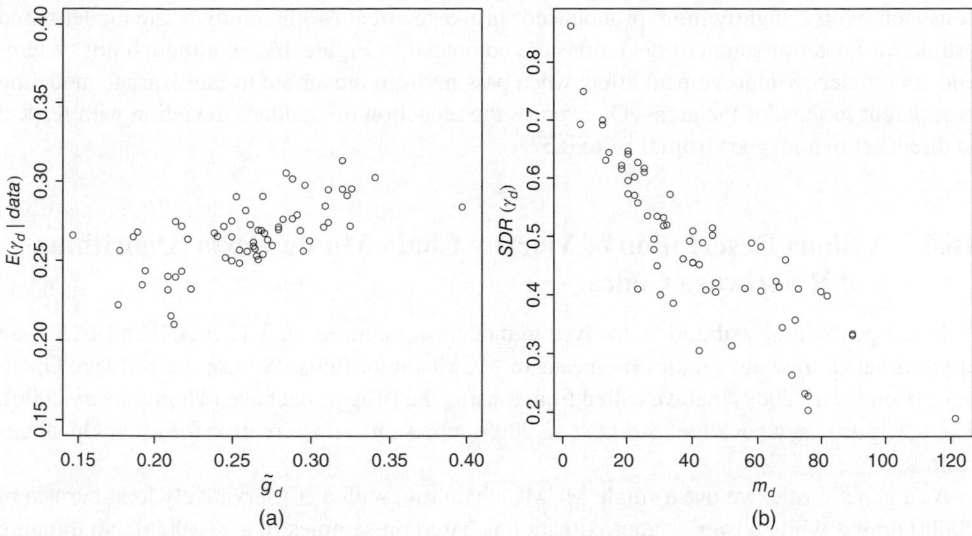

Figure 16.4 Direct versus model-based estimates based on the *mm* model for the Gini index γ_d

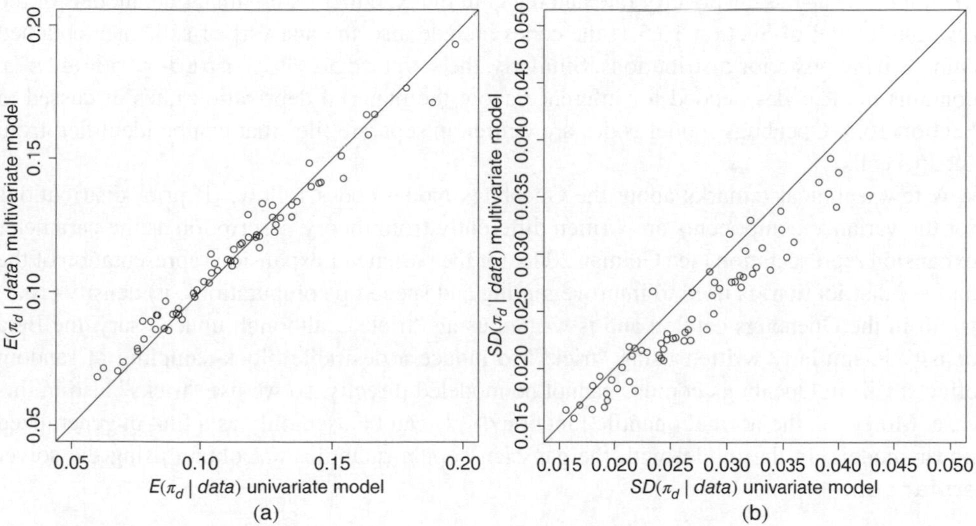

Figure 16.5 Comparing model-based estimates of π_d obtained using univariate and multivariate models

economies of the Emilia-Romagna and Toscana regions than that of the likely overdispersed direct estimates. Figure 16.4b shows the improvements in precision with a formula defined in the same way as (16.12); on average the reduction in the standard deviation is 0.48.

In Figure 16.5 we compare estimates of π_d obtained under the univariate and the multivariate (at both levels) models. Figure 16.5a shows that the two sets of direct estimates are largely consistent with a slightly more pronounced shrinkage effect for the multivariate model-based estimates. Posterior standard deviations are compared in Figure 16.5b: although not systematic, an efficiency improvement effect when passing from univariate to multivariate modeling is apparent in most of the areas. On average the reduction of standard deviation with respect to direct estimators goes from 0.53 to 0.57.

16.6 A Short Description of Markov Chain Monte Carlo Algorithms and R Software Codes

Relevant posterior distributions involved in models of Sections 16.3.1, 16.4.1, and 16.5.1 are approximated drawing samples by means of MCMC algorithms. We use the software Openbugs (Lunn *et al.* 2009) that we called from `R` using the `BRugs` package (Thomas *et al.* 2006). The `R2Openbugs` package (Sturtz *et al.* 2005) represents an alternative for using Openbugs from `R`.

As a general rule, we use a single MCMC chain run, with a conservatively long burn-in of 20000 draws, while posterior approximation is based on samples of size 10000. No thinning of the chains is needed for the models and data considered in this chapter. Convergence is checked using visual inspection of chain tracks and dynamic quantile plots.

The `R` script `code for hr gini models.r` contains the codes needed for the estimation of the at-risk-of-poverty rate and the Gini index, either using univariate models or the bivariate model of Section 16.5.1; the codes include also the analysis of estimates obtained summarizing posterior distributions. Similarly, the script `code for matdep models.r` contains the `R` codes needed for inference about the material deprivation rates discussed in Section 16.4. Openbugs model codes are written in separate files that can be identified from `BRugs` calls.

A few technical remarks about the Openbugs model codes follow: (i) prior distributions for the variance components are written differently from theory description as the parameter expansion representation (see Gelman 2006, for the parameter expansion representation of the half − t distribution) is used to improve mixing and speed up computations: ii) density (16.8) is not in the Openbugs catalog and is written using "tricks"; although unnecessary, the Beta density is similarly written using "tricks" to induce a desirable block-sampling of random effects; (iii) in Openbugs copulas cannot be modeled directly, so we use "tricks" also in this case. Moreover, the normal quantile function $\Phi^{-1}()$ can be used only as a link in generalized linear models; in the model with the gaussian copula quantiles are obtain using the solver `equation`.

16.7 Concluding Remarks

In this chapter, we discussed the application of Beta regression models to the estimation of poverty and inequality related small area parameters adopting area level models and approximated bayesian inference relying on MCMC algorithms.

None of the problems we consider can be solved using standard Beta regression; in the univariate case we introduced a mixture Beta likelihood that allows to accomodate direct estimates equal to 0 that are likely to occur when estimating a relatively rare characteristic (such as poverty) using small samples. We presented two alternative multivariate extensions of this model: in the first case we considered two rates based on increasing thresholds, a situation that drives our modeling of the correlation structure between the parameters. In the second, more general example, we considered two possibly correlated parameters with support in (0, 1); in this case the recourse to copulas is the solution we suggest to overcome the essential lack of a simple multivariate generalization of the Beta distribution with a flexible correlation structure.

We tried to illustrate the models using simple examples; the univariate model can be easily extended to cover the case of direct estimates equal to 1, while for multivariate models, in both cases extensions to more than two parameters are simple in principle, but would have led to more complicated theory and `R` codes.

References

Aitchinson J and Shen M (1980), Logistic-normal distribution: some properties and uses, *Biometrika*, **67**, 261–272.

Alfons A and Kraft S (2012), *simPopulation*: Simulation of Synthetic Populations for Surveys Based on Sample Data, https://cran.r-project.org/src/contrib/Archive/simPopulation/. [accessed 5-Sep-15]

Alfons A and Templ M (2013), Estimation of social exclusion indicators from complex surveys: The R package `laeken`, *Journal of Statstical Software*, **54**, 15.

Bayes CL, Bazàn JL, and Garcìa C (2012), A new robust regression model for proportions, *Bayesian Analysis*, **7**, 771–796.

Branscum AJ, Johnson WO, and Thurmond MC (2007), Bayesian beta regression: applications to household expenditure data and genetic distance between foot-and-mouth disease viruses, *Australian and New Zealand Journal of Statistics*, **49**, 287–301.

Clemen RC and Reilly T (1999), Correlations and copulas for decision and risk analysis, *Management Science*, **45**, 208–224.

Da Silva CQ, Migon HS, and Correia LT (2011), Dynamic Bayesian beta models, *Computational Statistics and Data Analysis*, **55**, 2074–2089.

Davidson R (2009), Reliable inference for the Gini index, *Journal of Econometrics*, **150**, 30–40.

Deltas G (2003), The small-sample bias of the Gini coefficient: results and implications for empirical research, *The Review of Economics and Statistics*, **85**, 226–234.

Elfadaly FG and Garthwaite PH (2013), Eliciting Dirichlet and Gaussian copulas prior distributions for multinomial models, http://mcs-brains.open.ac.uk/elicitation/Copula%20Paper.pdf. accessed also 7-Oct-15

European Parliament and Council (2003), Regulation (EC) No. 1177/2003 of June 16, 2003 concerning Community statistics on income and living conditions (EU-SILC), *Official Journal of the European Union*, 3.7.2003 L 165/1.

Fabrizi E, Ferrante MR, and Pacei S (2008), Measuring sub-national income poverty by using a small area multivariate approach, *The Review of Income and Wealth*, **54**, 597–615.

Fabrizi E, Ferrante MR, Pacei S, and Trivisano C (2011), Hierarchical Bayes multivariate estimation of poverty rates based on increasing thresholds for small domains, *Computational Statistics and Data Analysis*, **55**, 1736–1747.

Ferrari SLP and Cribari-Neto F (2004), Beta regression for modelling rates and proportions, *Journal of Applied Statistics*, **31**, 799–815.

Ferrari SLP and Pinheiro EC (2012), Improved likelihood inference in beta regression, *Journal of Statistical Computation and Simulation*, **81**, 431–443.

Foster J, Greer J, and Thorbecke E (1984), A class of decomposable poverty measures, *Econometrica*, **52**, 761–776.

Fusco A, Guio AC, and Marlier E (2010), Chracterizing the income poor and the materially deprived in European countries, in *Income and Living Conditions in Europe,* Atkinson A. and Marlier E. (eds), Publication Office of the European Union, Luxembourg.

Gelman A (2006), Prior distributions for variance parameters in hierarchical models, *Bayesian Analysis*, **1**, 515–533.

Huang A and Wand MP (2013), Simple marginally noninformative prior distributions for covariance matrices, *Bayesian Analysis*, **8**, 439–452.

Jasso G (1979), On Gini's mean difference and Gini's index of concentration, *American Sociological Review*, **44**, 867–870.

Langel M and Tillé Y (2013), Variance estimation of the Gini index: revisiting a result several times published, *Journal of the Royal Statistical Society, Series A*, **7**, 521–540.

Liu B, Lahiri P, and Kalton G (2014), Hierarchical Bayes modeling of survey weighted small area proportions, *Survey Methodology*, **40**, 1–13.

Lunn D, Spiegelhalter D, Thomas A, and Best N (2009), The BUGS project: Evolution, critique and future directions, *Statistics in Medicine*, **28**, 3049–3067.

Kieschnick R and McCullough BD (2003), Regression analysis of variates observed on (0,1): percentages, proportions and fractions, *Statistical Modelling*, **3**, 193–213.

Nelsen R.B. (1999), *An introduction to copulas*, Springer, New York.

Ospina R and Ferrari SLP (2012), A general class of zero-or-one inflated beta regression models, *Computational Statistics and Data Analysis*, **56**, 1609–1623.

Rao. NK (2003), *Small Area Estimation*, John Wiley & Sons, Inc., Hoboken.

Sturtz S, Ligges U, and Gelman A (2005), `R2WinBUGS`: A package for running `WinBUGS` from `R`, *Journal of Statistical Software*, **12,** 1–16.

Thomas A, O'Hara B, Ligges U, and Sturtz S (2006), Making BUGS Open, *R news*, **6**, 12–17.

Wieczorek J and Hawala S (2011), A Bayesian zero-one inflated Beta model for estimating poverty in U.S. counties, *JSM Proceedings, Section on Survey Research Methods*, 2812–2822.

You Y and Rao JNK (2002), Small area estimation using unmatched sampling and linking models, *Canadian Journal of Statistics*, **30**, 3–15.

17

Empirical Bayes and Hierarchical Bayes Estimation of Poverty Measures for Small Areas

Jon N. K. Rao[1] and Isabel Molina[2]

[1]*School of Mathematics and Statistics, Carleton University, Ottawa, Canada*

[2]*Department of Statistics, Universidad Carlos III de Madrid, Madrid, Spain*

17.1 Introduction

Measurement of poverty for local areas, such as municipalities in Mexico and Brazil and counties and school districts in the United States, has received a lot of attention in recent years. Poverty measures, such as poverty rate or incidence and more complex measures, at the local level are needed for formulating and implementing policies, allocation of funds and other purposes. Large-scale income and consumer expenditure sample surveys measure welfare variables, such as income and expenditure, of sample households, and the data are used to produce reliable direct estimators of poverty measures for domains (or subpopulations) with sufficiently large sample sizes, using domain-specific sample data. Reliability is often judged from the size of the coefficient of variation of the estimates. However, those surveys may not support reliable estimation at the level of a local area because area-specific sample sizes are often too small to provide direct estimates with acceptable coefficients of variation (CVs); some area sample sizes may be even zero in which case no direct estimation is possible. For example, the Spanish Survey of Income and Living Conditions (SSILC) can produce reliable estimates of poverty incidence at the level of Spanish Autonomous Communities (large regions) but not at the level of Spanish provinces due to small SSILC sample sizes for some of the provinces. We call a domain or local area as a small area if the area-specific sample size is too small to provide a reliable direct estimator.

Analysis of Poverty Data by Small Area Estimation, First Edition. Edited by Monica Pratesi.

Companion Website: www.wiley.com/go/pratesi/poverty

Because of difficulties with direct estimation for a small area, it becomes necessary to borrow information from related areas through a recent census and current administrative records providing auxiliary variables related to the welfare variable of interest and thus increase the *effective* area sample size. This is accomplished through modeling the welfare variable as a function of the auxiliary variables and a random area effect that can account for the residual variability. Resulting model-based estimators for small areas are called *indirect* estimators and such estimators can lead to large gains in efficiency over the direct estimators and in turn acceptable CVs because they are not area-specific. The book by Rao and Molina [1] gives an extensive account of model-based estimation for small areas, particularly for simple parameters such as area means, totals and proportions.

The focus of this chapter is on model-based estimation of small area poverty measures which are typically non-linear functions of the welfare variable. In particular, we consider empirical Bayes (EB) and hierarchical Bayes (HB) methods for handling non-linear parameters under area level and unit level models. Area level models relate the direct area estimators to area level covariates and random area effects. On the other hand, unit level models relate unit values of the welfare variable (or a suitable function of it) to unit level covariates and random area effects, in particular through a nested error linear regression model. Area level and unit level models have been used extensively for estimating area means and totals.

For poverty measures, we develop EB estimators and associated estimators of mean squared error (MSE) through an assumed nested error linear regression model. HB estimation further assumes that the model parameters are random and specifies a prior distribution for the model parameters, unlike the EB method. Posterior distributions of the area parameters are obtained conditional on the data and used to make inferences; in particular the mean of the posterior distribution is used as the point estimator, posterior variance as measure of variability of the estimator and posterior credible intervals for interval estimation. We focus on the Foster–Greer–Thorbecke (FGT) family of poverty measures [2] which includes poverty rate and poverty gap. FGT measures have received a lot of attention and are extensively used by the World Bank to produce poverty maps for many developing countries using the ELL method proposed by Elbers, Lanjouw, and Lanjouw [3], see Section 17.4.1. Both EB and HB methods are applicable to more complex poverty measures, such as the fuzzy monetary and supplementary indices based on ranking the individuals with respect to their level of poverty or welfare. The latter indicators do not require the specification of a poverty line, unlike the FGT family of measures. EB and HB methods for poverty measures are briefly described in Sections 17.4.2 and 17.4.3, respectively.

17.2 Poverty Measures

Suppose that the population consists of N population units with N_i units belonging to area $i\ (= 1, \ldots, M)$. We denote the welfare variable for the unit $j\ (= 1, \ldots, N_i)$ in area i by E_{ij}. The family of FGT measures for area i may be expressed as

$$F_{\alpha i} = N_i^{-1}\sum_j \{(z - E_{ij})/z\}^{\alpha} I(E_{ij} < z)\} \equiv N_i^{-1}\sum_j F_{\alpha ij}, \alpha \geq 0 \tag{17.1}$$

where z is a specified poverty threshold and $I(E_{ij} < z) = 1$ if E_{ij} is below z and zero otherwise. Poverty rate, poverty gap and poverty severity are obtained by taking $\alpha = 0$, $\alpha = 1$ and $\alpha = 2$, respectively in (17.1). It is clear from (17.1) that $F_{\alpha ij}$ is a non-linear function of E_{ij}. As noted earlier, a drawback of FGT measures is that they depend on z.

The Fuzzy Monetary Index (FMI), denoted by F_{Mi}, is obtained by replacing $F_{\alpha ij}$ in (17.1) by F_{Mij} which is the product of two terms: term 1 is the proportion of individuals in the population that are less poor than the individual ij raised to power $(\gamma - 1)$ and term 2 is the share of the total welfare for all individuals that are less poor than individual ij, where γ is a positive constant (see Ferretti and Molina [4] for details). The second term is equal to $1 - L(E_{ij})$ where $L(E_{ij})$ is the population Lorenz curve evaluated at E_{ij}. The Fuzzy Supplementary Index (FSI) is given by $F_{Si} = N_i^{-1} \sum_{j \in s(i)} F_{Sij}$ where F_{Sij} is obtained from F_{Mij} by replacing E_{ij} by a score S_{ij} obtained by applying a multi-dimensional approach. It is clear that both FMI and FSI are much more complex than FGT measures because FM_{ij} and FS_{ij} depend on the welfare values of all other population individuals unlike $F_{\alpha ij}$ which depends only on E_{ij}. Other indices of poverty include the quintile share, Gini coefficient, Sen index, Theil index, and generalized entropy [5].

For the area level model, we assume that area level covariate vectors x_i are available for all the areas, sampled and non-sampled. Similarly, for the unit level model we assume that unit level covariate vectors x_{ij} are available for all the population units; in the case of estimating means, it is sufficient to know only the population area means $\overline{X}_i$.

We suppose that $m(\leq M)$ of the areas are sampled and that independent samples $s(i)$ of sizes n_i are drawn from the sampled areas. In the case of unit level models, we observe E_{ij} and associated x_{ij} for $j \in s(i)$ and $i \in s$, where s is the sample of areas. In the case of area level models, for estimating area mean $\overline{E}_i$ it is sufficient to know only direct estimators of $\overline{E}_i$ and area level covariates x_i, but for complex parameters like $F_{\alpha i}$ we need the sample E_{ij} in order to compute direct estimators $\widehat{F}_{\alpha i}$ unless the latter are reported, where $\widehat{F}_{\alpha i} = N_i^{-1} \sum_{j \in s(i)} w_{ij} F_{\alpha ij}$ and w_{ij} are the design weights. Direct estimators of FMI and FSI are much more complex because we need to estimate F_{Mij} and F_{Sij} from the sample.

17.3 Fay–Herriot Area Level Model

An area level model, proposed by Fay and Herriot [6], is extensively used in model-based small area estimation of means, totals, or proportions. The main advantage of the Fay–Herriot (FH) model is that it uses only area level auxiliary information as covariates and direct estimates of the variable of interest for sampled areas. Direct area estimates take account of the sampling design through design weights. However, the FH model may not hold for the sample of areas if not all areas are sampled. This happens when the design used to sample the areas is informative in the sense that the design weights are related to direct estimates even after conditioning on the covariates. Here we assume that the sampling of areas is not informative so that the same model holds for the sampled and the non-sampled areas. Inference under informative sampling is more complex and beyond the scope of this chapter. The reader is referred to Rao and Molina [1], chapter 6 for inference on area means or totals under the FH model.

Molina and Morales [7] applied the FH model to estimate the FGT poverty measures. In this case the FH model consists of two components: (i) a sampling model $\widehat{F}_{\alpha i} = F_{\alpha i} + e_i$, $i = 1, \ldots, m$, where the e_i are sampling errors assumed to be independent with mean 0 and known variance ψ_i; and (ii) a linking model on the unknown $F_{\alpha i}$ that links the areas through the covariates x_i and random area effects v_i. It is given by $F_{\alpha i} = x_i^T \beta + v_i$, $i = 1, \ldots, M$ where the v_i are independent and identically distributed (i.i.d.) with mean 0 and unknown variance σ_v^2.

Combining (17.1) and (17.2) we get the FH model for the sampled areas:

$$\widehat{F}_{\alpha i} = x_i^T \beta + v_i + e_i, i = 1, \ldots, m \tag{17.2}$$

Model (17.2) is a special case of a linear mixed model and hence standard results for linear mixed models can be applied to get the best linear unbiased prediction (BLUP) of $F_{\alpha i}$ for sampled areas as

$$\widehat{F}_{\alpha i}^{BLUP}(\sigma_v^2) = \gamma_i \widehat{F}_{\alpha i} + (1-\gamma_i) x_i^T \widetilde{\beta}(\sigma_v^2), i = 1, \ldots, m \tag{17.3}$$

where $\widetilde{\beta}(\sigma_v^2)$ is the generalized least squares (GLS) estimator of β and $\gamma_i = \sigma_v^2/(\sigma_v^2 + \psi_i)$. It follows from (17.3) that the BLUP is a weighted combination of the direct estimator $\widehat{F}_{\alpha i}$ and the regression synthetic estimator $x_i^T \widetilde{\beta}(\sigma_v^2)$ and that it approaches the direct estimator when the sampling variance ψ_i is small relative to the between area variance σ_v^2 and to the regression synthetic estimator when ψ_i is large relative to σ_v^2. The latter feature of BLUP is practically appealing.

In practice, σ_v^2 is not known and we need to replace it by a suitable estimator $\widehat{\sigma}_v^2$ in (17.3) to get an empirical BLUP (EBLUP) given by $\widehat{F}_{\alpha i}^{EBLUP} = \widehat{F}_{\alpha i}^{BLUP}(\widehat{\sigma}_v^2)$. EBLUP does not require normality assumption provided a suitable moment estimator of σ_v^2 is used. For the non-sampled areas, we use the regression synthetic estimator $\widehat{F}_{\alpha i}^{SYN} = x_i^T \widehat{\beta}$, $i = m+1, \ldots, M$, where $\widehat{\beta} = \widetilde{\beta}(\widehat{\sigma}_v^2)$. If normality of sampling errors e_i and random effects v_i is also assumed, then the EBLUP is identical to the EB estimator.

A drawback of the FH model is the assumption of known sampling variance ψ_i. In practice, an estimator $\widehat{\psi}_i$ obtained from unit level data is often used as a proxy for ψ_i by ignoring the error due to estimating ψ_i. This approach can lead to substantial underestimation of the true MSE. Alternatively, the estimators $\widehat{\psi}_i$ are smoothed using some models connecting the estimators to suitable covariates and then treating the smoothed estimator $\widehat{\psi}_{iS}$ as a proxy for ψ_i.

The EBLUP is effective only when the covariates are related to $F_{\alpha i}$ for the chosen values of α. It may be difficult to find suitable covariates related to poverty measures regardless of the chosen value of α. On the other hand, it is usually easier to find suitable covariates that are related to direct estimators of area means of the welfare variable.

The MSE under the assumed model is used as a measure of variability of the EBLUP. A second-order correct estimator of MSE of $\widehat{\theta}_i^{EBLUP}$, assuming normality of the random effects and sampling errors and estimating σ_v^2 by the restricted maximum likelihood (REML) method, is given by

$$\text{mse}\,(\widehat{\theta}_i^{EBLUP}) = g_{1i}(\widehat{\sigma}_v^2) + g_{2i}(\widehat{\sigma}_v^2) + 2g_{3i}(\widehat{\sigma}_v^2) \tag{17.4}$$

The three terms on the right-hand side are obtained by replacing σ_v^2 by $\widehat{\sigma}_v^2$ in the following formulae:

$$g_{1i}(\sigma_v^2) = \gamma_i \psi_i, \qquad g_{2i}(\sigma_v^2) = (1-\gamma_i)^2 x_i \left(\sum_i \gamma_i x_i x_i^T \right)^{-1} x_i^T \tag{17.5}$$

and

$$g_{3i}(\widehat{\sigma}_v^2) = (1-\gamma_i)^2 (\sigma_v^2 + \psi_i)^{-1} \overline{V}(\widehat{\sigma}_v^2), \tag{17.6}$$

where $\overline{V}(\widehat{\sigma}_v^2)$ is the asymptotic variance of the REML estimator $\widehat{\sigma}_v^2$ which can be readily obtained from standard software such as SAS or R (see Rao and Molina [1], chapter 6 for

Table 17.1 Estimated coefficients of variation (CV%) of the direct, EBLUP, EB and HB estimators of poverty incidence for selected provinces by gender

Province	Gender	Sample size	Obs. poor	CV(direct)	CV(EBLUP)	CV(EB)	CV(HB)
Soria	F	17	6	51.9	25.4	16.6	19.8
Tarragona	M	129	18	24.4	20.1	14.9	12.3
Córdoba	F	230	73	13.0	10.5	6.2	6.9
Badajoz	M	472	175	8.4	7.6	3.5	4.2
Barcelona	F	1483	191	9.4	7.8	6.5	4.5

further details). The terms g_{2i} and g_{3i} in (17.4) are of lower order than the leading term g_{1i} and account for the estimation of β and σ_v^2, respectively. The leading term g_{1i} shows that the reduction in MSE of the EBLUP over the direct estimator is large when γ_i is small (or ψ_i is large relative to σ_v^2).

Molina and Morales [7] applied the FH model to the 2006 SSILC estimate poverty rates for the 52 Spanish provinces by gender. In the application, the estimated sampling variances of the direct estimators have been treated as the true variances. Domain proportions of individuals with Spanish nationality, in different age groups and in several employment categories were used as covariates. Table 17.1 reports the CVs of direct estimators of poverty rates, together with CVs of EBLUP estimators obtained using the MSE estimator given by (17.4), for a selection of domains (crossings province–gender). Concretely, we list the results for the domains with minimum and maximum sample sizes, and sample sizes closest to first quartile, median, and third quartile. It can be seen that the CVs of EBLUP estimators are smaller than those of direct estimators, especially for the smallest domain (Soria-Female). In fact, the results of Molina and Morales [7] indicated gains in efficiency of the EBLUPs based on the FH model with respect to the direct estimators for most of the domains, but in general the gains in efficiency are modest.

17.4 Unit Level Nested Error Linear Regression Model

In this section we present the ELL/World Bank method and the EB and HB methods for estimating area level poverty measures, all based on unit level nested error linear regression models. All three methods can be used to estimate general poverty indicators, but we focus on the FGT poverty indicators given by (17.1).

17.4.1 ELL/World Bank Method

The ELL method is based on the following data: (i) unit level auxiliary information $\{x_{ij}; j = 1, \ldots, N_i, i = 1, \ldots, M\}$ from recent census or other administrative sources; and (ii) survey data on the welfare variable of interest, E_{ij}, and the same auxiliary variables x_{ij}. Sampling design uses a two-stage design with clusters as first stage units, but for comparability with the EB and HB methods we take the clusters as small areas here. The welfare variable E_{ij} is first transformed to $y_{ij} = \log(E_{ij})$ and then a nested error model is assumed for the sample data

$\{(y_{ij}, x_{ij}), j = 1, \ldots, n_i; i = 1, \ldots, m\}$:

$$y_{ij} = x_{ij}^T \beta + v_i + e_{ij},\ j = 1, \ldots, n_i; i = 1, \ldots, m \tag{17.7}$$

where the area effects v_i are independent with mean 0 and variance σ_v^2 and independent of the unit errors e_{ij} which are assumed to be independent with mean 0 and variance σ_e^2. In the ELL method, unit errors are allowed to have unequal variances but for comparability with the other methods we assume a constant variance.

ELL uses ordinary least squares (OLS) to estimate the residuals $u_{ij} = v_i + e_{ij}$ and split the estimated residual $\hat{u}_{ij} = y_{ij} - x_{ij}^T \hat{\beta}_{OLS}$ into components $\hat{v}_i = \hat{u}_{i+} = n_i^{-1} \sum_j \hat{u}_{ij}$ and $\hat{e}_{ij} = \hat{u}_{ij} - \hat{u}_{i+}$. A simulated census of the welfare variable may then be generated as $\{E_{ij}^* = \exp(y_{ij}^*), j = 1, \ldots, N_i; i = 1, \ldots, m\}$, where y_{ij}^* is taken as $x_{ij}^T \hat{\beta}_{OLS} + v_i^* + e_{ij}^*$ by generating v_i^* and e_{ij}^* randomly from the fitted values $\{\hat{v}_i, i = 1, \ldots, m\}$ and $\{\hat{e}_{ij}, j = 1, \ldots, n_i; i = 1, \ldots, m\}$, respectively. Replacing E_{ij} in (17.1) by the simulated census value E_{ij}^*, the FGT poverty measure is calculated and denoted as $F_{\alpha i}^*$.

This process of generating a simulated census is repeated a large number of times, say $A = 100$, to obtain A values of the poverty measure, denoted by $F_{\alpha i}^{*(a)}, a = 1, \ldots, A$. The ELL estimator of $F_{\alpha i}$ is taken as

$$\hat{F}_{\alpha i}^{ELL} = A^{-1} \sum_{a=1}^{A} F_{\alpha i}^{*(a)},\ i = 1, \ldots, M. \tag{17.8}$$

The ELL method implicitly assumes that the model (17.7) fitted to the sample data also holds for the population. This assumption is valid under non-informative sampling or absence of sample selection bias.

The ELL estimator of MSE of $\hat{F}_{\alpha i}^{ELL}$ is simply obtained as

$$\text{mse}(\hat{F}_{\alpha i}^{ELL}) = A^{-1} \sum_{a=1}^{A} (F_{\alpha i}^{*(a)} - \hat{F}_{\alpha i}^{ELL})^2. \tag{17.9}$$

However, the ELL estimator of MSE can lead to significant underestimation of the true MSE because the model parameters are not re-estimated in each bootstrap replicate of the sample. Also, the ELL method can be inefficient for populations exhibiting significant area effects because it is not area-specific. For example, the ELL estimator of the area mean $\overline{Y}_i$ is approximately equal to the regression synthetic estimator $\overline{X}_i^T \hat{\beta}$ which is less efficient than the EB estimator of $\overline{Y}_i$ unless σ_v^2 is small. Here $\overline{X}_i$ is the known area mean of the x_{ij} values in area i.

An advantage of the ELL method is that it can be implemented without linking the sample to the population because a simulated census generates values for all the population units including the sampled units. Also, in the application of the ELL method to developing countries, the number of sampled areas is typically small compared with the number of areas in the population which means that for most of the areas no sample observations are available. The EB and HB methods are effective for areas with sample observations, and they also use regression synthetic type estimators for the non-sampled areas. However, in other applications, especially for poverty estimation in European countries, many areas will have sample observations and hence methods that take account of the sample data are more appropriate than the ELL method.

We refer the reader to Chapter 18 for details of the ELL method and some modifications.

17.4.2 Empirical Bayes Method

The EB_MR method of Molina and Rao [8] uses the nested error model (17.7) for the sample data with the further assumption that the random effects v_i and the unit errors e_{ij} are normally distributed. The poverty measure $F_{\alpha i}$ is expressed in terms of the transformed values y_{ij} as $F_{\alpha i} = N_i^{-1}\sum_j h_\alpha(y_{ij})$ using $E_{ij} = \exp(y_{ij})$ in (17.1). The N_i vector y_i of population values y_{ij} is partitioned into the vector of sample values y_{is} and the vector of out-of-sample values y_{ir}. The best predictor of $F_{\alpha ij} = h_\alpha(y_{ij})$ for $j \in r(i)$ is obtained as its expectation with respect to the conditional distribution of y_{ij} given the sample y_{is}, where $r(i)$ denotes the set of non-sampled units in area i. A closed form expression for the best predictor for general α does not exist, but it can be approximated by a Monte Carlo approximation by generating values $y_{ij}^{(l)}, l = 1, \ldots, L$ from the conditional distribution of the vector of out-of-sample values y_{ir} given y_{is}. The Monte Carlo approximation to the best predictor of $F_{\alpha ij}$ for $j \in r(i)$ is given by

$$\hat{F}_{\alpha ij}^{B} \approx L^{-1}\sum_{l=1}^{L} h_\alpha(y_{ij}^{(l)}),\ j \in r(i) \tag{17.10}$$

The best estimator (17.10) depends on the model parameters β, σ_v^2 and σ_e^2. Replacing the parameters in (17.10) by suitable estimators $\hat{\beta}, \hat{\sigma}_v^2$ and $\hat{\sigma}_e^2$, such as the maximum likelihood (ML) or REML estimators, we get the EB_MR (Empirical Bayes Molina Rao) estimator $\hat{F}_{\alpha ij}^{EB}$ of $F_{\alpha ij}$.

Under normality of the random effects and the unit errors in the model (17.7), the desired values $y_{ij}^{(l)}$ can be generated from a univariate normal distribution [8]. In particular, $y_{ij}^{(l)}$ for $j \in r(i)$ can be generated as $\hat{\mu}_{ij|s} + u_i^{(l)} + \varepsilon_{ij}^{(l)}$, where $\hat{\mu}_{ij|s} = x_{ij}^T\hat{\beta} + \hat{\gamma}_i(\bar{y}_i - \bar{x}_i^T\hat{\beta})$ is the jth element of the estimated mean vector of the conditional distribution of y_{ir} given y_{is}, and the random area effects $u_i^{(l)}$ and the errors $\varepsilon_{ij}^{(l)}$ are independently generated from $N[0, \hat{\sigma}_v^2(1-\hat{\gamma}_i)]$ and $N(0, \hat{\sigma}_e^2)$, respectively. It now follows that the EB_MR estimator of $F_{\alpha i}$ is given by

$$\hat{F}_{\alpha i}^{EB} = N_i^{-1}\left(\sum_{j\in s(i)} F_{\alpha ij} + \sum_{j\in r(i)} \hat{F}_{\alpha ij}^{EB}\right) \tag{17.11}$$

where $F_{\alpha ij}$ for $j \in s(i)$ is computed from the sample data and where $s(i)$ is the set of sampled units in area i. The above method of finding EB_MR estimators is valid for general non-linear functions $h(y_i)$ of the transformed values y_{ij}; it is not necessary to have the separable form of FGT measures. Hence, it can be applied to other poverty measures mentioned in Section 17.1. The EB_MR method is significantly more efficient than the ELL method in terms of MSE for populations exhibiting significant area effects.

The EB_MR estimator (17.11) requires the linking of the sample to the population to identify the non-sampled units. If this is not possible, we can predict all the values $F_{\alpha ij}$, as in the ELL/WB method, and use the Census EB (CEB) given by

$$\widehat{F}_{\alpha i}^{CEB} = N_i^{-1}\left(\sum_j \widehat{F}_{\alpha ij}^{EB}\right), \tag{17.12}$$

where the summation is over all the units j in area i. The CEB estimator is less efficient than the EB_MR estimator when the linking of the sample to the population is feasible.

Analytical approximations to the MSE are difficult to derive in the case of complex parameters such as the FGT poverty measures. Molina and Rao [8] used a parametric bootstrap method of estimating MSE of the EB_MR estimator. This method is computer-intensive but performed well in tracking the true MSE in empirical studies.

For the non-sampled areas i, the values $y_{ij}^{(l)}$ for $j = 1, \ldots, N_i$ are generated by bootstrap from $y_{ij} = x_{ij}^T\widehat{\beta} + v_i^* + e_{ij}^*$, where v_i^* is drawn from $N(0, \widehat{\sigma}_v^2)$ and e_{ij}^* from $N(0, \widehat{\sigma}_e^2)$. Formula (17.10) is then used to get the EB_MR estimator $\widehat{F}_{\alpha ij}^{EB}$ of $F_{\alpha ij}$ and the EB_MR estimator of $F_{\alpha i}$ using formula (17.12). In this case, the EB_MR estimator is essentially a synthetic estimator since sample observations are not available from a non-sampled area i. The EB_MR estimator is in fact essentially equivalent to the ELL estimator for a non-sampled area. However, Molina and Rao [8] obtain a proper MSE estimator of the EB_MR estimator through parametric bootstrap.

Ferretti and Molina [4] applied the EB_MR method to more complex poverty measures, in particular, the fuzzy monetary and the fuzzy supplementary poverty indicators. They developed a fast EB_MR method based on replacing the true value of the poverty indicator in simulated population l by a design-based estimator based on a sample drawn from this population. Samples are drawn independently across the simulated populations. Simulation results indicated that the fast EB_MR method is almost as efficient as the full EB_MR method. The fast EB_MR method is particularly useful for large populations in terms of computational efficiency.

17.4.3 Hierarchical Bayes Method

The HB approach can also be used to make inferences on poverty measures for small areas. We generate a large sample of values $\{F_{\alpha i}^{(h)}, h = 1, \ldots, H\}$ from the posterior distribution of $F_{\alpha i}$ based on a prior distribution on the model parameters. The HB estimator, $\widehat{F}_{\alpha i}^{HB}$, under squared error loss, is the posterior mean obtained by averaging $F_{\alpha i}^{(h)}$ over $h = 1, \ldots, H$. As a measure of variability, we use the posterior variance obtained as the variance of the $F_{\alpha i}^{(h)}$ values. Thus,

$$\widehat{F}_{\alpha i}^{HB} = E(F_{\alpha i}|y_s) \approx H^{-1}\sum_{h=1}^{H} F_{\alpha i}^{(h)}, \quad V(F_{\alpha i}|y_s) = H^{-1}\sum_{h=1}^{H}\left(F_{\alpha i}^{(h)} - \widehat{F}_{\alpha i}^{HB}\right)^2 \tag{17.13}$$

where y_s denotes the sample of y values.

In order to generate $F_{\alpha i}^{(h)}$, Molina, Nandram and Rao [9] reparametrized the model, which together with non-informative priors provides a proper posterior and at the same time enables simulating directly from the posterior distribution and avoiding Markov chain Monte Carlo (MCMC) sampling that is commonly used in HB inference. The reparametrization is based on expressing the model in terms of the intra-cluster correlation $\rho = \sigma_v^2/(\sigma_v^2 + \sigma_e^2)$ with $0 < \rho < 1$. A non-informative prior, $\pi(\beta, \sigma_e^2, \rho) \propto \sigma_e^{-2}$, on the model parameters β, σ_e^2 and ρ is used. MCMC is avoided by representing the joint posterior distribution of $\theta = (v^T, \beta^T, \sigma_e^2, \rho)$, where $v = (v_1, \ldots, v_m)^T$, as

$$\pi(\theta|y_s) = \pi(v|\beta, \sigma_e^2, \rho, y_s)\pi(\beta|\sigma_e^2, \rho, y_s)\pi(\sigma_e^2|\rho, y_s)\pi(\rho|y_s) \tag{17.14}$$

where the first three conditional distributions on the right-hand side of (17.14) have closed forms but not $\pi(\rho|y_s)$. However, random values from $\pi(\rho|y_s)$ can be easily generated using either a grid method or an accept–reject algorithm. Using (17.14) we can generate a random value from the posterior distribution of θ by first drawing a ρ from $\pi(\rho|y_s)$, then a σ_e^2 from $\pi(\sigma_e^2|\rho, y_s)$, then a β from $\pi(\beta|\sigma_e^2, \rho, y_s)$, and finally a v from $\pi(v|\beta, \sigma_v^2, \rho, y_s)$. This procedure is repeated a large number , H, of times to get a random sample $\{\theta^{(h)}, h = 1, \ldots, H\}$ from the joint posterior $\pi(\theta|y_s)$. Then, for each generated $\theta^{(h)}$ we can draw out-of-sample values $\{y_{ij}^{(h)}, j \in r(i)\}, i = 1, \ldots ., m$, noting that conditional on θ, the y_{ij} for $j \in r(i)$ are independent of y_s and

$$y_{ij}|\theta \sim_{\text{ind}} N(x_{ij}^T\beta + v_i, \sigma_e^2) \tag{17.15}$$

Thus we construct the full population vector $y_i^{(h)}$ by putting together the sample vector y_{is} and the generated non-sampled vector $y_{ir}^{(h)}$. Then using $y_i^{(h)}$ we can calculate the poverty measure $F_{\alpha i}^{(h)}$. For the non-sampled areas, we generate the whole population vector $y_i^{(h)}$ using (17.15) and $\theta^{(h)}$.

The above HB method avoids difficulties in MSE estimation associated with the EB_MR estimator based on bootstrap sampling. It is computationally very fast compared with the EB_MR method and also provides credible intervals on $F_{\alpha i}$ using the generated values $F_{\alpha i}^{(h)}$ from the posterior distribution of $F_{\alpha i}$. Molina, Nandram and Rao [9] provide HB methods for model checking.

17.5 Application

Molina, Nandram and Rao [9] used data from the 2006 SSILC to estimate poverty rates and poverty gaps for Spanish provinces by gender using the EB and the HB methods. Direct estimators are not reliable at the province level for several of the 52 provinces in Spain. They considered the nested error linear regression model for the log-equivalized disposable income and used the following explanatory (auxiliary) variables: indicators of five age groups, of having Spanish nationality, of 3 education levels, and of the labor force status (unemployed, employed or inactive). For each auxiliary variable, one of the categories was considered as base reference, omitting the corresponding indicator and then including an intercept term in the model. EU-SILC used a two-stage design with stratification of the first-stage units. Residuals from the fitted model were plotted against the sampling weights for each gender which indicated no evidence of informative sampling in this application. Several HB model diagnostics were also applied.

Table 17.1 above reports the CVs of the direct estimator, the EBLUP estimator based on the FH model, and the EB and HB estimators of poverty incidence based on the unit level model for selected provinces by gender. For the HB estimator the CV is computed from the posterior variance. It is clear from Table 17.1 that both EB and HB have much smaller CVs relative to the direct estimator and also smaller CVs than the EBLUPs based on the FH model, especially for provinces with small sample sizes (e.g., Soria).

17.6 Concluding Remarks

This chapter presented a brief account of efficient model-based methods for small area estimation of poverty measures, in particular focusing on the EB and the HB methods. A limitation of the proposed methods is the assumption of normality. Diallo and Rao [10] relaxed the assumption of normality by assuming the family of skew normal (SN) distributions for the random effects v_i and/or the errors e_{ij}. The SN family includes the normal distribution and it can represent a variety of skew distributions. Diallo and Rao [10] derived EB estimators of small area poverty indicators under the SN model. Their results indicate that the assumption of normality on v_i is robust to deviations from normality provided e_{ij} remains normal. On the other hand, under SN errors e_{ij}, normality-based EB estimators can induce significant bias and may not perform well compared with SN-based EB estimators. Van der Weide and Elbers [11] studied normal mixture models on the area effects v_i but assuming normality on the e_{ij}. Their results are in agreement with Diallo and Rao [10] in the sense that the normality-based EB method is robust provided e_{ij} remains normal.

Alternative approaches to estimation of small area poverty measures, based on the M-quantile method, are reviewed in Chapter 9.

References

[1] Rao, J.N.K., and Molina, I. (2015) *Small Area Estimation, Second Edition*. John Wiley & Sons, Inc., Hoboken.

[2] Foster, J. Greer, J., and Thorbecke, E. (1984) A class of decomposable poverty measures. *Econometrica*, **52**, 761–766.

[3] Elbers, C., Lanjouw, J.O., and Lanjouw, P. (2003) Micro-level estimation of poverty and inequality. *Econometrica*, **71**, 355–364.

[4] Ferretti, C. and Molina, I. (2012) Fast EB method for estimating complex poverty indicators in large populations. *Journal of the Indian Society of Agricultural Statistics*, **66**, 105–120.

[5] Neri, L., Ballini, F., and Betti, G. (2005) Poverty and inequality in transition countries. *Statistics in Transition*, **7**, 135–157.

[6] Fay, R.E. and Herriot, R.A. (1979) Estimation of income from small places: An application of James-Stein procedures to census data. *Journal of the American Statistical Association*, **74**, 269–277.

[7] Molina, I. and Morales, D. (2009) Small area estimation of poverty indicators. *Boletin de Estadistica e Investigacion Operativa*, **25**, 218–225.

[8] Molina, I. and Rao, J.N.K. (2010) Small area estimation of poverty indicators. *Canadian Journal of Statistics*, **38**, 369–385.

[9] Molina, I., Nandram, B., and Rao, J.N.K. (2014) Small area estimation of general parameters with application to poverty indicators: a hierarchical Bayes approach. *Annals of Applied Statistics*, **8**, 852–885.

[10] Diallo, M. and Rao, J.N.K. (2014) Small area estimation of complex parameters under unit-level models with skew-normal errors. *Proceedings of the Survey Research Section, American Statistical Association* http://www.amstat.org/sections/srms/proceedings/y2014/Files/311396_87439.pdf [accessed Sep-5-2015].

[11] Van der Weide, R. and Elbers, C. (2013) Estimation of normal mixtures in a nested error model with an application to small area estimation of welfare. Presented at the SAE Conference 2013, Bangkok, Thailand.

Part VI
Data Analysis and Applications

18

Small Area Estimation Using Both Survey and Census Unit Record Data

Links, Alternatives, and the Central Roles of Regression and Contextual Variables

Stephen J. Haslett[1,2]
[1] *Institute of Fundamental Sciences, Massey University, Palmerston North, New Zealand*
[2] *Statistical Consulting Unit, The Australian National University, Canberra, Australia*

18.1 Introduction

On discipline lines, there are at least three strands to techniques that generate estimates at small area level. In the main, these different methods have developed in parallel in the different disciplines, and the overlap is small.

The first strand is the now extensive statistical literature on small area estimation (SAE). The core reference remains Rao (2003) with its comprehensive coverage of statistics-based developments in SAE. More recent developments are extensively reviewed in Pfeffermann (2013), but because development remains rapid even this review is no longer comprehensive (see e.g. Chambers *et al.*, 2014). Interesting characteristics of this research are the distinction between design-based and model-based small area estimates, its focus not only on estimates at small area level but also their standard errors, the strong linear algebraic focus, and the very limited treatment in publications of issues such as data preparation and selection of the variables used to predict the small area variable(s) of interest. Design-based and model-based estimates generally both use models based on sample survey data for estimates at small area level, but differ

Analysis of Poverty Data by Small Area Estimation, First Edition. Edited by Monica Pratesi.

Companion Website: www.wiley.com/go/pratesi/poverty

in whether estimates of bias, variance, standard errors, mean square error, and so on are based on the randomisation structure inherent in the survey design, or on models conditional on the sample. The strong linear algebraic focus evident in Rao (2003) remains – see for example Chambers *et al.* (2014) – due to its utility in specifying models. How the comparative lack of explicit specification of data preparation methods or selection of the explanatory variables has affected the relationship with SAE related to developments in other disciplines is the topic of later sections. Suffice to say here that this reflects a general emphasis in the statistical literature on applications such as those in official statistics where the extent of improvement required from using SAE methods is more limited than sought in some other disciplines; usually in official statistics applications every small area contains a sub-sample so that SAE is feasible even when relatively few auxiliary regression variables and no unit record census data are available.

There have been essentially parallel developments in econometrics where the seminal paper is by Elbers, Lanjouw and Lanjouw (2003). The practicalities of implementing their method in its initial form, especially its use of contextual variables, are however more explicit in other papers and reports of the period, for example Elbers, Lanjouw and Lanjouw (2002), or more particularly Fujii (2003), and Haslett and Jones (2004). The method, commonly abbreviated to ELL, models sample survey data (using variables shared by survey and census, plus aggregated area-based contextual variables from the census), then predicts at household or individual level using unit record census data. Predictions are aggregated to form small area estimates. Repetition of the prediction to form a collection of pseudo-censuses, using pooled residuals from the estimated variance components and incorporating the variance of the regression parameter estimates, allows standard errors conditional on the model to be estimated. The extensions since 2003 are wide ranging, as Pfeffermann (2013) notes of the parallel developments in the statistical literature since Rao (2003). These further developments, as discussed later, are part of the reason the methods of Elbers, Lanjouw and Lanjouw (2003) cannot be adequately subsumed into those developed in Molina and Rao (2010) or those outlined in Chapter 17, and why further rapprochement is still needed between SAE models used in statistics and in econometrics (especially in SAE of poverty in the third world). The more complete summary of the ELL method and its importance are given in a later section. Since 2003, it has been core to SAE of poverty in over 50 countries and its estimates have been central to the allocation of billions of dollars of aid. In standard applications conforming to a particular form of ELL with two error components, it is often implemented using the software package PovMap (Zhao, 2006; Zhou and Lanjouw, 2009). Because ELL, like spatial microsimulation, was developed largely outside the statistical mainstream, its similarities, strengths and weaknesses have been the topic of much debate among statisticians, especially since the method could in principle be applied in a wide range of other application areas. Particular recent criticisms relate to bias (because there are no area-specific predictors of random effects), and to understated estimated standard errors. In this context, a focus of this chapter is the central role played in SAE by choice of contextual variables to control area-specific bias and regression variables to reduce standard errors. The marked reduction in standard errors in models such as ELL which use both unit record survey and census data, when they are compared with some other SAE methods, reflects in part an apparent focus in the published statistical literature on predicting rather than minimising random effects and standard errors. As outlined below, it also highlights the central importance of sound and careful choice of regressors and contextual variables to minimise standard error and to control bias in any good model, whatever the choice of SAE method.

In geography and public policy there has been a further development: spatial microsimulation. The ELL method and spatial microsimulation have a common origin. See, for example, Bramley (1991a, b, 1992); Bramley and Smart (1995, 1996); and Bramley and Lancaster (1998). In spatial microsimulation (as for ELL-type methods) a detailed synthetic pseudo-census dataset is generated that describes household characteristics at a local level. The technique combines sample survey datasets, and aggregate census data [as in Molina and Rao (2010), rather than unit record census data used in ELL], to get area-based estimates. The census information contains more households than the samples (even though in aggregated from), but not as many variables of interest. The analysis of spatial variation in such simulated data, it is claimed, can be used to study the impact and future impact of social policy initiatives, for example in health, justice and education. In essence the method includes techniques that may add census averages to household data where those averages are taken over neighbourhoods, some types of sample reweighting techniques, and prediction based on what is usually an implicit and untested statistical model. Standard errors are seldom estimated (cf. Pudney and Sutherland, 1994) and, when scenarios are used, the initial estimates at small area or local level are then often projected far into the future without the benefit of an explicit time series model (e.g. Stillwell *et al.*, 2004).

Small area estimates have a spatial dimension, particularly when estimates are mapped as in poverty mapping. SAE models, whether statistical or econometric, may or may not include a spatial component. Geographical Information System (GIS) software tends to focus on such spatial models, treated as continuous across space. These also form a strong component within the statistical mainstream [which is covered as part of, e.g., Cressie (1993) and Cressie and Wikle (2011)]. Some such continuous spatial models (e.g. kriging, geographically weighted regression) are not usually described as SAE, because they are based on data linked to particular sampled co-ordinate locations rather than being (or being treated as representing) unknown constant values within pre-specified administrative boundaries. This distinction is sometimes blurred by using centroids of administrative areas as point locations representing the whole of each small area [such as in Brunsdon, Fotheringham and Charlton (1998, p. 440)]. GIS co-ordinate information is increasingly being collected at household level, rather than locations in sample surveys being defined only via design information such as stratification and clustering, and such techniques are likely to become increasingly relevant as household geocodes or GPS co-ordinates are made available. The very broad topic of spatial data from known sampled or pre-specified locations is outside the scope of this chapter. What will nevertheless be discussed in a later section are the advantages and disadvantages of supplementing non-spatial small area models with models defined by areas' contiguity or adjacency, because such proximity models can smooth small area estimates, which nevertheless remain constant over administrative areas.

18.2 The ELL Implementation Process and Methodology

18.2.1 ELL: Implementation Process

The steps usually necessary when using ELL methodology for SAE of poverty and malnutrition are summarised below. A number of these steps would be required for any SAE research project that needs to assess feasibility for, or to produce, small area estimates. The steps particular to ELL, not required for SAE methods that do not use unit record census as well as unit record survey data, relate principally to matching variables and matching area codes between census and survey. Whatever the SAE technique used, there is extensive model searching required for making an optimal choice of regression variables and any contextual variables.

The steps involved are:

- Identification and examination of relevant data sources and reports to determine which show potential for use in SAE of SAE indicators.
- Identification and listing of questions asked in the census and in the selected surveys that *prima facie* are similar enough to be used for SAE. This investigation is usually based on English versions of questionnaires where available, and requires strong local involvement because questionnaires are seldom conducted in English and matching of questions in English translation is insufficient. Census only variables can also be added in aggregate form for context in the survey-based regression model (to limit the size of variance components) if area code matching is possible. However even without such contextual variables, the matching of variables between survey and census is required because ELL can base the non-contextual part of its survey-based regression model only on variables common to survey and census.
- Preliminary matching, for those questions apparently common to census and survey, of the census response categories with those of the corresponding survey question. This matching needs to be further examined via direct statistical comparison of estimates of each potential matching variable before the variable is usable in models for final SAE estimates.
- Investigating the matching of the various data sources via geocodes and/or survey design variables (such as stratification and clustering), and correcting these to ensure matching where necessary. This can be an extensive process where area codes have been changed for or after a census but older area codes are used for a survey, or vice versa. Matching of location at cluster level can be difficult for administrative reasons, and matching at household level is almost never possible (and, even if it were, could make no allowance for household mobility).
- Merging and cleaning data from each selected survey to create a dataset at household level (for expenditure poverty or kilocalorie consumption) or child level (for child under nutrition). For each observation, the dataset needs to contain the matched variables, suitably re-coded for census matching, together with the relevant survey design variables and regional, small area, cluster, and other area indicators.
- Creating cluster level means, and other area level means as contextual variables, from the population and housing census and any other available sources, and merging these with the survey data on an area basis applied at household or child level.
- Developing and testing preliminary statistical regression models with census means as contextual variables, including estimation of variance components, for expenditure per capita or kilocalories per capita or adult equivalent (using household level data), or – for children under 5 years of age – standardised height-for-age (for stunting prevalence), standardised weight-for-age (for underweight prevalence), and standardised weight-for-height (for wasting prevalence). Note that other individual level or household level variables (e.g. related to health or food security) may be and have been considered for SAE.
- Often at this stage, a preliminary feasibility report based on the survey model is written with recommendations, before any small area estimates have been produced, and even before census matching is completed, and a decision is made by the funding agency whether to proceed to the final phase.

- If training is specified or required, assessing the existing capacity of the government and agency staff to implement SAE by identifying capacity-building needs before providing formal training in SAE is advisable. Because time and resource required to ensure local autonomy in sound use of SAE techniques before research project completion is almost always insufficient, it is best that the principal researchers remain responsible for producing any final small area estimates, as well as for the formal training itself.
- Given there is to be a final phase, further cleaning of available census data for compatibility with the final survey dataset to be used for further model development and testing. At this stage the unit record level census dataset with matched and contextual variables needs to be finalised, as this (along with various model residuals from the survey) is required for application of the survey-based model to the census data to generate small area estimates and their estimated standard errors conditional on the model.
- Trial production of preliminary estimates of relevant SAE variables using available survey data in conjunction with census-based predictions amalgamated to small area level.
- Assessment of the quality and precision of the preliminary SAE estimates.
- Extensive, time consuming, further model development and production of small area estimates with maps for each SAE variable, based on choice from the many options for including or excluding candidate explanatory variables and their interactions plus contextual variables in the final survey model.
- Briefing the technical committee on the results and the level of aggregation possible based on acceptable accuracy at small area level.
- Producing a draft of the final SAE report and circulating it to stakeholders for comment.
- Incorporating the comments in the revision of the draft, and iterating as required.
- Producing a final SAE report and circulating it to stakeholders.
- Publishing the report and disseminating the findings at a national workshop/public seminar.

18.2.2 ELL Methodology: Survey Regression, Contextual Effects, Clustering, and the Bootstrap

This subsection gives a brief overview of the ELL (Elbers, Lanjouw and Lanjouw, 2003) SAE method.

Let $\boldsymbol{y}$ be a target (i.e. SAE) variable for which we want estimates for a number of small areas or subpopulations. As with other SAE applications, subpopulations are usually small geographical areas, but can be small domains made up of different subpopulations that are not collocated. The ELL method was originally applied to economic poverty measures, with $\boldsymbol{y}$ being log-transformed per capita income or expenditure. For extension to under-nourishment, log kilocalorie intake per person or per adult equivalent is used. For stunting, underweight and wasting in children under 5 years of age, $\boldsymbol{y}$ is standardised height-for-age, weight-for-age and weight-for-height, respectively.

Even in those small areas that contain sample, direct estimates (that are based only on data within each small area) generally have standard errors too large to be useful. So, in SAE, the central idea is that information available from parts of the sample outside a given small area, and from other sources such as a census even for unsampled parts of the population, can be used to improve on the direct estimates. Indeed, because models are sufficiently detailed that standard errors may in principle allow many small areas with estimates of sufficient accuracy, most small areas for which ELL generates an estimate contain no sample.

In the ELL method, unlike many other small area methods, the auxiliary or regression information used for survey respondents, denoted as $\boldsymbol{X}$, contains only variables that can be measured for the *entire* population, either by a census, through an administrative source, or via a GIS database. The variables in $\boldsymbol{X}$ fall into two categories, those at unit record level in both survey and census which can be adequately matched in aggregate, and census means (usually at cluster level in the survey) which are only available in the census and serve as contextual variables. This use of contextual variables, although not mentioned explicitly in Elbers, Lanjouw and Lanjouw (2002) or Elbers, Lanjouw and Lanjouw (2003) has been a feature of the method in applications since its inception. See, for example, expenditure modelling in Fujii (2003, p. 39, Table 9) which uses Cambodian village level variables, and Haslett and Jones (2004, p. 30, Table B1) which uses Bangladesh upazila and union level variables. When survey data are being fitted, each row in $\boldsymbol{X}$ then corresponds to a household or individual in the survey, and contains both matched and contextual variables.

Elbers, Lanjouw and Leite (2008) comment that 'Demombynes *et al.* (2006) demonstrate that performance of the ELL approach depends crucially on the ability to incorporate into the basic consumption or income model, locality-level explanatory variables inserted into the household survey data from outside datasets such as the census and/or other ancillary databases'. This statement supports the later conclusions of Molina and Rao (2010). However, Molina and Rao (2010) go further and claim their empirical best predictor (EBP) is superior to ELL, by comparing their method with a non-standard specification of ELL containing no contextual variables. The model that Molina and Rao use for comparison then is not ELL, but one which Elbers, Lanjouw and Leite (2008) have previously stated is a poor small area estimator. The superiority of EBP over ELL remains unproven.

The full specification of ELL with contextual variables also contrasts with the methods outlined in Rao (2003), where unit record census data is not used or not available. There $\boldsymbol{X}$ could be (but is usually not) more extensive information from the sampled units only, since it does not need to be matched at unit record level. However, unlike ELL in its original form, the type of statistical model can be more extensive, for example nonlinear or generalized linear models can be used to model the variable of interest directly.

In contrast to Rao (2003), ELL instead develops 'estimators of population parameters which are non-linear functions of the underlying variable of interest ... by deriving them from the full unit level distribution of that variable' (Elbers, Lanjouw and Lanjouw, 2003, p. 355). In their example from Ecuador, Elbers, Lanjouw and Lanjouw (2003) note that the population size in small areas for estimation of poverty incidence (which would more properly be called poverty prevalence) may be as small as 15 000 households. This is a rather smaller area than the methods in Rao (2003) are capable of producing reliable small area estimates for.

For ELL, the underlying model is a linear regression-type relationship between $\boldsymbol{y}$ and $\boldsymbol{X}$ namely:

$$\boldsymbol{y} = \boldsymbol{X}\boldsymbol{\beta} + \boldsymbol{u} \tag{18.1}$$

which is estimated from the survey data supplemented by the contextual variables, and where n, the sample size, is the number of rows in $\boldsymbol{y}$, $\boldsymbol{X}$ and $\boldsymbol{u}$. In equation (18.1), $\boldsymbol{\beta}$ represents the regression coefficients estimating the effect of the $\boldsymbol{X}$ variables on $\boldsymbol{y}$, and $\boldsymbol{u}$ is a random error.

The structure of $\boldsymbol{X}$ as a mixture of survey variables matched with the census plus contextual variables in the form of census means, and $\boldsymbol{u}$ as a set of random components, is central to the small area estimates and estimates of their standard error, respectively.

The core idea in ELL that can considerably reduce the standard errors for the small area estimates (or allow the small areas to be much smaller than is possible with many other SAE methods) is that if we can assume that this same survey-based relationship applies to the entire population, it can be used to predict $\boldsymbol{y}$ for *every* unit in the census population. These census unit record level predictions can and usually do contain substantial prediction error at household or child level, but the prediction error reduces substantially when they are averaged over the subpopulations or areas of interest. The small area estimates based on these predicted $\boldsymbol{y}$ values have smaller standard errors than any available direct estimates, even given the uncertainty in the unit record predicted values, because they are based on an entire census which is almost invariably several orders of magnitude larger than the sample. Although superficially different, the same central idea is pursued in Molina and Rao (2010), except that there, because by construction all small areas contain sample and are therefore much larger, and there are no contextual variables available, prediction of non-zero random effects replaces ELL's use of contextual variables.

For ELL then:

$$\boldsymbol{X\beta} = \boldsymbol{X}_s\boldsymbol{\beta}_s + \boldsymbol{X}_c\boldsymbol{\beta}_c \tag{18.2}$$

where the subscript s denotes a selection of the survey and census variables that match at unit record level, and the subscript c denotes contextual variables derived from census, GIS or administrative sources usually at survey cluster level.

The model of equation (18.2) (when it is applied to the full census data by redefining $\boldsymbol{X}$ to have N rows, where N is the population size) implicitly involves summing of contextual variables to higher (more aggregated) levels in the model, so that the estimate of $\boldsymbol{X}_c\boldsymbol{\beta}_c$ provides non-zero estimates of area level effects at higher levels than household, including small area level. Consequently $\boldsymbol{X}_c\boldsymbol{\beta}_c$ can be decomposed as

$$\boldsymbol{X}_c\boldsymbol{\beta}_c = \left(\boldsymbol{X}_{ca(1)}\,\boldsymbol{X}_{ca(2)}\ldots.\,\boldsymbol{X}_{ca(k)}\right)\begin{pmatrix}\boldsymbol{\beta}_{ca(1)}\\ \boldsymbol{\beta}_{ca(2)}\\ \ldots\\ \boldsymbol{\beta}_{ca(k)}\end{pmatrix} \tag{18.3}$$

where $a(1)$, $a(2)$, … $a(k)$ denote successively finer area effects in the model down to cluster level. The hierarchy is usually defined by administrative area boundaries, and can in principle be extended to levels even higher than small area level, such as district or regional level. Estimation can proceed directly through linear combinations which are the means of $\boldsymbol{X}_c\hat{\boldsymbol{\beta}}_c$ for each area, beginning with the most aggregated set, then proceeding in turn to each level directly below. It is possible, once $\boldsymbol{\beta}_c$ has been estimated from the survey data, to use the census data to improve on the averages of the contextual area level estimates, since these include all areas rather than only those in the sample. Producing explicit estimates of $\boldsymbol{X}_c\boldsymbol{\beta}_c$ for each area in the hierarchy, while useful, has not however formed a standard part of the ELL method. Interestingly the $N \times 1$ estimate of $\boldsymbol{X}_s\boldsymbol{\beta}_s$ can also contribute to contextual effects, because in ELL-type methods once $\boldsymbol{\beta}_s$ is estimated from the survey data it is applied to the full census data, so its estimates also vary at different levels of aggregation (e.g. small area, cluster, household) even if there is no sample in those areas.

In equation (18.1), the non-zero area effects within ELL small area estimates are incorporated within $\boldsymbol{X\beta}$ as fixed effects, rather than via random effects $\boldsymbol{u}$ as in Molina and Rao (2010). Because they use contextual variables via equations (18.1), (18.2) and (18.3) as well as unit record census data, ELL small area estimates are then not strictly synthetic estimates, in the sense that for ELL area specific effects are included in small area estimates via equation (18.1). This is contrary to the view in Molina and Rao (2010) summarised in Pfeffermann (2013), who do not recognise ELL's contextual effects. Of course, whether ELL uses a synthetic estimator depends on whether a synthetic estimator under equation (18.1) is defined via $\hat{\boldsymbol{y}} = \boldsymbol{X}\hat{\boldsymbol{\beta}}$ or by considering only the regression component without the contextual effects, that is $\hat{\boldsymbol{y}}_s = \boldsymbol{X}_s\hat{\boldsymbol{\beta}}_s$. Either way however, ELL does have non-zero estimated area specific effects. Further, ELL has the major advantage over Molina and Rao (2010) that it can estimate a non-zero area effect specific to a small area even if there is no sample in that particular small area, which means that generally it can provide estimates for much smaller small areas.

In ELL, these area effects, both at each area level in the model and in total via $\boldsymbol{X}_c\hat{\boldsymbol{\beta}}_c$ (where $\hat{\boldsymbol{\beta}}_c$ is the estimate of $\boldsymbol{\beta}_c$), can be and are estimated for all census records, and *every* small area not just the sampled ones, so that the final small area estimates include non-zero area effects individually tailored to each small area.

Estimates at finer effects (such as household, or within household) are modelled in ELL via $\boldsymbol{u}$ in equation (18.1), which also contains zero mean variance components at the higher levels. As a consequence of the contextual variables, the higher order effects in $\boldsymbol{u}$ are zero mean and comparatively small because the non-zero mean component of these higher level effects has been modelled and included in the small area estimates via the contextual effects.

18.2.3 Fitting Survey-based Models

Except in a strictly model-based environment, standard methods of fitting regression models cannot be used to fit equation (18.1), even if the software being used allows for both fixed and random effects, because unless the survey design uses an equi-probability selection method (and hence is called epsem), the survey data need to be weighted. Survey data are also generally collected from a sample that is both stratified and clustered, and this too needs to be accommodated in the analysis, even if only for robustness.

The clustering is not only a sample design property. The population is also grouped geographically or administratively into clusters containing units with similar characteristics, which when amalgamated form the small areas of interest. Households in close proximity tend to be similar, which is why even sampled households are not independent. When there is such structure in the population, and there are clusters in the sample (often called primary sampling units, or psu in design terms), the regression model (18.1) needs to be fitted using specialised methods and software rather than ordinary least squares (OLS). This issue, especially as it relates to ELL, is discussed in detail in Haslett, Isidro and Jones (2010), as the original specification of the estimation procedure for equation (18.1) in ELL is not entirely correct. This issue is also discussed in Haslett, Isidro and Jones (2010) and Haslett (2013). For an econometric perspective see Tarozzi and Deaton (2009). More general references on fitting statistical models to survey data include Skinner, Holt and Smith (1989), Chambers and Skinner (2003) and Lehtonen and Pahkinen (2003). Suitable software packages for fitting statistical models to complex survey data include Sudaan, Stata, SPSS Complex Surveys, and the survey package in R. There are also specialised but more limited routines in SAS.

18.2.4 Residuals and the Bootstrap

From the fitting of equation (18.1), the residuals (or estimated model errors) $\hat{\boldsymbol{u}}$ can be decomposed into residuals at all the hierarchical area levels in the model. Elbers, Lanjouw and Lanjouw (2003) decompose only at cluster and household level. Letting i denote cluster and j household within cluster, so that the ijth element of $\hat{\boldsymbol{u}}$ is $\hat{u}_{ij}$, the cluster level residuals are $\hat{c}_i = \hat{u}_{i\cdot}$, where the dot denotes averaging over j, and the household level residuals $\hat{e}_{ij} = \hat{u}_{ij} - \hat{c}_i$. Although not part of the original ELL method, since it focused on household expenditure, for models fitted at individual level such as stunting, underweight and wasting in children, an individual level random effect also needs to be fitted. In the original ELL specification, random effects at levels of aggregation higher than cluster level were not included. This has led to strong criticism that the model should also contain small area level random effects, since otherwise standard errors for the SAEs will be underestimated. However in a range of small area applications of ELL-type methods, for example in Bangladesh, Nepal, Philippines and Cambodia, the estimated small area level variance component (while difficult to estimate) has been negligible, because there is a non-zero small area specific effect for each small area already included in the model via the contextual effects. See Haslett and Jones (2004, 2005, 2006) and Haslett, Jones and Sefton (2013) for some further detail. ELL also models heteroscedasticity of household level residuals, although in the four applications of ELL-type methods above, the model for heteroscedasticity has had an R^2 of around 0.03, so that heteroscedasticity has been negligible. Details of the heteroscedasticity model, along with discussion of the role of R^2 in the fitted model for equation (18.1), variance components, and the importance ratios of variance components can be found in Haslett and Jones (2010) and Haslett (2013).

Bootstrapping (Efron and Tibshirani 1993), while not required for estimation of the small area estimates themselves, is one option for estimating standard errors of the ELL small area estimates. In the case of poverty mapping, bootstrapping ELL generates not only

$$\hat{\boldsymbol{y}} = \boldsymbol{X}\hat{\boldsymbol{\beta}} = \boldsymbol{X}_s\hat{\boldsymbol{\beta}}_s + \boldsymbol{X}_c\hat{\boldsymbol{\beta}}_c \tag{18.4}$$

but also a large number, B, usually $B = 100$, of alternative predicted values:

$$\hat{\boldsymbol{y}}^b = \boldsymbol{X}\hat{\boldsymbol{\beta}}^b + \hat{\boldsymbol{u}}^b = \boldsymbol{X}_s\hat{\boldsymbol{\beta}}_s^b + \boldsymbol{X}_c\hat{\boldsymbol{\beta}}_c^b + \hat{\boldsymbol{u}}^b \quad b = 1, \ldots B \tag{18.5}$$

(with $\hat{\boldsymbol{y}}$ and $\hat{\boldsymbol{y}}^b$ both $N \times 1$), to estimate the accuracy of the predictor (4), or aggregations of it over each small area.

Because $\hat{\boldsymbol{\beta}}$ is an unbiased estimator of $\boldsymbol{\beta}$ with estimated variance $\hat{\boldsymbol{V}}_{\hat{\beta}}$, for each $b = 1, \ldots, B$ we can draw $\hat{\boldsymbol{\beta}}^b$ independently from a multivariate normal distribution with mean $\hat{\boldsymbol{\beta}}$ and variance matrix $\hat{\boldsymbol{V}}_{\hat{\beta}}$. Putting aside for now (as ELL does), the need, even with contextual variables in the model, to test for and if necessary include small area level random effects, the cluster level effects can be drawn without replacement from their zero mean empirical distribution based on the sample only, possibly rescaled or 'unshrunk' so that their variance matches that of the corresponding estimated variance component. Taken over all B bootstraps, even for a given small area the mean of the cluster level predictions will be approximately zero, which is why Molina and Rao (2010), and Pfeffermann (2013) in his review, state ELL to be a synthetic estimation method. They have however not recognised that in ELL, properly specified, there are non-zero contextual effects via $\boldsymbol{X}_c\hat{\boldsymbol{\beta}}_c^b$ for every small area.

We could, as Molina and Rao (2010) suggest is optimal, use the actual residual for those small areas, clusters, and/or households that are in the sample. However in practice, the benefit is negligible even when operationally feasible, because the number of sampled clusters is almost always a very small percentage of the total number of clusters in the population from which they have been sampled, and this percentage (i.e. the sampling fraction) is smaller again for households.

Each $N \times 1$ bootstrap value $\hat{\boldsymbol{y}}^b$ in ELL for a fixed value of b yields a set of small area estimates by averaging of its relevant elements of $\hat{\boldsymbol{y}}^b$ for each small area, and the variance of the set of these for $b = 1, 2, \ldots., B$ gives standard errors for each small area. Usually however, neither $\hat{\boldsymbol{y}}$ nor $\hat{\boldsymbol{y}}^b$ are used directly.

Poverty incidence, gap and severity are defined using the FGT measures (Foster, Greer and Thorbeck, 1984):

$$P_\alpha = \frac{1}{N}\sum_{k=1}^{N}\left(\frac{z - E_k}{z}\right)^\alpha \cdot I(E_k < z) \tag{18.6}$$

where N is the population size of the area, E_k is the expenditure of the kth individual, z is the poverty line and $I(E_k < z)$ is an indicator function (which is equal to 1 when expenditure is below the poverty line, and 0 otherwise). Poverty incidence, gap and severity then correspond to $\alpha = 0, 1$ and 2, respectively. Individual expenditure is estimated by exponentiating predicted log expenditure for each household, and allocating it to the household members. Note that exponentiating must be the preliminary step. Poverty incidence, gap and severity are then usually mapped at small area level. See, for example, Figure 18.1.

To summarise, for ELL for each of the B predictions for each person in each census household, the poverty incidence, gap and severity are calculated using equation (18.5), then aggregated to small area level. The variation in these B small area estimates in each small area gives an estimate of the standard error for that small area conditional on the model being correct.

The integrity of the small area estimates and their standard errors depend on the model fitted being sound, meaning the fit is sufficiently good and unbiased, and the model applies to the census population in the same way that it applies to the sample. To have the same model apply to survey and census relies on non-informative sampling [see Pfeffermann (2013) for details and references], and good matching of survey and census variables used in $\boldsymbol{X}_s$ to provide auxiliary regression information. Even the census questionnaire (though generally much shorter than the survey questionnaire) contains many questions however, especially when categories or levels of categorical variables are involved, so extensive checking is necessary to avoid including relationships in ELL models which do not hold in the population. Such problems can be exacerbated by splitting the sample into subsamples and fitting separate models to each part (as is too often done), choosing variables from a very large set of possibilities (such as all available matching variables, plus contextual variables, and their interactions to high order), or by using final models with too many variables. While these strategies can lead to models with higher R^2 and small area estimates with apparently small standard errors, estimated standard errors are then underestimates, perhaps severely, even in the presence of contextual variables. Even when models are fitted using better methods, field verification as the final step

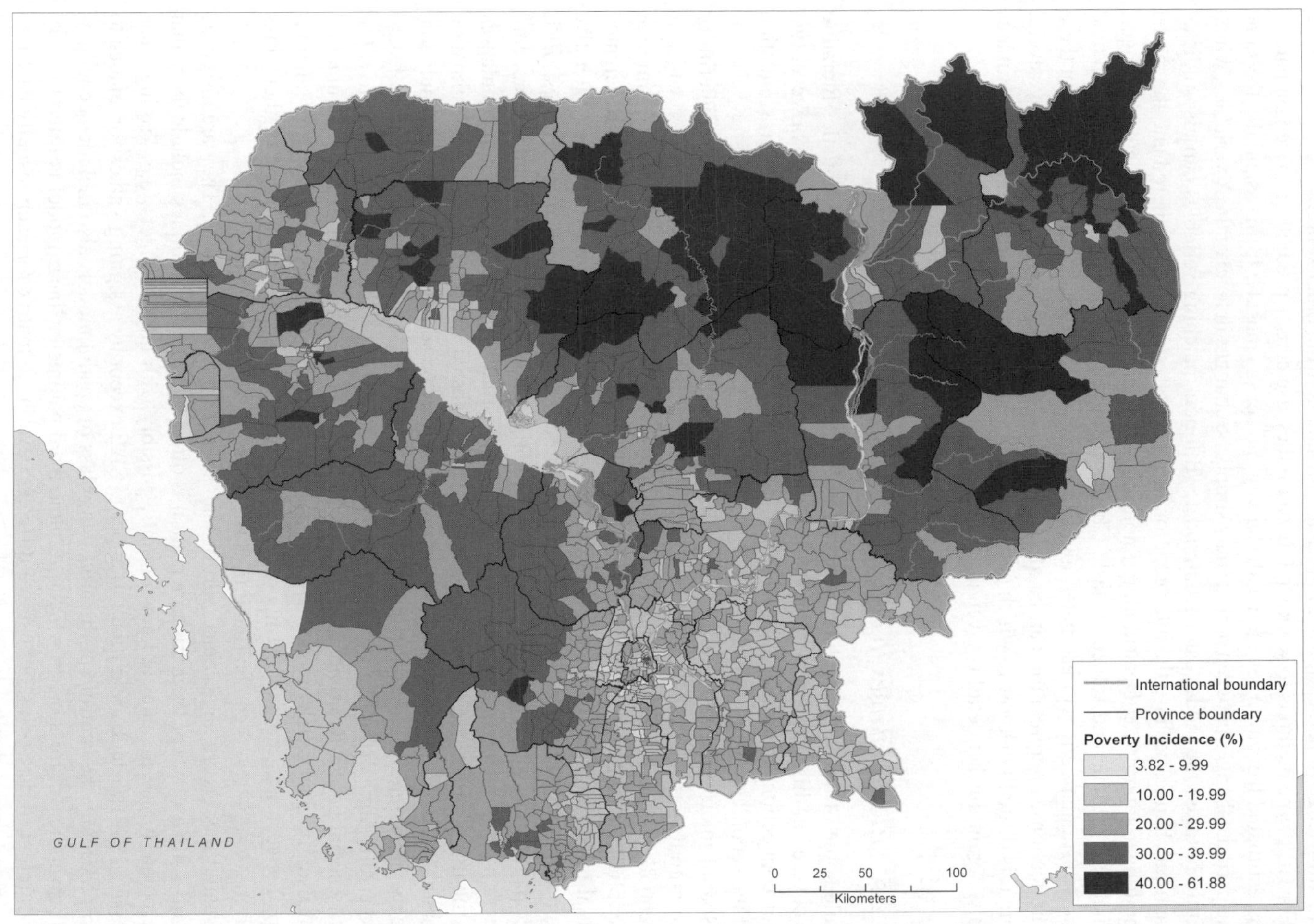

Figure 18.1 Poverty incidence (P0) in Cambodia at commune level, 2008/2009. *Source*: Reproduced with permission of the National Institute of Statistics, Cambodia

in poverty mapping is advisable, especially for small areas that have estimates out of line with local knowledge or perceptions or information available from other sources.

For ELL-type methods to work well, because they use census predictions based on parameters estimated from a survey model, the survey and census must be from essentially the same time period. Changes between periods may simply be changes in level of the X variables, which (within bounds) the model can accommodate. But non-matching variables can also indicate structural change, because the interpretation of particular variables has altered and the model is no longer appropriate. Determining if non-matching variables are due to structural changes, or not, is a difficult and subjective task. Changes of level and changes of structure both tend to add to standard errors of estimates, and in some cases to introduce bias. Even without these complications, interpretation can be affected. If date of survey and census differ by several years it is difficult to know to which period or date even otherwise sound small area estimates apply. Again, further detail is given in Haslett (2013).

18.2.5 ELL: Linkages to Other Related Methods

While some of the extensive links between ELL and SAE methods in the statistics literature are covered in this chapter, the topic is too broad to be covered in full. There are however two other connections that deserve further mention. The first of these is spatial microsimulation and the second is spatial statistics.

Spatial microsimulation and ELL have in common the intention (either as an interim or final output) to produce a rectangular dataset (or datasets) without missing values, which is essentially a pseudo-census created by substitution of missing information using an implicit or explicit statistical model. (The exception is the subclass of deterministically reweighted spatial microsimulation methods, which are reweighting techniques based on use of iterative proportional fitting or its equivalent used to calibrate to various census and other tables.) ELL and spatial microsimulation also have a common origin. Indeed ELL's contribution is not construction of pseudocensus which is also part of its historical link to spatial microsimulation, but the use of unit record census data, an explicit and tested statistical model, and estimation of standard errors. The latter two aspects could usefully be added to spatial microsimulation. The two techniques also differ in that spatial microsimulation, even without the benefit of standard errors, often extrapolates many years into the future to produce scenarios used to set or justify social policy, based not on a time series but on a single period of data and an untested (and usually unspecified) model. An extensive examination and discussion of the relationships between ELL and spatial microsimulation (which includes examination of their links with mass imputation) can be found in Haslett *et al.* (2010).

Spatial models are an accepted part of mainstream statistical methods. They are also used in GIS software, where their implementation is often occluded. Techniques include those that interpolate between values at known points, possibly with nugget effects [e.g. kriging – see Cressie (1993)], which are difficult to apply to SAE except by assigning values for an area to a point, usually the centroid of the area. This class of spatial models also includes geographically weighted regression, in which the same fixed parameter linear model is fitted to every geo-referenced data point, but using a different covariance matrix for each, usually based on a weighted kernel defined by the distance between points. Because covariance structures at different points are different, so are the parameter estimates. The effects of the covariance structure, and conditions under which parameter estimates are equal, is a long standing topic. See for example Rao (1971), Puntanen and Styan (1989) and Haslett *et al.* (2014). Geographically weighted regression in GIS software nevertheless does not usually fit a smooth surface across a map; rather it sets a constant value across each pre-specified administrative area. The other type of spatial model, and the type more often used in SAE, is the class of adjacency models. Adjacency, proximity or contiguity may be defined in a number of ways (e.g. by a rook's or by a queen's moves in chess) and may include not only nearest neighbours but also those two or more steps away. An example is a conditional autoregressive (CAR) or a simultaneous autoregressive (SAR) model. For a more general discussion of CAR models see Gelfand and Vounatsou (2003). Spatial models have been advocated as part of ELL. There has even been an argument advanced that they are an essential part (Tarozzi and Deaton, 2009).

However ELL, like many small area models, is essentially non-spatial, in the sense that none of the regression variables indicate geographical location or proximity. Small area models fitted without using geospatial models, when mapped provide important information about model fit. If, even without specifying that they should be via a spatial model, adjacent areas are generally similar in colour or shading rather than being a fine mosaic of colours or shades, this is a strong diagnostic in support for the underlying non-spatial model. In contrast, if the underlying small area model is inadequate, adding geographical smoothing via an adjacency model can only disguise (rather than rectify) the fact. The cautionary note is that, rather than improving models, using spatial information such as adjacency can disguise underlying model inadequacy.

18.3 ELL – Advantages, Criticisms and Disadvantages

Evidence that ELL can work well is given in Elbers, Lanjouw and Leite (2008) and in Haslett *et al.* (2010). See also Demombynes *et al.* (2006). The first of these papers details a study undertaken in Brazil which, rather unusually (since income and expenditure is difficult to collect because there are many questions involved to cover all its components), has census level information on income. This allowed ELL-type small area estimates, derived without use of the known census-based income information, to be compared with the known census values at small area level. The second is the simulation study in Haslett *et al.* (2010) in which an ELL-type model was used to generate survey and census data, and ELL models fitted and compared with the generating model. Perhaps unsurprisingly, model recovery was good. Small area estimates were also good, with accurate estimated standard errors for the small area estimates. These results support the view that biased small area estimates are not a necessary consequence of using ELL-type methods, despite the view in Molina and Rao (2010), reiterated in Pfeffermann (2013). Indeed the results in Elbers, Lanjouw and Leite (2008) and Haslett *et al.* (2010) are corroborating evidence that the contextual effects in ELL provide the set of non-zero estimated area effects required to adjust for bias of synthetic estimators, and support the view that (because ELL uses area specific contextual effects) ELL is not a synthetic estimator which relies only on a standard fixed effect regression model through $\boldsymbol{X}_s\hat{\boldsymbol{\beta}}_s$ in equation (18.4).

ELL methods tend to use more detailed regression models than most SAE methods in the statistical literature. This is possible because, remarkably perhaps, survey response rates for income and expenditure surveys and for health surveys in excess of 95% are common in developing countries, though are lower in transition countries. See, for example, United

Nations (2005, p. 502, item 29 and pp. 584–585, Tables XXXV.3 and XXV.4). Data quality can also be good, although not universally; for some details see United Nations (2005, p. 240, Table XXIII.7).

Nevertheless, Molina and Rao (2010) and Pfeffermann (2013) are not the first to point out possible complications with ELL-type methods. Their objections are worth discussing in more detail.

First however note that, with two main modifications, Molina and Rao's empirical best predictor (EPB) is the ELL estimator, and their parametric bootstrap mean squared error estimator is the ELL estimator of variance.

The first difference is that ELL (pseudo) census values do not contain the observed sample data. In practice however, sample sizes are in the thousands or tens of thousands, and populations in the millions or tens of millions, so the benefit of including sampled observations in pseudo censuses is negligible, even if all households or individuals could be matched by location. Molina and Rao claim, based on their simulation, that EBP is considerably better than ELL because they incorporate the sample into every pseudo census, but in their simulation (unlike real world applications) sample sizes are substantial, $n = 4000$, in comparison with their census size, $N = 20\,000$, and their census is remarkably small.

The second difference is more substantive. The EPB predicts non-zero area level random effects, and ELL does not. This means that, ELL's contextual variables aside, the Molina and Rao method differs essentially from ELL only if there is sample in a small area. Molina and Rao's simulation study reflects this. All their small areas have sample and, because there are no contextual effects used in their simulation, ELL then becomes a synthetic estimator, and EPB appears to perform much better. To understand why this matters, consider the contextual effects in the ELL model. Combining equations (18.1) and (18.2) gives the ELL model for the sampled observations, namely

$$y = X_s\beta_s + X_c\beta_c + u \tag{18.7}$$

ELL tends to have greater model richness via X_s, but assuming for now a specification that favours EBP, namely that the corresponding Molina and Rao (2010) EBP model has no contextual effects and the same rather than fewer non-contextual effects than ELL gives:

$$y = X_s\beta_s + u_0 \tag{18.8}$$

So, comparing EBP with ELL:

$$u_0 = X_c\beta_c + u \tag{18.9}$$

Molina and Rao, by stipulating the problem as predicting the non-sampled observations from the sampled ones, are able to predict non-zero random effects for non-sampled observations that, to the extent possible, mirror the effect of the contextual variables available for non-sampled observations in ELL.

Ideally, but not necessarily, as in other mixed model contexts u_0 substitutes for missing regressor variables at the appropriate level in the EBP model. Excepting the comparatively small number of sampled values, since EPB and ELL treat those slightly differently, we would then want:

$$E_{\xi,EBP}(\hat{u}_{0\bar{s}}) = X_{c\bar{s}}E_{\xi,ELL}(\hat{\beta}_{c\bar{s}}) \tag{18.10}$$

where E_ξ denotes model expectation, and $\bar{s}$ denotes the units not in the sample. Note that, for a given set of small areas, S, although for the EBP $E_{\xi,EBP}(\hat{u}_{0\bar{s}}|S) \neq 0$, in ELL $E_{\xi,ELL}(\hat{u} \mid S)$ can be zero and there will still be area level effects via $X_{c\bar{s}} E_{\xi,ELL}(\hat{\beta}_{c\bar{s}})$. Equation (18.10) does not imply that the estimated variances of EBP and ELL are equal. ELL has the advantage that the basis of its area level effects is explicit, rather than being a substitute via random effects for the missing regression variables in equation (18.10). A more technical discussion of links between best linear unbiased estimations (BLUEs) and best linear unbiased predictions (BLUPs), with possibly singular covariance and/or X matrices, is given in Haslett *et al.* (2014).

Returning to model richness, the simulation in Molina and Rao (2010) contains only two explanatory (or auxiliary) regression variables, both Bernoulli (i.e. 0 or 1), and an intercept, but no contextual variables. In their example using data from Spain based on the EU Statistics on Income and Living Conditions (EU-SILC) 2006, as auxiliary regression variables they used the indicators of five age groups; a Bernoulli indicator of Spanish nationality; indicators of three levels of education level; and indicators of three employment categories namely unemployed, employed and inactive, making four variables with ten linearly independent categories in all, plus an intercept. Both models then contain few terms in comparison with those routinely used in ELL, even putting aside the contextual effects that are in ELL, but not in EBP as implemented in Molina and Rao (2010). This is one reason EPB appears better that ELL in Molina and Rao (2010). Without contextual effects and with regression models containing a limited number of auxiliary variables, random area level effects are more marked in these particular EBP models than they would be in an ELL implementation with its richer regression models and its contextual variables. This is the reason that EBP can be used by Molina and Rao for small areas only at Spanish provincial level (of which there are 52, all containing sample) for males and females (making 104 small areas or domains in all, based on a population of around 47 million). Molina and Rao (2010) use no contextual variables, essentially because they do not have access to unit record or cluster level census data, but instead reconstruct the census from a limited set of multiway tables (which in turn restricts how rich their regression model can be in terms of candidate regression variables). In comparison, ELL-type methods, with unit record survey and census data are freed by use of contextual effects from the need to have sample in every small area to avoid area level bias, and by using richer regression models, are often able to get reliable small area estimates at a much finer level. See for example Haslett, Jones and Sefton (2013) where there are small area estimates for Cambodia for approximately 1600 communes and a population of around 14 million at the time of the 2008 census. Effort and expertise put into deciding what should and should not be included in the auxiliary variable matrix X of equation (18.1) really does matter.

The derivation of the EPB in Molina and Rao (2010) appears to make stronger assumptions than ELL about the distribution of the census data, as it utilises multivariate normality of the census data, and makes reference to maximum likelihood (ML) and restricted maximum likelihood (REML) for linear models. In implementing EBP however, this assumption is not required, as even without the normality assumption, EBP-type methods can utilise BLUE of fixed effects and BLUP of random effects, or use the empirical analogues of these methods. In this sense, EBP and ELL are not different. EBP and ELL may differ in the choice of method used to fit the underlying model to the survey data [for ELL model fitting see Haslett, Isidro and Jones (2010)], but this need not be so.

There is however one very important difference between EPB and ELL. Contextual effects, where available and adequate, mean that non-zero random predictions based on the random effects at small area level area are not necessary. ELL can then provide much finer level estimates (or sound estimates for many more small areas) than can EBP without contextual variables but with non-zero random effect predictions.

The wider set of issues that have been raised about ELL can be summarised as:

1. Stringent data requirements.
   (a) The survey and census must be essentially contemporaneous, otherwise the period to which the small area estimates apply is not well defined. The maximum allowable time between survey and census is however difficult to specify, as it depends on the stability of each country's political and administrative structure.
   (b) Area codes for survey and census (and other data sources such as GIS) must match at least to cluster level. Area codes are needed for the survey to permit estimation of variance components at various levels in the survey-based model, and so that the analysis can properly incorporate the survey design. Area codes are needed in the census to specify small areas. Matching of these area codes is required for contextual effects from the census to be merged with the survey data for model fitting, and for them to be applied as required at the prediction stage to produce the pseudo censuses and small area estimates. Matching to household level is seldom feasible, both for administrative reasons and due to mobility, but is not essential as sampled households are few in comparison with the census (cf. Molina and Rao, 2010).
   (c) It is not generally advisable to fit small area models to only part of a census, for example in cities (such as has been proposed by the World Bank for Dhaka in Bangladesh and Kuala Lumpur in Malaysia) using subsamples from the survey data. This is especially so if the unit record census data available are a subsample, particularly a clustered one.
   (d) Variables and categories used or derived from survey and census must match. This first requires the matching of questions from the questionnaires. Particular care, field manuals, and local knowledge are required especially if the questionnaires are being compared in translation. Note that, because census information is usually collected on and for a particular day of the year and surveys can be carried out over an annual cycle to better account for seasonality, even the same question may elicit different responses. Different definitions in survey and census (such as for head of household), and eligibility criteria for completion (such as definition of residence) also affect matching. For statistical comparisons, z-scores for test of difference in mean are often used, but power of the test will depend on whether tests used are for the data *in toto* or for subgroups from split samples.
2. Difficult to find sound models.
   (a) There are many possible models since there are many matching variables. A soundly based and extensive search procedure, often taking months of research, is required.
   (b) With even two-way interactions, this adds many more model possibilities to be checked, and the number of such models can be significant relative to sample size.
   (c) Criteria for selecting models are important. F-tests are not sufficient as some effects, for example child age in stunting models, may give highly significant F-values, but have essentially the same distribution in every small area and hence almost no effect on small area estimates.
   (d) Diagnostics are important. Using only R^2, for example, can be highly misleading, especially if in order to improve it, the sample is split by region (or even province

within region) and separate models fitted to each split. R^2 is not entirely appropriate for mixed models such as (18.1). It also depends on the level at which a model is fitted, for example at household or individual level, so comparison of household and individual level models via R^2 is not useful.

3. Where survey-based models with weighting are used, models are not adequately fitted to allow for survey design.
   Ordinary least squares is not an adequate model fitting method for survey data. Methods that recognise the sample design are required. For a discussion see Haslett, Isidro and Jones (2010). This has at times been an issue in use of ELL.
4. Sensitive to model misspecification.
   (a) Because models used are difficult to fit well, extensive checking is required. Note however that the parameters estimates for particular variables used in the model are not central if the small area estimates are not affected by them, even if their regression parameter estimates are significant using an F-test.
   (b) Because the small area estimates are generally sensitive to choice of model, running short courses with comparatively inexperienced staff, who then use PovMap, is inadvisable.
   (c) Calibration and benchmarking of small area estimates to direct estimates at higher levels of aggregation is a possibility, but care is needed if bias is not to be disguised rather than controlled. See Pfeffermann (2013, section 6.3) for a useful general summary.
5. Absence of random effects at small area level in models
   See however comments above about contextual effects, which do mean that non-zero small area level effects are used extensively in ELL.
6. No specific non-zero area level effects.
   In ELL these are modelled instead by contextual effects.
7. Models assume homogeneity within regions.
   See further comments below. Also note that contextual effects need not be homogeneous.
8. Lack of recognition that cluster effects are spatially correlated.
   See earlier comments about spatial models. Also see below.
9. Models involve mass imputation.
   Both ELL and EBP use residuals from the survey data, via the bootstrap, for all census observations. This is however rather different from generating $\boldsymbol{X}$ as well as $\boldsymbol{y}$ for many of the non-sampled (or even for sampled) observations in equation (18.1) when it is applied for prediction to the $N \times 1$ census data. For a more extensive discussion, see Haslett *et al.* (2010, pp. 61–62).
10. Understatement of standard errors.
    (a) In ELL, as in other forms of SAE, standard errors are conditional on the model being correct. Standard error is controlled not just by regression or auxiliary variables known for each census observation, but also via the area specific contextual effects. When bootstrapping as ELL does to estimate standard errors for small area estimates, allowance has to be made (as in Section 18.2.4) for the standard errors of the estimated parameters in the regression as well as the variation due to the random effects.
    (b) Other methods for estimating standard errors have also been used for ELL. See for example, Hentschel *et al.* (2000) and Minot, Baulch and Epprecht (2008).

11. Can only be carried out every 10 years.
    Without another census, ELL cannot be used again. This severely limits use of ELL to variables that tend to change only gradually in each small area. Updating methods, given a new survey but no new census, are discussed in Christiaensen *et al.* (2010), Isidro (2010) and Isidro, Haslett and Jones (2010).

Tarozzi and Deaton (2009; already available in 2007 as a report) suggest that ELL will often yield an overly optimistic assessment of the precision of its small area estimates. They argue that ELL relies on crucial assumptions which they claim likely fail in most real settings. First, they state that a household survey-based model of income or expenditure at regional level is not likely to be good enough to predict welfare at the level of a small area, unless the region is essentially homogenous. Secondly, they claim that the assumption of homoscedastic and independent and identically distributed cluster random effects is a very strong one, because within a region, sub-regional areas are likely to be integrated. They note this could result in spatial correlation of residual cluster effects if regressors did not sufficiently capture such integration. Part of Elbers, Lanjouw and Leite's (2008) reply has been quoted in Section 18.2.2. In essence they argue that contextual effects, as in equations (18.1) and (18.2) address both issues, which reinforces the view (now almost universally held), that contextual effects are essential to proper implementation of ELL. One useful diagnostic for assessing whether there is unmodelled spatial variation, is to map residuals from survey-based non-spatial models such as ELL at a higher level of aggregation where it is possible to check for area-based bias since average residuals should be approximately zero within these higher levels if the model is not spatially biased.

The purpose of this section has not been to argue ELL is best, but instead that it is and remains a viable candidate method, especially in poverty related applications in the developing world. If no contextual variables are available, SAE must be limited to small areas that all contain sample, because SAE for areas without sample tend to be biased, and a technique such as EPB is needed to correct for this bias via predictions of non-zero area specific random effects. But if unit record data are available for both survey and census, and good matching can be achieved along with a good model, then ELL is able to provide sound, unbiased small area estimates of an acceptable accuracy at a much finer level, even for non-sampled small areas.

18.4 Conclusions

The importance of any SAE method is to provide estimates at as fine a level as possible, subject to acceptable accuracy. A central question is, 'How accurate is necessary?' A second question is, 'How small can the areas be that can meet the accuracy criterion?' A third is, 'What is the best method to use?' Most of the published literature (as distinct from the reports on particular applications of SAE) focuses on the third question.

The first question needs to be answered before the second. Strict criteria are difficult to formulate, so that answers include rules of thumb, for example that a given percentage of the small area estimates must have a standard error of less than 5% for a proportion (such as poverty incidence, or percentage of children under 5 years who are stunted, underweight or wasted). All sound answers recognise that not all the small area estimates will be equally accurate, whether because of the data available, the type of model fitted, or the size of the small area.

The second question needs to follow the first rather than precede it, because there is almost inevitably pressure, often political, to produce estimates at a finer level than the data and analysis methods permit. Extreme examples have included published small area maps for a proportion in which estimated standard errors exceeded 40% in many of the small areas. So the political pressures to produce small area estimates at ever lower levels of administrative structure are real enough. But whatever small area method is used, results must be fit for purpose, especially if results are to be used for aid allocation to small areas with the highest estimated levels of deprivation. Aid allocation often utilises ranks of small areas by deprivation. Standard errors are too often forgotten or ignored, and statistical methods more suitable for ranking are often not used.

The level at which small area estimates are useful and can be mapped depend on the variable of interest. Some measures are considerably more difficult to model well. Examples include Gini coefficients (since they depend on the full distribution of income or expenditure and can be very much affected by predictions that are outliers within small areas) and food consumption score (which measures the number of times a particular food item has been eaten over a period, and is non-additive when food groups are aggregated).

Accuracy of small area estimates would be improved if sample surveys used for modelling were designed with small area estimates in mind. For example, if variance components are required at various administrative levels the design would best reflect that. For example, in ELL (or other SAE methods), if both small area level and cluster level variance components are required, then it is best if a reasonable proportion of those small areas which are sampled contain more than one sampled cluster. Further discussion of sample design for SAE can be found in Haslett (2012).

Application areas, and variables used for SAE in general, and ELL in particular have broadened since Rao (2003) and Elbers, Lanjouw and Lanjouw (2003). Food security measures include food expenditure and food poverty, kilocalories, and the problematic food consumption score (which cannot be compared across surveys with different questionnaires, let alone between countries). Economic measures again include food poverty (via its incidence, gap and severity), and Gini coefficients plus other measures of welfare via expenditure.

There are issues around the setting of poverty lines and their use in modelling. Nevertheless, whatever method is used, relative poverty is largely unaffected, and it is relative poverty that is the usual basis of aid allocation. Use of poverty lines in models nevertheless requires care when regional prices differ – it is better to scale local expenditures to the same scale and use a single poverty line. Then (given the relative index is sound) household measures used in the model are on the same scale.

Finally, principal components type measures as portmanteau statistics may be intuitively appealing but the same principal components will not apply in different countries, so using a standard linear combination in different countries has little meaning. For example, in Nepal stunting and underweight are positively correlated but stunting and wasting are negatively correlated [stunting can occur more in the mountains and wasting more on the plains or Terai, or in the west – see Haslett and Jones (2006) for further details]. In Cambodia however stunting and underweight are again positively correlated, but SAEs for wasting is difficult [as it differs only a relatively small amount over the whole country – see Haslett, Jones and Sefton (2013)]. Correlation between stunting and wasting would be much closer to zero in Cambodia than in

Nepal and this would mean the principal components for stunting, underweight and wasting taken together would also be different.

Atlases that contain many maps are a better way of incorporating and considering the multiple dimensions of poverty. See for example Bangladesh Planning Commission and the UN World Food Programme (2005) and Nepal Development Research Institute and UN World Food Programme (2010).

References

Bangladesh Planning Commission and the UN World Food Programme (2005) *The Food Security Atlas of Bangladesh 2004: Towards a poverty and hunger free Bangladesh*, ed. S. Hollema, Bangladesh Planning Commission and the United Nations World Food Programme, First Edition, March 2005.

Bramley, G. (1991a) *Bridging the Affordability Gap in 1991: An Update of Research on Housing Access and Affordability*, BEC Publications, Birmingham.

Bramley, G. (1991b) *Bridging the Affordability Gap in Wales: A Report of Research on Housing Access and Affordability*, House Builders Federation/Council of Welsh Districts, Cardiff.

Bramley, G (1992) Homeownership affordability, in England, *Housing Policy Debate* **3**(3), 143–182.

Bramley, G. and Lancaster, S. (1998) Modelling local and small-area income distributions in Scotland, *Environment and Planning C,* **16**, 681–706.

Bramley, G. and Smart, G. (1995) Rural Incomes and Housing Affordability, Rural Research Report No. 2, Rural Development Commission, London/Salisbury.

Bramley, G. and Smart, G. (1996). Modelling local income distributions in Britain, *Regional Studies,* **30**(3), 239–255.

Brunsdon, C., Fotheringham, S. and Charlton, M. (1998) Geographically weighted regression-modelling spatial non-stationarity, *Journal of the Royal Statistical Society: Series D*, **47**(3), 431–443.

Chambers, R., Chandra, H., Salvati, N. and Tzavidis, N. (2014) Outlier robust small area estimation, *Journal of the Royal Statistical Society: Series B*, **76**(1), 47–69.

Chambers, R.L. and Skinner, C.J. (2003) *Analysis of Survey Data*, John Wiley & Sons, Inc., Hoboken.

Christiaensen, L., Lanjouw, P., Luoto, J. and Stifel, D. (2010) *The reliability of small area estimation prediction methods to track poverty*, United Nations University, Working Paper No. 2010/99.

Cressie, N.A.C. (1993) *Statistics for Spatial Data,* Revised Edition, John Wiley & Sons, Inc., New York.

Cressie, N.A.C. and Wikle, C.K. (2011) *Statistics for Spatio-temporal Data,* John Wiley & Sons, Inc., Hoboken.

Demombynes, G., Lanjouw, J.O., Lanjouw, P. and Elbers, C. (2006) *How good a map? Putting small area estimation to the test*, The World Bank, Policy Research Working Paper No. 4155.

Efron, B., R.J. Tibshirani (1993), An introduction to the bootstrap, Chapmann & hall, London.

Elbers, C., Lanjouw, J.O. and Lanjouw, P. (2002) *Micro-level estimation of welfare*, Development Research Group, The World Bank, Policy Research Working Paper 2911.

Elbers, C., Lanjouw, J.O. and Lanjouw, P. (2003) Micro-level estimation of poverty and inequality, *Econometrica* **71**, 355–364.

Elbers, C., Lanjouw, P. and Leite, P.G. (2008). *Brazil within Brazil: testing the poverty map methodology in Minas Gerais*, Development Research Group, The World Bank, Policy Research Working Paper 4513.

Foster J.E., Greer J. and Thorbeck E. (1984) A class of decomposable poverty measures, *Econometrica,* **52**, 761–766.

Fujii, T. (2003) *Commune-Level Estimation of Poverty Measures and Its Application in Cambodia*, March 2003, http://siteresources.worldbank.org/INTPGI/Resources/342674-1092157888460/Fujji.Commune-LevelCambodia .pdf. [accessed 5-Sep-15]

Gelfand, A. E. and Vounatsou, P. (2003) Proper multivariate conditional autoregressive models for spatial data analysis, *Biostatistics* **4**(1), 11–25.

Haslett, S. (2012) Practical guidelines for design and analysis of sample surveys for small area estimation, Special Issue on Small Area Estimation, *Journal of the Indian Society of Agricultural Statistics*, **66**(1), 203–212.

Haslett. S. (2013) Small area estimation of poverty using the ELL/PovMap method, and its alternatives, Chapter 12 in *Poverty and Social Exclusion: New Methods of Analysis*, eds G. Betti and A Lemmi, Routledge, New York.

Haslett, S., Isidro, M. and Jones, G. (2010) Comparison of survey regression techniques in the context of small area estimation of poverty, *Survey Methodology*, **36**(2), 157–170.

Haslett, S., Isotalo, J., Liu, Y. and Puntanen, S. (2014) Equalities between OLSE, BLUE and BLUP in the linear model, *Statistical Papers*, **55**, 543–561.

Haslett, S. and Jones, G. (2004) *Local Estimation of Poverty and Malnutrition in Bangladesh*, Bangladesh Bureau of Statistics and UN World Food Programme, May 2004, http://documents.wfp.org/stellent/groups/public/documents/liaison_offices/wfp120022.pdf. [accessed 5-Sep-15]

Haslett, S. and Jones, G. (2005) *Estimation of Local Poverty in the Philippines*, Philippine National Statistics Coordination Board/World Bank, November 2005, http://www.nscb.gov.ph/poverty/sae/NSCB_LocalPovertyPhilippines.pdf. [accessed 5-Sep-15]

Haslett, S. and Jones, G. (2006) *Small Area Estimation of Poverty, Caloric Intake and Malnutrition in Nepal*, Nepal Central Bureau of Statistics/UN World Food Programme/World Bank, September 2006, http://documents.wfp.org/stellent/groups/public/documents/ena/wfp110724.pdf. [accessed 5-Sep-15]

Haslett, S. and Jones, G. (2010) Small area estimation of poverty: the aid industry standard and its alternatives, *Australian and New Zealand Journal of Statistics*, **52**(4), 341–362.

Haslett, S., Jones, G., Noble, A. and Ballas, D. (2010) More for Less? Using Existing Statistical Modeling to Combine Existing Data Sources to Produce Sounder, More Detailed, and Less Expensive Statistics, Official Statistics Report Series, 5, Statistics New Zealand, http://www.statisphere.govt.nz/official-statistics-research/series/2010/page1.aspx. [accessed 5-Sep-15]

Haslett, S., Jones, G. and Sefton, A. (2013) *Small-area Estimation of Poverty and Malnutrition in Cambodia*, Cambodia Institute of Statistics, UN World Food Programme, April 2013, http://www.wfp.org/content/cambodia-small-area-estimation-poverty-and-malnutrition-april-2013. [accessed 5-Sep-15]

Hentschel, J., Lanjouw, J., Lanjouw, P. and J. Poggi, J. (2000) Combining census and survey data to trace the spatial dimensions of poverty: A case study of Ecuador. *World Bank Economic Review,* **14**(1), 147–165.

Isidro, M. (2010) *Intercensal Updating of Small Area Estimates*, Unpublished PhD thesis, Massey University, New Zealand.

Isidro, M., Haslett, S. and Jones, G. (2010) Extended Structure-Preserving Estimation Method for Updating Small-Area Estimates of Poverty, *Joint Statistical Meetings, Proceedings of the American Statistical Association 2010,* Vancouver, Canada, http://www.amstat.org/meetings/jsm/2010/onlineprogram. [accessed 5-Sep-15]

J. N. K. Rao (2003) Small Area Estimation, Wiley, New York, http://eu.wiley.com/WileyCDA/Section/id-302479.html?query=J.+N.+K.+Rao.

Lehtonen, R. and Pahkinen, E. (2003) *Practical Methods for Design and Analysis of Complex Surveys*, 2nd Edition, John Wiley & Sons, Inc., Hoboken.

Minot, N., Baulch, B. and Epprecht, M. (2008) Poverty and Inequality in Vietnam: Spatial Patterns and Geographic Determinants, Research Report 148, International Food Policy Research Institute, Washington, DC.

Molina, I. and Rao, J.N.K. (2010) Small area estimation of poverty indicators, *Canadian Journal of Statistics*, **38**(3), 369–385.

Nepal Development Research Institute and UN World Food Programme (2010) *The Food Security Atlas of Nepal*, ed. S. Hollema, Food Security Monitoring Task Force, National Planning Commission, Government of Nepal and UN World Food Programme, First Edition, July 2010.

Pfeffermann, D. (2013) New important developments in small area estimation, *Statistical Science,* **28**(1), 40–68.

Pudney, S. and Sutherland, H. (1994) How reliable are microsimulation results? An analysis of the role of sampling error in a UK tax-benefit model, *Journal of Public Economics*, **53**, 327–365.

Puntanen, S. and Styan, G.P.H. (1989) The equality of the ordinary least squares estimator and the best linear unbiased estimator, *The American Statistician*, **43**(3), 153–161.

Rao, C.R. (1971), Unified theory of linear estimation (Corr: 72V34 P194; 72V34 P477), *Sankhyä, A*, **33**, 371–394.

Skinner, C.J., Holt, D. and Smith, T.M.F. (1989) *Analysis of Complex Survey Data*, John Wiley & Sons, Inc., New York.

Stillwell, J., Birkin, M., Ballas, D., Kingston, R. and Gibson, P. (2004) Simulating the city and alternative futures, in *Twenty-first Century Leeds: Contemporary Geographies of a Regional City*, eds R. Unsworth and J. Stillwell, Leeds University Press, Leeds, pp. 345–364.

Tarozzi, A. and Deaton, A. (2009) Using census and survey data to estimate poverty and inequality for small areas, *The Review of Economics and Statistics*, **91**, 773–792.

United Nations (2005) *Household Sample Surveys in Developing and Transition Countries*, Department of Economic and Social Affairs: Statistics Division, Studies in Methods, ST/ESA/STAT/SER.F/96, https://unstats.un.org/unsd/hhsurveys/pdf/Household_surveys.pdf. [accessed 5-Sep-15]

Zhao, Q. (2006) *User Manual for PovMap*, The World Bank, http://iresearch.worldbank.org/PovMap/PovMap2/PovMap2Manual.pdf. [accessed 5-Sep-15]

Zhou, Q. and Lanjouw, P. (2009) *PovMap2: A User's Guide*, The World Bank, http://go.worldbank.org/QG9L6V7P20. [accessed 5-Sep-15]

19

An Overview of the U.S. Census Bureau's Small Area Income and Poverty Estimates Program[1]

William R. Bell, Wesley W. Basel and Jerry J. Maples
U.S. Census Bureau, Washington, DC, USA

19.1 Introduction

The U.S. Census Bureau's Small Area Income and Poverty Estimates (SAIPE) program provides annual estimates of income and poverty statistics for all states, counties, and school districts of the U.S. A key SAIPE product is poverty estimates for school-age (5–17) children for all U.S. school districts. These estimates are used by the U.S. Department of Education in determining annual allocations of federal funds to school districts. The allocations amounted to more than 14 billion in fiscal year 2013.

Before the creation of the SAIPE program, income and poverty estimates for small (sub-state) geographic areas in the U.S. were produced only from the decennial census long form sample. Concerns about the 10-year gap between censuses led to the development, by 1993, of a coalition of five U.S. federal government agencies to fund Census Bureau research on developing postcensal small area estimates of income and poverty.[2] In addition, the Statistics of Income Division of the Internal Revenue Service (IRS, the U.S. government agency charged

[1]**Disclaimer**: This chapter is published to inform interested parties of research and to encourage discussion. The views expressed on statistical, methodological, technical, or operational issues are those of the authors and not necessarily those of the U.S. Census Bureau.

[2] These five U.S. government agencies were: Department of Agriculture, Food and Nutrition Service; Department of Education, National Center for Education Statistics; Department of Health and Human Services, Head Start Program; Department of Housing and Urban Development, Office of Policy Development and Research; and Department of Labor, Employment and Training Administration.

Analysis of Poverty Data by Small Area Estimation, First Edition. Edited by Monica Pratesi.

Companion Website: www.wiley.com/go/pratesi/poverty

with collecting federal income taxes) agreed to participate in the project by providing data that could be used in developing the estimation approaches.

In September 1994 the U.S. Congress passed the Improving America's Schools Act, whose Title I stated that the distribution of federal funds to school districts should be based largely on "the number of children aged 5 to 17, inclusive, from families below the poverty level on the basis of the most recent satisfactory data … available from the Department of Commerce." The Census Bureau, an agency of the Department of Commerce, was charged with producing the required poverty estimates. The law further required the Secretary of Education to use the updated Census Bureau estimates in 1997 for counties and, beginning in 1999, for school districts, unless the Secretaries of Education and Commerce determined the estimates to be "inappropriate or unreliable" for use in the fund allocations. To this end, the law directed the Secretary of Education to fund a U.S. National Academy of Sciences panel to advise the Secretaries on the suitability of the Census Bureau estimates, hereafter the SAIPE estimates, for determining the fund allocations.

The National Academy's *Panel on Estimates of Poverty for Small Geographic Areas* reviewed the SAIPE research and resulting poverty estimates, and issued their analyses of this work and their recommendations in a series of reports (National Research Council, hereafter NRC, 1997–2000a,b). In the first report (NRC 1997), the panel expressed a desire for further evaluations of the SAIPE estimates, and so recommended that 1997 allocations use county poverty estimates obtained by averaging the SAIPE school-age county poverty rates for 1993 with school-age county poverty rates from the 1990 U.S. census. The second and third reports (NRC 1998,1999) reviewed revised models and additional evaluations produced by the SAIPE program. NRC (1998) recommended that estimates from the revised county model (again for 1993) be used in making 1998 fund allocations, while NRC (1999) recommended that 1999 allocations be based on the 1995 SAIPE school district estimates. NRC (2000a) combined and updated the three earlier reports, and NRC (2000b) examined some issues for the users of SAIPE estimates, and suggested areas where more research and development could be done to improve the SAIPE models.

The original Census Bureau commitment to fulfill the requirements of Title I was to provide updated poverty estimates every 2 years starting with estimates for 1993. To improve the timeliness of the estimates, SAIPE quickly moved to releasing annual estimates. These are available for states starting in 1995, for counties starting in 1997, and for school districts starting in 1999. For states and counties SAIPE provides estimates of:

- the total number of people in poverty;
- the number of children under age 5 in poverty (for states only);
- the number of children age 5–17 in families in poverty;
- the number of children under age 18 in poverty; and
- median household income.

For school districts SAIPE provides estimates of:

- total population;
- the number of children age 5–17; and
- the number of children age 5–17 in families in poverty (the estimates used in determining the Title I fund allocations).

The remainder of this chapter provides a technical overview of the SAIPE program, summarizing the data, models, and methods used, and discussing some issues that have arisen for implementing the methods in a production environment. Thus, Section 19.2 discusses the data sources used by SAIPE,[3] while Section 19.3 reviews the poverty models used at the state and county levels, the estimation procedures used at the school district level, and the major changes SAIPE has made to these over time. Section 19.4 discusses some recent research aimed at improving the SAIPE models, data, and methods, and Section 19.5 offers some conclusions. Given the important role among the SAIPE products of the age 5–17 poverty estimates, much of the attention in this chapter will be focused on the data, models, methods, and results related to these estimates.

A key theme running throughout the chapter is the importance to small area estimation in practice of having data of good quality from which to develop small area models and predictions. This is well-illustrated by SAIPE results presented at all three geographic levels: state, county, and school district. Very limited data, or data of poor quality, will limit the improvements achieved over the direct survey estimates, thus compromising the goal of small area estimation, regardless of how sophisticated the models are that are being used.

19.2 U.S. Poverty Measure and Poverty Data Sources

19.2.1 Poverty Measure and Survey Data Sources

The official U.S. poverty measure used by the Census Bureau compares a family's total income to one of 48 "poverty thresholds" that vary by family size and composition. If the family's income is below its threshold, then all members of the family are considered to be in poverty. Family income includes money income before taxes, but not capital gains or noncash benefits such as public housing subsidies, Medicaid, and Supplemental Nutrition Assistance Program benefits. The poverty thresholds were originally developed in 1963–1964 to reflect the costs to families of various sizes of using an economy food plan developed by the U.S. Department of Agriculture (USDA), with the costs scaled up by a factor of three to account for the average share of total family expenditures coming from food costs. The thresholds are updated for inflation using the U.S. Consumer Price Index (CPI–U), but they do not vary geographically.[4] More information on the poverty measure can be found at http://www.census.gov/hhes/www/poverty/about/overview/measure.html.

Poverty status cannot be determined for some individuals who are thus excluded from the "poverty universe." For example, children under age 15 who are not living with an adult family member (such as foster children) cannot be assigned a poverty status because survey income questions are asked of people age 15 and older, so for these children there is no income to use to determine their poverty status. Other individuals also excluded from the poverty universe are people living in institutional group quarters (such as prisons or nursing homes), college dormitories, military barracks, and in other living situations without conventional housing.

Formally, the poverty estimates used in the Title I fund allocations restrict consideration to "children age 5–17 in families." This differs only slightly from the concept of poverty for all children age 5–17 because the group providing the difference, unrelated children ages 15–17,

[3] Some of the input data for the SAIPE models is publicly available and may be downloaded from the SAIPE web site at http://www.census.gov/did/www/saipe/data/model/tables.html. NRC (2000b, pp. 34–43) also discusses various U.S. data sources for poverty estimation.

[4] See Fisher (1997) for a detailed discussion of the development and history of the poverty thresholds. A short summary of his paper is available at http://www.census.gov/hhes/povmeas/publications/povthres/fisher4.html

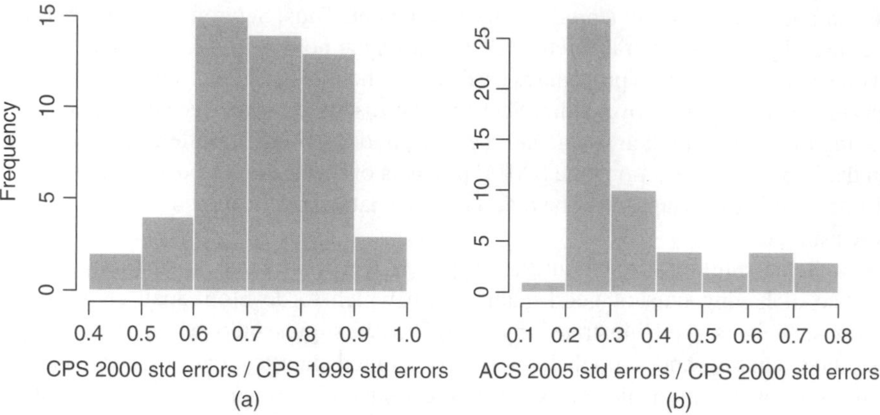

Figure 19.1 Histograms comparing standard errors of direct survey state poverty ratio estimators: CPS 2000 standard errors / CPS 1999 standard errors (a) and ACS 2005 standard errors / CPS 2000 standard errors (b)

is small.[5] We shall thus take the "in families" qualifier as implied, so our discussion of models and results for poverty of children age 5–17 should be understood to refer to "poverty of age 5–17 children in families", except where otherwise noted.

Census Bureau surveys that collect household income data include the CPS (Current Population Survey), ACS (American Community Survey), and, through 2000, the decennial census long form. The CPS provided the official direct state level poverty estimates through 2004, and still provides the official national level poverty estimates. The ACS started providing official direct subnational poverty estimates, with far greater geographic detail than the CPS, in 2005. Thus, SAIPE modeled CPS estimates through 2004 and ACS estimates starting in 2005.

The CPS, sponsored jointly by the U.S. Census Bureau and the U.S. Bureau of Labor Statistics, is the labor force survey of the U.S., providing monthly estimates of employment, unemployment, and other labor force characteristics. The CPS produces poverty estimates from its Annual Social and Economic Supplement (ASEC), which collects household income data in February, March, and April, asking questions about receipt of more than 50 different income types, with a reference period of the previous calendar year. Since 2001 the ASEC annual sample has consisted of approximately 100 000 addresses; prior to this the sample varied around 50 000–60 000 addresses, with data collected only in March (with the supplement then referred to as the March supplement). The 2001 sample expansion substantially reduced the sampling error in the direct CPS poverty estimates used in the SAIPE state and county models starting with the estimates for 2000 (since income data collected in 2001 was for income received in 2000). Figure 19.1a gives a histogram of the ratios of the sampling standard errors of the CPS direct estimates of the state 5–17 poverty ratios for 2000 to those for 1999, with the estimates of the standard errors taken from the generalized variance function model of Otto and Bell (1995), rather than from the noisier direct standard error estimates. The average reduction in the standard errors from 1999 to 2000, and the median reduction, are both about 27%.

[5] "Related children" in a household include sons and daughters, including stepchildren and adopted children, of the householder, as well as all other children under 18 years old if they are related to the householder by birth, marriage, or adoption. All other children in a household are "unrelated children." While SAIPE produces poverty estimates for all children 5–17 at the state level (but not at the county or school district levels), these are just done as an intermediate step to producing the state poverty estimates for all children under 18 and for persons of all ages.

The ACS is a nationwide survey designed to provide annual estimates of a broad range of demographic, social, economic, and housing characteristics at detailed levels of geography including states, counties, and school districts. Starting with 2010, the U.S. decennial census collects only basic demographic and housing information, and ACS has effectively replaced the long form sample. ACS produces poverty estimates using data collected on eight types of income (eight much broader categories than the 50 used in the CPS ASEC) using questions similar to those previously used for the census long form sample.

The ACS annual sample size was roughly 3 million addresses from 2005 to 2010, and increased to beyond 3.5 million addresses in 2013.[6] The sample is selected from all counties and county equivalents in the U.S. Single-year direct survey estimates are published for all states, and for counties and other places with population size 65 000 or larger. Published estimates for smaller places are produced by pooling 5 years of ACS sample data.[7] For input to the state and county models, SAIPE uses ACS poverty estimates produced from a single year of ACS data. This means that for counties with population less than 65 000, the direct ACS estimates used by SAIPE are unpublished. While sampling variances of these direct ACS single-year poverty estimates are large for the smallest counties, the use of single-year ACS estimates for all counties achieves two objectives. First, it uses estimates that are consistent and comparable across all counties, an important property for data being used in modeling. Secondly, it uses the most timely ACS estimates that could be used, something particularly important since the lack of timeliness of the census long form poverty estimates was a key concern that originally motivated the SAIPE project.

There are several important differences in the collection of income data by ACS versus the CPS ASEC that lead to differences in their poverty estimates. The ACS collects data on a monthly basis from one-twelfth of its annual sample, with income reported by respondents for the previous 12 months (at time of response), not for the previous calendar year as is the case for the CPS ASEC. This means that the effective reference period for single-year ACS income and poverty estimates extends across two successive calendar years plus 1 month, roughly centering them around the first sample month, which is January of the estimation year. The ACS estimates are thus shifted about 6 months forward and back, respectively, from the two closest CPS ASEC estimates.[8]

The much larger sample of ACS compared with CPS ASEC (recently about 3.5 million addresses versus 100 000), leads to much smaller sampling standard errors for the ACS direct estimates. This is shown in Figure 19.1b, which presents the histogram of the ratios of the standard errors of the ACS direct state estimates of the age 5–17 poverty ratios for 2005 to

[6] Due to subsampling of nonrespondents after the initial two phases of data collection (mail and telephone interviewing), the number of completed final ACS interviews is substantially less, around 2 million for 2005–2010, and increasing to about 2.5 million by 2012.

[7] For comparability, five-year estimates are produced for all places including those with populations over 65,000. Through 2013 ACS also provided three-year pooled sample estimates for places with populations larger than 20,000, but these estimates were discontinued for 2014 and, as of this writing, their future status is uncertain due to budgetary considerations.

[8] For each month's sample, ACS uses three different modes of data collection successively over 3 months: mail, telephone, and computer assisted personal interviewing (CAPI). The ACS data collection scheme thus results in a sequence of 14 overlapping 12-month income reference periods. For example, for 2013 poverty estimates the earliest such interval (January–December of 2012) occurred for mail respondents from the January 2013 sample, and the latest such interval (February 2013–January 2014) occurred for CAPI respondents from the December 2013 sample (who responded in February 2014). This effectively centered the resulting 1-year poverty estimate around January 2013. In contrast, the poverty estimates from the 2013 CPS ASEC sample are for 2012 (centered around July 1, 2012), while those from the 2014 CPS ASEC are for 2013 (centered around July 1, 2013). Previous discussions of this point generally refer to a 23-month period, which accounts for the use of the 12 monthly samples, but not for the 3-month periods of data collection for each sample.

those for 2000 from the CPS (the same CPS standard errors as those used in Figure 19.1a). The mean reduction in state age 5–17 poverty ratio standard errors from CPS 2000 to ACS 2005 is about 64%, and the median reduction is about 72%.

Data from the previous decennial census long form sample are also used in the SAIPE state and county poverty models by providing regression covariates, as discussed in Sections 19.3.1 and 19.3.2. Through 1998 the "previous census" data used by SAIPE were obtained from 1990 census results, but from 1999 on SAIPE switched to using results from Census 2000. (The Census 2000 poverty estimates relate to income received in 1999.) Previous census long form estimates of poverty at the school district level have played an even more fundamental role in the SAIPE school district poverty estimates, as discussed in Section 19.3.3.

More discussion of the various survey sources of income and poverty estimates can be found on the Census Bureau poverty web page at https://www.census.gov/hhes/www/poverty/about/datasources/description.html.

19.2.2 Administrative Data Sources Used for Covariate Information

19.2.2.1 U.S. Federal Income Tax Data

Each year the Census Bureau receives, from the IRS, selected data items from all U.S. individual federal income tax returns.[9] This includes: counts of "exemptions," which correspond to the set of people for which the tax return is being filed; the adjusted gross income (AGI) for the persons represented on the return; and the home address of the filer. SAIPE assigns a "pseudo-poverty status" to each tax return by comparing its AGI to the poverty threshold appropriate for the set of persons (exemptions) on the return, treating this set as if it constitutes a family. While neither age nor birth date is given for the exemptions, the returns do identify child exemptions[10] and exemptions for persons over age 65. We can thus tabulate, for all states and counties, total exemptions, child exemptions, age 65 and over exemptions, and, by subtraction, age 0–64 exemptions, all of these by pseudo-poverty status (poor versus not poor). Pseudo-poverty ratios, such as poor child exemptions divided by total child exemptions, can then be constructed.

The tax data have certain limitations for providing measures of poverty. First, the exemptions on a return may not correspond to the family unit for which poverty status would actually be defined. For example, a return may list financially dependent children no longer resident in the filer's household, such as college students. Secondly, about 12% of U.S. residents are not represented on tax returns, largely because many low income households do not owe any taxes, and so are not required to file a tax return. Finally, the address may not readily identify the relevant school district(s) for the child exemptions, a limiting factor for using the tax data in the school district poverty estimation, as discussed in Sections 19.3.3 and 19.4.

Due to their limitations, these pseudo-poverty measures cannot be taken as estimates of true poverty. Still, they have potential for having strong relationships to true poverty, and so to

[9] These data are kept in the strictest confidentiality, consistent with the requirements in IRS Publication 1075, "Tax Information Security Guidelines For Federal, State and Local Agencies" (http://www.irs.gov/pub/irs-pdf/p1075.pdf), as well as with the Census Bureau's own confidentiality standards.

[10] Child exemptions are generally children (or other descendents, such as grandchildren) of the filer(s) who are either age 18 or younger, or 23 or younger and still a student, or any age and permanently and totally disabled. They thus do not exactly correspond to a specific age range.

provide good covariates for small area models. Experience has shown, in fact, that covariates generated from the tax data are often the most valuable in the SAIPE state and county models. Also, in 2005 SAIPE started using tabulations of tax data for school districts in a limited way in the estimation of school district poverty (see Section 19.3.3), and current research discussed in Section 19.4 is attempting to improve the assignment of tax returns to school districts and other subcounty areas.

19.2.2.2 Supplemental Nutrition Assistance Program Participants Data

The Supplemental Nutrition Assistance Program (SNAP) provides subsidies to low income households for food purchases. These subsidies are now given in the form of electronic benefit cards, but previously they were given in a paper form of "stamps" or coupons bound into booklets, and so the program was originally (until October 1, 2008) known as the federal food stamp program. SNAP has the same eligibility requirements and benefit levels across all states except for Alaska and Hawaii. While the determination of SNAP eligibility differs from the determination of poverty status, SNAP eligibility is determined by a standard tied to the poverty level. This makes it likely that variations in SNAP participants across geographic areas could show a strong relationship to corresponding geographic variations in poverty, with potential for defining useful poverty model covariates. This has mostly been the case, with some qualifications to be discussed.

Each year SAIPE obtains monthly time series of the state numbers of SNAP participants from the USDA Food and Nutrition Service (USDA/FNS), which provides the SNAP funding. For most states, USDA/FNS also provides numbers of the county SNAP participants, but only for the months of July and, usually, January. We obtain county SNAP participants data for the remaining states directly from the individual state offices. SNAP participants data are not tabulated below the county level, and so are not available for use in estimating school district poverty.

We make two adjustments to the SNAP participants data. We adjust the data for Alaska and Hawaii by subtracting estimates of their additional SNAP recipients who would not be eligible under the criteria used in the rest of the U.S. These adjustments come from the "SNAP Household Characteristics" survey of the USDA/FNS.

We also adjust the SNAP time series for outliers arising from the issuance of emergency SNAP benefits (previously called "disaster relief food stamps") following major natural disasters.[11] Of these, the most important by far are major hurricanes. [Other natural disasters (floods, tornadoes, etc.) generally affect far fewer people and so have a much smaller effect on state SNAP participant counts.] Figure 19.2 shows the huge increase in monthly SNAP participants for the state of South Carolina following Hurricane Hugo, which struck South Carolina on September 21–22, 1989. Because emergency SNAP benefits are issued under a different set of eligibility requirements focused on persons directly affected by the natural disaster, such increases in a state's SNAP participants are not expected to reflect a commensurate increase in the number of persons in poverty. SAIPE thus adjusts for these outliers using the outlier detection and adjustment capabilities of the X-12-ARIMA seasonal adjustment program (Findley *et al.* 1998). For the example of Figure 19.2, adjustment of the series

[11] The USDA/FNS web site provides further information on SNAP benefits for disaster relief—see http://www.fns.usda.gov/disaster/food-assistance-disaster-situations.

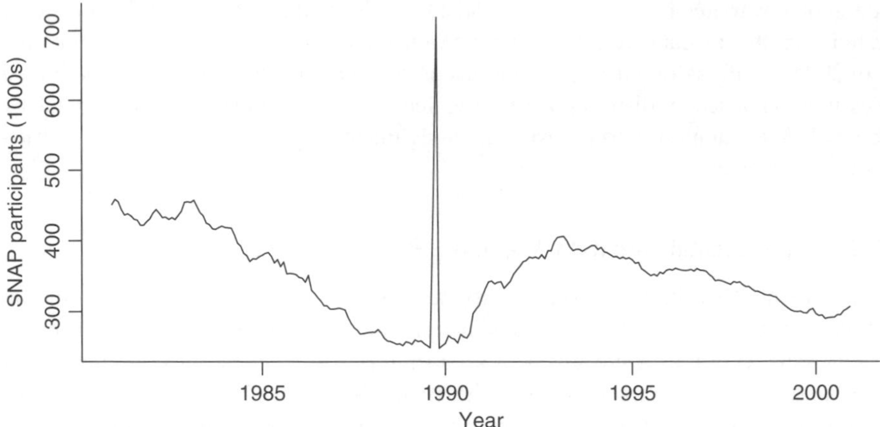

Figure 19.2 South Carolina SNAP participants, January 1981 to December 2000

for the September and October 1989 outliers essentially interpolates between the August and November values, removing the peak.

We cannot adjust county level SNAP data for the issuance of emergency SNAP benefits since we have neither monthly SNAP time series data for counties nor data on issuance of emergency SNAP benefits by county.

19.2.2.3 Other Administrative Data Sources for Model Covariates

For the 65 and over state poverty ratio model, the SNAP participants ratio is replaced by a Supplemental Security Income (SSI) recipiency rate. The federal SSI program provides monthly cash assistance to low income elderly persons and to low income disabled persons. The Social Security Administration tabulates various measures of state SSI participation.

Data from the Census Bureau's Population Estimates Program also play key roles in the SAIPE state and county models. In the state models, state population estimates for the relevant age groups are used to construct the denominators of several of the regression covariates. They also multiply the model predicted state poverty ratios to convert them to model-based predictions of numbers of persons in poverty. In the county models, the log of county population (either total or under 18) serves as a regression covariate, and the county population estimates are used to convert county model predictions of numbers in poverty for the various age groups into estimates of county poverty rates by age. For this purpose, the county population estimates are converted into estimates of the county poverty universes.

19.3 SAIPE Poverty Models and Estimation Procedures

The production SAIPE state and county poverty models follow the general area level model of Fay and Herriot (1979), which can be written

$$y_d = Y_d + e_d \qquad d = 1, \ldots, D \tag{19.1}$$

$$Y_d = \mathbf{x}_d' \beta + u_d \tag{19.2}$$

where Y_d is the population characteristic of interest for area (state or county) d, y_d is the direct survey estimate of Y_d, e_d is the sampling error in y_d, $\mathbf{x}_d$ is a $p \times 1$ vector of values of regression variables for area d, β is the corresponding vector of regression parameters, and D is the number of small areas. The sampling errors e_d are assumed to be distributed independently over d as $N(0, v_d)$. The area random effects u_d (also called "model errors") are assumed to be independent and identically distributed $N(0, \sigma_u^2)$ and distributed independently of the e_d. In this model the sampling variances v_d are typically treated as known although, in reality, they are generally estimated from the survey microdata and thus subject to errors. This point, as it applies to the various SAIPE models, is discussed in the following subsections and also in Section 19.4.

The unknown parameters of the model given by (19.1) and (19.2) are β and σ_u^2. Given a value for σ_u^2, β can be estimated by weighted least squares regression of y_d on $\mathbf{x}_d$ for $d = 1, \ldots, D$ using weights $(\sigma_u^2 + v_d)^{-1}$, giving $\hat{\beta} = (\mathbf{X}'\mathbf{V}_y^{-1}\mathbf{X})^{-1}\mathbf{X}'\mathbf{V}_y^{-1}\mathbf{y}$ and $\text{Var}(\hat{\beta}) = (\mathbf{X}'\mathbf{V}_y^{-1}\mathbf{X})^{-1}$, where $\mathbf{y} = [y_1, \ldots, y_D]'$, $\mathbf{X}$ is the $D \times p$ regression matrix with rows $\mathbf{x}_d'$, and $\mathbf{V}_y$ is the $D \times D$ diagonal covariance matrix of $\mathbf{y}$, which has diagonal elements $\sigma_u^2 + v_d$. There are various ways of estimating σ_u^2 such as method of moments, maximum likelihood (ML), or restricted maximum likelihood. One can apply these in an iteration that alternates estimation of β given σ_u^2 with estimation of σ_u^2 given β. Alternatively, one can use a Bayesian approach to deal with β and σ_u^2 (see, e.g. Berger 1985, pp. 190–193).

If σ_u^2 and the v_d are known, standard best linear unbiased prediction (BLUP) results give the predictors $\hat{Y}_d$ and their error variances:

$$\hat{Y}_d = h_d y_d + (1 - h_d)\mathbf{x}_d'\hat{\beta} \tag{19.3}$$

$$\text{Var}(Y_d - \hat{Y}_d) = \sigma_u^2(1 - h_d) + (1 - h_d)^2\mathbf{x}_d'\text{Var}\left(\hat{\beta}\right)\mathbf{x}_d \tag{19.4}$$

where $h_d = \sigma_u^2/(\sigma_u^2 + v_d)$. In practice, we substitute $\hat{\sigma}_u^2$ and the estimated v_d into the above formulas and into the expression for h_d. Asymptotic corrections to $\text{Var}(Y_d - \hat{Y}_d)$ are available to account for error in the estimation of σ_u^2 (Prasad and Rao 1999; Datta and Lahiri 2000). One can also use a Bayesian approach to develop posterior distributions of the unobserved population characteristics Y_d. In particular, the posterior means and variances, $E(Y_d|\mathbf{y})$ and $\text{Var}(Y_d|\mathbf{y})$, can be obtained by appropriately integrating (19.3) and (19.4) with respect to the posterior distribution of σ_u^2, either numerically or by Monte Carlo simulation (Bell 1999). See Rao (2003) for more details on estimation and prediction for the Fay–Herriot model.

Sections 19.3.1 and 19.3.2 discuss the specific models and methods used for estimation and prediction in the SAIPE state and county poverty models. Section 19.3.3 describes the estimation procedure used to produce school district poverty estimates, which does not rely on a formal statistical model. Finally, Section 19.3.4 summarizes the major changes that have been made over time in the SAIPE models and procedures.

19.3.1 State Poverty Models

The SAIPE state poverty models, developed originally by Fay and Train (1997), apply the Fay–Herriot model (19.1) and (19.2) separately to direct survey estimates of state poverty ratios for four age groups: 0–4, 5–17, 18–64, and 65 and over.[12] Note that these differ from

[12] Separate models of the same form are used for poverty of total children age 5–17 and for poverty of children age 5–17 in families, whose distinction is discussed in Section 19.2.1.

the publication age groups of 0–4, 5–17, 0–17, and all ages. The four modeled age groups are modeled separately to capture the different behavior of poverty across ages, as doing so has some potential to improve model-based predictions. The most dramatic difference across ages is that poverty for age 65 and over is much more stable over time than is poverty for the other three age groups. This fact is particularly relevant to the use of data from the previous census long form sample in the 65 and over poverty models.

The state poverty *ratios* are defined as the number in poverty divided by the population, each for the given age group. Poverty *rates* change the denominators to the corresponding *poverty universes*, which, as discussed in Section 19.2.1, exclude a small part of the total population (e.g. foster children). Differences between the ratios and the rates are generally small and are unimportant for modeling. We model the poverty ratios rather than the rates because the estimated ratios can be readily converted to estimates of the numbers of persons in poverty for the four age groups by multiplying them by the corresponding state population estimates. Estimates of the number in poverty for the other age groups, 0–17 and all ages, are then obtained by simple aggregation. Estimates of the poverty universe are not a standard Census Bureau product, but are produced for the publication age groups (0–4, 5–17, 0–17, and all ages) to permit estimates of the number in poverty to be converted to corresponding estimates of poverty rates.

As discussed in Section 19.2.1, through 2004 the direct survey estimates of the state poverty ratios (y_d in the models) came from the CPS, while from 2005 on these have come from the ACS. Sampling variances (v_d) of the CPS poverty ratio estimates were obtained using a generalized variance function (GVF) model with random effects developed by Otto and Bell (1995) to improve the direct CPS sampling variance estimates. With the switch in 2005 to the use of data from ACS, which has much larger state sample sizes than CPS, direct sampling variance estimates have been used.

Regression variables ($\mathbf{x}_d$) in the models for age 5–17 include an intercept term and, in most years:

- The "tax return poverty rate" for children, defined as the number of child tax exemptions in poverty divided by the total child exemptions.
- The under age 65 "tax nonfiler rate," defined as the difference between the estimated population under age 65 and the number of tax exemptions under age 65, expressed as a percentage of the population under age 65.
- The SNAP participation rate (formerly known as the Food Stamp Program participation rate)—the outlier adjusted, monthly average number of participants of all ages receiving SNAP benefits over a 12-month period (whose timing is discussed below), expressed as a percentage of the total state population.
- Either the previous census estimated age 5–17 poverty ratios, or the residuals from a regression of the previous census age 5–17 poverty ratios on the values of the above variables for the previous census income year (1989 for the 1990 census and 1999 for Census 2000).[13]

[13] Bell (1999) notes that, for the state models, use of census residuals as a regressor is essentially equivalent to using a bivariate model that combines the regression equation for the census estimates with the model given by (19.1) and (19.2) for the CPS (or ACS) estimates. This result holds for the state models but not the county models, since it depends on there being negligible sampling error in the census poverty estimates (obtained from the long form sample), something that was true for states but not for counties. For discussion of the bivariate Fay–Herriot model, see Huang and Bell (2012).

Regression variables for the age 0–4 and age 18–64 models include the under 65 tax nonfiler rate and SNAP participation rate, as well as the under 65 tax poverty rate and the previous census poverty ratios or residuals specific to their age ranges. For age 65 and over, the variables refer specifically to the age 65 and over population, and the SNAP participation rate is replaced by the SSI recipiency rate, which is defined as the 12-month average number of state SSI recipients age 65 years and over for the estimation year divided by the demographic estimate of the state 65 and over population for the following year. See any of the links to state model documentation on the SAIPE web site at http://www.census.gov/did/www/saipe/methods/statecounty/index.html for more information.

For models of CPS data (i.e., through the 2004 estimates), the tax return poverty rate and nonfiler rate used are for the same year as the CPS estimates. Also, the 12 months for the SNAP participants averages cover July of that year through June of the following year, and the "total population" in the denominator of the SNAP participation rate is actually the average of the population estimates for the 2 years involved.[14] Because of the 6-month timing shift of the ACS poverty estimates relative to those from CPS, the appropriate choice of year for the regression variables in models of ACS poverty estimates is less clear. SAIPE chose to use the earlier year regression variables (e.g. regression variables for 2004 when modeling the 2005 ACS poverty estimates) rather than the later year versions (2005 regression variables for modeling 2005 ACS estimates). One reason was that the data for the earlier year regression variables are, of course, available 1 year earlier, facilitating timeliness of the production schedule.

Estimation of the SAIPE state poverty ratio models was initially by ML. Starting with the models for 1996, however, the models have been given a Bayesian treatment with a flat prior on the model error variance σ_u^2. The motivation for this change was that ML estimates of σ_u^2 were frequently zero in the SAIPE state models for the earlier years to which the models were applied. While $\hat{\sigma}_u^2 = 0$ is not necessarily a bad (highly inaccurate) estimate, it leads to some undesirable results, including empirical Bayes predictions from (19.3) that equal the regression fitted values $(\mathbf{x}_d'\hat{\beta})$, giving no weight to the direct survey estimates y_d. For the SAIPE state models it also led to prediction error variances from (19.4) that were clearly too low for many states, and that followed implausible patterns over the states. The Bayesian approach with a suitable prior effectively prevents estimating $\sigma_u^2 = 0$ and so addresses these concerns. Bell (1999) discusses this issue in more detail. Gelman (2006) discusses some alternative priors for variance component models.

The choice between use of either previous census poverty ratio estimates or "census residuals" is an interesting feature of the SAIPE state models. This choice has been guided by empirical comparisons of model fit using the information criterion AIC (Akaike 1973), though since these comparisons are of models with the same number of parameters, the AIC penalty term drops out of the AIC comparisons, which then reduce to log–likelihood comparisons. We have not, however, always picked the best fitting model, because if we did so then minor AIC differences could produce frequent model changes over time and unnecessarily inconsistent choices over the age groups. We instead have looked for evidence of consistency in model preference over time and across the age groups below 65. We gave separate consideration to the 65 and over age group because it shows different behavior from that of the other three age groups.

[14] The timing of the interval used to define the SNAP participation rate was determined empirically, though shifting it forward or back several months did not materially affect the model fits. Also, note that population estimates for a given year refer to July 1 of that year.

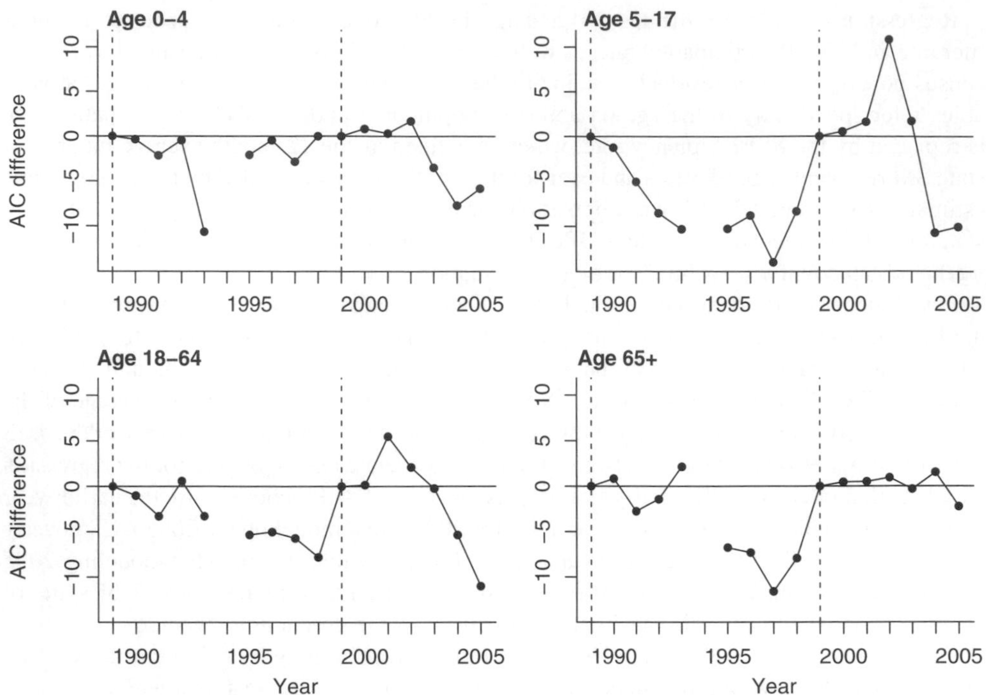

Figure 19.3 SAIPE state poverty ratio models with CPS data: AIC differences for models with census residuals versus census estimates (negative AIC differences favor models with census residuals)

Figure 19.3 shows the AIC comparisons.[15] Values of the AIC difference below zero favor use of previous census residuals in the model; values above zero favor use of previous census estimates. (In 1989 and 1999, the income years for the 1990 and 2000 censuses, the two models can be shown to be equivalent, hence all AIC differences for these two years are zero.) Figure 19.3 shows that in the early years of the 1990s (1990–1993) the AIC differences are mostly (13 of 16) negative, favoring the model with census residuals. Age 65 and over has two positive and two negative AIC differences in this interval, and so provides something of an exception. In the later years of the decade (1995–1998) almost all the AIC differences (except for the two values close to zero for age 0–4 in 1996 and 1998) favor the model with census residuals.

Following the 2000 census, the pattern is different. In the early years of that decade (2000–2002) the AIC differences are all positive, favoring the model using census estimates. Those for 2003 are equivocal (one positive, one negative, and two approximately zero), but for 2004 and 2005 all but one AIC difference favors census residuals. The results for age 65 and over again stand out as somewhat different, not necessarily recommending either model over the other.[16]

[15] No AIC differences are shown for 1994 because estimates were not produced that year due to technical issues with producing sampling variance estimates caused by the phased introduction of a new CPS sample that year.

[16] Notice that the 2005 AIC differences shown on the graphs come from models fitted to CPS state poverty ratio estimates for 2005, not to the ACS 2005 estimates used in SAIPE production. AIC differences from models fitted to ACS data for 2005 through 2009 were also computed (though not shown in Figure 19.3), and these also favored models with census residuals. These AIC differences were even larger, which may be partly due to the greater time separation from the census, and partly due to the much lower level of sampling error in the ACS estimates.

We conjecture that if the distribution of poverty across states in the year being modeled is similar to that in the census income year, something that is more likely when modeling data for years close to the census income year, then the model using previous census estimates may fit better. This seems to have been the case for the early years of the 2000s, and models with census residuals were not clearly favored until 2004. It was apparently not the case in the 1990s, when models with census residuals were favored almost immediately. At the beginning of both decades, results for age 65 and over stand out as somewhat different, showing more durable support for models with census estimates.

The original production models, which started in 1993 (results for 1990–1992 were developed later for testing purposes), used census residuals for ages 0–4, 5–17, and 18–64, and used census estimates for 65 and over. We later started making the AIC comparisons and in 1998 switched the 65 and over model to use census residuals. Following the 2000 census we switched to using models with previous census estimates for all age groups, but in 2004 switched to using census residuals for ages below 65. Since the 2010 census omitted the long form sample of earlier censuses, and so did not collect income data, we have not had the same decision process since the 2010 census. Instead, because the Census 2000 estimates are becoming more and more distant from our current modeling years, we are now considering replacing use of the previous (2000) census data in our models with 5-year ACS estimates prior to our current modeling year. See Section 19.4 for a brief discussion and Huang and Bell (2012) for an investigation.

Results related to the SNAP participation rate variable in the 0–4, 5–17, and 18–64 poverty ratio models illustrate the importance to small area models of using variables that are defined consistently across all the small areas. Figure 19.4 shows t-statistics for the regression coefficients on the SNAP participation rate in the poverty ratio models fitted to CPS data from 1989 through 2005.[17] Notice that the coefficients are fairly consistently statistically significant (t-statistics usually exceed 1.96) until 1997, at which point they become statistically insignificant for 4 years, returning to statistical significance in 2001. This behavior was not entirely surprising. In 1997 new legislation (the "welfare reform act") went into effect that changed how certain U.S. federal benefit programs were administered, including giving individual states more freedom to decide how they would administer the food stamp program (and renaming the program as SNAP). As different states started administering this program in different ways, this change had the potential to alter participation in the program in ways that could differ across states. The apparent consequence of this to the SAIPE state models was that the coefficients on the SNAP participation rate became insignificant starting in 1997. Since we wanted clear evidence that things had changed for this variable before changing our models, we waited one more year to check the coefficients for 1998. When the coefficients were again insignificant, we removed the SNAP variable from the models. We continued to monitor fits of models with the SNAP variable, intending to reintroduce the variable into the models if and when its coefficients again became consistently significant. Though in Figure 19.4 this occurs in 2001, the figure shows results for all years from the models with census residuals, whereas from 2000 to 2003 we were instead using census estimates in the models. In these models the coefficients on the food stamp variable were usually not significant. In 2004 we switched back to using models with census residuals, so we then reinstated the SNAP variable in the models.

[17] For comparability over time, models of the same form were used across all years shown in Figure 19.4, although this meant that they disagreed with the SAIPE production models for some years. In particular, all the models used census residuals, rather than census estimates. Also, these results were obtained from models fitted by ML. Differences between these t-statistics and analogous quantities obtained from a Bayesian approach are small.

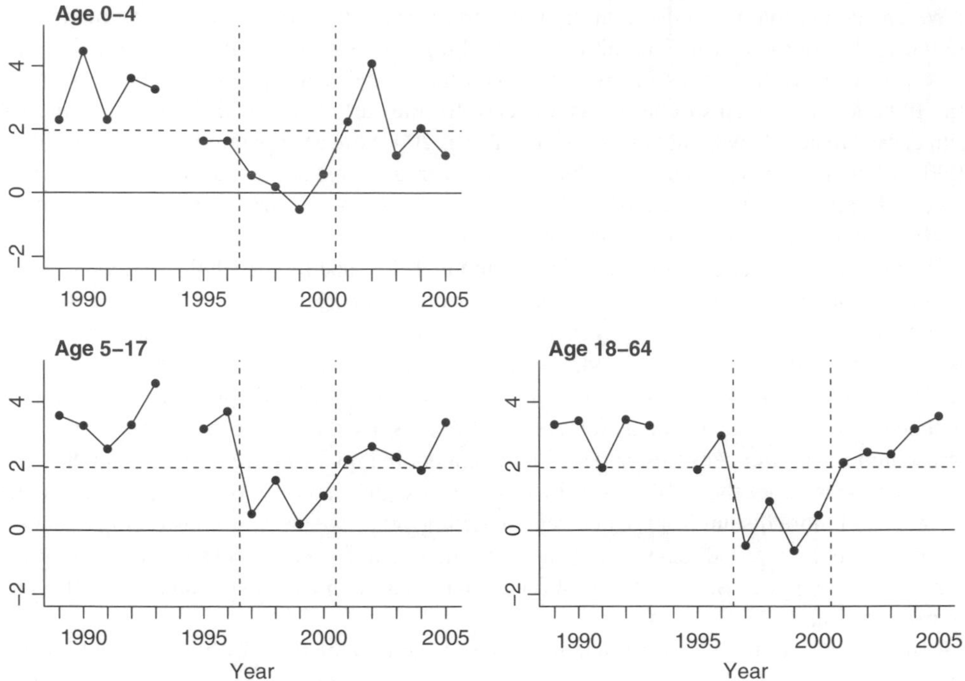

Figure 19.4 t-statistics of the regression coefficients on the SNAP participation rate variable in the state poverty ratio models with CPS data for ages 0–4, 5–17, and 18–64, plotted over time. Points above the horizontal dotted line indicate significant positive t-statistics (exceeding 1.96)

Modeling of the CPS data achieved substantial and important reductions in the standard errors of the direct poverty estimates for virtually all states. In contrast, standard errors of the direct ACS state estimates of the poverty ratios for the age groups used by SAIPE are sufficiently low that modeling to improve them is less important. In fact, for some states modeling the ACS estimates has little effect: the smoothing weights h_d in (19.3) are fairly close to 1 and the predictions $\widehat{Y}_d$ are close to the direct ACS estimates. Modeling is still important and beneficial for improving the direct ACS estimates for the smallest 15 or so states, as Figure 19.5 illustrates.

Figure 19.5a compares prediction standard errors from the CPS 2005 5–17 poverty ratio model with the sampling standard errors of the CPS direct estimates. It shows dramatic improvements from modeling, with standard error reductions for all but a few states exceeding 50% (points below the dashed line, which is $y = x/2$). Figure 19.5c shows corresponding results from the ACS 2005 model. While the modeling shows standard error reductions for all states, for the 30 or so states with the smallest ACS sampling standard errors the reductions are not so dramatic and the sampling standard errors are fairly small anyway. (Note that the axis ranges in this plot cover less than half the axis ranges in Figure 19.5a.) For the 15 or so states with the largest sampling standard errors, the improvements from modeling are larger and more important.

Figure 19.5b shows that prediction standard errors from the CPS 2005 model are higher for most states than the sampling standard errors of the ACS 2005 direct estimates. Figure 19.5d

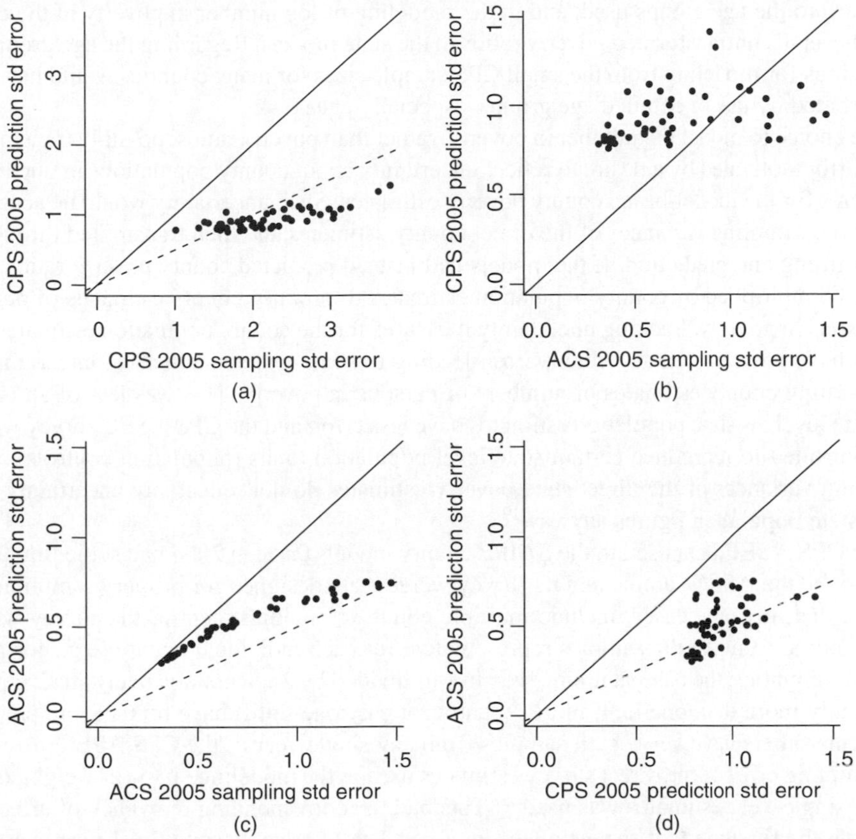

Figure 19.5 Comparing standard errors of various estimators and predictors for state 5–17 poverty ratios: ACS 2005 and CPS 2005 equations. Dashed lines are at $y = x/2$, showing where standard errors of the predictor on the y-axis would be one-half the standard errors of the predictor on the x-axis

shows the not surprising result that prediction standard errors from the ACS 2005 model are substantially lower than those from the CPS 2005 model. Many of the standard errors from the ACS model are only about one half or less of those from the CPS model (points lying near to or below the dashed line, which again is $y = x/2$).[18]

19.3.2 County Poverty Models

SAIPE county poverty estimation applies the Fay–Herriot model (19.1) and (19.2) separately to logarithms of direct survey estimates of the number of persons in poverty in counties for three age groups: 0–17, 5–17, and all ages. The county models differ from the state models

[18] The discussion of CPS-ACS differences in Section 19.2.1 noted that differences in the timing of their income reports effectively shifts the ACS estimates 6 months earlier than the CPS estimates for a given year. While this may raise questions about possible effects of this difference on the comparisons made in Figure 19.5, if we replace the CPS 2005 results in the graphs with corresponding CPS 2004 results, which are roughly 6 months earlier than the ACS 2005 results, the general appearance of the plots is very similar.

in regard to the age groups used, and in the modeling of log number in poverty in the county models versus untransformed poverty ratios in the state models. Regarding the age groups, the limitations for modeling from the small CPS sample sizes for many counties would have been exacerbated by use of detailed age groups, especially ages 0–4.

The choice to model log number in poverty, rather than poverty ratios, possibly transformed, was partly motivated by a desire to reflect uncertainty about county populations in uncertainty measures for the model-based county poverty estimates. Such uncertainty would be accounted for by the sampling variances of the direct county estimates and then transmitted through the model fitting and prediction. If the models had instead predicted county poverty ratios, these would be multiplied by county population estimates to produce county estimates of numbers of persons in poverty. Lacking uncertainty measures for the county population estimates, they would have been treated as without error, leading to some understatement of uncertainty for the resulting county estimates of numbers of persons in poverty. This was less of an issue at the state level, as state population estimates have less error, and the CPS ASEC survey weights are controlled to reproduce certain state level population totals (population controls), so the sampling variances of the direct state poverty estimates do not reflect any uncertainty about these state population figures anyway.[19]

The CPS ASEC data used in the SAIPE county models through 2004 had some limitations. First, since the ASEC sample and its survey weights are designed for primary sampling units (PSUs) that, in some cases, include multiple counties, an adjustment to the survey weights was required to make the samples representative for each individual county (e.g., for a PSU with two counties, the base sampling weight was divided by 2). Secondly, many small counties (originally more than one-half, more recently roughly one-third) have no CPS sample at all, and many others have very small samples from any single year of the CPS ASEC. To reduce the sampling error of the direct survey estimates used in the modeling, a 3-year weighted average of single-year estimates was used.[20] This had the corresponding drawback of effectively reducing the timeliness of the estimates by about 1 year. Also, despite the 3-year averaging, the samples for some small counties contained no 5–17 year old children in poverty, leading to direct estimates of zero. Since logs could not be taken of these zero estimates, the data from these counties were dropped from the model fitting, and for prediction purposes the counties were treated as if they had no survey data.[21]

Another limitation of the CPS ASEC data used through 2004 was that direct estimates of the county sampling error variances were not produced because replicate weights used for variance estimation were developed for state level estimates but not for county estimates. This limitation required a work-around for estimation of the county Fay–Herriot models. The approach

[19] A similar situation now exists at the county level, since ACS estimation now uses population controls for substate weighting areas consisting of either individual counties (for the larger counties) or small groups of counties. See the ACS web site at http://www.census.gov/programs-surveys/acs/methodology.html. The choice to model log number in poverty also depended on some evaluations SAIPE did of various alternative county models, including comparisons of model predictions against 1990 census long form estimates. As reported in NRC (1998, chapter 4), the evaluations found the models for untransformed poverty rates did not fare as well, while those for logs of either number in poverty or poverty rates performed comparably.

[20] The direct estimates modeled were defined as log[(3-yr weighted average poverty rate) × (3-yr weighted average poverty universe)]. For the age 5–17 estimates, the weights in each year for a given county were proportional to the numbers of interviewed housing units in the county containing at least one child age 5–17. For counties with CPS sample in only one or two of the 3 years, the values for only that year, or for the analogous 2-year average, were used.

[21] Any observations of zero counts for the unlogged covariate data are imputed to be positive, so logs can be defined. In this way, a regression prediction for county level poverty can always be obtained.

Table 19.1 Comparison of county level coverage, CPS ASEC and ACS

	CPS ASEC 2004	ACS 2012
Counties whose sample includes no age 5–17 children	1604	3
Percentage among all counties	51.1	0.1
Counties whose sample includes no age 5–17 children in poverty	273	108
Percentage among counties with age 5–17 children in sample	17.8	3.4

used was to fit a model of the form of (19.1) and (19.2) to the corresponding previous census estimates (for which sampling error variance estimates were available), and to then use the resulting estimate of σ_u^2 in the model fitted to the CPS ASEC data for all years through the decade. A very simple GVF model was specified for the sampling variances in this model. Initially, the GVF model $v_d = \gamma/n_d$, where γ is an unknown parameter and n_d is the sample size, was used. With σ_u^2 held fixed, the parameter γ could be estimated jointly with the regression parameters β in fitting the model to the CPS ASEC data. [For some analyses of such models, see Slud (2000).] Starting in 2000, the GVF was changed to $v_d = \gamma/\sqrt{n_d}$, as Asher and Fisher (2000) found this reduced heteroskedasticity in the regression residuals of the fitted Fay–Herriot model, and yielded model predictions that performed favorably in certain evaluations that compared model predictions with 1990 census long form estimates.

As noted in Section 19.2.1, starting with the estimates for 2005, the SAIPE county models have used ACS single-year direct estimates of log poverty instead of CPS ASEC estimates. The much larger sample size of ACS conveys substantial advantages. First, ACS has sample in all U.S. counties, and the occurrences of zero estimates (requiring that a county's data be left out of the model fitting) are greatly reduced. Table 19.1 illustrates these results for the county 5–17 poverty models. For the CPS ASEC in 2004, a little over half the counties had no 5–17 children in sample, and thus no direct estimate of poverty, while for the 2012 ACS, this occurred for only three counties. (The U.S. has about 3142 counties, with slight variations in the exact number over time.) The proportion of counties with sample but with zero poverty estimates is also much higher for the CPS estimates compared with the ACS estimates.[22]

Another advantage to using ACS data is that sampling variances of the direct log poverty estimates are available for all counties from a successive difference replicate weight approach [extending that described in Fay and Train (1995)].[23] These direct sampling variance estimates are used in the ACS-based SAIPE county models, permitting the standard approach to model estimation where the sampling variances are held fixed and σ_u^2 is estimated.

Details on the sources of covariate data are given in Section 19.2.2. The following focuses on the specific concepts and timing of the covariates included in the county level models.

- U.S. federal tax information: The SAIPE county model uses logarithms of county level tabulations of total exemptions on tax returns, and of exemptions on returns in poverty (as defined in Section 19.2.2). For the 0–17 and 5–17 models, the exemption counts and poverty

[22] Note that for both CPS ASEC and ACS the proportion of the total sample that gets dropped from the model fitting is much lower than these figures since the counties with no poor 5–17 year olds in sample are invariably small counties with small samples.

[23] Coincidentally, replicate weights to facilitate county variance estimates were added as part of CPS ASEC processing around 2005, but at that point SAIPE had switched to using ACS poverty estimates.

exemption counts are limited to dependents identified as children on the return. For the all ages model, all tax exemptions are included except for nonchild dependents.

- SNAP participants: The covariate based on SNAP participants is the log of the number of county participants in the SNAP program for the July of the year preceding the SAIPE estimation year. The county participant counts are benchmarked to the adjusted state total (the numerator of the SNAP participant ratio used in the state model).
- Population estimates: For scaling effects relative to the other variables, the SAIPE county models include logs of county population estimates from the Census Bureau's Population Estimates Program. The vintage and reference year of these estimates matches the SAIPE estimation year. For the 0–17 and 5–17 poverty models, the log of the population estimate for ages 0–17 is used as a covariate.[24] For the all ages model, the log of the all ages population estimate is used.
- Prior census long form poverty estimates: The SAIPE county models include logs of the long form estimated county number in poverty for each age-specific modeled domain. Through 1998, the prior census estimates came from the 1990 census; from 1999 on, they have come from Census 2000. Although, without a long form in the 2010 census, the prior census poverty estimates are becoming further and further removed from the advancing SAIPE estimation years, this variable still displays significant correlation in the regression models, at least through the SAIPE 2012 county models.

Estimation of the county models' variance parameters was by method of moments for the CPS-based model (estimating γ) and by ML for the ACS-based model (estimating σ_u^2). The estimation of these parameters iterates with estimation of the regression parameters β as noted at the beginning of Section 19.3. County model estimation has not encountered problems with estimating zero for σ_u^2 (or for γ) as was the case with the state model. Though a Bayesian treatment of the county model was examined, results were essentially the same as for ML.

Given a fitted model, the BLUP predictor and corresponding prediction error variance are obtained in the log scale using the formulas in equations (19.3) and (19.4). Asymptotic adjustments to (19.4) as in Prasad and Rao (1990) and Datta and Lahiri (2000) were examined but found to be small. For the CPS-based models, the GVF sampling variances for the 3-year average estimates were generally high relative to the estimate of σ_u^2, and thus the shrinkage weights h_d applied to the direct estimates of log poverty y_d were minimal for all but the handful of largest counties. Figure 19.6 displays the distribution of these shrinkage weights for both the 2004 CPS-based shrinkage estimates and the 2012 ACS-based shrinkage estimates. From the CPS-based distribution, more than 75% of the counties have a weight on the direct survey estimate of less than 0.10. Only the top 1% (30 counties) have shrinkage weights higher than 0.20. In contrast, with the ACS-based model the shrinkage weights exceed 0.10 for at least half the counties, and exceed 0.20 for more than 25% of the counties. The top percentile of the shrinkage weights, those which apply to the largest 30 counties, exceeds 0.85 for the ACS-based predictions.

Prediction results on the log scale are transformed back to the native scale using properties of the lognormal distribution. Letting $Z_d = \exp(Y_d)$ denote the population number in poverty

[24] The log of the age 0–21 population estimate was also tried. However, spikes in the 0–21 population for counties with high populations of college students are not matched by corresponding spikes in child tax exemptions, which led to biased predictions for these counties (NRC 1998, pp. 58–61).

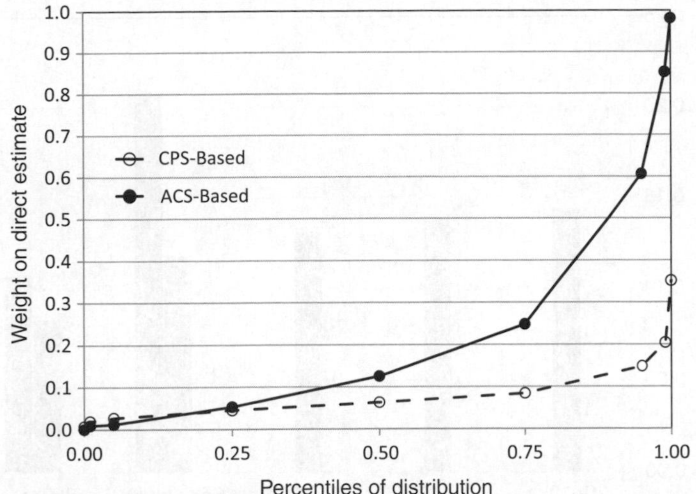

Figure 19.6 Percentiles of the distributions of the shrinkage weights applied to the direct survey 5–17 poverty estimates for age 5–17 in families: ACS-based SAIPE 2012 model (solid) and CPS-based SAIPE 2004 model (dashed)

for county d, the prediction of Z_d and the estimated prediction error variance are given by:

$$\widehat{Z}_d = \exp\left\{\widehat{Y}_d + [\widehat{\sigma}_u^2 - \mathrm{Var}(\widehat{Y}_d)]/2\right\}$$

$$\mathrm{Var}(Z_d - \widehat{Z}_d) = \widehat{Z}_d^2\left[e^{\mathrm{Var}(Y_d - \widehat{Y}_d)} - 1\right]$$

The expression for $\widehat{Z}_d$ produces an unbiased predictor $[E(\widehat{Z}_d) = E(Z_d)]$, incorporating an adjustment to the standard lognormal results to account for the error in estimating β.

The resulting predictions of county numbers of persons in poverty are ratio adjusted (raked) so their aggregations for each state agree with the corresponding predictions from the state models. To also account for the effect of estimation error in the state model, a Taylor series approximation is applied to generate the standard errors of the final, raked estimates.

Relative standard errors for the final estimates from the SAIPE CPS-based 2004 and ACS-based 2012 age 5–17 poverty models are displayed in Figure 19.7, which shows the medians of the relative standard errors for groups of counties defined by total population size. The relative standard errors are defined as the standard errors of the final raked estimates divided by the corresponding raked point predictions. For comparison, the median sampling coefficients of variation (CVs) of the direct ACS estimates are given for those county groups for which single-year direct estimates are published, that is for counties with over 65 000 in total population.

For all population size groups, the ACS-based predictions are substantially more precise than those from the CPS-based models. The medians across counties of the CPS-based relative standard errors decline from about 20 to 18% from the smallest to largest size groups, while the ACS-based versions decline from about 14 to 8%. The decline is more pronounced for the ACS-based SAIPE estimates because the much larger ACS sample size leads to higher

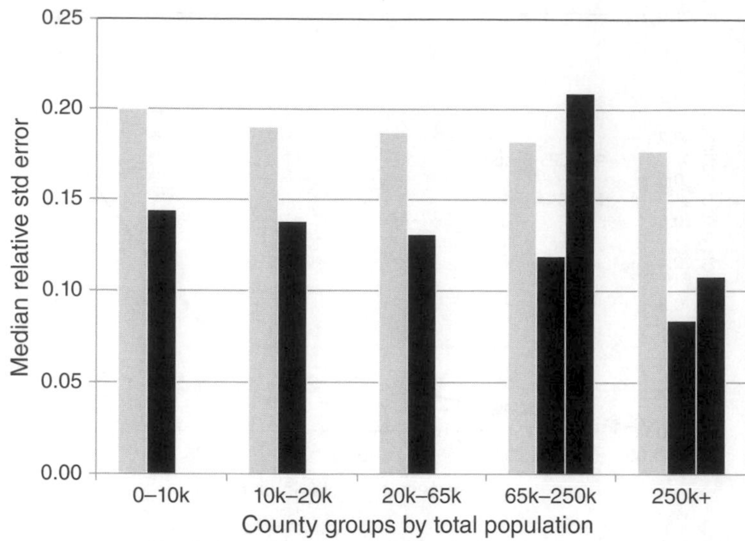

Figure 19.7 Median relative standard errors of various county poverty estimates for age 5–17 in families: SAIPE CPS-based 2004 model (left); SAIPE ACS-based 2012 model (middle); and ACS 2012 direct estimates (right)

shrinkage weights h_d (Figure 19.6), allowing the predictions to benefit more from the contribution of the more reliable ACS direct estimates for the larger counties. Figure 19.7 also shows that median relative standard errors for the direct ACS estimates are about 21 and 11% for the two largest size groups compared with about 12 and 8% for the ACS-based models. The difference is smaller for the largest counties because their model-based predictions and direct estimates are more similar due to the large values of the shrinkage weights h_d. Apart from the largest 7–10 counties, the ACS-based model predictions are always more precise than the direct estimates, given the model assumptions.[25]

19.3.3 School District Poverty Estimation

U.S. school districts differ from states and counties in several respects that create the following issues for estimating the poverty of school-age children.

Many school districts are very small. Of roughly 13 500 school districts in the U.S. in 2014, Census 2010 data showed 29% had child populations less than 500, and 64% had child populations less than 2000. While most school districts are small, most school-age children live in the larger districts—roughly two-thirds live in districts with total populations over 40 000.

Some counties contain multiple school districts and some school districts cross county boundaries. While in some states counties are the school districts, in other states counties may contain multiple districts, and in still other states there is essentially no relation between

[25] For the shrinkage estimates in the log-scale, the model predictions are more precise than the survey estimates for every county, matching the empirical BLUP theory. The ratio-adjustment procedure generates additional uncertainty, however, from the state-level predictions, and thus the largest 7–10 counties display slightly higher standard errors for the SAIPE estimates relative to the direct estimates.

county and school district boundaries. NRC (1998, p. 41), noted that in 1990 27% of school districts crossed county lines. To maintain consistency between school district and county estimates, SAIPE splits districts that cross county boundaries into school district pieces corresponding to the parts of the districts that overlap each county. Poverty estimates for the pieces within each school district are then summed to create the school district poverty estimates. This facilitates controlling the SAIPE school district poverty estimates to agree with the SAIPE county poverty estimates, but also requires the production of more estimates (over 20 000 compared to about 13 500) to cover the many small school district pieces.

Some school districts cover only selected grade ranges (e.g. elementary school districts may cover only up to grade 5 or 6, while separate secondary school districts cover grades from this level up through grade 12). Tabulating survey or other data sources for these districts requires that either the data provide grades of the school-age children in the files, or that the grades be assigned based on age or some other data. The 1990 and 2000 census long forms collected data used to determine grades within given ranges.[26] Individual grades of children could then be approximated using also the reported data on their ages. For the 2010 census, lacking a long form, SAIPE approximated grades from reported ages by assigning age 5 children to kindergarten, age 6 children to first grade, and so on. The ACS asks respondents to report the highest grade completed, by single-year grades, for all persons at the address.

School district boundaries may change over time, whereas county boundaries change only rarely. In addition to boundaries being moved, new school districts are formed while others go out of existence (as when multiple school districts are consolidated into one larger district). This issue has been addressed by having a school district boundary survey done every 2 years. Retabulation of the data sources, particularly previous census data, to account for the boundary changes, has been a critical aspect of updating the school district poverty estimates.

Originally, the only data source able to provide poverty estimates for school districts was the census long form. The CPS ASEC sample was far too small for this purpose: most school districts would have had no CPS sample, while many others would have had very small samples. The tax data and SNAP participants data were not tabulated for school districts. An administrative data source available specifically for school districts, data on participation in the free and reduced price lunch program of the USDA/FNS, was rejected due to serious concerns about its incompleteness and perceived variable quality across districts—see NRC (2000b, pp. 137–142) and Cruse and Powers (2006). This lack of usable current data precluded any modeling at the school district level. Given this, SAIPE employed a "census poverty shares" approach to produce updated school district poverty estimates.

The *census poverty shares approach* formed long form estimates of the ratio of the number of age 5–17 children in poverty in each school district piece to the number of age 5–17 children in poverty in the corresponding county. We call these ratios the "(school district piece to county) census poverty shares." To form current-year estimates of the number of age 5–17 children in poverty for the school district pieces, the poverty share estimates from the most recent census long form were multiplied by the corresponding SAIPE county model estimates of the number of age 5–17 children in poverty for the current year. Since the estimated poverty shares for given district pieces would remain fixed until the next census long form, changes in the poverty estimates for school districts were thus driven entirely by (i) changes in the

[26] Census 2000 collected reports of children attending preschool, kindergarten, or grades in the ranges 1–4, 5–8, and 9–12 (plus college undergraduate or graduate school). The 1990 census collected reports of highest grade completed, expanding 9–12 to the individual grade years, and also collected more detail about college education.

county poverty estimates obtained from the county model; and (ii) changes in school district boundaries (which forced retabulations of the previous census data to provide poverty share estimates for the redefined school district pieces).

This approach was used with 1990 census poverty shares through 1998, and from 1999 to 2004 with Census 2000 poverty shares. Starting with the 2005 estimates, a new approach was implemented that combined census poverty shares with tabulations of tax return data for school district pieces. The use of these tabulations posed some new challenges, however.

Nearly all tax return addresses can be directly assigned to a county. The Census Bureau's standard approach to geographic assignment of addresses to subcounty areas such as school district pieces first goes through an intermediate step called "geocoding," in which the address is assigned to a census block. Unfortunately, not all tax return addresses can be geocoded, leaving pools for many counties of nongeocoded tax exemptions not assigned to any of its school district pieces. This compromises tabulations of the tax data for the school district pieces of such counties to an extent depending on their amounts of nongeocoded exemptions. In 2012, approximately 8% of all tax returns nationally could not be geocoded. Non-city style addresses (those which lack a house number/street identification) are a major reason why certain tax returns are not geocoded, so that nongeocoded rates in rural counties can be much higher than elsewhere, sometimes exceeding 50%. In fact, for 2012, there were 34 counties for which over 90% of the tax returns lacked identification of subcounty locations.

A second issue with using the tax data is that tax returns do not include any age information about child exemptions. Such information is needed to determine the number of county poor child exemptions corresponding to the grade range of each of its school district pieces, whether the range is the full K-12 (age 5–17) or is limited. Fortunately, county age distributions for children by single years of age are available each year from county population estimates, and these can be used to make age-grade range adjustments to the data. To do this we multiply the number of geocoded poor child exemptions assigned to the geographic area for each school district piece by the proportion from the population estimates of the age 0–17 children in the county who are in that piece's age-grade range. The county's nongeocoded poor child exemptions are similarly age-adjusted to reflect the age 5–17 population. For counties that contain only unified school districts (covering all grades) the grade assignments do not matter and the age adjustments ultimately cancel out. In 2014, this was the case for 2891 (92%) of the 3142 U.S. counties. Only for the remaining 251 counties (8%) would the age-grade adjustments differentially affect the school district piece poverty estimates.

Beginning with the 2010 estimates, SAIPE has also adjusted the geocoded poor child exemptions to account for the difference in poverty rates between different age groups. This adjustment multiplies the geocoded poor child exemptions for any school district piece having a restricted age range by the ratio of the state poverty rate for that age range to the age 5–17 state poverty rate, using state-level ACS 5-year poverty rate estimates for this purpose.

In addition to these adjustments, the nongeocoded poor child exemptions must be allocated to the school district pieces before poverty shares can be constructed. While this allocation might be done according to the county distributions of either population or geocoded poor child exemptions, we had concerns that, in some counties, the true distribution of the nongeocoded exemptions across school district pieces might be very different. The Minimum Change (MC) algorithm (Maples and Bell 2007) was developed to address this concern. The idea was to use the previous census long form poverty shares except where the geocoded tax exemptions for a county provide definite evidence to contradict the previous census results. The MC algorithm achieves this by allocating the nongeocoded poor child exemptions for each county to make the resulting within county shares of poor child exemptions as close as possible, in some sense, to the corresponding shares from the most recent census. For details, see Maples and Bell (2007). If the number of nongeocoded poor child exemptions in a county is large, it is likely they can be allocated so that the resulting shares of the poor child tax exemptions for the county's school district pieces will reproduce the corresponding census shares. If so, the tax data for the county has no effect on the estimates for its school district pieces. If a county has a small number of nongeocoded poor child exemptions, then its child tax poverty shares after allocation of the nongeocoded exemptions will remain close to those from just the geocoded exemptions, which may differ substantially from the census poverty shares. If all tax returns for a county are geocoded, then the child poverty shares used are those from the tax data and the previous census results have no effect.[27]

After allocating the nongeocoded poor child exemptions through the MC algorithm, we tabulate the total poor child exemptions for each school district piece. An estimate of poor age 5–17 children for the district piece is then its share, s_{dj}, of the total county poor child exemptions times the SAIPE county model estimate of children age 5–17 in poverty (C_d):

$$\frac{\text{Adjusted poor child exemptions}_{dj}}{\sum_j \text{Adjusted poor child exemptions}_{dj}} \times \text{SAIPE county}_d \text{ poor children age 5–17} = s_{dj}C_d. \tag{19.5}$$

We refine (19.5) as follows. First, we make age-grade range adjustments to total child exemptions as was done for the poor child exemptions. We then use the MC algorithm with previous census population shares to allocate nongeocoded total child exemptions to school district pieces. We thus construct school district piece population estimates analogously to (19.5), replacing there the county child poverty estimate by a demographic estimate of county age 5–17 population. We then compute poverty rates for the school district pieces, and multiply these by school district piece population estimates to get refined child poverty estimates for each piece. The latter are then raked to agree with the county poverty estimates C_d. See Maples (2008, p. 2) for details.

Maples and Bell (2007) report results of some evaluations of the census shares approach and the MC algorithm. Estimates were formed for 1999 from the two approaches using county model predictions for that year and taking the "previous census poverty shares" from the 1990 census results. The estimates were compared with direct school district poverty estimates from Census 2000. This process was also reversed in time, forming estimates for 1989 using poverty shares from the Census 2000 results and the 1989 county age 5–17 poverty model estimates. These estimates were compared with the 1990 census long form estimates. The evaluation results showed relative errors of the estimators tended to increase with the nongeocoding rate, and tended to decrease with increasing population size. They also showed some advantage to the MC approach, which tended to perform better than the census shares approach when the nongeocoding rate was low, and similarly to the census shares approach when the nongeocoding rate was high.

Maples (2008) took this evaluation approach further, making stronger assumptions to obtain approximations to the mean squared errors of the school district poverty estimators. For the

[27] Starting in 2010, the most recent ACS 5-year estimates of poverty shares have replaced the previous census long form poverty shares for use in the MC algorithm.

Table 19.2 Median RRMSEs (within school district size groups) for 1999 school district age 5–17 poverty estimates (evaluations against Census 2000 estimates)

Estimator	School district population size					All districts
	<500	500–1K	1K–2K	2K–5K	5000+	
Census 2000 long form	0.32	0.24	0.21	0.16	0.09	0.22
MC shares	0.77	0.44	0.35	0.29	0.21	0.44
Census shares	0.70	0.50	0.44	0.43	0.35	0.52
Number of districts	4418	2356	2636	2936	1988	14 334

Results taken from Maples (2008, Table 9).

MC estimators this involved using squared differences between the predicted 1999 poverty shares [s_{dj} in (19.5)] and the Census 2000 estimated poverty shares (allowing for their sampling variances) to estimate a parametric error variance function. Analogous calculations were done for the estimator using the 1990 census poverty shares, and two parameterizations of the variance function were tried. Table 19.2 contains an excerpt of his results, taken from Maples (2008, Table 9). The results are expressed as relative root mean squared errors (RRMSEs) of the school district estimates of number of age 5–17 children in poverty [labeled “CVs” in Maples (2008)]. Sampling CVs of the corresponding direct Census 2000 long form estimates are included for comparison. We see that, with the exception of the smallest size group, the RRMSEs of the MC approach are smaller than those of the census shares approach, though both sets of RRMSEs are substantially larger than the CVs of the direct long form estimates (which is not surprising). We also note that the RRMSEs of the MC and census shares estimators are rather high, exceeding 20% even for the largest size category, and being much larger for the smallest size categories.[28] Compare these to the results in Figure 19.7 showing that the highest median RRMSEs for the county models, which occurred for the smallest counties, were 20% for the CPS 2004 model and 14% for the ACS 2012 model.

19.3.4 Major Changes Made in SAIPE Models and Estimation Procedures

An ongoing program of production of model-based small area estimates provides opportunities to make improvements over time by refining models and incorporating new or updated data sources. It also dictates responsibilities to take corrective actions in response to any problems discovered that may surface due to changing circumstances. Since its inception, the SAIPE program has made many changes to its models and estimation procedures, and to how it has used its data sources. Many of these changes, such as slight refinements in the definition of a regression variable, were relatively minor and had only minor effects on the estimates produced. But some changes have had substantial impacts. The most important changes are listed below. For changes previously discussed, we simply note in which section(s) they are discussed. For changes not elsewhere discussed, we provide brief explanations of why they were

[28] Results are shown for Maples's variance parameterization A. Results for his other parameterization B show somewhat lower RRMSEs, but the pattern of the results comparing the MC and census shares approaches is very similar. Maples also provides analogous results for comparing “predictions” for 1989 to the 1990 census results. These results are very similar to those shown, except that the RRMSEs for the smallest size group also favor the MC estimator over the census shares approach.

important. Detailed documentation of model changes is provided in the annual documentation on the SAIPE web site at http://www.census.gov/did/www/saipe/methods/index.html.

- *Bayesian treatment of the SAIPE state model (Section 19.3.1)*
- *Change to using census estimates as a regressor and then a change back to using census residuals in the state model (Section 19.3.1)*
- *Dropping, then later reinstating, the SNAP participation rate variable in the state models (Section 19.3.1)*
- *Replacing 1990 census data with data from Census 2000*

  The Census 2000 long form poverty estimates, which were for 1999, were available early enough to be used in the SAIPE models for that year. This updating of the census data had important effects on the SAIPE estimates, particularly at the school district level, for which they were a key component of a limited set of input data. (See Section 19.3.3.)
- *Switch from use of CPS data to use of ACS data (Sections 19.2.1, 19.3.1, and 19.3.2)*
- *Changes to the SAIPE production schedule*

  The SAIPE production schedule has been advanced several times to improve the timeliness of the estimates, a key aspect of their quality. In particular, starting in 2004 the production schedule was accelerated by a full year, which allowed SAIPE to release estimates for both 2002 and 2001 that year. The new schedule put the SAIPE estimates at just a 1-year lag from the corresponding release of the CPS ASEC direct national poverty estimates. The change was made possible by a successful effort to acquire the SNAP participants data from the USDA/FNS that are used in the county poverty models a year earlier than in previous years. SAIPE also changed to using the first vintage of the Census Bureau's population estimates for constructing regression variables for the state and county models, and for constructing school district population estimates. For more on this, see http://www.census.gov/did/www/saipe/methods/02change.html.
- *Implementing the MC algorithm to use subcounty tabulations of tax data in updating school district poverty estimates (Section 19.3.3)*
- *Developing margins of error for intertemporal and cross-sectional comparisons*

  Many users of SAIPE estimates are interested in making statistical comparisons of estimates across time or space. The prediction intervals SAIPE provides for the estimates do not address this because prediction errors in the estimates are correlated across time and space. For intertemporal statistical comparisons of SAIPE county estimates, Basel *et al.* (2010) developed a specification of the intertemporal distribution of the error components, with the sampling errors assumed independent across any given pair of years.[29] The covariance of the random effects is then estimated using a method of moments approach, and the resulting covariance between the errors in pairs of estimates is then adjusted for all the steps involved in the estimation: regression, shrinkage, transformation to the native scale, and ratio adjustment. A similar development yields prediction standard errors for cross-sectional comparisons, that is for comparing estimates for two counties in a given year.
- *Replacing Census 2000 estimates by corresponding ACS 5-year estimates in the MC algorithm used to produce school district poverty estimates (Section 19.3.3)*

[29] This assumption fails for the CPS data, hence this methodology applies only to the ACS-based SAIPE estimates, not to the CPS-based estimates.

19.4 Current Challenges and Recent SAIPE Research

Despite the improvements made over time in the SAIPE data sources, models, and prediction methods, several challenges remain: with no 2010 long form, the "previous census estimates" used in SAIPE models (from Census 2000) are going further out of date; single-year direct ACS sampling variance estimates are imprecise for many small counties; the county model still requires dropping zero poverty estimates from the model fitting (though this is a minor problem with the ACS data); and we do not have a formal model for predicting school district poverty. Recent SAIPE research has sought to address these challenges.

A natural option for addressing the challenge of the previous census estimates going out of date is to replace them in the SAIPE state and county models by 5-year ACS estimates from the 5 years just prior to the current estimation year. Huang and Bell (2012) investigated this option using a bivariate version of the Fay–Herriot model, which repeats equations of the form of (19.1) and (19.2) for a second set of direct survey poverty estimates y_{2d}, with the population quantities Y_{1d} and Y_{2d} allowed to be correlated through their model errors u_{1d} and u_{2d}. Huang and Bell (2012) compared posterior (prediction error) variances of Y_{1d} (current year poverty) from using Census 2000 estimates versus using prior ACS 5-year estimates as y_{2d} in the bivariate model. They found no disadvantage to replacing the Census 2000 estimates by the prior ACS 5-year estimates, and some indications that this change could produce a slight improvement.

Rather than incorporating prior ACS 5-year estimates into the modeling, an alternative is to jointly model current and some number (> 1) of prior ACS 1-year estimates by specifying a time series model for the model errors u_{dt}, the t subscript indexing time. Taciak and Basel (2012) investigated models with first-order autoregressive [AR(1)] random effects for application to logs of ACS county age 5–17 poverty estimates, while Hawala and Lahiri (2012) investigated AR(1) models for application to untransformed ACS estimates of county age 5–17 poverty rates.

As noted in Section 19.3.1, when the SAIPE state poverty rate models were based on CPS data, the direct CPS variance estimates were improved using a GVF model with random state effects (Otto and Bell 1995). Maples (2011) investigated similar models for variance estimates of ACS county poverty estimates, refining the GVFs for this application and using random county effects.

While the switch from CPS to ACS data greatly reduced the occurrences of zero poverty estimates for small counties, this issue can potentially be eliminated by replacing the linear Fay–Herriot model of log poverty with a suitable generalized linear mixed model (GLMM). Such models are discussed by Ghosh *et al.* (1998) and Rao (2003, sections 5.6 and 10.11). Slud (2000, 2004) did several analyses comparing results from GLMM models with results from models similar to the SAIPE county production model (in the form applied to CPS data). Franco and Bell (2013) used a binomial/logit normal model for the direct estimates of county numbers in poverty, with a logistic regression model for the binomial "success probability" p_d (the true poverty rate). They used GVF estimates of sampling variances to define an "effective sample size" for the binomial distribution to account for effects of the ACS sample design and survey weighted estimation. They then extended the model to a bivariate version applied to ACS 1-year and prior ACS 5-year county poverty estimates.

As an alternative, Wieczorek *et al.* (2012) investigated the *zero-one inflated beta model* (ZOIB model). This model allows positive probabilities that the direct survey estimate of the poverty rate for an area is either zero or one. Conditional on this not occurring, the estimated rate is assumed to follow a beta distribution with a mean whose logit follows a normal distribution. In a small simulation study, Wieczorek *et al.* reported some favorable results for the ZOIB model in regard to bias, mean squared error, and confidence interval coverage, in comparison with estimates from a linear Fay–Herriot model for poverty rates, and to direct survey estimates.

The switch to ACS data provided direct survey estimates of school district poverty, creating possibilities for modeling at the school district level. To address the lack of suitable covariate data for school districts, recent SAIPE research has sought to extend the geocoding process so that all tax returns can be assigned to school districts. The research has extended an approach used by other Census Bureau programs to impute missing geocodes using the zip code[30] information from address records. The imputations come from an empirical distribution of the populations of subcounty areas (such as school districts) that intersect each given zip code in a county. This distribution is constructed using data from the geocoded tax returns, and from the Census Bureau's Master Address File. A progression of successively more detailed zip code distributions is used, starting with the standard five-digit zip but, when available, using the complete nine-digit zip.

19.5 Conclusions

We have reviewed the data sources, models, and prediction methods used by the SAIPE program to produce poverty estimates for states, counties, and school districts of the U.S. While all these things are important to the success of the program, here we have paid particular attention to discussing our data sources, as their quality is so critical. Note that of the nine major SAIPE changes listed in Section 19.3.4 five (the third through sixth and the last) specifically involved changes to or addressing issues with, data sources, while another (the seventh) involved both new data tabulations and a change in methodology. Also, the part of the SAIPE program with the most potential for improvement is the school district poverty estimation, where the data sources are the most limited. The efforts mentioned at the end of Section 19.4 to improve the assignment of tax returns to subcounty areas to produce better school district tabulations of tax data are key to improving the school district estimates.

While some of the discussion here of data sources and even of models may be particular to SAIPE, we suspect that some issues similar to those SAIPE has faced will arise in other applications. We thus hope that this chapter provides a useful illustration of how such problems and challenges may be dealt with in a production environment for small area estimation. One challenge that we believe holds generally is that the smaller the areas are for which estimates are sought, the more limited and problematic the potential data sources are likely to be. Thus, data problems tend to present the most difficulties for the implementation of small area models at the smallest geographic levels. While outright solutions to this problem seem unlikely, it may at least be helpful to know where the most difficult challenges can be expected to lie.

[30] "Zip codes" are the U.S. system of numeric postal codes. The basic zip code consists of five digits, but the full zip code of nine digits provides even more detailed geographic identification. Many, though not all, tax return addresses include nine-digit zip codes.

References

Akaike H 1973 Information theory and an extension of the maximum likelihood principle. In *the* 2nd International Symposium on Information Theory (eds Petrov BN and Czaki F), Budapest: Akademia Kiado, 267–287.

Asher JL and Fisher RC 2000 Alternate CPS sampling variance structures for constrained and unconstrained county models. SAIPE Technical Report No. 4, U.S. Census Bureau, available online at http://www.census.gov/did/www/saipe/publications/files/tech.report.4.pdf. [accessed Sep-5-15]

Basel WW, Hawala S, and Powers D 2010 Serial comparisons in small domain models: A residual-based approach. SAIPE Technical Report, U.S. Census Bureau, available online at http://www.census.gov/did/www/saipe/publications/files/BaselHawalaPowers2010asa.pdf. [accessed Sep-5-15]

Bell WR 1999 Accounting for uncertainty about variances in small area estimation. *Bulletin of the International Statistical Institute*, 52nd Session, Helsinki.

Berger JO 1985 *Statistical Decision Theory and Bayesian Analysis*. Springer-Verlag.

Cruse C and Powers D 2006 Estimating school district poverty with free and reduced-price lunch data. *Proceedings of the American Statistical Association, Section on Government Statistics* [CD-ROM], Alexandria, VA: American Statistical Association, available online at https://www.census.gov/did/www/saipe/publications/conference.html. [accessed Sep-5-15]

Datta GS and Lahiri P 2000 A unified measure of uncertainty of estimated best linear unbiased predictors in small-area estimation problems. *Statistica Sinica* **10**, 613–628.

Fay RE and Herriot R 1979 Estimation of income from small places: An application of James-Stein procedures to census data. *Journal of the American Statistical Association* **74**, 269–277.

Fay RE and Train GF 1995 Aspects of survey and model–based postcensal estimation of income and poverty characteristics for states and counties. *Proceedings of the American Statistical Association, Section on Government Statistics*, Alexandria, VA: American Statistical Association, 154–159, available online at http://www.census.gov/did/www/saipe/publications/files/FayTrain95.pdf. [accessed Sep-5-15]

Fay RE and Train GF 1997 Small domain methodology for estimating income and poverty characteristics for states in 1993. *Proceedings of the American Statistical Association, Social Statistics Section*, 183–188, available online at http://www.census.gov/did/www/saipe/publications/files/FayTrain97.pdf. [accessed Sep-5-15]

Findley DF, Monsell BC, Bell WR, Otto MC, and Chen B-C 1998 New capabilities and methods of the X-12-ARIMA seasonal adjustment program (with discussion). *Journal of Business and Economic Statistics* **16**, 127–177.

Fisher GM 1997 The development of the Orshansky poverty thresholds and their subsequent history as the official U.S. poverty measure. Available online at http://www.census.gov/hhes/povmeas/publications/orshansky.html. [accessed Sep-5-15] A condensed earlier version of this paper appeared in 1992 as 'The development and history of the poverty thresholds' *Social Security Bulletin* **55**(4), 3–14.

Franco C and Bell WR 2013 Applying bivariate binomial/logit normal models to small area estimation. *Proceedings of the American Statistical Association, Section on Survey Research Methods*, 690–702, available online at http://www.amstat.org/sections/srms/Proceedings/. [accessed Sep-5-15]

Gelman A 2006 Prior distributions for variance parameters in hierarchical models (Comment on Article by Browne and Draper). *Bayesian Analysis* **1** (3), 515–534.

Ghosh M, Natarajan K, Stroud TWF, and Carlin BP 1998 Generalized linear models for small area estimation. *Journal of the American Statistical Association* **93**, 273–282.

Hawala S and Lahiri P 2012 Hierarchical Bayes estimation of poverty rates. *Proceedings of the American Statistical Association, Survey Research Methods Section*, 3410–3424, available online at http://www.census.gov/did/www/saipe/publications/files/hawalalahirishpl2012.pdf. [accessed Sep-5-15]

Huang ET and Bell WR 2012 An empirical study on using previous American Community Survey data versus Census 2000 data in SAIPE models for poverty estimates. Research Report RRS2012/04, Center for Statistical Research and Methodology, U.S. Census Bureau, available online at http://www.census.gov/srd/papers/pdf/rrs2012-04.pdf. [accessed Sep-5-15]

Maples JJ 2008 Calculating coeffcient of variation for the minimum change school district poverty estimates and the assessment of the impact of nongeocoded tax returns. Research Report RRS2008/10, Center for Statistical Research and Methodology, U.S. Census Bureau, available online at https://www.census.gov/srd/papers/pdf/rrs2008-10.pdf. [accessed Sep-5-15]

Maples JJ 2011 Using small. area modeling to improve design-based estimates of variance for county level poverty rates in the American Community Survey. Research Report RRS2011/02, Center for Statistical Research and Methodology, U.S. Census Bureau, available online at http://www.census.gov/srd/papers/pdf/rrs2011-02.pdf. [accessed Sep-5-15]

Maples JJ and Bell WR 2007 Small area estimation of school district child population and poverty: Studying the use of IRS income tax data. Research Report RRS2007/11, Center for Statistical Research and Methodology, U.S. Census Bureau, available at online http://www.census.gov/srd/papers/pdf/rrs2007-11.pdf. [accessed Sep-5-15]

National Research Council 1997 *Small Area Estimates of Children in Poverty, Interim Report 1, Evaluation of 1993 Estimates for Title I Allocations* (eds Citro CF, Cohen ML, Kalton G, and West KK), *Panel on Estimates of Poverty for Small Geographic Areas, Committee on National Statistics*, Washington, DC: National Academy Press.

National Research Council 1998 *Small Area Estimates of Children in Poverty, Interim Report 2, Evaluation of Revised 1993 Estimates for Title I Allocations* (eds Citro CF, Cohen ML, and Kalton G), *Panel on Estimates of Poverty for Small Geographic Areas, Committee on National Statistics*, Washington, DC: National Academy Press.

National Research Council 1999 *Small Area Estimates of Children in Poverty, Interim Report 3, Evaluation of 1995 County and School District Estimates for Title I Allocations* (eds Citro CF and Kalton G), *Panel on Estimates of Poverty for Small Geographic Areas, Committee on National Statistics*, Washington, DC: National Academy Press.

National Research Council 2000a *Small-Area Estimates of School-Age Children in Poverty: Evaluation of Current Methodology* (eds Citro CF and Kalton G), *Panel on Estimates of Poverty for Small Geographic Areas, Committee on National Statistics*, Washington, DC: National Academy Press.

National Research Council 2000b *Small Area Income and Poverty Estimates: Priorities for 2000 and Beyond* (eds Citro CF and Kalton G), *Panel on Estimates of Poverty for Small Geographic Areas, Committee on National Statistics*, Washington, DC: National Academy Press.

Otto MC and Bell WR 1995 Sampling error modelling of poverty and income statistics for states. *Proceedings of the American Statistical Association, Social Statistics Section*, 160–165, available online at http://www.census.gov/did/www/saipe/publications/files/BellOtto95.pdf. [accessed Sep-5-15]

Prasad NGN and Rao JNK 1990 The estimation of mean squared errors of small area estimators. *Journal of the American Statistical Association* **78**, 47–59.

Rao JNK 2003 *Small Area Estimation*. Hoboken: John Wiley & Sons, Inc.

Slud EV 2000 Models for simulation and comparison of SAIPE analyses. SAIPE Technical Report, U.S. Census Bureau, available online at http://www.census.gov/did/www/saipe/publications/files/saipemod.pdf. [accessed Sep-5-15]

Slud EV 2004 Small area estimation errors in SAIPE using GLM versus FH models. *Proceedings of the American Statistical Association, Section on Survey Research Methods*, 4402–4409, available online at http://www.amstat.org/sections/srms/Proceedings/. [accessed Sep-5-15]

Taciak J and Basel W 2012 Time series cross sectional approach for small area poverty models. *Proceedings of the American Statistical Association, Section on Government Statistics*, available online at http://www.census.gov/did/www/saipe/publications/files/JTaciakBaseljsm2012.pdf. [accessed Sep-5-15]

Wieczorek J, Nugent C, and Hawala S 2012 A Bayesian zero-one inflated beta model for small area shrinkage estimation. *American Statistical Association, Proceedings of the Survey Research Methods Section*, 3896–3910, available online at http://www.census.gov/did/www/saipe/publications/files/wieczoreknugenthawalajsm2012.pdf. [accessed Sep-5-15]

20

Poverty Mapping for the Chilean Comunas

Carolina Casas-Cordero Valencia[1], Jenny Encina[2] and Partha Lahiri[3]
[1]*Instituto de Sociología y Centro de Encuestas y Estudios Longitudinales, Universidad Católica de Chile, Santiago, Chile*
[2]*Inter-American Development Bank, Washington, DC, USA*
[3]*Joint Program in Survey Methodology and Department of Mathematics, University of Maryland, College Park, USA*

20.1 Introduction

The eradication of poverty has been at the center of various public policies in Chile and has guided public policy efforts. The nationwide survey estimate of the poverty rate has declined since the early 1990s suggesting some progress towards this goal. While this result is encouraging, erratic time series patterns have emerged for small *comunas* – the smallest territorial entity in Chile. Moreover, for a handful of extremely small comunas, survey estimates of poverty rates are unavailable for some or all time points simply because the survey design, which traditionally focuses on precise estimates for the nation and large geographical areas, excludes these comunas for some or all of the time points. In any case, direct survey estimates of poverty rates typically do not meet the desired precision for small comunas and thus the assessment of implemented policies is not straightforward at the comuna level. In order to successfully monitor trends, identify influential factors, develop effective public policies, and eradicate poverty at the comuna level, there is a growing need to improve on the methodology for estimating poverty rates at this level of geography.

Chile's official data source for poverty statistics is the National Socioeconomic Characterization Survey (Casen) – a survey sponsored every 2 or 3 years by the Ministerio de Desarrollo Social (henceforth referred to as the *Ministry*) since 1987 with sample in most of the comunas. The demand for various socioeconomic data at the comuna level is relevant for the design and

Analysis of Poverty Data by Small Area Estimation, First Edition. Edited by Monica Pratesi.

Companion Website: www.wiley.com/go/pratesi/poverty

evaluation of public policies, especially because municipalities[1] are the first level of contact for the Chileans with their (local) government.

In Chile, direct design-based estimates of poverty rates were routinely released for the nation, all regions and all the self-representing comunas in each Casen sample.[2] In 1999, the Programa de las Naciones Unidas para el Desarrollo (PNUD)[3] and the Ministry used direct design-based survey estimates as input data for producing the PNUD's Human Development Index (HDI)[4] for the 72 self-representing comunas in the Casen 1990 and 1998 samples. In 2000, comuna level estimates were produced as inputs for the HDI for 333 comunas using a mix of standard design-based and a Ministry-PNUD synthetic method using Casen 1998 data.[5] In 2005, an updated version of the comuna level HDI was produced for 334 comunas using Casen 1994 and 2003 data.

The need for socioeconomic data at lower levels of geography also found its way into the Chilean legislation in 2007 when an amendment to the law of the *Fondo Común Municipal* (FCM) established a new set of indicators for its fund allocation algorithm among comunas.[6] The regulation passed in 2009 required the Ministry to provide poverty rate estimates for all comunas in Chile.[7] In 2010, the Ministry produced for the first time poverty rate estimates for all 345 comunas in Chile using both standard design-based and the Ministry-PNUD synthetic method.

In 2010, the Ministry convened an Expert Commission to evaluate the design of the Casen survey and to make recommendations for future innovations.[8] Regarding the production of comuna level estimates, the experts raised concerns because of (i) the significant costs associated with sampling almost all comunas in the country, and (ii) the relatively low precision for some comuna level estimates making the planned comparison among comunas and/or across time useless. The Commission recommended to (i) reduce the overall sample significantly, (ii) stop the production of comuna level estimates, and (iii) search for alternative data sources such as administrative records or develop a new data collection effort specifically designed for comuna level representation of social indicators of interest for various public policies.

The Commission's concerns motivated the Ministry to convene a Ministry-PNUD working group in 2011 with the goal of developing a new methodology aimed at improving the precision of comuna level estimates. After reviewing the literature on small area estimation (SAE), the

---

[1] A municipality is a decentralized institution responsible for the local administration of each comuna. Ministerio del Interior [2006].

[2] Self-representing comunas were those considered with large enough sample sizes for the purpose of producing direct design-based estimates. The number of self-representing comunas increased from 48 in Casen 1987 to 335 in 2006. Ministerio de Desarrollo Social [2012a].

[3] In English, the United Nations Development Programme (UNDP).

[4] The HDI is a statistical tool used to measure a country's overall achievement in its social and economic dimensions. United Nations Development Programme (UNDP) [1990].

[5] The method was developed with the aim at producing estimates of poverty rates for comunas without sample in the Casen survey. The estimate corresponds to the average poverty rate calculated within a group of comunas with similar characteristics. Groups of comunas with similar characteristics were defined using Principal Component Analysis (PCA). Programa de las Naciones Unidas para el Desarrollo (PNUD) Chile y Ministerio de Planificación de Chile [2000].

[6] Ministerio del Interior [2007].

[7] Ministerio del Interior [2009].

[8] Comisión de Técnicos de la Encuesta Casen [2010].

Ministry decided to develop a methodology that could take advantage of the Casen data and relevant administrative records available in Chile. The advice of an international expert (Partha Lahiri) was sought and a model-based method was developed using Casen 2009 data and administrative data (henceforth referred to as the *SAE method*). In 2011, estimates of comuna level poverty rates were obtained using the SAE method for all comunas with sample in Casen 2009 and the synthetic Ministry-PNUD method for comunas without sample in Casen 2009. In 2012, 2013, and 2014 the same strategy was followed using Casen 2011 data.

The SAE method implemented in Chile consists of an array of different methodologies and thus a variety of choices had to be made at different stages of the production process. The purpose of this chapter is to shed some light on these issues by describing and discussing the application of the SAE method implemented in Chile using the Casen 2009 data. The sections that follow present a description of the poverty measures and the data sources used in Chile (Section 20.2), the data preparation process, SAE estimation method, and the results of the implementation of the SAE method (Sections 20.3, 20.4, and 20.5), and a discussion with directions for future research (Section 20.6).

20.2 Chilean Poverty Measures and Casen

A measure of *absolute poverty*[9] based on income has been used continuously since the 90's to assess one of the multiple dimensions of welfare of the Chilean population. The data used for estimating poverty has always been the Casen survey. In this section, we briefly describe the poverty measure used in Chile and the design of the Casen survey. In particular, we describe the design of Casen 2009 since this dataset was used for the development of the SAE method presented in this chapter.

20.2.1 The Poverty Measure Used in Chile

In Chile, poverty is measured using the poverty rate, also known as Headcount Index, defined as the proportion of households[10] with *income* below the *poverty threshold* or *poverty line*.

The first ingredient of the poverty rate is the *poverty line*. For most Latin American countries, the poverty line is the cost of a basket of essential food and non-food items.[11] This poverty line is expressed in per-capita terms. The methodology for estimating Chile's poverty line was developed by the Comisión Económica para América Latina y el Caribe (CEPAL)[12] in 1990

[9] "Absolute poverty lines represent the cost of buying a basket of essential items that allows one to meet the absolute thresholds of satisfying certain basic needs. The definition of the normative basket should therefore entail first, deciding on absolute threshold for each of the basic needs; second, defining the types and quantities of the goods and services that are necessary to meet each of those standards; and third, pricing those goods and services". United Nations Expert Group on Poverty Statistics (Rio Group) [2006, p. 53].

[10] In the context of the Casen survey, households are *consumption/economic units* defined as the group of residents, within a housing unit, that share the costs of food consumption. Ministerio de Desarrollo Social [2012a].

[11] For a review of methodologies for income-based measures of poverty used in Latin America see United Nations Expert Group on Poverty Statistics (Rio Group) [2006].

[12] In English, United Nations Economic Commission for Latin America and the Caribbean (ECLAC).

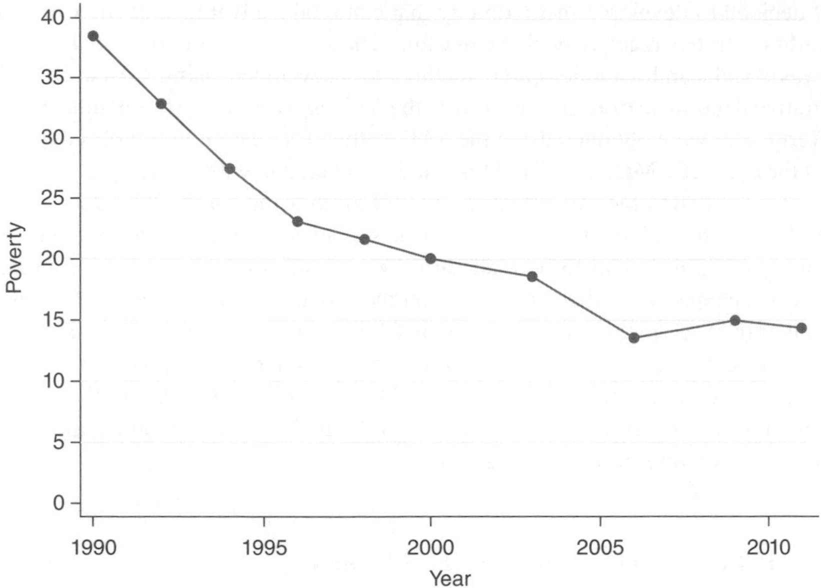

Figure 20.1 Estimates of national poverty rates in Chile, by year. *Source*: Data from Ministerio de Desarrollo Social [2012b].

and has remained almost unaltered since then.[13] Data from the Chilean expenditure survey *Encuesta de Presupuestos Familiares 1987–1988* was used to estimate the value of the food basket. Two different poverty lines were derived from the food basket – one for rural areas and the other for urban areas.[14] The second ingredient of the poverty rate, the *per-capita income,* is the ratio of the *total household income* and the *household size*. Households whose per-capita income falls below the poverty line are considered in poverty. The poverty rate is then the percent of households in each region/comuna that are in poverty.

In Chile, national and regional estimates of per-capita income and poverty rates are produced from the Casen survey using standard design-based methods. The official national level poverty rate estimates are published every 2 or 3 years following the release of each Casen data. Figure 20.1 displays estimates of poverty rates for the period 1990–2011, which reveals a declining trend over time.

20.2.2 The Casen Survey

The Casen survey is a cross-sectional multipurpose household survey designed to understand the socioeconomic conditions of the population and the evaluation of social programs. The survey has been fielded regularly every 2 or 3 years since 1987. Over time, Casen has become

[13] Comisión para la Medición de la Pobreza [2014].

[14] For a more comprehensive review of the methodology for the estimation of the poverty rate in Chile see Comisión Económica para América Latina y el Caribe (CEPAL) [1990].

the main data source for monitoring the performances of poverty and inequality indicators such as the poverty rate and the Gini coefficient.[15]

The survey's stratified, multistage, area probability sample design has evolved to allow for changes in the sampling frame and the sampling method used. The sample size and geographic coverage have expanded to accommodate the increasing demands of representation from local areas. The mode of data collection – in-person paper-and-pencil interview completed by a household informant (18+ years old) on behalf of all household members – has not changed since its inception. The questionnaire has increased in length and scope reflecting mostly changes in national social programs and the emergence of new topics of social interest.

The 2009 Casen survey collected data from 246 924 persons in 71 460 households, representing a total of 16 607 007 persons living in private dwellings in Chile in November 2009. The sampling design used was as follows: the target population was defined to cover 334 out of the 345 comunas in the country,[16] which were defined as the smallest domains for publication of official statistics. Samples were drawn independently from 602 sampling strata formed by the comuna's urban/rural subdivisions. Using a two-stage sampling design, small geographic entities, known as *secciones*, were sampled at the first stage (Primary Sampling Units, PSUs) and housing units were sampled at the second stage (Secondary Sampling Units, SSUs) within each sampling strata. The PSUs were selected with probability proportional to size, measured in terms of the number of occupied housing units. A variable number of SSUs were selected with equal probability using a systematic sampling algorithm with a random start within each selected PSU. Within each housing unit interviews were attempted with all households (i.e., no subsampling was implemented beyond the selection of the housing units).

The 2009 Casen microdata contain both comuna and region level weights. Both sets share the same base weights and nonresponse adjustments, but differ at the calibration stage.[17] In this application, regional weights were used to produce thresholds or cutoffs for the design-based weight trimming procedure (section 20.3.1), developing synthetic estimates of comuna level design effects (section 20.3.1) and setting the control for implementing raking adjustments (section 20.3.5). Comuna specific weights, on the other hand, were trimmed (and re-calibrated) to produce direct, standard survey-weighted, estimates of the comuna level poverty rates (section 20.3.1). Section 20.3 elaborates more on these procedures.

20.3 Data Preparation

The basic idea for the Chilean official small area methodology is to borrow strength from various relevant data sources using an appropriate comuna level model. In 2011, a major effort

[15] Comisión de Técnicos de la Encuesta Casen [2010].

[16] The 11 comunas excluded correspond to a few isolated rural areas, located in 4 of the 15 regions of the country that are not covered by the sampling frame maintained by the National Institute of Statistics. Ministerio de Desarrollo Social [2012a].

[17] Region weights are calibrated using 30 strata given by region and urban/rural subdivision, whereas comuna weights are calibrated using 602 strata given by comuna and urban/rural subdivision. Both sets of weights, when aggregated over all sampled respondents within a region, yield the population estimate of all persons living in private dwellings in the comunas included in the 2009 sample for that region. The region weights are used to produce national and regional level estimates whereas the comuna weights are only used for comuna level estimates. Ministerio de Planificación [2010].

was undertaken by a Ministry-PNUD working group to understand the relevance of different data sources for the production of comuna level poverty rates and to prepare data to be fed into the methodology described in Section 20.4. We now describe the data preparation steps.

20.3.1 Comuna Level Data Derived from Casen 2009

Extreme weights generally yield unreliable estimates. In order to reduce the effects of extreme comuna weights, the working group decided to trim these weights using a design-based method suggested by Potter [1993]. Since only a small number of PSUs in small comunas were available, the design-based mean squared error (MSE) estimates needed for obtaining cutoffs by the Potter method were not reliable for the small comunas. Thus, a synthetic method that used the same cutoff for all comunas within a large group was used. For this application, 30 large groups were formed by a combination of the 15 regions in Chile and 2 zonal (urban/rural) classifications. For each of these 30 groups, the Potter method found a cutoff point by minimizing the design-based empirical MSE of the direct poverty rate computed using the regional weights. Using these cutoff points, trimmed weights were obtained for the comunas using the original comuna weights. Finally, the trimmed comuna weights were calibrated to the corresponding comuna population projections.

The trimming procedure resulted in trimming of approximately 7.2% of the original comuna weights for the Casen 2009 survey. Table 20.1 displays descriptive statistics for the original comuna weights, the cutting-off point used and the total number of sampling weights truncated in each of the 30 groups. Figure 20.2 plots the original and trimmed comuna weights. Although no large difference can be observed between the original and trimmed comuna weights for the majority of the cases, the plot helps to identify cases that are most affected by the weight trimming procedure. All cases with survey weights larger than 2000, for example, were trimmed to values close to 1500. Some of the most extreme cases occurred in the comuna of Las Condes, where weights as high as 4000 were truncated to 1500. In the rest of the chapter, we refer to the set of trimmed comuna weights as "trimmed weights" or "survey weights".

Using the trimmed weights, direct estimates of the poverty rates for all the comunas were produced. Let p_i denote the direct estimate of poverty rate for the ith comuna. Before application of the well-known two-level normal model, discussed in Section 20.4, direct poverty rate estimates were transformed using the arcsine transformation: $y_i = \sin^{-1}\sqrt{p_i}$. Modeling of survey-weighted proportions has been considered in other small area applications. For example, in producing state level poverty rates in the US Census Bureau's SAIPE program[18], normal distribution is used to model the survey-weighted proportions. Because of much smaller sample sizes for the Chilean comunas, the working group felt the need to apply a suitable variance stabilizing transformation before applying a normal model. For the county level estimation in the SAIPE program, the logarithms of survey-weighted counts are assumed to be normally distributed. While logarithmic transformation is often used before modeling counts, the working group did not consider this logarithmic transformation appropriate for modeling survey-weighted proportions for two reasons. First, comunas with zero survey-weighted proportions would be excluded in developing the estimates. Secondly, such

[18] Small Area Income and Poverty Estimates (SAIPE) program of the U.S. Census Bureau. Details about the SAIPE program can be found in Bell [1997], Bell *et al.* [2007] and National Research Council [2000].

Table 20.1 Descriptive statistics for the original survey weights, truncation point and total number of original comuna weights truncated, by region and zonal group. Casen 2009 data.

Truncation groups	Descriptive statistics original comuna weights			Truncation point	Number of original comuna weights truncated
	Average	Minimum	Maximum		
1	87.4902	5	501	137.6	748
2	10.1405	2	34	63.0	0
3	94.3949	3	692	672.7	32
4	4.5907	1	27	32.5	0
5	59.8353	6	1.363	731.7	12
6	14.3449	4	47	83.4	0
7	95.7634	7	558	524.6	82
8	27.3082	4	134	127.6	58
9	74.0826	5	1.020	112.6	4117
10	23.0577	3	100	75.7	105
11	53.7106	2	637	262.0	229
12	22.6640	2	110	87.7	171
13	69.0955	3	405	141.5	1471
14	26.0215	4	160	49.8	1760
15	66.7037	4	1095	346.3	105
16	20.2520	4	203	30.6	2929
17	60.0223	4	548	318.7	225
18	28.2342	6	183	37.2	2152
19	70.5936	2	693	118.5	1547
20	22.8171	3	461	67.9	673
21	37.9711	5	183	52.8	497
22	9.9334	3	34	11.8	328
23	85.7458	8	544	777.8	0
24	9.8642	2	41	14.1	67
25	147.5910	5	4103	1445.2	81
26	38.0404	2	1147	476.6	18
27	53.8487	6	520	287.1	86
28	31.1182	6	237	135.4	54
29	106.1280	1	475	133.2	268
30	14.3313	1	57	103.1	0
Total	—	—	—	—	17815

Source: Data from Ministerio de Desarrollo Social [2013a].

a transformation may result in a poverty rate estimate greater than 1. After some discussions, the working group settled on the arcsine transformation for this project because it stabilizes the sampling variances and at the same time results in a better normal approximation.

The arcsine transformation used for this application was motivated by Efron and Morris [1975] who found such transformation useful for estimating proportions in a number of small area related applications. The descriptive statistics on the number of respondents and households displayed in Table 20.2 suggest that our sample sizes are more than the sample sizes Efron and Morris used for their well-known baseball data analysis. Carter and Rolph [1974]

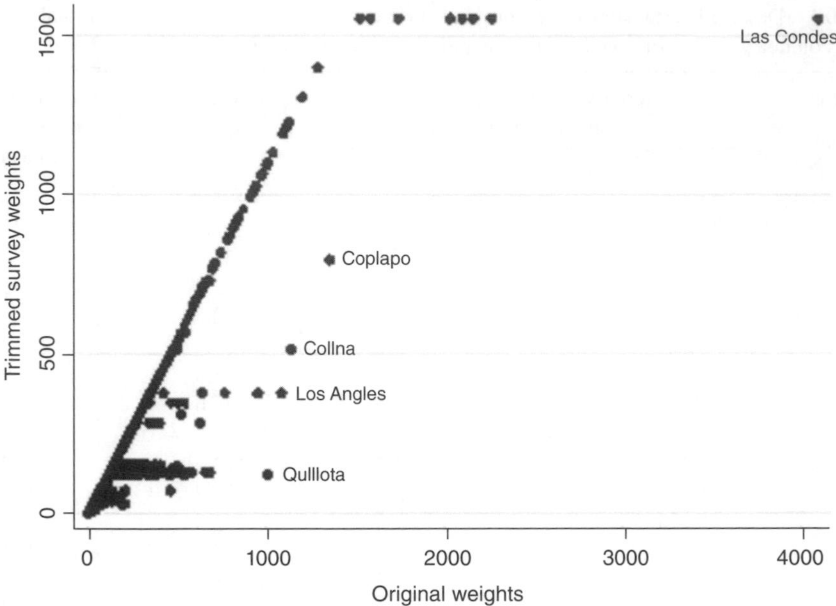

Figure 20.2 A plot of original weights (x-axis) and trimmed survey weights (y-axis) for all observations in Casen 2009. *Source*: Data from Ministerio de Desarrollo Social [2013a].

Table 20.2 Descriptive statistics of the number of cases at the respondent level and the household level. Casen 2009 data.

Quartiles of comunas respondent sample	Respondent level sample			Household level sample		
	Minimum	Mean	Maximum	Minimum	Mean	Maximum
1	53	491.0	610	20	152.7	198
2	612	654.9	692	155	195.7	239
3	693	752.2	864	177	214.1	265
4	873	1064.3	1608	211	294.4	409

Source: Data compiled by authors based on Casen 2009 data.

used arcsine transformation in estimating false fire alarm probabilities for small areas. In the context of SAE with complex survey data (Jiang *et al.* [2001], Raghunathan *et al.* [2007], Xie *et al.* [2007]), arcsine transformations of survey-weighted proportions were useful in modeling the sampling distributions.

Figure 20.3 plots poverty rate direct estimates using the trimmed and original survey weights for all comunas in Casen 2009. The two sets of estimates do not differ substantially. The plot also displays heterogeneity in poverty conditions among comunas in Chile with poverty rates ranging from 0 to 50%.

The sampling variance for the poverty estimate in the ith comuna was approximated by $D_i = 1/(4n_i)$, where n_i is the effective sample size obtained from the sample size divided by a

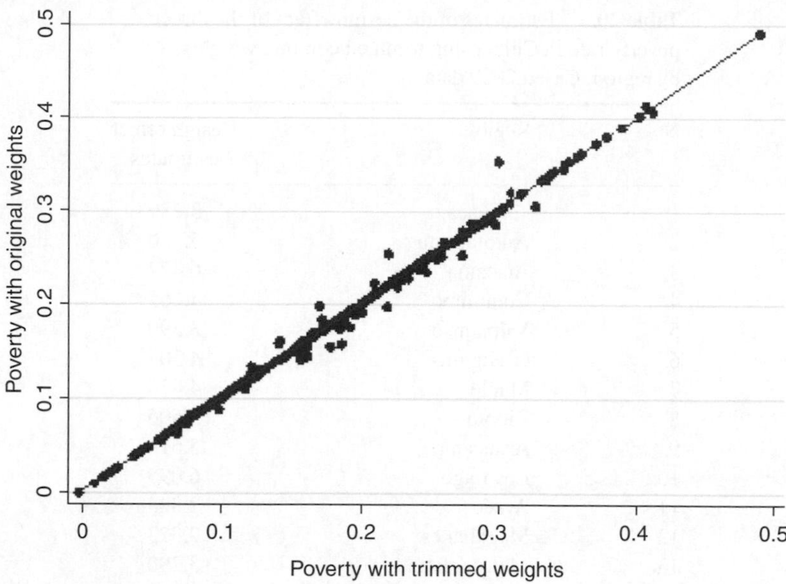

Figure 20.3 A plot of direct survey estimates with original survey weights (y-axis) and trimmed survey weights (x-axis) for comunas in Chile. Casen 2009 data. *Source*: Data from Ministerio de Desarrollo Social [2013a].

synthetic estimate of the design effect. In this application, the comuna level design effects are approximated by the standard design effect formula at the regional level. The heterogeneity in the sampling variance estimates across geographical areas is evident in Table 20.3, with regions in the north (e.g., Antofagasta, Atacama) and south (e.g., Los Lagos, Los Rios) of the country displaying more extreme estimates of the design effect than those in the center of the country.

Table 20.4 displays quartiles of sampling variance (D_i) estimates, where Group 1 provides results for those with the 25% lowest values and Group 4 those with the 25% highest values. Results show that the average sampling variance in the highest quartile (i.e., 0.0030616) is almost three times larger than that in the lowest quartile (i.e., 0.0009047). As shown in Figure 20.4, comunas with larger sample sizes in the Casen survey (y-axis) are associated with lower levels of sampling variability (x-axis) and vice versa. We can also observe some degree of variability in estimates of the sampling variances for comunas with relatively similar sample sizes.

20.3.2 Comuna Level Administrative Data

The initial task of identifying a wide range of auxiliary variables expected to be correlated with poverty was facilitated by the wealth of administrative data available within the Chilean government system. The initial list of auxiliary variables was reviewed by subject matter experts who helped narrowing down the list (Table 20.5). The selection of variables up to this point included those that were (i) associated with *social and economic conditions* (in time this will help understanding the evolution of the social and economic condition of the comunas

Table 20.3 Estimates of the design effect of the direct poverty rate in Chile using trimmed comuna weights, by region. Casen 2009 data.

No.	Region	Design effect estimates
1	Tarapacá	3.280
2	Antofagasta	5.750
3	Atacama	6.477
4	Coquimbo	4.665
5	Valparaíso	3.390
6	O'Higgins	4.307
7	Maule	4.870
8	Biobío	5.506
9	Araucanía	5.618
10	Los Lagos	6.095
11	Aysén	2.843
12	Magallanes	2.323
13	Metropolitana	3.290
14	Los Ríos	8.681
15	Arica y Parinacota	2.864

Source: Data from Ministerio de Desarrollo Social [2013a].

Table 20.4 Descriptive statistics of D_i and B_i by groups of comunas formed using quartiles of the distribution of the sampling variances (D_i). Casen 2009 data.

Quartiles of D_i	Descriptive statistics of D_i			Descriptive statistics of B_i		
	Mean	Minimum	Maximum	Mean	Minimum	Maximum
Group 1 (lowest 25% of D_i)	0.0009047	0.0004453	0.0012006	0.2765	0.1598	0.3390
Group 2	0.0014632	0.0012023	0.0016854	0.3836	0.3393	0.4186
Group 3	0.0019475	0.001686	0.0021987	0.4535	0.4186	0.4843
Group 4 (highest 25% of N_i)	0.0030616	0.0022082	0.0113181	0.5474	0.4854	0.8286
All	0.0018417	0.0004453	0.0113181	0.4150	0.1598	0.8286

Source: Data from Ministerio de Desarrollo Social [2013a].

in Chile), (ii) collected and administered by reliable sources (this helps to control measurement errors – an important aspect of the quality of estimators based on these data), and (iii) elaborated and published periodically (this helps to depict more accurately and timely the phenomenon under study).

At the outset, transformations were taken on all potential comuna level auxiliary variables. To be specific, arcsine transformation was applied to all auxiliary variables that were proportions and logarithmic transformation was taken on the rest of the auxiliary variables. The analysis was performed on a subset of comunas with population greater than 10 000

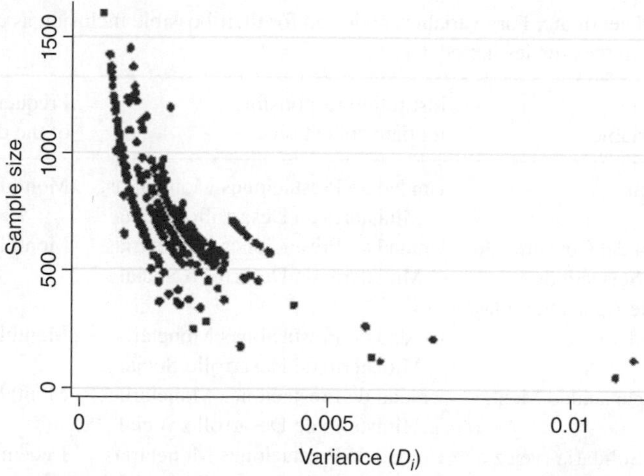

Figure 20.4 Comuna level sample size (y-axis) and comuna level estimate of variance (x-axis) of the direct estimate of the poverty rates. Casen 2009 data. *Source*: Data from Ministerio de Desarrollo Social [2013a].

inhabitants,[19] which had larger sample sizes, in order to reduce the potential effect of sampling errors on the dependent variable (the poverty rates) in the regression analysis.

To identify a parsimonious model for the poverty rates across the small areas, a data driven approach was followed using the stepwise procedure and model diagnostics available in Stata 11 [StataCorp, 2009]. Standardized residuals were examined for normality and homoskedasticity. The Q-Q plot displayed in Figure 20.5 and the residual plot displayed in Figure 20.6 do not provide any strong evidence of lack of normality. The Shapiro–Wilk statistics ($W = 0.99485$) further supports the lack of evidence against the normality assumption.

In order to investigate the validity of the homoskedasticity assumption, we examined residuals. Figure 20.7 displays the adjusted (x-axis) and the unadjusted (y-axis) residuals of the comuna level model. No pattern in the residuals emerges from the graph. This and the Spearman correlation coefficients, displayed in Table 20.6, do not support any evidence of lack of the homoskedasticity assumption.

The final set of auxiliary variables included in the small area model are: average wage for dependent workers,[20] percentage of rural population,[21] percentage of illiterate population,[22] percentage of school attendance,[23] the average of the comuna level poverty rates from Casen 2000, 2003, and 2006, and region level indicators for the 7th, 8th and 9th regions of the country. The comuna level model, fitted using ordinary least squares, yielded an adjusted R^2 coefficient of 67%. Table 20.7 displays the results for the model. The magnitude and sign of the coefficients were all consistent with our expectation. Comuna level poverty rates were positively

[19] In Chile 246 of the 345 comunas had 10 000 or more inhabitants, according to the population projections published by the National Institute of Statistics by June 30 of 2011.

[20] Corresponds to auxiliary variable 24 in Table 20.5.

[21] Corresponds to auxiliary variable 19 in Table 20.5.

[22] Corresponds to auxiliary variable 18 in Table 20.5.

[23] Corresponds to auxiliary variable 20 in Table 20.5.

Table 20.5 Initial set of auxiliary variables reviewed for their possible inclusion as comuna level auxiliary variables in the area level model.

Name of the auxiliary variable	Institution responsible for data collection	Frequency of publication of the data
1. Subsidio Familiar	Unidad de Prestaciones Monetarias, Ministerio de Desarrollo Social	Monthly and yearly
2. Subsidio al Pago del Consumo de Agua Potable y Servicio de Alcantarillado de Aguas Servidas	Unidad de Prestaciones Monetarias, Ministerio de Desarrollo Social	Monthly and yearly
3. Bono Chile Solidario	Unidad de Prestaciones Monetarias, Ministerio de Desarrollo Social	Monthly and yearly
4. Subsidio de Discapacidad Mental	Unidad de Prestaciones Monetarias, Ministerio de Desarrollo Social	Monthly and yearly
5. Pensión Básica Solidaria (vejez e invalidez)	Unidad de Prestaciones Monetarias, Ministerio de Desarrollo Social	December
6. Aporte Previsional Solidario (vejez e invalidez)	Unidad de Prestaciones Monetarias, Ministerio de Desarrollo Social	December
7. Bonificación al Ingreso Ético Familiar	Unidad de Prestaciones Monetarias, Ministerio de Desarrollo Social	Monthly and yearly
8. Beca de Apoyo a la Retención Escolar, BARE	Unidad de Prestaciones Monetarias, Ministerio de Desarrollo Social	Monthly and yearly
9. Afiliados Sistema de Capitalización Individual	Superintendencia de Pensiones	Monthly and yearly
10. Matrícula	Ministerio de Educación	Yearly
11. Rendimiento	Ministerio de Educación	Yearly
12. SIMCE	Ministerio de Educación	Yearly or every 2 years
13. Titulados Educación Superior	Ministerio de Educación	Yearly
14. Índice de Vulnerabilidad del Establecimiento (IVE-SINAE)	Junta Nacional Escolar y Becas (Junaeb)	Yearly
15. Situación Nutricional estudiantes básica y media	Junta Nacional Escolar y Becas (Junaeb)	Yearly
16. Población beneficiaria Fonasa	Ministerio de Salud	Yearly
17. Atenciones sector privado	Ministerio de Salud	Yearly
18. Razón de analfabetos respecto a la población de 10 y más años en la comuna	CENSO, INE	Every 10 years
19. Porcentaje de Población Rural	CENSO, INE	Every 10 years
20. Porcentaje de Asistencia Escolar Comunal	SINIM	Monthly
21. Tamaño promedio del hogar	CENSO, INE	Every 10 years
22. Tasa de pobreza histórica	CASEN	Every 2 or 3 years
23. Contribuciones de Vivienda	SII (http://www.sii.cl/avaluaciones/estadisticas/estadisticas_bbrr.htm#2)	Yearly
24. Remuneraciones promedio de los trabajadores dependientes		Yearly

Source: Data from Ministerio de Desarrollo Social [2013a].

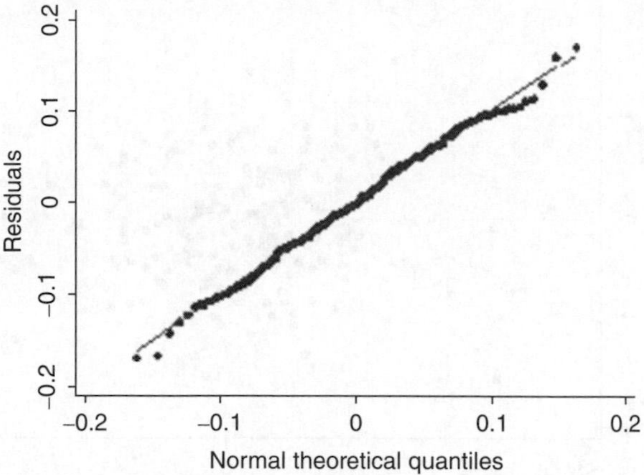

Figure 20.5 Q-Q plot of the standardized residuals. *Source*: Data from Ministerio de Desarrollo Social [2013a].

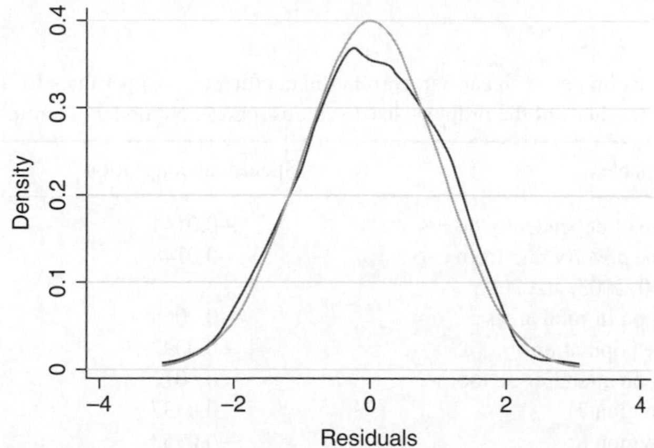

Figure 20.6 Distribution of the standardized residuals (black line). *Source*: Data from Ministerio de Desarrollo Social [2013a].

associated with the 3-year average poverty rates in the same area, the percentage of illiterate population and the indicator variables for the 7th, 8th and 9th regions of the country. Negative associations were observed with the average income for dependent workers, percentage of school attendance, and the percentage of rural population in the area.

20.4 Description of the Small Area Estimation Method Implemented in Chile

Modeling is needed to borrow strength from different relevant databases discussed in Section 20.3.2. Since this was the first time a model-based method was implemented in Chile for

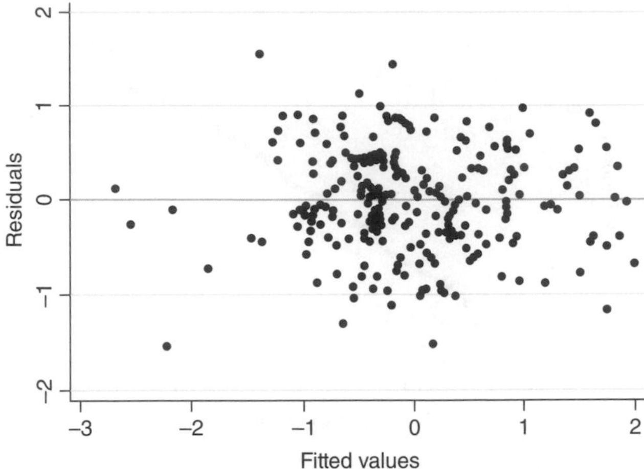

Figure 20.7 Plot of standardized residuals against fitted values. *Source*: Data from Ministerio de Desarrollo Social [2013a].

Table 20.6 Estimates of Spearman correlation coefficients and p-values for the squared standardized residuals of the ordinary least squares regression model in Table 20.7.

Auxiliary variable	Spearman correlation	p-values
Average wage of dependent workers	−0.0144	0.8264
Average of the poverty rate from Casen 2000, 2003, and 2006	−0.0148	0.8214
% of population in rural areas	−0.0065	0.9214
% of illiterate population	−0.0092	0.8882
% of population attending school	0.0072	0.9126
Dummy for region 7	0.0337	0.607
Dummy for region 8	−0.095	0.1467
Dummy for region 9	−0.0061	0.9256

Source: Data from Ministerio de Desarrollo Social [2013a].

the production of official statistics, the Ministry-PNUD working group preferred a relatively simple working model over a more elaborate model and hoped to correct gross model misspecification by a variety of adjustments. In developing a SAE method for producing official poverty rates for Chilean comunas, the Ministry-PNUD working group was guided by the following four basic simple guidelines to reduce the effect of model misspecification:

1. The method must use the Casen survey data directly to the extent possible since these are the largest data that collect information on most current poverty related variables.
2. Poverty rate estimates should be close to the survey-weighted direct estimates for comunas with reasonably large samples.

Table 20.7 Output of regression analysis based on comunas with population more than 10 000 inhabitants (dependent variable: arcsine transformed direct survey estimate of the poverty rate with original and trimmed weights; independent variables: a set of variables used in the comuna level model with arcsine transformation for proportions and logarithmic transformation for the rest).

Independent variables	Regression coefficient estimate (t-statistics): original comuna weights	Regression coefficient estimate (t-statistics): trimmed comuna weights
Average wage of dependent workers (log)	−0.09575646 (3.52**)	−0.21927953 (3.52**)
Average of the poverty rate from Casen 2000, 2003, and 2006 (arcsin)	0.49548266 (7.92**)	0.48474029 (7.92**)
% of population in rural areas (arcsin)	−0.13409847 (4.96**)	−0.39252745 (4.96**)
% of illiterate population (arcsin)	0.40349163 (2.57*)	0.25176513 (2.57*)
% of population attending school (arcsin)	−0.21883535 (2.23*)	−0.0938032 (2.23*)
Dummy for region 7 (=1)	0.03442978 (2.11*)	0.08671043 (2.11*)
Dummy for region 8 (=1)	0.03882056 (2.67**)	0.12474226 (2.67**)
Dummy for region 9 (=1)	0.105632 (6.04**)	0.28328927 (6.04**)
Constant	1.61477028 (4.24**)	−0.00203088 (0.06)
Number of observations	235	235
Adjusted R^2	0.67	0.67

*Statistically significant at the 5% level;
**statistically significant at the 1% level.
Source: Data from Ministerio de Desarrollo Social [2013a].

3. The method must not produce poverty rate estimates that considerably deviate from the corresponding direct survey estimates even for small comunas.
4. Poverty count estimates, when aggregated over all the comunas in a given region, must produce the official survey-weighted count for that region.

The Ministry-PNUD working group discussed at length SAE methods that have been implemented (Elbers *et al.* [2003, 2008], National Research Council [2000]) and not implemented (Molina and Rao [2010]) in the production of official poverty statistics for possible adaptation in estimating poverty rates for Chilean comunas. The main idea behind the method of Elbers *et al.*, also known as the ELL method, is simple and innovative. The method essentially fits a model, however complex, at the unit level (e.g., person or household level) using household survey data that relate the welfare variable (e.g., income) to a set of related auxiliary variables. A good working model can often be developed using the auxiliary variables collected in the survey data, but in order to gain the maximum benefit from the ELL method, one needs these strong auxiliary variables in the population frame as well so that the welfare variable can be imputed for all units of the frame. The poverty measures, however complex, can then be estimated since the imputed welfare variable is available for all units of the population frame. Clearly, the ELL method does not comply with Guideline 1 listed above – the poverty rate estimates would be entirely based on imputed data and Casen data would be used only to fit the model. For a possible implementation of the ELL method in the Chilean poverty mapping application of 2011, the Census 2002 would have had to be used as the population frame – a dataset that was considered neither rich in terms of the availability of the strong auxiliary variables found in Casen nor recent for obtaining reliable poverty rates for the comunas in 2009.[24]

The empirical Bayes (EB) method proposed by Molina and Rao meets Guideline 1. Their method is based on a nested error model on the logarithm of the welfare variable. In reality, as ELL argue, one observes heteroskedasticity in the error term even for the log-transformed variable. In theory, the Molina–Rao method could be extended to include this heterogeneity, but such a method was not available at the time of developing methodology for Chilean poverty mapping. Other than this modeling issue, it is not possible to gain the maximum benefit from the Molina–Rao method since the possible population frame is old (Census 2002) and does not contain the same strong auxiliary variables contained in the Casen survey. Moreover, it would be difficult, if not impossible, to match the sampled units in Casen 2009 with the 2002 Census as required by Molina and Rao [2010]. We also note that Molina and Rao did not incorporate the survey weights in their EB method. Given these considerations and tight time line for developing a poverty mapping system in Chile, the Ministry-PNUD working group decided not to pursue the ELL or Molina–Rao method.

The working group considered the SAIPE program of the U.S. Census Bureau the most promising reference for a variety of reasons.[25] First, the SAIPE program, like the Chilean case, focusses on the estimation of poverty rates and counts and has been successfully implemented in a major government program for fund allocation purposes for small areas across the United States. Secondly, it conforms with Guidelines 1, 2, and 4 listed above. Subsequently, the group developed its methodology with SAIPE as the baseline with the addition of recent best practices available in the literature.

20.4.1 Modeling

At the core of the Chilean poverty mapping methodology is the following two-level normal model, commonly referred to as the Fay–Herriott [1979] model in the SAE literature:

For $(i = 1, \dots, m)$,
Level 1 (Sampling model): Given θ_i, y_i's are independent with $y_i \sim N(\theta_i, D_i)$;
Level 2 (Linking model): θ_i's are independent with $\theta_i \sim N(x_i'\beta, A)$,

where m is the number of comunas in Chile covered by Casen; $\theta_i = \sin^{-1}\sqrt{P_i}$; P_i is the true poverty rate; $x_i' = (x_{i0}, \cdots, x_{is-1})$ is a $s \times 1$ vector of s known fixed comuna specific auxiliary variables with $x_{i0} = 1$; $\beta = (\beta_0, \cdots, \beta_{s-1})$ is a $s \times 1$ column vector of unknown regression coefficients where β_0 denotes the intercept; A is the unknown model variance.

[24] For applications of the ELL method using Casen 2003 data and Census 2002 data, see Agostini *et al.* [2008, 2010].
[25] For a recent review on small area estimation methods, see Pfeffermann [2013] and Rao and Molina [2015].

20.4.2 *Estimation of* A *and* β

In an effort to reduce the effect of sampling variability, the model parameters A and β were estimated using data for the 235 comunas with more than 10 000 inhabitants because these comunas have relatively larger sample sizes in the Casen 2009 survey.

The model variance A was estimated by the adjusted profile likelihood method of Li and Lahiri [2010], which is obtained by maximizing the following adjusted profile log-likelihood with respect to A in the range $[0, \infty)$:

$$l_{adj}(A) = c - (1/2)(\log|\Sigma| - y'Py) + \log(A),$$

where c is a normalizing constant free of A; $\Sigma = \text{diag}(A + D_1, \cdots, A + D_m)$ is a diagonal matrix of dimension m; $X' = (x_1, \cdots, x_m)$ is a $s \times m$ known fixed matrix; and $P = \Sigma^{-1} - \Sigma^{-1}X(X'\Sigma^{-1}X)^{-1}X'\Sigma^{-1}$ is the usual $m \times m$ projection matrix. The adjusted maximum profile likelihood estimator of A, say $\hat{A}$, is strictly positive and enjoys the higher order asymptotic properties of the profile maximum likelihood estimator.

The estimator of β is given by: $\hat{\beta} = (X'\hat{\Sigma}^{-1}X)^{-1}X'\hat{\Sigma}^{-1}y$, where $\hat{\Sigma} = \text{diag}(\hat{A} + D_1, \cdots, \hat{A} + D_m)$. This is the standard weighted least squares estimator of β with estimated A under the marginal model for y (i.e., the model obtained from the Fay–Herriot model by integrating out the random effects θ_i's).

20.4.3 *Empirical Bayes Estimator of* θ_{i}

The Bayes estimator of θ_i, under the squared error loss function and the Fay–Herriot model, is given by: $\hat{\theta}_i^B = (1 - B_i)y_i + B_i x_i'\beta$, where $B_i = D_i/(A + D_i)$, known as the shrinkage factor. The Bayes estimator shrinks the direct estimator y_i towards the regression mean $x_i'\beta$.

An EB estimator of θ_i is obtained from $\hat{\theta}_i^B$ when the model parameters are replaced by their respective estimators and is given by: $\hat{\theta}_i^{EB} = (1 - \hat{B}_i)y_i + \hat{B}_i x_i'\hat{\beta}$, where $\hat{B}_i = D_i/(\hat{A} + D_i)$.

The weight the EB estimator puts on the direct estimator y_i depends on the ratio $\hat{A}/D_i$. The choice of the adjusted maximum profile likelihood estimator of A over the usual residual maximum likelihood (REML) estimator was intentional and was used to assign more weight on the direct estimator since the adjusted profile likelihood tends to have more upward bias than the REML. Since the adjusted maximum profile likelihood estimator is strictly positive, it avoids the common problem of the full shrinkage (i.e., $\hat{B}_i = 1$) that is often encountered with the REML-based EB estimator of θ_i. In theory, EB estimates can go out of the admissible range $[0, \pi/2]$. Thus, $\hat{\theta}_i^{EB}$ is truncated to 0 if $\hat{\theta}_i^{EB}$ is negative and to $\pi/2$ if $\hat{\theta}_i^{EB}$ is greater than $\pi/2$.

20.4.4 *Limited Translation Empirical Bayes Estimator of* θ_{i}

Efron and Morris [1972] proposed a limited translation EB estimator, which essentially limits the deviation of the EB estimator from the corresponding direct estimator. This conservative adjustment, also known as Winsorization in the statistics literature, protects EB estimates from extreme effects. Fay and Herriot used this adjustment in their census application. The working group found this adjustment to the EB estimator useful to limit extreme effects of model-based estimation in conforming with Guideline 3 listed above. The limited translation EB estimator of θ_i is given by:

$$\widehat{\theta}_i^{LT} = \begin{cases} \widehat{\theta}_i^{EB} & \text{if } \; y_i - k\sqrt{D_i} \le \widehat{\theta}_i^{EB} \le y_i + k\sqrt{D_i}, \\ y_i - k\sqrt{D_i} & \text{if } \; \widehat{\theta}_i^{EB} \le y_i - k\sqrt{D_i}, \\ y_i + k\sqrt{D_i} & \text{if } \; \widehat{\theta}_i^{EB} \ge y_i + k\sqrt{D_i}, \end{cases}$$

where k is a constant. For this application, $k = 1$ as in Fay and Herriot [1979].

20.4.5 Back-transformation and raking

The limited translation estimator $\widehat{\theta}_i^{LT}$ is back-transformed to the original scale to produce the following estimator of poverty rate for comunas with sample in the Casen 2009 survey: $\widehat{P}_i = \sin^2 \widehat{\theta}_i^{LT}$. For a few comunas with no sample in the Casen 2009 survey, the estimates of the poverty rate were computed using the Ministry-PNUD synthetic method cited in Section 20.1. Whether a comuna is in the Casen sample or not, the final official raked SAE estimates of poverty rates for all the comunas that belong to the rth region are given by: $\widehat{P}_i^{SAE} = \widehat{P}_i \times R_r$, where $R_r = p_r^{regn} N_r^{regn} / \sum_{i=1}^{m_r^*} \widehat{P}_i N_i$ is the raking factor common to all comunas in the region r; m_r^* is the total number of comunas in region r; p_r^{regn} is the direct design-based estimate of the regional level poverty rate using the original regional weights; N_i is an estimate of the population projection in comuna i belonging to region r; and N_r^{regn} is an estimate of the population projection in region r. We have $N_r^{regn} = \sum_{i=1}^{m_r^*} N_i$.

20.4.6 Confidence intervals for the poverty rates

The working group adapted the parametric bootstrap confidence interval proposed by Li and Lahiri [2010] (see also Chatterjee *et al.* [2008]) to construct confidence intervals for the poverty rates. The procedure involves the following steps:

Step 1: Generate R independent parametric bootstrap samples $\{(y_i^{(r)}, \theta_i^{(r)}),\ i = 1, \cdots m\}$, $r = 1, \cdots, R$ as follows: $\theta_i^{(r)} \sim N(x_i^T \widehat{\beta}, \widehat{A})$, $y_i^{(r)} | \theta_i^{(r)} \sim N(\theta_i^{(r)}, D_i)$, $i = 1, \ldots, m$.

Step 2: Produce estimates $\widehat{A}^{(r)}$, $\widehat{B}_i^{(r)}$ and $\widehat{\beta}^{(r)}$ by replacing the original data with the parametric bootstrap samples generated in Step 1. Repeat this step R times.

Step 3: For each bootstrap sample, calculate the following pivotal quantity: $t_i^{(r)} = (\theta_i^{(r)} - \widehat{\theta}_i^{EB(r)}) / \sqrt{D_i(1 - \widehat{B}_i^{(r)})}$, where $\widehat{\theta}_i^{EB(r)} = (1 - \widehat{B}_i^{(r)}) y_i^{(r)} + \widehat{B}_i^{(r)} x_i' \widehat{\beta}^{(r)}$.

Step 4: For comuna i, obtain q_{1i} and q_{2i}, the $100\alpha/2$ and $100(1 - \alpha/2)$ percentiles of $\{t_i^{(r)},\ r = 1, \cdots, R\}$.

Step 5: For comuna i, an approximate $100(1 - \alpha)\%$ confidence interval for θ_i is obtained as: (L_i, U_i), where $L_i = \widehat{\theta}_i^{EB} + q_{1i}\sqrt{D_i(1 - \widehat{B}_i)}$, and $U_i = \widehat{\theta}_i^{EB} + q_{2i}\sqrt{D_i(1 - \widehat{B}_i)}$. Note that the admissible range for θ_i is $[0, \pi/2]$. Thus, L_i is truncated to 0 if L_i is negative and U_i is truncated to $\pi/2$ if U_i is greater than $\pi/2$. The probability that (L_i, U_i) is not contained in $(0, \pi/2)$ is expected to be negligible unless $4n_i$ is very small. The truncated confidence interval for θ_i is denoted by (L_i^*, U_i^*).

Step 6: Finally, the lower and upper limits of the confidence interval (L_i^*, U_i^*) in Step 5 are back-transformed to yield the following approximate $100(1 - \alpha)\%$ confidence interval of the poverty rate P_i: $(\sin^2 L_i^*,\ \sin^2 U_i^*)$. Note that the parametric bootstrap confidence interval for any one-to-one transformed parameter can be easily obtained using the simple back-transformation. In our case, the motivation for this back-transformed confidence interval comes from the fact that for any $0 < p < 1$ and $0 < \theta < \pi/2$, $\sin^{-1}\sqrt{p}$ and $\sin^2\theta$ are monotonically increasing functions of p and θ, respectively.

20.5 Data Analysis

In this section, we report the data analysis conducted when implementing the SAE methodology described in Section 20.4 using the comuna level reduced data from Casen 2009 and the set of selected comuna level auxiliary data described in Section 20.3.

The adjusted profile maximum likelihood method yielded an estimate 0.00234134 of the model variance A, which, in turn, produced estimates of the shrinkage factors B_i for all comunas. Some descriptive statistics of the estimated shrinkage factors are reported in Table 20.4 for four different groups of comunas created by classifying all comunas in a given quartile of the sampling variances (D_i's). That is, Group 1 and Group 4 provide descriptive statistics for comunas with the 25% lowest and 25% highest values of D_i, respectively. The average B_i over all comunas was 42%. Thus, on the average, EB estimates of θ_i assigned 42% weight to the synthetic estimate and the remaining 58% to the direct survey estimate. Table 20.4 shows that the average B_i (0.5474) in the highest quartile (Group 4) was approximately two times larger than that (0.2765) in the lowest quartile (Group 1). Overall, for only a quarter of the comunas the EB estimates assigned a weight of 49% or more to the synthetic estimates of the comuna level poverty rates. The EB estimates were always within the admissible range so truncation was not needed.

The limited translation EB procedure resulted in the truncation of EB estimates in 45 of the 334 comunas: 21 comunas where the EB estimates were above the upper cutoff points and 24 comunas below the lower cutoff points. Figure 20.8 presents results of the Winsorization process, where the x-axis displays the limited translation EB estimates sorted by level of the poverty rates (y-axis). The figure also shows the upper and lower cutoff points: comunas with EB estimates above the upper cutoff point are identified by a circle at the upper cutoff point and comunas below the lower cutoff points are identified by a cross at the lower cutoff point. For most of the comunas with EB estimates beyond the cutoff points, the differences between the EB estimates and the cutoff points are negligible. For a few cases, however, the differences were large. The three comunas most affected by the limited translation procedure had their EB estimates cut down by approximately 10 percentage points. After Winsorization, the limited translation EB estimates deviated from the corresponding direct estimates by approximately 5 percentage points for these three comunas.

Table 20.8 displays the raking factors for the 15 regions of the country. On average, the raking factor was approximately 1.01, which means that on the average the comuna level estimates required only a minor fine-tuning to converge to the regional level estimate. For two regions (Tarapacá and Los Ríos), the raking factors were more than 1.08.

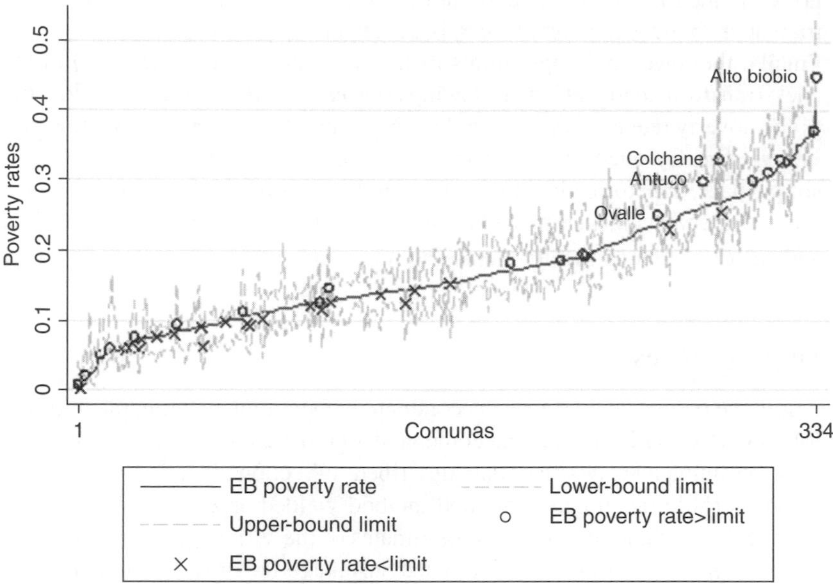

Figure 20.8 Limited translation empirical Bayes (EB) estimates of the comuna level poverty rates, and the upper and lower thresholds. Comunas on the x-axis sorted by the estimate of the poverty rate. *Source*: Data from Ministerio de Desarrollo Social [2013a].

Table 20.8 Raking factors used so that the total of the model-based estimates of the number of poor within each region matches the standard design-based official estimate for the region, by region.

No.	Region	R_r
1	Tarapacá	1.12172
2	Antofagasta	0.97455
3	Atacama	1.06685
4	Coquimbo	1.04309
5	Valparaíso	1.00387
6	O'Higgins	1.00430
7	Maule	1.05292
8	Biobío	0.99010
9	Araucanía	1.01628
10	Los Lagos	1.04088
11	Aysén	1.06255
12	Magallanes	0.97368
13	Metropolitana	0.97765
14	Los Ríos	1.08572
15	Arica y Parinacota	0.99486

Source: Data from Ministerio de Desarrollo Social [2013a].

Let us now turn our attention to the construction of the parametric bootstrap confidence intervals for the comuna level poverty rates. First of all, we note that in this application, (L_i, U_i) was always found to be well within $(0, \pi/2)$ so the truncation step was not needed. In general, as Figure 20.9 shows, parametric bootstrap confidence intervals are shorter (i.e., more precise) than the corresponding direct confidence intervals. The parametric bootstrap confidence intervals, however, are not necessarily symmetric. Figure 20.10 plots the histograms of the pivot for three comunas. For Puchuncavi the distribution of the pivot is approximately symmetric around zero. The histograms for Providencia and Peumo are, however, very asymmetric. These three histograms suggest that parametric bootstrap methods based on confidence intervals are likely to be more appropriate than those based on normal or t-distribution and justify the parametric bootstrap confidence interval methodology.

20.6 Discussion

In this chapter, we present an overview of different efforts undertaken in Chile to measure poverty and describe the recently developed statistical methodology for producing the official poverty rates for all comunas in Chile. Since this is the first attempt to use an explicit statistical model in producing official estimates of the comuna level poverty rates in Chile, a conservative small area model-based technique was adopted so that the poverty rates were not markedly different from the design-based poverty rates. While it may be possible to improve on the current methodology using alternative models and estimation methods such as the one based on an alternative arcsine transformation (see, e.g., Brown *et al.* [2001]), there is no denying the fact that this is a first solid step in moving away from the direct design-based methods, which are known to be unreliable for estimating poverty rates for small comunas. We present below some directions for future research.

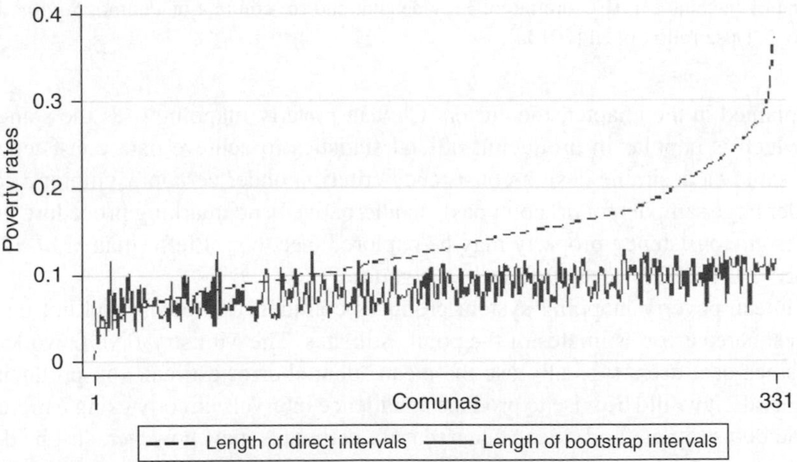

Figure 20.9 Length of the direct and parametric bootstrap confidence intervals of the comuna level poverty rates for comunas sorted by the limited translation empirical Bayes estimates of the poverty rate. The three comunas with the largest estimates of the length of the direct confidence interval were excluded from the graph. *Source*: Data from Ministerio de Desarrollo Social [2013b].

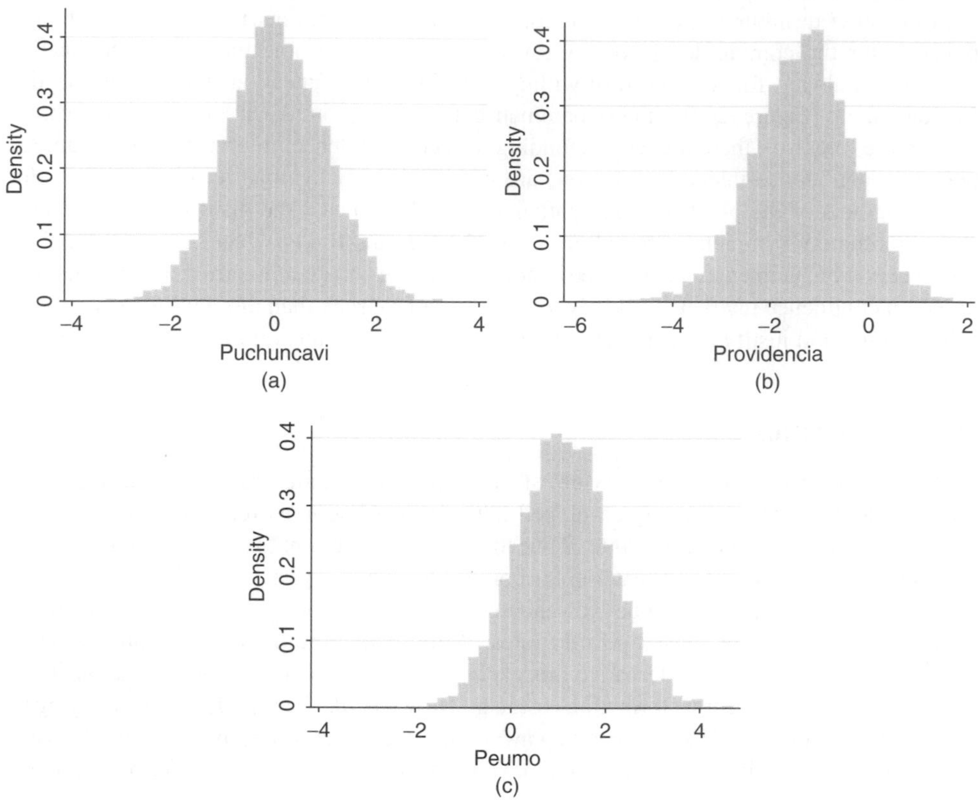

Figure 20.10 Histograms of pivots in the parametric bootstrap method with 5000 bootstrap samples: (a) comuna of Puchuncavi; (b) comuna of Providencia; and (c) comuna of Peumo. *Source*: Data from Ministerio de Desarrollo Social [2013a].

As explained in the chapter, the current Chilean poverty mapping uses the standard ratio raking, which is popular in producing official statistics to achieve data consistency. While this may satisfy a desirable design-consistency criterion under certain asymptotic conditions (e.g., under large samples for *all* comunas), an alternative benchmarking procedure that has an inbuilt design-consistency property may be explored (see, e.g., Pfefferman *et al.* [2014], Ha and Lahiri [2014]). This needs further investigation.

The Chilean poverty mapping system produces confidence intervals and not the customary mean squared error estimates of the point estimates. The Ministry-PNUD working group felt that since data users typically use the mean squared error estimates in producing confidence intervals, it would be wise to produce confidence intervals directly using a more efficient parametric bootstrap method. Mean squared error estimates may, however, still be desired by some data users. At the time of developing the Chilean poverty mapping system, there was no method available in the literature for computing mean squared error estimates of a raked back-transformed translation invariant SAE estimator. The parametric bootstrap method proposed by Chatterjee and Lahiri [2013] is promising to solve the problem and may be tried in the Chilean poverty mapping system in the future.

Poverty rates using the SAE method were estimated only for the comunas with sample cases in the Casen survey. As requested by the Ministry, the Ministry-PNUD synthetic method was used to estimate the poverty rates of the comunas without sample. In the future, the use of the Ministry-PNUD synthetic method versus alternative options, such as predictions from the SAE area level model used for comunas with sample, should be evaluated explicitly.

In 2014 a new Expert Commission[26] recommended the Ministry to complement the current income-based measure of poverty with a multidimensional indicator of social needs for the Chilean households covering education, health, employment and social security, housing, and environment and social networks. Looking forward, the Commission encouraged development of a single aggregated measure, at the housing unit level, using the method of Alkire and Foster [2007] where the five dimensions would be weighted equally (i.e., 20% each). The index would have a range of 0–100%, and a household would be considered under social welfare need if they achieve at least 33%. For the present time, however, the Commission suggested the development of separate indicators for the different social needs dimensions, using the Casen data for the national estimate and administrative records for comuna level estimates. In the future, it would be interesting to study the feasibility of developing a SAE technique such as the one described in this chapter to address the problem of estimating a single aggregated multidimensional poverty measure.

Moving away from the standard design-based method to a model-based method for the production of official statistics, such as the poverty rates, was a big challenge for the Ministry as data producer. According to Statistics Netherlands ([2008], pp. 3–4): "NSIs (National Statistical Institutes) need to play safe in the production of official statistics and therefore do not want to rely on model assumptions, particularly if they are not verifiable. Another factor is that the methodology of advanced model-based techniques, used for example in the context of small area estimation, is intellectually and practically inaccessible (Heady and Ralphs [2005]). Indeed the statistical theory is rather complex and the available software at NSIs is often not suitable to conduct the required calculations in a straightforward manner in a production environment. This hampers the implementation in survey processes to produce timely official statistics". The change is also challenging for the public and other stakeholders for whom the new estimates are not possible to replicate and are much more difficult to comprehend. Additional resources need to be allocated to adequately communicate and educate stakeholders on new statistical methods.

Before ending the chapter, we reiterate that while this is a very first step to improve on statistical methodology for estimating poverty rates for comunas in Chile, researchers should be encouraged to think about evaluating new methods as they become available in the continually evolving small area methodology. We hope that this chapter will encourage researchers in other government statistical organizations to consider model-based SAE techniques to improve the small area poverty statistics.

Acknowledgements

The authors would like to thank the Ministerio de Desarrollo Social for authorizing us to write this chapter in order to explain the Ministry's experience in implementing the SAE method for the estimation of comuna level poverty rates in Chile. We would also like to thank Snigdhansu

[26] Comisión para la Medición de la Pobreza [2014].

Chatterjee, Stephanie Eckman, Osvaldo Larrañaga, and Isabel Millán for their reviews and comments on earlier versions of this chapter. Of course, any errors or omissions that remain are the sole responsibility of the authors. All three authors were part of the group that worked on the development of the SAE method for the comuna level poverty rates in Chile.

References

Ministerio del Interior, Subsecretaría de Desarrollo Regional y Administrativo (2006). Ley 18.695 Art.1. *Fija el Texto Refundido, Coordinado y Sistematizado de la Ley No. 18.695, Orgánica Constitucional de Municipalidades*. Santiago, Chile, Publicación July 26, 2006.

Ministerio de Desarrollo Social (2012a). *Metodología del Diseño Muestral y Factores de Expansión Encuesta de Caracterización Socioeconómica Nacional*. Serie de Documentos Metodológicos No. 1, Observatorio Social.

United Nations Development Programme (UNDP) (1990). *Human Development Report 1990*. Oxford University Press, New York.

Programa de las Naciones Unidas para el Desarrollo (PNUD) Chile y Ministerio de Planificación de Chile (2000). *Desarrollo Humano en las comunas de Chile*. Santiago, Chile.

Ministerio del Interior, Subsecretaría de Desarrollo Regional y Administrativo (2007). Ley 20.237 Art.1 (1). *Modifica el Decreto Ley No. 3.063, de 1979, Sobre Rentas Municipales la Ley No. 18.695, Orgánica Constitucional de Municipalidades, y Otros Cuerpos Legales, en Relación con el Fondo Común Municipal y Otras Materias Municipales*. Santiago, Chile, Publicación December 17, 2007.

Ministerio del Interior, Subsecretaría de Desarrollo Regional y Administrativo (2009). Decreto 1293 Art. 13(f). *Reglamento para la Aplicación Del Artículo 38 Del Decreto Ley No. 3.063, De 1979, Modificado por el Artículo 1° de la Ley No. 20.237*. Santiago, Chile, Publicación January 2, 2009.

Comisión de Técnicos de la Encuesta Casen (2010). *Informe de la Comisión de Técnicos sobre la Encuesta Casen*. Santiago, Chile.

United Nations Expert Group on Poverty Statistics (Rio Group) (2006). *Compendium of Best Practices in Poverty Measurement*. United Nations Economic Commission for Latin America and the Caribbean (ECLAC) and Brazilian Institute for Geography and Statistics (IBGE). Rio de Janeiro, September 2006.

Comisión para la Medición de la Pobreza (2014). *Informe Final de la Comisión para la Medición de la Pobreza*. Santiago, Chile.

Comisión Económica para América Latina y el Caribe (CEPAL) (1990). *Una estimación de la medición de pobreza para Chile, 1987*. Santiago, Chile.

Ministerio de Desarrollo Social (2012b). *Indicadores de Pobreza Casen 2011*. Power point slides retrieved from http://observatorio.ministeriodesarrollosocial.gob.cl/layout/doc/casen/pobreza_casen_2011.pdf

Ministerio de Planificación (2010). *Informe Metodológico Casen 2009*. División Social, Santiago, Chile.

Potter, F.J. (1993). *The Effect of Weight Trimming on Nonlinear Survey Estimates*. American Statistical Association, San Francisco, CA.

Bell, W. (1997). *Models for County and State Poverty Estimates*. Preprint, Statistical Research Division, U. S. Census Bureau, Washington, DC.

Bell, W., Basel, W., Cruse, C., Dalzell, L., Maples, J., O'Hara, B., and Powers, D. (2007). *Use of ACS Data to Produce SAIPE Model-Based Estimates of Poverty for Counties*, Census Report.

Citro, C. and Kalton, G. (eds) (2000). *Small-area Estimates of School-age Children in Poverty: Evaluation of Current Methodology*. National Academy Press, Washington, DC.

Efron, B. and Morris, C. (1975). Data analysis using Stein's estimator and its generalizations. *Journal of the American Statistical Association* **70**, 311–319.

Carter, G. and Rolph, J. (1974). Empirical Bayes methods applied to estimating fire alarm probabilities. *Journal of the American Statistical Association* **69**, 880–885.

Jiang, J., Lahiri, P., Wan, S., and Wu, C. (2001). Jackknifying in the Fay-Herriot Model with an example. *Proceedings of the Seminar on Funding Opportunity in Survey Research*, Council of Professional Associations on Federal Statistics.

Raghunathan, T.E., Xie, D., Schenker, N., Parsons, V.L., Davis, W.W., Dodd, K.W., and Feuer, E.J. (2007). Combining information from two surveys to estimate county-level prevalence rates of cancer risk factors and screening. *Journal of the American Statistical Association* **102**, 474.

Xie, D., Raghunathan, T.E., and Lepkowski, J.M. (2007). Estimation of the proportion of overweight individuals in small areas – a robust extension of the Fay-Herriot model. *Statistics in Medicine* **26**, 2699–2715.

StataCorp (2009). *Stata Statistical Software: Release 11*. StataCorp LP, College Station, TX.

Elbers, C., Lanjouw, J., and Lanjouw, P. (2003). Micro-level estimation of poverty and inequality. *Econometrica* **71**, 355–364.

Elbers, C., Lanjouw, P., and Leite, P. (2008). *Brazil within Brazil: Testing the Poverty Map Methodology in Minas Gerais*. World Bank Policy Research Working Paper No. 4513.

Molina, I. and Rao, J.N.K. (2010). Small area estimation of poverty indicators. *Canadian Journal of Statistics* **38**, 369–385.

Agostini, C., Brown, P., and Gongora, D. (2008). Distribución Espacial de la Pobreza en Chile. *Estudios De Economía* **35**, 79–110.

Agostini, C., Brown, P., and Roman, A. (2010). Estimando Indigencia y Pobreza Indígena Regional con Datos Censales y Encuestas de Hogares. *Cuadernos de Economía* **47**, 125–150.

Rao, J.N.K. and Molina, I. (2015). *Small Area Estimation*. Second Edition. John Wiley & Sons, Inc., Hoboken.

Pfeffermann, D. (2013). New important developments in small area estimation. *Statistical Science* **28**, 40–68.

Fay, R.E. and Herriot, R.A. (1979). Estimates of income for small places: An application of James_Stein procedure to census data. *Journal of the American Statistical Association* **74**, 269–277.

Li, H. and Lahiri, P. (2010). An adjusted maximum likelihood method for solving small area estimation problems. *Journal of Multivariate Analysis* **101**, 882–892.

Efron, B. and Morris, C. (1972). Limiting the risk of Bayes and empirical Bayes estimators-Part II: the empirical Bayes case. *Journal of the American Statistical Association* **67**, 130–139.

Chatterjee, S., Lahiri, P., and Li, H. (2008). Parametric bootstrap approximation to the distribution of EBLUP, and related prediction intervals in linear mixed models. *The Annals of Statistics* **36**, 1221–1245.

Brown, L.D., Cai, T.T. and DasGupta, A. (2001). Interval estimation for a binomial proportion. *Statistical Science* **16**, 101–133.

Pfeffermann, D., Sikov, A., and Tiller, R. (2014). Single and two-stage cross-sectional and time series benchmarking procedures for small area estimation. *Test*, DOI: 10.1007/s11749-014-0398-y.

Ha, N. and Lahiri, P. (2014). Comment: Single and two-stage cross-sectional and time series benchmarking procedures for small area estimation. *Test*, DOI: 10.1007/s11749-014-0400-8.

Chatterjee, S. and Lahiri, P. (2013). A Simple Computational Method for Estimating Mean Squared Prediction Error in General Small-Area Model. Paper presented at the SAE 2013, Bangkok.

Alkire, S. and Foster, J. (2007). *Counting and Multidimensional Poverty Measurement*. OPHI Working Paper Series, No. 7.5.

Statistics Netherlands (2008). *Model-Based Estimation for Official Statistics*. Discussion Paper 08002, The Hague, Netherlands.

Heady, P. and Ralphs, M. (2005). EURAREA: an overview of the project and its findings. *Statistics in Transition* **7**, 557–570.

Ministerio de Desarrollo Social (2013a). *Procedimiento de cálculo de la Tasa de Pobreza a nivel Comunal mediante la aplicación de Metodología de Estimación para Áreas Pequeñas (SAE)*. Serie de Documentos Metodológicos No. 1, Observatorio Social.

Ministerio de Desarrollo Social (2013b). *Incidencia de la Pobreza a nivel Comunal, según Metodología de Estimación para Áreas Pequeñas. Chile 2009 y 2011*. Serie de Informes Comunales No. 1, Observatorio Social.

21

Appendix on Software and Codes Used in the Book

Antonella D'Agostino[1], Francesca Gagliardi[2] and Laura Neri[2]
[1]*Department of Business and Quantitative Studies, University of Naples "Parthenope", Naples, Italy*
[2] *Department of Economics and Statistics, University of Siena, Siena, Italy*

21.1 Introduction

Empirical applications of small area estimation (SAE) methods need access to statistical software in both data management and estimation phases of the specified models. In this chapter we briefly present two software that, in our experience, are suitable for this specific task, that is R and SAS. This list of software is, obviously, not intended to be comprehensive and there is no doubt many other packages that would have similar facilities (e.g., STATA, SPSS, S-PLUS). In addition, we provide key information on the use of *custom-made* scripts written in R or SAS that implement the advanced methods covered in the book chapters. These script files contain *custom-made* functions necessary for performing the analytic steps in R or SAS and are available on the book's website. These functions are a genuine contribution to the literature and the authors may aim at condensing them into a proper R package or SAS macro soon (see also Sections 21.2 and 21.4).

It is worth noting, that, while most of the book chapters (5–14, 16–18) present an empirical analysis, only Chapters 6–10, 12–14, and 16 are supported by *custom-made* scripts written in SAS or R. The applications presented in Chapters 5, 11, 17, and 18 need standard R packages specifically designed for SAE analyses, the details of which can be easily found on the R website (http://www.r-project.org); a brief description of these packages is presented in Section 21.2. With regards to the remaining Chapter 18, the analysis is performed by using PovMap,

Analysis of Poverty Data by Small Area Estimation, First Edition. Edited by Monica Pratesi.

Companion Website: www.wiley.com/go/pratesi/poverty

a package developed by the World Bank (http://econ.worldbank.org/external/default/main?theSitePK=477894&contentMDK=22717057&menuPK=546584&pagePK=64168182&piPK=64168060). In an attempt to make as homogeneous as possible the information provided for each chapter, we proceed chapter-by-chapter, always presenting the software, the standard package or the *custom-made* script used for performing the analyses discussed in the corresponding book chapter. Last but not least, the description has been not structured as a traditional user's guideline, where syntax is displayed with commands and specific functions that need to be performed are described in detail. Rather, we give more general guidelines. In particular, we provide a quick guide that includes a description of the key aspects that users have to know in order to replicate the empirical analysis of the chapter concerned. We recommend reading these quick guides which provide a useful and very concise documentation especially for beginners in SAE methods. In addition, each custom-made script available on the book's website is well structured with enough comments explaining the script step-by-step. Comments in R are inserted after the "#" sign and in SAS after "*". Moreover, some meta-information about the script is inserted at the beginning of each script. However, to use these advanced scripts the user should be familiar with features common to all estimation commands and should be able to prepare the data for analysis using R or SAS. Indeed, this chapter does not discuss the creation of the dataset in the proper format (SAS or R). This type of practice can be easily acquired studying the tutorials available on the Web. Once users have their data in R and/or SAS, they can start using the scripts. Furthermore, most of the analyses conducted in this book are based on sensible data or data that cannot be delivered for free, in such cases datasets are not available to users. In any case, the practitioners can actually implement the methods covered in the book, provided that they have data required for the specific analysis. To sum up, this chapter is structured into three distinct parts. The first part (Section 21.2) gives a brief explanation on general features of R and SAS software and also outlines the available tools for SAE methodologies. The second part (Section 21.3) concerns the EU Statistics on Income and Living Conditions (EU-SILC) survey, because this database is used in several empirical analyses presented; it should be very useful for practitioners not familiar with these data. The section briefly describes EU-SILC data and provides guidance on the data managing generally requested for organizing datasets. Finally, the third part (Section 21.4) provides a description of the key aspects of each chapter for which scripts are available; a common structure of presentation is used.

21.2 R and SAS Software: a Brief Note

This section is designed to introduce the general aspects of R and SAS and to provide a brief and general reference for R and SAS concerning SAE methodologies. Both R and SAS software allow several statistical analyses and offer tools from basic statistics to more advanced methods. To use R packages or SAS procedures for SAE one should be familiar with features common to all estimation commands and able to prepare the data for analysis.

R is a free software environment for statistical computing and graphics. It compiles and runs on a wide variety of UNIX platforms, Windows and MacOS. R can be downloaded, free of charge, from the web. The CRAN and R homepage (http://www.r-project.org/) is the R's core reference, giving information on the R project and everything related to it. It acts as the download area, carrying the software itself, available packages and PDF manuals. The core of R is an interpreted computer language, most of its packages dealing with statistics, data analysis, and graphics are provided and updated by volunteer researchers. A package specification allows the production of loadable modules for specific purposes, currently the CRAN package repository features 7154 available packages and several contributed packages are made available through the CRAN sites (https://cran.r-project.org/web/packages/). Many resources on getting started with R are available on the web. For example, a couple of R Tutorials are:

- An Introduction to R (http://www.stat.ufl.edu/~presnell/Courses/sta4504-2000sp/R/R-intro.pdf) by the R Core Development Team.
- Data Analysis and Graphics Using R – Introduction, Code and Commentary (http://maths-people.anu.edu.au/~johnm/r/usingR.pdf) by John Maindonald. There are also some accompanying datasets and scripts.

In addition, users can edit the R scripts also in a user-friendly script editor such as TINN-R, Rstudio, and/or JGR, or use the package R commander (`Rcmdr`), which has a user-friendly graphical interface. This would help beginners to get some first ideas about the R syntax. R offers various packages for SAE. The following gives a brief description of those packages. Package "`sae`" (Molina and Marhuenda, 2015; http://journal.r-project.org/archive/2015-1/molina-marhuenda.pdf) provides functions for estimation in domains with small sample sizes, including functions for mean squared error (MSE) estimation. Basic estimators include direct, post-stratified synthetic and sample size dependent. Model-based estimators include the empirical best linear unbiased prediction (EBLUP) based on a Fay–Herriot model and the EBLUP based on a unit level nested error model. Estimators obtained from spatial and spatio-temporal Fay–Herriot models and the empirical best method based on the unit level nested error model for estimation of general non-linear parameters are also included. Package "`sae2`" (Fay and Diallo, 2015) provides two primary functions, `eblupRY` and `eblupDyn`, to fit non-spatial time-series small area models to area level data. The function `mvrnormSeries` provides simulated data under either model. Fitting time series models for SAE based on area level models, this package supplements the functionality of the package sae, specifically EBLUP fitting of the original Rao–Yu model and also of a modified ("dynamic") version. Both univariate and multivariate applications are supported. Package "`JoSAE`" (Breidenbach, 2015) implements the unit level EBLUP (Battese *et al.*, 1988) and generalized regression (GREG) (Sarndal, 1984) estimators as well as their variance estimators. Package "`mme`" (Lopez-Vizcaino *et al.*, 2013a) implements three multinomial area level mixed effects models for SAE. The first model (Model 1) is based on the area level multinomial mixed model with independent random effects for each category of the response variable (Lopez-Vizcaino *et al.*, 2013a). The second model (Model 2) takes advantage of the availability of survey data from different time periods and uses a multinomial mixed model with independent random effects for each category of the response variable and with independent time and domain random effects. The third model (Model 3) is similar to the second one, but with correlated time random effects. To fit the models, the penalized quasi-likelihood (PQL) method, introduced by Breslow and Clayton (1993) for estimating and predicting the fixed and random effects, is combined with the residual maximum likelihood (REML) method for estimating the variance components. In all models of the package the MSE is estimated by using the analytical expression and then by bootstrap techniques. Package "`rsae`" (Schoch, 2015) computes robust basic unit and area level and predicts area-specific means It estimates SAE models that have been set up by `saemodel` (or synthetic data generated by `makedata`) by various (robust) estimation methods. Dissemination actions and training on SAE has been conducted within the framework of some projects: the BIAS project (2008–2011, http://www.bias-project.org.uk/), the EU SAE project (2009–2012, http://www.cros-portal.eu/content/sae-finished), the ESSnet project on SAE (http://www.essnet-portal.eu/project-information/sae/), the InGRID project (2013–2017, http://www.ingridproject.eu/), and the EMOS Spring School project (2015, http://www.ec.unipi.it/news/item/586-emos-spring-summer.html).

SAS is general-purpose software for a wide variety of statistical analyses, data management, and predictive analytics. Detailed information on SAS is available on the official website http://www.sas.com. The first development of SAS began in 1976 and it has been developed over the years up to version 9.4 which is supported on many operating systems and platforms. SAS is not free of charge, it needs a license that should be renewed annually; the licence agreement includes some basic modules, specific modules should be licensed. The SAS language is a computer programming language used in the SAS System. The SAS language runs under compilers that can be used on Microsoft Windows, Linux, and various other UNIX and mainframe computers. Official statistical groups and institutions such as Eurostat or national statistical institutes (NSIs) use the SAS System for its power with very large datasets and its high performances in computations. SAS provides a graphical point-and-click user interface for non-technical users and more advanced options through the SAS programming language. SAS programs have a `DATA step`, which retrieves and manipulates data, usually creating a SAS dataset, and a `PROC step`, which analyses the data. Small area estimation requires to go beyond the survey data analysis methods that are available in the SAS/STAT® survey procedures; the main procedure for this kind of analysis is the `MIXED`. The `MIXED` procedure provides parameter estimates for unit level models and the small area predictions. Small area predictions are based on EBLUP, and the mean squared error of predictions (MSEP) measures the precision of the predictions. Moreover the `MIXED` procedure also allows to fit area level models; then, the EBLUPs and the MSEPs are computed with the IML procedure. The Interactive Matrix Language (IML), namely SAS/IML (http://www.sas.com/en_us/software/analytics/iml.html), is a high-level matrix programming language that enables natural mathematical syntax to be used to write custom algorithms and to compute statistics that are not built into any SAS procedure. Standard linear mixed model theory cannot be applied to unmatched sampling and linking models. Instead, a hierarchical Bayes (HB) approach to estimation can be applied. This approach uses the `PROC MCMC` to estimate the means and variances of the posterior distributions of the small area parameters of interest. With regards to SAE and SAS, it is worth noting that within the EURAREA project (http://www.ons.gov.uk/ons/guide-method/method-quality/general-methodology/spatial-analysis-and-modelling/eurarea/index.html) the estimation software in small areas has been developed using `SAS/IML`. The project, carried out within the 5th Framework Programme of R&D launched by the EU from 2000 to 2004, aimed at evaluating the effectiveness of standard estimation techniques for small areas (synthetic, GREG and composite estimators). All the theory developed has been implemented in SAS. Moreover, within the framework of the EURAREA project has been developed the SAS macro `EBLUPGREG` for SAE using unit level data. The program includes several estimators from the GREG estimator to the linear mixed models, which may contain spatial and time correlation structures. The aim is to use sample data estimate totals or means of a continuous response variable and with the help of population data yield predicted values for the study variable over geographical regions. The `EBLUPGREG` code has been written in SAS macro language. Macros consist of collections of regular SAS program statements, macro variables, macro language statements, and macro functions contained within a `%MACRO` and a `%MEND`. The `%MACRO` statement includes a name and the macro is called using the macro's name preceded by % (http://www2.sas.com/proceedings/sugi29/243-29.pdf).

21.3 Getting Started: EU-SILC Data

The data source used in several of the empirical analyses presented in this book is the EU-SILC survey. EU-SILC was launched in 2003 on the basis of a gentlemen's agreement between Eurostat and six Member States (Austria, Belgium, Denmark, Greece, Ireland, and Luxembourg) and Norway. It was formally launched in 2004 in 15 countries and expanded in 2005 to cover all of the EU-25 Member States, together with Norway and Iceland. Bulgaria launched EU-SILC in 2006 while Romania, Switzerland, and Turkey introduced the survey in 2007. EU-SILC provides two types of annual data:

- cross-sectional data pertaining to a given time or a certain time period with variables on income, poverty, social exclusion, and other living conditions;
- longitudinal data pertaining to individual level changes over time, observed periodically over a 4-year period.

It covers many topics such as living conditions, income, social exclusion, housing, work, demography, and education. The standard database available for Users is organized into four (cross-sectional or longitudinal) files named, respectively: R-file, D-file, H-file, and P-file. The R- and P-files are at individual level; R-file covers all individuals, while P-file contains only individuals aged 16+. The D- and H-files are at household level. Access to EU-SILC data is not free, data can be released by Eurostat or NSIs for research purposes. Some countries are changing their privacy policy and they have started to upload free on the web the national EU-SILC datasets (e.g., Spain; http://www.ine.es/prodyser/microdatos.htm). Users not familiar with this data source could have difficulty in understanding some of the programs provided for the empirical analysis, since much of the information from this survey is assumed to be known. This brief section aims to introduce some basic information on the most common variables used in the book chapters. The variable used for monetary poverty analysis is the *Household Disposable Income* (*`HY020`*) or its equivalized version, *Equivalised Disposable Income* (*`HX090`*), based on the modified-OECD equivalent scale. A detailed description of the EU-SILC variables can be found in the Eurostat official document "Description of target variables: cross-sectional and longitudinal", known also as EU-SILC doc. 065 (https://circabc.europa.eu/sd/a/1ad4dc02-7695-4765-b6db-c609acb1a162/SILC065%20operation%202011%20VERSION%20MAY%202011.pdf).

Other useful documents that describe the variables and the construction of the main official indicators are: the Eurostat working paper with the description of the "Income and living conditions dataset" (http://epp.eurostat.ec.europa.eu/portal/page/portal/income_social_inclusion_living_conditions/documents/tab/Tab/Working_paper_on_EU_SILC_datasets.pdf) and the EU-SILC metadata web page (http://ec.europa.eu/eurostat/cache/metadata/FR/ilc_esms.htm).

In some book chapters the indicators for the Europe 2020 strategy, based on EU-SILC data, are used. These indicators are included also in the EU-SILC dataset, starting from the 2010 users' database release. In particular, the severe material deprivation indicator has been used in Chapter 16. This indicator characterized individuals being at the state of enforced inability to pay for at least four of the following nine material deprivation items: (1) to pay their rent, mortgage or utility bills; (2) to keep their home adequately warm; (3) to face unexpected expenses; (4) to eat meat, fish, or a protein equivalent every second day; (5) to enjoy a week of holiday away from home once a year; (6) to have a color television; (7) to have a washing machine; (8) to have a car; and (9) to have a telephone. In the users' database the definition of this indicator is "Severely Materially Deprived (`SEV_DEP`)". For more details see the above mentioned Eurostat working paper or the web page: http://ec.europa.eu/eurostat/statistics-explained/index.php/EU_statistics_on_income_and_living_conditions_(EU-SILC)_methodology.

21.4 A Quick Guide to the Scripts

21.4.1 Basics of the Scripts

In R and SAS the canonical way to document functions and make them easily accessible to others (especially, beginner users) is to make an R package and a SAS macro, respectively. Indeed, an R package or a SAS macro passes the build checks if developers provide totally general codes in a user-friendly format, and supply sufficiently detailed help files for each of their functions/datasets. The *custom-made* scripts available on the website are not yet properly either R packages or SAS macros. Rather they are advanced program files that need at least the knowledge of basic tools of SAS and R to be understood; therefore less skilled users have to refer to the relative SAS or R documentation for details on the standard procedures recalled in these scripts (see Section 21.2). In particular, some of the custom-made scripts need datasets already rearranged and managed in the form required from the scripts (Chapters 5, 8–11, 13, 14, 17, and 18). Whereas other scripts also provide syntax for a general procedure to follow in order to prepare the data for the analysis (Chapters 6, 7, 12, and 16). Moreover, the syntax used in each script includes the application of standard commands/functions of SAS or R rearranged in more than one block of syntax, so that each instruction block produces a particular output. For this reason, R and SAS beginners need to learn basic tools of R and SAS before running these scripts to be familiar with features common to all estimation commands and to be able to prepare the data for analysis. Basically, each script assumes that your data are stored in your own working path and combines basic documentation and code in a single file (i.e., above each code block there is a text which explains what it is going to do). In an attempt to make as homogeneous as possible the quick guide, we use a common structure that includes only key aspects of the script concerned. For this reason, each quick guide is very simple and it is composed of the following five arguments: (i) analysis provided by the script; (ii) required software and packages; (iii) required data and their structure; (iv) how to use and run the script available on the website; and (v) output provided by the script. These arguments are useful, especially for first-time users of SAE methods, and are designed for users that are not expert in R and/or SAS. The first argument (*Analysis provided by the script*) summarizes the main analysis that can be performed using the script; in this manner users have a direct link between the chapter concerned and the script available on the website. In explaining this argument, we always refer to the estimation of quantities of interest for subpopulations (domains). Domains are defined by any characteristics disaggregating the population into a set of mutually exclusive subpopulations. Clearly this field is not applicable for chapters 5, 11, 17 and 18. Indeed, the scripts associated with these chapters are not available on the website because standard R packages have been used. The second argument (*Required software and packages*) indicates if the script runs in R or SAS and gives guidelines on the degree of difficulty that beginners can meet running the script. The third argument (*Required data and their structure*) summarizes the data used in the empirical analysis of the corresponding chapter and contains additional details on the particular data structure necessary for running the code. Accordingly, users find the following list of information.

1. *Sampling data* (basically it reports the name of the sample survey used for the analysis). Obviously, users can also run the script using a different sample survey, nevertheless users (especially beginners in SAE methods) have to check if their sampling data contain the variables necessary for the analysis (the *core variable* and the *auxiliary variables)* before running the *custom-made* script.
2. *Core variable* (i.e., the variable of interest necessary for implementing the analysis). This variable must be included in the sampling data. Across the empirical analyses presented in the book's chapters, it is generally the equivalized household income.
3. *Auxiliary data* (i.e., an additional source of data or a variety of external sources of information to be used for obtaining small area estimates). This is a very important argument in the SAE settings. In detail, auxiliary data can be other sample surveys, population data, or auxiliary information on domains (usually it corresponds to small geographical areas or small subpopulations). Thus, this argument will specify chapter-by-chapter the additional sources of data.
4. *Auxiliary variables* (i.e., the variables users need for implementing estimation at small area level). Obviously auxiliary variables are included in auxiliary data and some of them also in sampling data. The lists of the specific auxiliary variables will be provided as they have been referred to in the chapter concerned and script.
5. *Link variable between sampling and auxiliary data* (i.e., the variable used for linking sources of data used in the analysis). The link variable is crucial in the context of SAE and it depends on the specified model: unit level models relate the unit values of the core variable to unit-specific auxiliary variables; area level models relate area-specific direct survey estimates to area-specific auxiliary data.

For chapters 5, 11, 17 and 18 this field is not appliacable because the required data and their structure follow the guidelines for the standard R packages used. The fourth argument (*How to use and run the script available on the website*) includes a description of how to use the materials provided on the website. These materials are organized in several folders named the same as each corresponding book's chapter. It is worth noting that the R and SAS scripts are provided as *txt*, so that users can easily open them. Nevertheless in the quick guides we refer to them as R or SAS files. We remember that only Chapters 6–10, 12–14, and 16 have a *custom-made* script associated with each of them. Therefore, this argument is not pertinent for Chapters 5, 11, 17, and 18 and we simply list the specific standard packages used for the given task. For the former set of chapters, the *custom-made* script can be organized in a set codes that have to be run in sequence. In these cases we provide both the description of each single script and the order that has to be followed for obtaining the output. While we attempt to provide as homogenous as possible quick guide, the materials available can be different in each folder.

These differences depend on the data files that are needed for replicating the empirical analysis. Indeed, as previously explained, some chapters use data not available for users or not free of charge. For this reason, where EU-SILC data and/or Census have been used for the empirical analysis we only provide a simple *.txt file where we summarize the structure of the data used for the analysis. Thus, the information reported in this argument vary a lot across chapters. Finally, the fifth argument (*Output provided by the script*) is a simple description of the target variable(s) (i.e., indicators) that users can obtain running the script concerned. Obviously, for chapters 5, 11, 17 and 18 this field is not applicable because the output is the one described in the standard R packages used.

21.4.2 A Quick guide to Chapter 5 (Impact of Sampling Designs in Small Area Estimation with Applications to Poverty Measurement)

21.4.2.1 Analysis Provided by the Script

Not applicable.

21.4.2.2 Required Software and Packages

The analysis uses R software. In particular, the function `glm` and the packages "`sae`" and "`nlme`" have been used.

21.4.2.3 Required Data and their Structure

Not applicable.

21.4.2.4 How to Use and Run the Script Available on the Website

Not applicable. Users can use the guidelines provided by the R packages listed in Section 21.4.2.2.

21.4.2.5 Output Provided by the Script

Not applicable.

21.4.3 A Quick guide to Chapter 6 (Model-assisted Methods for Small Area Estimation of Poverty Indicators)

21.4.3.1 Analysis Provided by the Script

The script provides design-based estimates for poverty and inequality indicators (poverty rate and Gini coefficient) for planned domains (districts). Statistical properties (design bias and accuracy) of the methods are assessed by design-based simulation experiments In particular, GREG and model calibration (MC) estimators have been used, while the Horvitz–Thompson (HT) estimator acts as a reference.

21.4.3.2 Required Software and Packages

All the programs provided in this chapter run on R software. The R packages "`Matrix`", "`lme4`" and "`nlme`" are required. The users have to be also confident with programming in R because the codes are organized using functions written by the authors.

21.4.3.3 Required Data and their Structure

The data structure for estimation purposes of this chapter consists of sample survey data (typically based on a national income survey) and administrative register data on variables that are related to the income variable. Assuming that the sample survey data and the register data can be linked uniquely at the micro-level the predicted values from the fitted model can be calculated at population unit and successively incorporated in the construction of the estimators of the desired indicators for the domains of interest. Data used in the empirical investigation are simulated data (see Section 6.2.2). Details on data are as follows:

Sampling data: s.txt (available on the website) is the sample dataset, $n = 2000$ drawn from the population dataset (*u.txt* on the website). It includes the domain variable (the district DIS), the design weight, the equivalized income, and other variables explained in the script "*Lehtonen-Veijanen example of code use.txt*".

Core variable: equivalized disposable income (`income`).

Auxiliary data: u.txt on the website. The auxiliary data is unit level population based on register data from Statistics Finland). The population consisted of 1 million people in 36 NUTS 4 regions in Western Finland. Moreover, another additional data source is used for labor force status and the socio-economic status; these variables were obtained from a household survey for the household head and imputed for the other members of each household.

Auxiliary variables: the labor force status and the socio-economic status.

Link variable between sampling and auxiliary data: personal identification number.

21.4.3.4 How to Use and Run the Script Available on the Website

The authors include a text file "Lehtonen–Veijanen examples of code use.txt" and "RDomest.r"; both describe step-by-step the R functions defined in the scripts.

21.4.3.5 Output Provided by the Script

Output of this script is composed of two parts: the output of the simulation experiments; and the output of the empirical investigation. Concerning the design-based estimation of the Gini coefficient, the percentile-adjusted predictor and the HT direct estimator are compared by mean absolute relative error (MARE) and mean coefficient of variation (MCV) of the estimators, over domain classes defined by sample size (see Table 6.2). Concerning the estimation of the poverty rate the authors compare the direct estimator (HT) versus the indirect estimators computed with different assisting models: MARE of the estimators in a SRSWOR (simple random sampling without replacement) sample over NUTS 4 regions are provided (see Table 6.4).

21.4.4 *A Quick Guide to Chapter 7 (Variance Estimation for Cumulative and Longitudinal Poverty Indicators from Panel Data at Regional Level)*

21.4.4.1 Analysis Provided by the Script

The scripts provided the estimation of three years' averaged poverty measures at national and NUTS 2 regional level. The poverty measures chosen are the "at-risk-of-poverty rate" (head count ratio, HCR) using the national poverty line, the "S80/S20" ratio of income shares of the percentiles and the Gini index (see Chapter 7 for definitions). For all the measures, standard errors and confidence intervals are computed. The methodology used for variance estimation is the resampling method known as Jack-knife Repeated Replication (JRR).

21.4.4.2 Required Software and Packages

The scripts are in SAS software. Only the Base SAS software is needed. The users have to be also confident with the SAS programming language because the script includes SAS macros, developed inside the program, written by the authors.

21.4.4.3 Required Data and their Structure

The analysis conducted in this chapter is based on EU-SILC data, and therefore the datasets are not available for the users. Nevertheless, the required data for the analysis are summarized in order to help readers to understand the structure of the input files as follows:

Sampling data: UDB (Users Data Base) EU-SILC, cross-sectional data for 2009, 2010, and 2011 for Spain.

Core variable: equivalized disposable income (HX090) through which the reference measures are constructed (HCR with poverty line as 60% of median national poverty line; ratio of income shares of the percentiles, S80/S20; the Gini index).

Auxiliary data: not applicable.

Link variable between sampling and auxiliary data: not applicable.

Auxiliary variables: EU-SILC 2009, 2010, and 2011 non public variables such as DB040 (region code at NUTS 2), DB050 (strata code) and DB060 [primary sampling unit (PSU) code], available for Spain. It is worth noting that in order to run this program it is necessary that the strata (DB050) and PSUs (DB060) can be linked from one year to the others, across the cross-sectional data.

21.4.4.4 How to Use and Run the Script Available on the Website

The authors use only one script. The program can be divided into three parts. In the first part a common structure of Strata and PSUs is created for the 3 years (see Chapter 7). In the second part the datasets for the 3 years are prepared with all the needed variables. The third part is the core of the program: the macros for the estimation of the indicators and of their variances are developed. There are three macros for the estimation of the poverty measures (`stat1`, `stat10`, `stat11`).The macro `perc_bound` allows any kind of weighted percentile of a distribution to be calculated. The computed value is the linear interpolation of the percentile. It means that if any real value of the distribution lies in this percentile, a value between the two nearest values – above and below the percentile – is interpolated. It uses information from individual datasets (PID).

The macro `jrr_var` implements the JRR methodology (for additional details see Chapter 7). In order to estimate the standard errors, the measures are estimated inside the replications. In fact, inside this macro there is the computation of the "`stat&k`". Inside the replications we also reallocate the weights. Once a PSU is deleted, its weights are assigned to the other PSUs in the same stratum of the deleted PSU, so that the total sum of the weights does not change.

Then there are two further macros to create a cycle: the macro waves repeats the loop for each year; and the macro `sub_ciclo` where the indicator to be computed can be chosen and the number of PSUs of the data are specified.

21.4.4.5 Output Provided by the Script

The output provided is a table with the estimate of the reference measures, their standard errors, and the confidence interval bounds.

21.4.5 A Quick Guide to Chapter 8 (Models in Small Area Estimation when Covariates are Measured with Error)

21.4.5.1 Analysis Provided by the Script

The scripts provided the Gibbs sampling algorithm for estimating the model in Arima *et al.* (2015).

21.4.5.2 Required Software and Packages

The code provided is written in R. The users also have to be confident with programming in R because this code is composed of two functions written by the authors.

21.4.5.3 Required Data and their Structure

The analysis conducted in this chapter is based on the 1979 U.S. Current Population Survey (CPS) whose data files are not available for the users. The data structure is described in the file "structure.txt", no data has been provided; the authors just give the list of the input information required by the script "GIBBS.r". For this reason we do not give information in a list of arguments as provided for the other chapters.

21.4.5.4 How to Use and Run the Script Available on the Website

The authors use the R code "GIBBS.r" for their empirical analysis. Users load their empirical data at the beginning of the code. The script is composed of two functions written by the authors: `gibbs.meas` and `gibbs.meas.noerr`. The first function (`gibbs.meas`) implements the Bayesian method for FME (functional measurement error) models described in Section 8.2.1. In particular, `gibbs.meas` implements the Gibbs sampling algorithm for estimating the model in Arima *et al.* (2014). In the function, the full conditional distributions are specified in order to obtain samples from the posterior distributions. The function returns a list of values defining the posterior distribution of the parameters of interest (for details, see the GIBBS.r code on the website). The second function (`gibbs.meas.noerr`) implements the Bayesian method for the Fay–Herriot model. In the function, the full conditional distributions are specified in order to obtain samples from the posterior distributions. The same arguments of the `gibbs.meas` function are required with the exception of the "`c.x.`" argument. The same output is returned. Users load their empirical data at the beginning of the code.

21.4.5.5 Output Provided by the Script

Only posterior distributions are returned by the R code. The table and figures reported in the Chapter are not supported by the GIBBS.r code.

21.4.6 *A Quick Guide to Chapter 9 (Robust Domain Estimation of Income-based Inequality Indicators)*

21.4.6.1 Analysis Provided by the Script

The scripts provided estimates of poverty indicators for unplanned domains . The term unplanned is used to denote domains that do not feature in the design and allocation of the EU-SILC sample. For the application these domains are defined by provinces cross-classified by the gender of the head of the household.

21.4.6.2 Required Software and Packages

All the codes provided are in R. Users must download the "`MASS`", the "`nlme`" and the "`nl`" packages before running the R codes. If users are interested in the estimation of the Gini coefficient the "`ineq`" package also has to be downloaded. Users also have to be confident with programming in R because the codes are organized using functions written by the authors.

21.4.6.3 Required Data and their Structure

The analysis conducted in this chapter is based on EU-SILC survey and Population Census therefore data files are not available for the users. Nevertheless, the required data for the analysis have been summarized in order to help the readers to organize their input files as follows:

Sampling data: Italian EU-SILC 2008.

Core variable: household equivalized income.
Auxiliary data: Italian Population Census 2001.
Auxiliary variables: the auxiliary variables for sampling data and for Population data (harmonized with respect to those of the sampling data) are: marital status of the head of the family, the employment status of the head of the household, the years of education of the head of the household, the mean house surface at municipality level and the number of household members (arguments `my.x.s` and `my.x.r` in the defined R functions).
Link variable between sampling and auxiliary data: unique code at level of region (arguments `myregioncode.s` and `myregioncode.r` in the defined R functions).

21.4.6.4 How to Use and Run the Script Available on the Website

There are four R codes used for obtaining the results reported in this chapter. The R code "func-poverty.R" contains instruction for estimating the HCR, the poverty gap (PG) and S80/S20 ratio. Users have to run this R code at the beginning. Doing this, functions to compute HCR, PG and S80/S20 are loaded in the R environment and can be passed as argument to the SAE function (argument `FUN` in the defined R function `MQ.SAE.MC` or `MQ.SAE.MC.NLAV`). The two R codes MQ.SAE.MC.R and MQ.SAE.MC.NoLinkAuxVar.R contain a given number of R functions written by the authors. The main functions `MQ.SAE.MC` and `MQ.SAE.MC.NLAV` provide point estimates. Indeed, both of them return objects (for instance, the small area point estimates, the vector of the residuals, etc.) required by the R code "M-quantileBootMC.R" that contains the bootstrap function needed for the root MSE (RMSE) estimate. The `MQ.SAE.MC.NLAV` is an alternative to `MQ.SAE.MC` because it has to be used to obtain the point estimate when the linkage between sample and census data is not possible. The same principle applies to the RMSE estimate.

21.4.6.5 Output Provided by the Script

At the beginning of each R code, the specific output obtained is described in detail.

21.4.7 A Quick Guide to Chapter 10 (Nonparametric Regression Methods for Small Area Estimation)

21.4.7.1 Analysis Provided by the Script

The script provided estimates of the average household per-capita consumption expenditures at district level using alternative nonparametric methods and then they compare these alternative approaches.

21.4.7.2 Required Software and Packages

All the programs provided in this chapter run on R software and on three specific R packages. Therefore, the packages "`MASS`", "`mgcv`", and "`nlme`" must be downloaded before running the R codes.

21.4.7.3 Required Data and their Structure

The analysis conducted is based on two data sources: the Living Standards Measurement Study (LSMS) survey; and the Population and Housing Census. Consequently, data are not available for the users. Nevertheless, the R code can be easily run once the input files have been organized following the structure described "*file_structure.txt*" (available on the web). The main characteristics of the required data for the analysis are summarized as follows:

Sampling data: LSMS conducted in Albania in 2002.
Core variable: per-capita consumption expenditure ("`consumption`" variable in the sampling data file).
Auxiliary data: Albanian Population and Housing Census carried out in 2001.
Auxiliary variables: the auxiliary variables for sampling data and population data (harmonized with respect to those of the sampling data) are described in detailed in Section 10.3.
Link variable between sampling and auxiliary data: unique code for districts (variable "`id_district`").

21.4.7.4 How to Use and Run the Script Available on the Website

Both sampling (albania_sample.txt) and census data (AlbaniaCensus.txt) are loaded. Indeed, the users cannot directly run the R code because, as explained above, these files are not available for the users. There are three main R codes used for estimation that must to be applied in sequence: (i) a_EBLUP_NPEBLUP.R; (ii) b_NPMQ.R; and (iii) c_NPMBDE.R. The first one contains instructions for the estimation of both a nested error unit level linear regression model (EBLUP) and a nested error unit level nonparametric regression model (NPEBLUP). R codes (MSE_EBLUP.R, MSE_NPEBLUP.R, MSE_NPEBLUP_COND.R) are, respectively, called inside a_EBLUP_NPEBLUP.R using the function source. The second R code (b_NPMQ.R) includes scripts for the district level estimates using a nonparametric M-quantile model with a Chambers–Dunstan type bias correction (NPMQCD). At the beginning the R code "Mqnpar_functions.R" is loaded. This R code contains a collection of functions to estimate nonparametric M-quantile models as introduced in Pratesi *et al.* (2009). Finally, the third R code (c_NPMBDE.R) computes the NPMBDE by Salvati *et al.* (2010).

21.4.7.5 Output Provided by the Script

Table 10.1, Table 10.2, and Table 10.3 as well as Figure 10.1 and Figure 10.2 are generated by running the a_EBLUP_NPEBLUP.R program, whereas Table 10.4 can be achieved by running the other two main R codes.

21.4.8 A Quick Guide to Chapter 11 (Area-level Spatio-temporal Small Area Estimation Models)

21.4.8.1 Analysis Provided by the Script

Not applicable.

21.4.8.2 Required Software and Packages

The empirical analysis is implemented using R. In particular, the R package "`saery`" has been used (see Section 21.2 for further details on this R package).

21.4.8.3 Required Data and their Structure

Not applicable.

21.4.8.4 How to Use and Run the Script Available on the Website

Not applicable. Users can use the guidelines provided by "`saery`".

21.4.8.5 Output Provided by the Script

Not applicable.

21.4.9 A Quick Guide to Chapter 12 (Unit Level Spatio-temporal Models)

21.4.9.1 Analysis Provided by the Script

The script provided procedures for the estimation of unit level models with spatio-temporal random area effects and uncorrelated residual errors.

21.4.9.2 Required Software and Packages

This script is written using SAS/IML (see Section 21.2 for a brief description of SAS/IML).

21.4.9.3 Required Data and their Structure

The analysis conducted in this chapter is based on the EU-SILC survey. The EU-SILC data cannot be shared, so the provided code refers to properly simulated data; nevertheless, the required data structure have been introduced in order to help the readers to organize their input files.

Sampling data: repeated-by-time collection of 724 householders in the years 2005–2008 ($T = 4$ time instants) from IT-SILC data, leaving out all the units with partially missing information.

Core variable: equivalized disposable income.

Auxiliary data: administrative data sources and statistical registers at the unit level.

Auxiliary variables: education level and age of the householder (see Section 12.4 for further details).

Link variable between sampling and auxiliary data: personal identification number.

21.4.9.4 How to Use and Run the Script Available on the Website

The SAS code is divided into seven parts. In Part 1 an example for generating simulated data is provided. The numerical values of the repeated samples (v), the number of areas in study (m), the number of time instants (t), the coding time vector (time) and the number of subjects in each area (sub) must be specified. Furthermore the SAS code calls a file "work.distances"; this file includes the matrix of the row standardized distances referred to small areas considered (in the code the number is fixed to 24). Users must create it before running the simulation exercise because it is called in the program by the "`proc iml`". The remaining part of the SAS code generates the core variable (Y) by means of a spatio-temporal unit level model and finally stores all the simulated data in the new work SAS file "work.simdata".

The code from Parts 2 to 7 goes through the standard estimation procedures, either for the time-varying effects unit level model and the "spatial" version of the state space unit level model. In particular, Parts 2–4 deal with a spatio-temporal unit level model with two random-area effects per time (one random-area effect, plus one random effect per area and time). Specifically, in Part 2 REML covariance parameters estimates are generated and in Part 3 EBLUP and MSE of EBLUP estimates are computed; in the case of estimation by simulated data, users can also reproduce the average relative bias of the EBLUP. In Part 4 the relative bias of covariance parameters is also provided if users apply the SAS code to simulated data. Parts 5–7 deal with the spatial state-space unit-level model (one random-area effect per time). Specifically, Part 5 generates REML covariance parameter estimates and Part 6 computes EBLUP and MSE of EBLUP estimates. Once again the average relative bias of the EBLUP is provided in the case of estimation by simulated data. Finally, Part 7 computes the relative bias of covariance parameters provided if users again apply the SAS code to simulated data.

21.4.9.5 Output Provided by the Script

The SAS code does not provide the estimation of all five models presented in the empirical analysis but just the estimation procedure for model 3 and model 5 (see Table 12.1). Parameter estimates computed by Part 2 and Part 5 of the SAS code are presented in Table 12.2. Table 12.3 and Table 12.4 present the EBLUP estimates and their average root MSE (Parts 3 and 6 of the SAS code).

21.4.10 A Quick Guide to Chapter 13 (Spatial Information and Geoadditive Small Area Models)

21.4.10.1 Analysis Provided by the Script

The script provided the estimate of the average household per-capita consumption expenditure at district level. In particular, a geoadditive SAE model has been applied.

21.4.10.2 Required Software and Packages

All the programs provided in this chapter run on R software. The R packages "`SemiPar`", "`mgcv`", "`msm`" and "`shapefiles`" are required. The R codes used in the scripts are organized using functions written by the authors in combination with specific packages available in R, so users need to be quite confident in the R environment.

21.4.10.3 Required Data and their Structure

The analysis conducted is based on the LSMS and the Population and Housing Census (PHC) plus geographical referencing data. These data files are not available for the users. The required data for the analysis have been summarized in the following in order to help the readers to organize their input files. Moreover in the "*structure_data.txt*" file a complete list of the variables included in both sampling and auxiliary data is reported (see the website).

Sampling data: LSMS 2002.

Core variable: household consumption expenditure per-capita ("`pcons3`" variable in the sampling data file).

Auxiliary data: PHC 2001.

Auxiliary variables: a detailed description of the auxiliary variables is provided in the chapter (see Section 13.5.1).

Link variable between sampling and auxiliary data: commune code (variable "`id_com`" in data files).

Sampling data and auxiliary data have been combined. Geographical location of each unit of analysis: the location of each household is available for the LSMS data (variable "`latitude`" and "`longitude`" in data file). For the auxiliary data, the geographical location is approximated with the centroid coordinates of the commune where the household is located.

21.4.10.4 How to Use and Run the Script Available on the Website

The authors use only the R code "GEOADDITIVE_SAE.r". First of all, thin plate spline-type Z matrices are created by the definition of the `Ztps` function. Once sampling data are loaded in R the "`shapefiles`" package is used to read the shape files included on the website. Users can now call the "`default.knots.2D`" internal function of the "`SemiPar`" package in order to plot knots and boundaries together with the data (see Figure 13.2). Then a geoadditive small area random effect model is estimated using the "`lme`" function. Then, the image plots are generated. Once auxiliary data are loaded, the unit coordinates are approximated with the commune centroids. Then the script calculates the empirical model-based model-calibrated (MBMC) weights that are used some lines below in order to compute the two types of MBDE model-calibrated estimators (the HT and the Hajek specification). Finally, robust MSE are calculated.

21.4.10.5 Output Provided by the Script

The R code provides all the figures presented in the chapter. The estimated parameters of the geoadditive SAE log-transformed model for the household per-capita consumption expenditure at district level Table 13.1. The district level estimates of the average household per-capita consumption expenditure, with estimated RMSE and coefficient of variation are shown in Table 13.2.

21.4.11 A Quick guide to Chapter 14 (Model-based Direct Estimation of a Small Area Distribution Function)

21.4.11.1 Analysis Provided by the Script

The script provided the estimate of the gender-wise distribution of equivalized income and HCR at province level. The estimator proposed is a model-based direct estimator (MBDE) of the small area distribution function. The analysis provides also the MSE estimation of the MBDE.

21.4.11.2 Required Software and Packages

The code is written in R. Users must download the "`MASS`" and "`nlme`" packages before running the R code. Users also have to be confident with programming in R because the code is organized using functions written by the authors.

21.4.11.3 Required Data and their Structure

The analysis conducted in this chapter is based on the EU-SILC survey and Population Census and the data files are not available for the users. The required data for the analysis have been summarized in order to help the readers to organize their input files as follows:

Sampling data: EU-SILC survey 2006 for Italy.
Core variable: equivalized disposable income (`y` variable in the sampling data file).
Auxiliary data: Italian Population Census 2001.
Auxiliary variables: age, gender, years of education, household size (variables `eta`, `sex`, `year_edu`, `hsize` in data files). More information on variables used in the analysis can be found in Section 14.3.
Link variable between sampling and auxiliary data: commune code (variable `PROV` in data files).

21.4.11.4 How to Use and Run the Script Available on the Website

The authors use only the R code "MBDE.r". In detail, it is important to highlight that this R code, in its current form, provides only the estimation of the HCR. It means that the R code predicts the value of the small area cumulative distribution function at t. For example the value of t can be set to 60% of the median equivalized income (poverty line). The R script loads three datasets named "toscana hh2new.txt", "lombardia hh2new.txt", and "campania hh2new.txt" that are sample data from EU-SILC and the dataset "cens_tosc_sll_new2.txt" referring to Census data 2010. Indeed, the users cannot run directly the R code because, as explained above, these files are not available for the users. Nevertheless users can easily run the R code once they have organized their files following the instruction above. On the website the "file_structure.txt" has the same structure as the input files used in this R code. The R code starts defining the "`MBDECDF`" function that contains instructions for obtaining the MBDE. Then data files are loaded in order to have the input information for the "`MBDECDF`" function. The remaining program lines include scripts for the robust MSE estimation.

21.4.11.5 Output Provided by the Script

The R code provides the SAEs of the HCR and their estimated MSEs. These results are written in an external data file at the end of the R code using write.table and they are reported in Table 14.2 by gender.

21.4.12 A Quick Guide to Chapter 16 (Bayesian Beta Regression Models for the Estimation of Poverty and Inequality Parameters in Small Areas)

21.4.12.1 Analysis Provided by the Script

This script provided small area estimates for parameters taking values in the (0,1) interval generally used for describing poverty, social exclusion, and inequality. In particular, Bayesian Beta regression models have been used. The specific issues considered are listed in Section 16.1.

21.4.12.2 Required Software and Packages

All the programs provided in this chapter run on R software. The specific R package "`BRugs`" must be installed before running the programs. Moreover, users need to also download the free `OpenBUGS` (version 3.2.3) software from a specific website (http://www.openbugs.net/w/FrontPage). The R scripts call this software using the "`BRugs`" package as it allows the connection between the two software. Additional details on Markov chain Monte Carlo (MCMC) algorithms can be found in Section 16.6.

21.4.12.3 Required Data and their Structure

The dataset used is a subset of the Italian 2010 sample of the EU-SILC survey and an arrangement of these data for the empirical analysis is provided by the authors on the website. Therefore users can reuse the R code to reproduce results. The target small areas are the 64 health districts belonging to two administrative regions.

Sampling data: not applicable. The script works on the data files available on the website.
Core variable: `hr60` [direct estimation at risk of poverty – see formula (16.1)], `gini` [direct estimation of the Gini index — see formula (16.4)], `depmatr4` (direct estimation of severe material deprivation rate), and `depmatr3` (direct estimation of material deprivation rate).
Auxiliary data: not applicable. The script works on the data files available on the website.

Auxiliary variables: `size` (household size in the district *i*), `Avgdich` (average taxable income claimed by private residents), `Freqpop` (percentage of residents filling tax forms), `DR` (dependency ratio), `quotastr` (percentage of resident immigrants), and `logdens` (log of population density).

Link variable between sampling and auxiliary data: not applicable. The script works on the data files available on the website.

21.4.12.4 How to Use and Run the Script Available on the Website

The two R codes used for estimation are "code for hr and gini models.r" and "code for matdep models.r". The three txt files "hr and g univ.txt", "hr and g multi.txt", and "hr and g univ IL multi IIL.txt" are the bugs code called, respectively, in sequence by the first R script above. The txt file "Bugs code for matdep.txt" is the bugs code called by the second R script above. Section 16.6 gives more information on these R codes.

21.4.12.5 Output Provided by the Script

Users obtain, respectively, Figure 16.1, Figure 16.4, Figure 16.5, and Table 16.1 by running the "code for hr and gini models.r" whereas the "code for matdep models.r" provides Figure 16.2 and Figure 16.3.

21.4.13 A Quick Guide to Chapter 17 (Empirical Bayes and Hierarchical Bayes Estimation of Poverty Measures for Small Areas)

21.4.13.1 Analysis Provided by the Script

Not applicable.

21.4.13.2 Required Software and Packages

The empirical analysis is implemented using R and in particular the package "`sae`" has been used. The following functions of "`sae`" are called:

`eblupFH`: calculates the EBLUPs of small area means under a Fay–Herriot model;

`mseFH`: gives the Prasad–Rao analytical MSE estimates of EBLUPs based on the Fay–Herriot model.

`ebBHF`: calculates empirical best estimates of general non-linear small area parameters when assuming the nested error model, also called the Battesse–Harter–Fuller (BHF) model, on a transformation of a continuous welfare variable, with normality of random effects and errors. The transformation can be chosen from the Box–Cox or the power families.

`pbmseebBHF`: gives parametric bootstrap estimates of the MSE of empirical best estimates of general parameters under the BHF model for a transformation of the welfare variable.

21.4.13.3 Required Data and their Structure

Not applicable.

21.4.13.4 How to Use and Run the Script Available on the Website

Not applicable. Users can use the guidelines provided by the R package listed in Section 21.4.13.2.

21.4.13.5 Output Provided by the Script

Not applicable.

21.4.14 A Quick Guide to Chapter 18 - (Small Area Estimation Using Both Survey and Census Unit Record Data: Links, Alternatives, and the Central Roles of Regression and Contextual Variables)

21.4.14.1 Analysis Provided by the Script

Not applicable. The script associated with this chapter is not available on the website because the World Bank software PovMap has been used. This software can be downloaded from the following link:

http://econ.worldbank.org/WBSITE/EXTERNAL/EXTDEC/EXTRESEARCH/EXTPROGRAMS/EXTPOVRES/0,,contentMDK:22717057~pagePK:64168182~piPK:64168060~theSitePK:477894,00.html. System requirements for using the package are: Microsoft Windows® NT or later; and a minimum memory requirement of 128M.

21.4.14.2 Required Software and Packages

See Section 21.4.14.1.

21.4.14.3 Required Data and their Structure

Not applicable.

21.4.14.4 How to Use and Run the Script Available on the Website

Not applicable. Users can use the guidelines provided by PovMap.

21.4.14.5 Output Provided by the Script

Not applicable.

References

Arima, S., Datta, G.S., and Liseo, B. 2015 Bayesian estimators for small area models when auxiliary information is measured with error. *Scandinavian Journal of Statistics* **42**, 518–529.

Battese, G. E., Harter, R. M. & Fuller, W. A. (1988), An error-components model for prediction of county crop areas using survey and satellite data, *Journal of the American Statistical Association*, **83**, 28–36.

Breidenbach J. (2015). Packae 'JoSAE'. (https://cran.r-project.org/web/packages/JoSAE/JoSAE.pdf).

Breslow N. E., Clayton D. G. (1993). Approximate Inference in Generalized Linear Mixed Models, *Journal of the American Statistical Association*, Vol. 88, No. 421, pp. 9–25.

Molina I and Marhuenda Y 2015 sae: An R package for small area estimation. *The R Journal* **7**(1), 81–98.

Fay R.E., Diallo M. (2015). Package 'sae2'. (https://cran.r-project.org/web/packages/sae2/sae2.pdf).

Lopez-Vizcaino E., Lombardia M.J. and Morales D. (2013a). Package 'mme'. (https://cran.r-project.org/web/packages/mme/mme.pdf).

Lopez-Vizcaino, ME, Lombardia, MJ. and Morales, D. (2013b). Multinomial-based small area estimation of labour force indicators. *Statistical Modelling*, **13**, 153–178.

Pratesi M, Ranalli MG, and Salvati N 2009 Nonparametric M-quantile regression using penalised splines. *Journal of Nonparametric Statistics* **21**(3), 287–304.

Salvati N, Chandra H, Ranalli MG, and Chambers R 2010 Small area estimation using a nonparametric model-based direct estimator. *Computational Statistics & Data Analysis* **54**(9), 2159–2171.

Sarndal, C. E. (1984). Design-consistent versus model-dependent estimation for small domains, *Journal of the American Statistical Association*, **79**, 624–631.

Schoch T 2015 rsae: Robust small area estimation. *R package version* 0.1-5.

Author Index

Analysis of Poverty Data by Small Area Estimation, First Edition. Edited by Monica Pratesi.

Companion Website: www.wiley.com/go/pratesi/poverty

Subject Index

Analysis of Poverty Data by Small Area Estimation, First Edition. Edited by Monica Pratesi.

Companion Website: www.wiley.com/go/pratesi/poverty

The Wiley Series in Survey Methodology covers topics of current research and practical interests in survey methodology and sampling. While the emphasis is on application, theoretical discussion is encouraged when it supports a broader understanding of the subject matter.

The authors are leading academics and researchers in survey methodology and sampling. The readership includes professionals in, and students of, the fields of applied statistics, biostatistics, public policy, and government and corporate enterprises.

ALWIN · Margins of Error: A Study of Reliability in Survey Measurement
BETHLEHEM · Applied Survey Methods: A Statistical Perspective
*BIEMER, GROVES, LYBERG, MATHIOWETZ, and SUDMAN · Measurement Errors in Surveys
BIEMER and LYBERG · Introduction to Survey Quality
BIEMER · Latent Class Analysis of Survey Error
BRADBURN, SUDMAN, and WANSINK · Asking Questions: The Definitive Guide to Questionnaire Design—For Market Research, Political Polls, and Social Health Questionnaires, Revised Edition
BRAVERMAN and SLATER · Advances in Survey Research: New Directions for Evaluation, No. 70
CALLEGARO, BAKER, BETHLEHEM, GORITZ, KROSNIK and LAVRAKAS (Editors) · Online Panel Research: A Data Quality Perspective
CHAMBERS and SKINNER (editors) · Analysis of Survey Data
COCHRAN · Sampling Techniques, Third Edition
CONRAD and SCHOBER · Envisioning the Survey Interview of the Future
COUPER, BAKER, BETHLEHEM, CLARK, MARTIN, NICHOLLS, and O'REILLY (editors) · Computer Assisted Survey Information Collection
COX, BINDER, CHINNAPPA, CHRISTIANSON, COLLEDGE, and KOTT (editors) Business Survey Methods
*DEMING · Sample Design in Business Research
DILLMAN · Mail and Internet Surveys: The Tailored Design Method
D'ORAZIO, DI ZIO and SCANU · Statistical Matching: Theory and Practice
FULLER · Sampling Statistics
GROVES and COUPER · Nonresponse in Household Interview Surveys
GROVES · Survey Errors and Survey Costs
GROVES, DILLMAN, ELTINGE, and LITTLE · Survey Nonresponse
GROVES, BIEMER, LYBERG, MASSEY, NICHOLLS, and WAKSBERG · Telephone Survey Methodology
GROVES, FOWLER, COUPER, LEPKOWSKI, SINGER, and TOURANGEAU · Survey Methodology, Second Edition
*HANSEN, HURWITZ, and MADOW · Sample Survey Methods and Theory, Volume 1: Methods and Applications
*HANSEN, HURWITZ, and MADOW · Sample Survey Methods and Theory, Volume II: Theory
HARKNESS, van de VIJVER, and MOHLER · Cross-Cultural Survey Methods
HEDAYAT and SINHA · Design and Inference in Finite Population Sampling
HUNDEPOOL, DOMINGO-FERRER, FRANCONI, GIESSING, NORDHOLT, SPICER and DE WOLF · Statistical Disclosure Control

KALTON and HEERINGA · Leslie Kish Selected Papers
KISH · Statistical Design for Research
*KISH · Survey Sampling
KORN and GRAUBARD · Analysis of Health Surveys
KREUTER (Editor) · Improving Surveys with Paradata: Analytic Uses of Process Information
LEPKOWSKI, TUCKER, BRICK, DE LEEUW, JAPEC, LAVRAKAS, LINK, and SANGSTER (editors) · Advances in Telephone Survey Methodology
LESSLER and KALSBEEK · Nonsampling Error in Surveys
LEVY and LEMESHOW · Sampling of Populations: Methods and Applications, Fourth Edition
LUMLEY · Complex Surveys: A Guide to Analysis Using R
LYBERG, BIEMER, COLLINS, de LEEUW, DIPPO, SCHWARZ, TREWIN (editors) Survey Measurement and Process Quality
LYNNE (editor) · Methodology of Longitudinal Surveys
MADANS, MILLER, MAITLAND and WILLIS · Question Evaluation Methods: Contributing to the Science of Data Quality
MAYNARD, HOUTKOOP-STEENSTRA, SCHAEFFER, VAN DER ZOUWEN · Standardization and Tacit Knowledge: Interaction and Practice in the Survey Interview
MILLER, CHEPP, WILSON and PADILLA (editors) · Cognitive Interviewing Methodology
PORTER (editor) · Overcoming Survey Research Problems: New Directions for Institutional Research, No. 121
PRATESI (editor) · Analysis of Poverty Data by Small Area Estimation
PRESSER, ROTHGEB, COUPER, LESSLER, MARTIN, MARTIN, and SINGER (editors) · Methods for Testing and Evaluating Survey Questionnaires
RAO and MOLINA · Small Area Estimation, Second Edition
REA and PARKER · Designing and Conducting Survey Research: A Comprehensive Guide, Third Edition
SARIS and GALLHOFER · Design, Evaluation, and Analysis of Questionnaires for Survey Research, Second Edition
SARNDAL and LUNDSTROM · Estimation in Surveys with Nonresponse
SCHWARZ and SUDMAN (editors) · Answering Questions: Methodology for Determining Cognitive and Communicative Processes in Survey Research
SIRKEN, HERRMANN, SCHECHTER, SCHWARZ, TANUR, and TOURANGEAU (editors) · Cognition and Survey Research
SNIJKERS, HARALDSEN, JONES and WILLIMACK · Designing and Conducting Business Surveys
STOOP, BILLIET, KOCH and FITZGERALD · Improving Survey Response: Lessons Learned from the European Social Survey
SUDMAN, BRADBURN, and SCHWARZ · Thinking about Answers: The Application of Cognitive Processes to Survey Methodology
UMBACH (editor) · Survey Research Emerging Issues: New Directions for Institutional Research No. 127
VALLIANT, DORFMAN, and ROYALL · Finite Population Sampling and Inference: A Prediction Approach
WALLGREN and WALLGREN · Register-based Statistics: Statistical Methods for Administrative Data, Second Edition